Introduction to Fluid Mechanics

Introduction to Fluid Mechanics

Second Edition

ROBERT W. FOX
ALAN T. McDONALD
School of Mechanical Engineering
Purdue University

JOHN WILEY & SONS
New York   Santa Barbara   Chichester   Brisbane   Toronto

Library of Congress Cataloging in Publication Data

Fox, Robert W   1934–
Introduction to fluid mechanics.

Includes bibliographical references and index.
1. Fluid mechanics.   I. McDonald, Alan T., joint
author.   II. Title.

TA357.F69 1978   532   77-20839
ISBN 0-471-01909-7

Preface

This textbook was written for an introductory course in fluid mechanics. In this, the second edition, we have used the international system of units (SI) throughout; SI units are used in approximately 70 percent of the example problems and problem exercises at the end of each chapter. (Approximately 30 percent of English engineering units have been retained to provide experience with this traditional system and to highlight methods of conversion among unit systems.)

We make no change in approach to the subject in the second edition. The physical concepts of fluid mechanics and methods of analysis beginning from basic principles are emphasized throughout. The primary objective of this book is to help students develop an orderly approach to problem solving. Thus we start from basic equations, state assumptions clearly, and relate results to expected physical behavior. The approach is illustrated by the 90 example problems in the text, whose solutions have been prepared to demonstrate good solution techniques and to explain troublesome points. The example problems are set apart from the text in format and type style, so that they are particularly easy to follow.

Complete explanations in the chapter discussions, together with the numerous detailed examples, have in our experience made the book understandable for students. This allows the instructor freedom to depart from conventional "lecture" teaching methods. Classroom time can be used to bring in outside material, expand on special topic areas (such as blood flow, non-Newtonian flow, and measurement methods), solve example problems, or explain the troublesome points of the assigned homework problems. Thus each class period can be used in the manner most appropriate to satisfy student needs.

The material has been carefully selected. There is a detailed presentation of a broad range of topics suitable for a one- or two-semester course in fluid mechanics at the junior or senior level. Desirable prerequisites are introductory courses in rigid-body dynamics, heat power or thermodynamics, and mathematics through integral calculus. The thermodynamics course may be taken concurrently.

The presentation is organized into five broad topic areas:

1. Introductory concepts, scope of fluid mechanics, and fluid statics (Chapters 1, 2, and 3).
2. Development and application of control volume forms of basic equations (Chapter 4).
3. Development and application of differential forms of basic equations (Chapters 5 and 6).
4. Dimensional analysis and correlation of experimental data (Chapter 7).
5. Applications for incompressible flow (Chapter 8) and one-dimensional compressible flow (Chapters 9 and 10).

"Summary Objectives" have been added to each chapter. They indicate specifically what students should be able to do on completion of their study of each chapter.

More than 150 new problems have been added for homework assignment and student exercises. The second edition now contains 651 problems, so that at least four semesters can be covered without repeating problem assignments. All problem statements have been revised to follow the 70/30 percent ratio of SI to traditional units.

The *Solutions Manual* that accompanies the book contains complete and detailed solutions for each of the 651 problems in the text; the same format used in the example problems has been followed. Solutions may be photocopied directly for library use or for student study. Thus the labor of problem solving is essentially eliminated for the instructor.

Many fine instructional films and film loops are available for further clarification and demonstration of basic principles in fluid mechanics. We refer to these films in the text, where their use is appropriate; a complete list of suppliers and titles is included in Appendix C.

When our students have finished this course, we expect them to be able to apply the basic equations to a variety of problems, including new problems that they have not encountered previously. We emphasize physical understanding throughout, to make students aware of the variety of phenomena that can occur in fluid flow situations. By minimizing the number of "magic formulas" and emphasizing the fundamental approach, we believe that students will feel confident in their ability to apply the material and will be able to reason out solutions to rather challenging problems.

The book is also well suited for independent study by students or practicing engineers. The readability and clear examples help to build student confidence. The summary objectives at the end of each chapter may be used for review and to assess achievement of educational goals.

We recognize that no single approach can satisfy all needs. We are grateful to students and faculty whose comments have helped us improve the second edition. We welcome criticisms and suggestions from interested readers or users of this book.

Alan T. McDonald
Robert W. Fox

Contents

1. **INTRODUCTION**   1

    1-1   Definition of a Fluid   1
    1-2   Scope of Fluid Mechanics   2
    1-3   Basic Equations   4
    1-4   Methods of Analysis   4

       1-4.1   System and Control Volume   5
       1-4.2   Differential versus Integral Approach   8
       1-4.3   Methods of Description   8

    1-5   Note to Students   11
    1-6   Dimensions and Units   13

       1-6.1   Systems of Dimensions   13
       1-6.2   Systems of Units   14
       1-6.3   Preferred Systems of Units   15

    Summary Objectives   17
    Problems   17

2. **FUNDAMENTAL CONCEPTS**   21

    2-1   Fluid as a Continuum   21
    2-2   Velocity Field   23

2-2.1	One-, Two-, and Three-Dimensional Flows	24
2-2.2	Pathlines, Streaklines, and Streamlines	26

2-3	**Stress Field**	**28**
2-3.1	Surface and Body Forces	28
2-3.2	Stress at a Point	30

2-4	**Newtonian Fluid: Viscosity**	**33**
2-4.1	Newtonian Fluid	33
2-4.2	Viscosity	35

2-5	**Description and Classification of Fluid Motions**	**38**
2-5.1	Viscous and Inviscid Flows	38
2-5.2	Laminar and Turbulent Flows	44
2-5.3	Compressible and Incompressible Flows	45

	Summary Objectives	**46**
	Problems	**46**

3.	**FLUID STATICS**	**53**
3-1	**The Basic Equation of Fluid Statics**	**53**
3-1.1	Pressure Variation in a Static Fluid	58

3-2	**The Standard Atmosphere**	**65**
3-3	**Absolute and Gage Pressures**	**66**
**3-4	Hydraulic Systems**	**67**
3-5	**Hydrostatic Forces on Submerged Surfaces**	**68**
3-5.1	Hydrostatic Force on Plane Submerged Surfaces	68
3-5.2	Hydrostatic Force on Curved Submerged Surfaces	76

**3-6	Buoyancy and Stability**	**86**
**3-7	Fluids in Rigid-Body Motion**	**89**
	Summary Objectives	**96**
	Problems	**97**

4.	**BASIC EQUATIONS IN INTEGRAL FORM FOR A CONTROL VOLUME**	**113**
4-1	**Basic Laws for a System**	**114**
4-1.1	Conservation of Mass	114
4-1.2	Newton's Second Law	114
**4-1.3	Moment of Momentum	114

** Sections so marked may be omitted without loss of continuity in the text material.

4-1.4 The First Law of Thermodynamics   115
4-1.5 The Second Law of Thermodynamics   115

4-2 Relation of System Derivatives to
     Control Volume Formulation   116

4-2.1 Derivation   117
4-2.2 Physical Meaning   122

4-3 Conservation of Mass   124

4-3.1 Control Volume Equation   124
4-3.2 Special Cases   125

4-4 Momentum Equation for
     Inertial Control Volume   133

4-4.1 Control Volume Equation   133
**4-4.2 Differential Control Volume Analysis   146
4-4.3 Control Volume Moving with Constant Velocity   149

4-5 Momentum Equation for Control Volume
     with Rectilinear Acceleration   153

4-5.1 Control Volume Equation   154

**4-6 Momentum Equation for Control Volume with
       Arbitrary Acceleration   162

4-6.1 Control Volume Equation   162

**4-7 Moment of Momentum   170

4-7.1 Equation for Fixed Control Volume   171
4-7.2 Application to Turbomachinery   172
4-7.3 Equation for Rotating Control Volume   181

4-8 The First Law of Thermodynamics   187

4-8.1 Rate of Work for a Control Volume   188
4-8.2 Control Volume Equation   190

4-9 The Second Law of Thermodynamics   195
Summary Objectives   198
Problems   199

5. INTRODUCTION TO DIFFERENTIAL ANALYSIS OF
    FLUID MOTION   223

5-1 Review of the Field Concept   223
5-2 The Continuity Equation   223

5-2.1	Rectangular Coordinate System	224
5-2.2	Cylindrical Coordinate System	227

**5-3 Stream Function for Two-Dimensional
Incompressible Flow ... 231
5-4 Motion of a Fluid Element (Kinematics) ... 235
**5-5 Fluid Rotation ... 237
**5-6 Irrotational Flow ... 239

5-6.1	Velocity Potential	240
5-6.2	Stream Function and Velocity Potential for Irrotational, Two-Dimensional, Incompressible Flow	241
5-6.3	Irrotational Flow and Viscosity	242

5-7 Momentum Equation ... 247

5-7.1	Acceleration of a Fluid Particle	248
5-7.2	Acceleration of a Fluid Particle in a Velocity Field	248
5-7.3	Formulation of Forces Acting on a Fluid Particle	253
5-7.4	Differential Momentum Equation	255

Summary Objectives ... 257
Problems ... 258

6. DYNAMICS OF INCOMPRESSIBLE INVISCID FLOW ... 265

6-1 Stress Field in an Inviscid Flow ... 265
6-2 Momentum Equation for Frictionless Flow:
Euler's Equations ... 266
**6-3 Euler's Equations for Fluids in
Rigid-Body Motion ... 269
6-4 Euler's Equations in Streamline Coordinates ... 270
6-5 Bernoulli Equation—Integration of Euler's
Equations along a Streamline for Steady Flow ... 272

6-5.1	Derivation Using Streamline Coordinates	273
**6-5.2	Derivation Using Rectangular Coordinates	274
6-5.3	Applications	276

6-6 Static, Stagnation, and Dynamic Pressures ... 283
6-7 Relation between the First Law of
Thermodynamics and the Bernoulli Equation ... 287
**6-8 Bernoulli Equation Applied to Irrotational Flow ... 292
**6-9 Unsteady Bernoulli Equation—Integration of
Euler's Equation along a Streamline ... 293

Summary Objectives 297
Problems 297

7. DIMENSIONAL ANALYSIS AND SIMILITUDE 305

7-1 Introduction 305
7-2 Nature of Dimensional Analysis 306
7-3 Buckingham Pi Theorem 307
7-4 Detailed Procedure for Use of
Buckingham Pi Theorem 309

7-4.1 Selection of Parameters 309
7-4.2 Procedure for Determining the Π Groups 309
7-4.3 Comments on the Procedure 313

7-5 Physical Meaning of Common Dimensionless
Groups 315

7-5.1 The Reynolds Number 316
7-5.2 The Mach Number 316
7-5.3 The Froude Number 317
7-5.4 The Euler Number (Pressure Coefficient) 317

7-6 Flow Similarity and Model Studies 317
7-7 Similitude Established from the
Differential Equations 321
Summary Objectives 322
Problems 322

8. INCOMPRESSIBLE VISCOUS FLOW 329

Part A. Introduction 329
8-1 Internal and External Flows 330
8-2 Laminar and Turbulent Flows 332

Part B. Fully-Developed Laminar Flow 333
8-3 Fully-Developed Laminar Flow between
Infinite Parallel Plates 334

8-3.1 Both Plates Stationary 334
8-3.2 Upper Plate Moving with Constant Velocity, U 341

8-4 Fully-Developed Laminar Flow in a Pipe 346

Part C. Flow in Pipes and Ducts 354
8-5 Velocity Profiles in Pipe Flow 355
8-6 Shear Stress Distribution in Fully-Developed
Pipe Flow 356

8-7 Energy Considerations in Pipe Flow 358

   8-7.1 Head Loss 359

8-8 Calculation of Head Loss 361

   8-8.1 Major Losses: Friction Factor 361
   8-8.2 Minor Losses 367

8-9 Solution of Pipe Flow Problems 376

   8-9.1 Single-Path Systems 376
   **8-9.2 Multiple-Path Systems 391
   **8-9.3 Noncircular Ducts 397

Part D. Boundary Layers 398
8-10 The Boundary-Layer Concept 398
8-11 Displacement Thickness 400
8-12 Momentum Integral Equation 404

   8-12.1 Application of the Basic Equations 404
   8-12.2 Special Case—Flow over a Flat Plate 409

8-13 Use of the Momentum Integral Equation for
Zero Pressure Gradient Flow 410

   8-13.1 Laminar Flow 411
   8-13.2 Turbulent Flow 413

8-14 Pressure Gradients in Boundary-Layer Flow 417

   8-14.1 Effect of Pressure Gradient on Flow: Separation 419
   8-14.2 Determination of Pressure Gradient 420

Part E. Fluid Flow about Immersed Bodies 424
8-15 Drag 424

   8-15.1 Flow over a Flat Plate Parallel to the Flow:
       Friction Drag 425
   8-15.2 Flow over a Flat Plate Normal to the Flow:
       Pressure Drag 426
   8-15.3 Flow over a Sphere and Cylinder: Friction
       and Pressure Drag 428
   8-15.4 Streamlining 434

8-16 Lift 436

Part F. Flow Measurement 447
8-17 Simple Methods 447
8-18 Flowmeters for Internal Flows 449

8-18.1	The Orifice Plate	452
8-18.2	The Flow Nozzle	453
8-18.3	The Venturi	455
8-18.4	The Laminar Flow Element	456

8-19 Mechanical Flowmeters 460
8-20 Traversing Methods 461
Summary Objectives 461
Problems 462
References 482

9. INTRODUCTION TO COMPRESSIBLE FLOW 485

9-1 Review of Thermodynamics 486
9-2 Propagation of Sound Waves 493

9-2.1 Speed of Sound 493
9-2.2 Types of Flow—The Mach Cone 498

9-3 Reference State: Local Isentropic
Stagnation Properties 500

9-3.1 Local Isentropic Stagnation Properties for the
Flow of an Ideal Gas 501
9-3.2 Critical Conditions 510

Summary Objectives 510
Problems 511

10. ONE-DIMENSIONAL COMPRESSIBLE FLOW 515

10-1 Basic Equations for Isentropic Flow 515
10-2 Effect of Area Variation on Flow Properties
in Isentropic Flow 520
10-3 Isentropic Flow of an Ideal Gas 523

10-3.1 Basic Equations 523
10-3.2 Reference Conditions for Isentropic Flow of
an Ideal Gas 524
**10-3.3 Tables for Computation of Isentropic Flow
of an Ideal Gas 528
10-3.4 Isentropic Flow in a Converging Nozzle 529
10-3.5 Isentropic Flow in a Converging-Diverging
Nozzle 536

10-4 Adiabatic Flow in a Constant Area Duct with
     Friction   544

    10-4.1   Basic Equations   544
    10-4.2   The Fanno Line   547
  **10-4.3   Tables for Computation of Fanno Line Flow
          of an Ideal Gas   552

10-5 Frictionless Flow in a Constant Area Duct
     with Heat Transfer   562

    10-5.1   Basic Equations   562
    10-5.2   The Rayleigh Line   565
  **10-5.3   Tables for Computation of Rayleigh Line
          Flow of an Ideal Gas   572

10-6 Normal Shocks   577

    10-6.1   Basic Equations   577
  **10-6.2   Tables for Computation of Normal Shocks
          in an Ideal Gas   586
    10-6.3   Flow in a Converging-Diverging Nozzle   592

    Summary Objectives   593
    Problems   594

APPENDIX A
    Fluid Property Data   609
APPENDIX B
    Tables for Computations in Compressible Flow   621
APPENDIX C
    Films and Film Loops for Fluid Mechanics   641
APPENDIX D
    Review of Vector Concepts and Operations   649
APPENDIX E
    SI Units, Prefixes, and Conversion Factors   661

    Answers to Even-Numbered Problems   665

    Index   677

Chapter 1

Introduction

In beginning the study of any subject, a number of questions come to mind immediately. Among those that a student in the first course in fluid mechanics may ask are the following:

What is fluid mechanics all about?
Why do I have to study it?
Why should I want to study it?
How does it relate to subject areas with which I am already familiar?

In this chapter we shall try to provide at least a qualitative answer to these and similar questions and thus an introduction to the subject.

1-1 DEFINITION OF A FLUID

Since fluid mechanics deal with the behavior of fluids at rest and in motion, it is logical to begin our study of the subject with a definition of the term *fluid*.

A fluid is a substance that deforms continuously under the application of a shearing (i.e. tangential) stress no matter how small the shearing stress.

Thus, according to the physical forms in which matter exists, fluids comprise the liquid and gas (or vapor) phases. The distinction between a fluid and the remaining possible state of matter (i.e. the solid state) is clear if one compares a fluid as defined above with the behavior of a solid. A solid is a substance that deforms when a shear stress is applied, but it does not continue to deform.

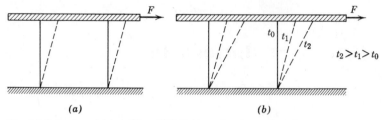

Fig. 1.1 Behavior of (a) solid and (b) fluid, under the action of a constant shear force.

In Fig. 1.1 the behavior of a solid (Fig. 1.1a) and a fluid (Fig. 1.1b) under the action of a constant shear force are contrasted. In Fig. 1.1a the shear force is applied to the solid through the upper of two plates to which the solid has been bonded. When the shear force is applied to the plate, the block is deformed as shown. From our previous work in mechanics, we know that, provided the elastic limit of the solid material is not exceeded, the deformation is directly proportional to the applied shear stress, τ, where $\tau = F/A$ and A is the area of the surface in contact with the plate.

Now let us repeat the experiment using a fluid between the plates. In order to observe the behavior of the fluid, we use a dye marker to outline a fluid element as shown by the solid lines (Fig. 1.1b). Upon application of the force, F, to the upper plate, we notice that the fluid element continues to deform as long as the force is applied. The shape of the fluid element, at successive instants of time, $t_2 > t_1 > t_0$, is shown by the dashed lines in Fig. 1.1b, which represent the positions of the dye markers. Note also that the fluid in direct contact with the solid boundary has the same velocity as the boundary itself; that is, there is no slip at the boundary. This is an experimental fact based on numerous observations of fluid behavior.[1]

Because the fluid motion continues under the application of a shear stress, we may alternatively define a fluid as a substance that cannot sustain a shear stress when at rest.

1-2 SCOPE OF FLUID MECHANICS

Having defined a fluid and noted the characteristics that distinguish it from a solid, we might ask the question: "Why study fluid mechanics?"

---

[1] The no-slip condition is demonstrated in the film loops S-FM003, *Shear Deformation of Viscous Fluids,* and S-FM006, *Boundary-Layer Formation.* These loops were produced by Educational Services, Inc., Watertown, Mass., and the National Committee for Fluid Mechanics Films. The films are distributed by Encyclopaedia Britannica Educational Corporation. (A complete list of fluid mechanics film titles and sources is given in Appendix C.)

Knowledge and understanding of the basic principles and concepts of fluid mechanics are essential in the analysis and design of any system in which a fluid is the working medium. The design of virtually all means of transportation requires an application of the principles of fluid mechanics. Included are aircraft for both subsonic and supersonic flight, ground effect machines, hovercraft (now in service for channel crossings between France and England), vertical takeoff and landing aircraft requiring minimum runway length, surface ships, submarines, and automobiles. In recent years automobile manufacturers have given more consideration to aerodynamic design. This has been true for some time for the designers of both racing cars and boats. The design of propulsion systems for space flight as well as for toy rockets is based on the principles of fluid mechanics. The collapse of the Tacoma Narrows Bridge some years ago is evidence of the possible consequences of neglecting the basic principles of fluid mechanics.[2] It is commonplace today to perform model studies to determine the aerodynamic forces on and flow fields around buildings and structures. These include studies of skyscrapers, baseball stadiums, smokestacks, and shopping plazas.

The design of all types of fluid machinery including pumps, fans, blowers, compressors, and turbines clearly requires a knowledge of the basic principles of fluid mechanics. Lubrication is an area of considerable importance in fluid mechanics. Heating and ventilating systems for private homes, large office buildings, and underground tunnels, and the design of pipeline systems are further examples of technical problem areas requiring a knowledge of fluid mechanics. The circulatory system of the body is essentially a fluid system. It is not surprising then that the design of artificial hearts, heart-lung machines, breathing aids, and other such devices must rely on the basic principles of fluid mechanics.

Even some of our recreational endeavors are directly related to fluid mechanics. The slicing and hooking of golf balls can be explained by the principles of fluid mechanics (although they can be corrected only by a golf pro!).

The list of applications of the principles of fluid mechanics could be extended considerably. Our main point here is that fluid mechanics is not a subject studied for purely academic interest; rather, it is a subject with widespread importance both in our everyday experiences and in modern technology.

Clearly, we cannot hope to consider in detail even a small percentage of these and other specific problems of fluid mechanics. Instead, the purpose

---

[2] For dramatic evidence of aerodynamic forces in action, see the Ohio State University film, *Collapse of the Tacoma Narrows Bridge*.

of this text is to present the basic laws and associated physical concepts that provide the basis or starting point in the analysis of any problem in fluid mechanics.

1-3 BASIC EQUATIONS

An analysis of any problem in fluid mechanics necessarily begins, either directly or indirectly, with statements of the basic laws governing the fluid motion. These laws, which are independent of the nature of the particular fluid, are:

1. Conservation of mass.
2. Newton's second law of motion.
3. Moment of momentum.
4. The first law of thermodynamics.
5. The second law of thermodynamics.

Clearly, not all of these laws are always required in the solution of any one problem. In some problems, it is necessary to bring into the analysis additional relations, in the form of constitutive equations describing the behavior of physical properties of fluids under given conditions.

It is obvious that the basic laws with which we shall deal are the same as those used in mechanics and thermodynamics. Our task will be to formulate these laws in forms suitable for the solution of fluid flow problems and to apply them to the solution of a wide variety of problems.

It should be emphasized that there are, as we shall see, many apparently simple problems in fluid mechanics that cannot be solved by totally analytical means. In such cases we must resort to experiments and experimental observations.

1-4 METHODS OF ANALYSIS

As we have indicated, the basic laws that are employed in the analysis of problems in fluid mechanics are the same ones that you have used previously in your earlier studies of thermodynamics and basic mechanics. From these earlier studies you will recall that the first step in solving a problem is to define the system that you are attempting to analyze. In basic mechanics, extensive use was made of the free body diagram. In thermodynamics you referred to the system under analysis as either a closed system or an open system. In this text we shall employ the terms *system* and *control volume*. The importance of defining the system or control volume to which the basic equations are to be applied in the analysis of a problem cannot be overemphasized. At this point it is wise to review the basic difference between a system and a control volume.

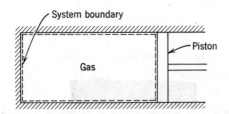

Fig. 1.2 Piston-cylinder assembly.

1-4.1   SYSTEM AND CONTROL VOLUME

A system is defined as a fixed, identifiable quantity of mass; the system boundaries separate the system from the surroundings. The boundaries of the system may be fixed or movable, but no mass crosses them.

In the familiar piston-cylinder assembly from thermodynamics, Fig. 1.2, the gas in the cylinder is considered as the system. If a high temperature source is brought in contact with the left end of the cylinder, the piston will move to the right; the boundary of the system thus moves. From our study in thermodynamics we know that heat and work may cross the boundaries of the system, but the quantity of matter within the system boundaries remains fixed, that is, there is no mass transfer across the system boundaries.

Example 1.1

A radiator of a steam heating system has a volume of 0.7 ft^3. When the radiator is filled with dry saturated steam at a pressure of 20 psia, all valves to the radiator are closed. How much heat will have been transferred to the room when the pressure of the steam is 10 psia?

Example Problem 1.1

GIVEN:

Steam radiator with volume, $\forall = 0.7$ ft^3, is filled with dry saturated steam at 20 psia and then the valves are closed. As a result of heat transfer, pressure in the radiator drops.

FIND:

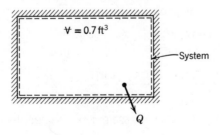

Amount of heat transferred when pressure reaches 10 psia.

SOLUTION:

We are dealing with a system.

At: State ① $\forall = 0.7$ ft³ $p_1 = 20$ psia   dry saturated steam

  State ② $\forall = 0.7$ ft³ $p_2 = 10$ psia

Basic equation: First law for the system: $Q_{12} + W_{12} = E_2 - E_1$

Assumptions: $W_{12} = 0$, since no shaft work is done and $\forall$ = constant

      $E = U$, since the system is stationary

Under these assumptions, the first law statement reduces to

$$Q_{12} = U_2 - U_1 = m(u_2 - u_1)$$

Thus, to determine Q_{12}, we must determine m, u_2, and u_1.

At state ① $p_1 = 20$ psia and from steam tables for dry saturated steam

$$u_1 = \frac{1081.9 \text{ Btu}}{\text{lbm}} \qquad v_1 = \frac{20.089 \text{ ft}^3}{\text{lbm}}$$

Since $v = \forall/m$, then

$$m_1 = \frac{\forall}{v_1} = 0.7 \text{ ft}^3 \times \frac{\text{lbm}}{20.089 \text{ ft}^3} = 0.0348 \text{ lbm}$$

For steam

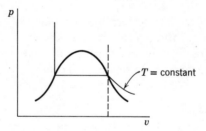

Consequently, when p is reduced at constant v, then state ② must lie in the wet region. Thus at state ②

$$v_2 = v_1 = v_{2f} + x v_{2fg} \qquad \text{and} \qquad x = \frac{v_1 - v_{2f}}{v_{2fg}}$$

From the steam tables at $p_2 = 10$ psia,

$$v_{2f} = \frac{0.017 \text{ ft}^3}{\text{lbm}}, \qquad v_{2fg} = \frac{38.40 \text{ ft}^3}{\text{lbm}}$$

$$x = \frac{v_1 - v_{2f}}{v_{2fg}} = \frac{20.089 - 0.017}{38.40} = 0.523$$

Then

$$u_2 = u_{2f} + x u_{2fg} = 161.14 + 0.523(911.1) = 638 \text{ Btu/lbm}$$

Finally,

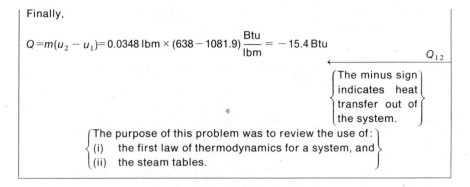

$$Q = m(u_2 - u_1) = 0.0348 \text{ lbm} \times (638 - 1081.9)\frac{\text{Btu}}{\text{lbm}} = -15.4 \text{ Btu}$$

$$\left.\begin{array}{l}\text{The minus sign} \\ \text{indicates heat} \\ \text{transfer out of} \\ \text{the system.}\end{array}\right\}$$

$$\left.\begin{array}{l}\text{The purpose of this problem was to review the use of:} \\ \text{(i)} \quad \text{the first law of thermodynamics for a system, and} \\ \text{(ii)} \quad \text{the steam tables.}\end{array}\right\}$$

In your earlier mechanics courses you made extensive use of the free body (system approach). This was logical because you were dealing with an easily identifiable rigid body. However in fluid mechanics, we are normally concerned with the flow of fluids through devices such as compressors, turbines, pipelines, nozzles, and so on. In these cases it is difficult to focus attention on a fixed identifiable quantity of mass. It is much more covenient, for purposes of analysis, to focus attention on a volume in space through which the fluid flows, that is, to use the control volume approach.

A control volume is an arbitrary volume in space through which fluid flows. The geometric boundary of the control volume is called the control surface. The control surface may be real or imaginary; it may be at rest or in motion. Figure 1.3 shows a possible control surface for use in analyzing the flow through a pipe. Here the inside surface of the pipe, a real physical boundary, comprises part of the control surface. However, the vertical portions of the control surface are imaginary—that is, there is no corresponding physical surface. These imaginary boundaries are selected arbitrarily for accounting purposes. Since the location of the control surface has a direct effect on the accounting procedure in applying the basic laws, it is

Fig. 1.3 Fluid flow through a pipe.

extremely important that the control surface be clearly defined before beginning any form of analysis.

1-4.2 DIFFERENTIAL VERSUS INTEGRAL APPROACH

The basic laws that we apply in our study of fluid mechanics can be formulated in terms of infinitesimal or finite systems and control volumes. As you might suspect, the equations will look different in each case. In the first case the resulting equations are differential equations; in the latter case we might call them global equations, that is, equations governing the gross behavior of the flow. Both approaches are important in the study of fluid mechanics and both will be developed in the course of our work.

If it is assumed that we can solve the equations, the differential approach (i.e. the use of differential equations governing the motion) provides a means of determining the detailed (i.e. point by point) behavior of the flow.

Frequently in the problems under study, the information sought does not require a detailed knowledge of the flow. We are often interested in the gross behavior of a device and hence can use the integral formulation of the basic laws. The integral formulation, that is, the use of finite systems or control volumes, is usually easier to treat analytically. Since mechanics and thermodynamics deal with the formulation of the basic laws in terms of finite systems, these formulations are the basis for deriving the control volume equations in Chapter 4.

1-4.3 METHODS OF DESCRIPTION

Mechanics and thermodynamics deal almost exclusively with systems. Thus you have made extensive use of the basic equations applied to a fixed, identifiable quantity of mass. Later, in attempting to analyze thermodynamic devices, you often found it necessary to go to a control volume (open system) analysis. Clearly, the type of analysis depends on the problem. Where it is easy to keep track of identifiable elements of mass (e.g. in particle mechanics), one utilizes a method of description that follows the particle. This is often referred to as the Lagrangian method of description.

Consider for example the application of Newton's second law to a particle of fixed mass, m. Mathematically, we can write Newton's second law for a system of mass, m, in one of the following forms:

$$\sum \vec{F} = m\vec{a} \tag{1.1a}$$

or

$$\sum \vec{F} = m\frac{d\vec{V}}{dt} = \frac{d}{dt}(m\vec{V}) \tag{1.1b}$$

or

$$\sum \vec{F} = m \frac{d^2\vec{r}}{dt^2} = \frac{d^2}{dt^2}(m\vec{r}) \tag{1.1c}$$

where $\sum \vec{F}$ is the sum of all external forces acting on the system, $\vec{a}$ the acceleration of the center of mass of the system, $\vec{V}$ the velocity of the center of mass of the system, and $\vec{r}$ the position vector of the center of mass of the system relative to a fixed coordinate system.

Thus, in describing the motion of a particle in a rectangular coordinate system:

$$\vec{F} = \hat{i}F_x + \hat{j}F_y + \hat{k}F_z \tag{1.2a}$$

$$\vec{a} = \hat{i}a_x + \hat{j}a_y + \hat{k}a_z \tag{1.2b}$$

$$\vec{V} = \hat{i}u + \hat{j}v + \hat{k}w \tag{1.2c}$$

$$\vec{r} = \hat{i}x + \hat{j}y + \hat{k}z \tag{1.2d}$$

where F_x, F_y, and F_z are the components of $\vec{F}$ in the x, y, and z directions, respectively; a_x, a_y, and a_z are the components of $\vec{a}$ in the x, y, and z directions, respectively; u, v, and w are the components of $\vec{V}$ in the x, y, and z directions, respectively; and x, y, and z are the coordinates of the particle.

Since the particle moves under the action of the force, $\vec{F}$, we recognize that its position, $\vec{r}$, velocity, $\vec{V}$, and acceleration, $\vec{a}$, are in general functions of time. That is,

$$\vec{r} = \vec{r}(t) \quad \text{and, hence,} \quad x = x(t), \quad y = y(t), \quad z = z(t)$$
$$\vec{V} = \vec{V}(t) \quad \text{and, hence,} \quad u = u(t), \quad v = v(t), \quad w = w(t)$$
$$\vec{a} = \vec{a}(t) \quad \text{and, hence,} \quad a_x = a_x(t), \quad a_y = a_y(t), \quad a_z = a_z(t)$$

From elementary mechanics we know further (as has already been implied in Eqs. 1.1) that

$$\vec{a} = \frac{d\vec{V}}{dt} \quad \text{and} \quad a_x = \frac{du}{dt}, \quad a_y = \frac{dv}{dt}, \quad a_z = \frac{dw}{dt}$$

$$\vec{V} = \frac{d\vec{r}}{dt} \quad \text{and} \quad u = \frac{dx}{dt}, \quad v = \frac{dy}{dt}, \quad w = \frac{dz}{dt}$$

Example 1.2
A ball is thrown vertically upward with an initial speed of 30 m/sec. Neglecting air resistance, determine the maximum height to which it will rise, and the time required for it to reach the maximum height.

Example Problem 1.2

GIVEN:

A ball thrown vertically upward.
At $t = 0$, $x = 0$

$$\vec{V} = u_0 \hat{i} = 30\hat{i} \text{ m/sec}$$

Neglect air resistance.

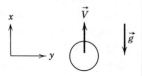

FIND:

(a)   The maximum height to which the ball will rise.
(b)   The time required to reach maximum height.

SOLUTION:

Basic equations: $\sum \vec{F} = m\vec{a}$

$$\sum F_x = ma_x \qquad a_x = \frac{d^2x}{dt^2} \qquad u = \frac{dx}{dt}$$

Drawing a free body diagram, we obtain

$$\vec{V} = u\,\hat{i}$$

$$\vec{W} = -mg\,\hat{i}$$

$$\sum F_x = ma_x$$

$$-W = -mg = ma_x = m\frac{d^2x}{dt^2}$$

$$\therefore \frac{d^2x}{dt^2} = -g$$

Integrating with respect to time between limits of 0 and t, we have

$$\frac{dx}{dt}\bigg]_t - \frac{dx}{dt}\bigg]_{t=0} = -gt \qquad \text{or} \qquad \frac{dx}{dt} = u_0 - gt$$

Integrating again with respect to time between limits of 0 and t, we have

$$x - x_0 = u_0 t - \tfrac{1}{2}gt^2 \qquad \text{or} \qquad x = u_0 t - \tfrac{1}{2}gt^2$$

The maximum height is reached when $u = dx/dt = 0$. At this condition

$$\frac{dx}{dt} = 0 = u_0 - gt$$

Thus the time required to reach the maximum height is given by

$$t = \frac{u_0}{g} = \frac{30}{\text{sec}}\ \frac{\text{m}}{9.81\ \text{m}} \times \frac{\text{sec}^2}{9.81\ \text{m}} = 3.06 \text{ sec} \qquad t_{x_{\max}}$$

The maximum height is obtained from

$$x = u_0 t - \frac{1}{2} g t^2 \qquad \text{with} \qquad t = \frac{u_0}{g}$$

$$\therefore x_{max} = u_0 \frac{u_0}{g} - \frac{1}{2} g \left(\frac{u_0}{g} \right)^2 = \frac{1}{2} \frac{u_0^2}{g} = \frac{1}{2} \times (30)^2 \frac{m^2}{sec^2} \times \frac{sec^2}{9.81 \ m}$$

$$x_{max} = 45.9 \ m \qquad \overset{x_{max}}{\longleftarrow}$$

This problem is intended as a reminder of the method of description used in particle mechanics. Note that the velocity, u, is a function of time in this method of description.

We may consider a fluid to be composed of a very large number of particles whose motion must be described; keeping track of the motion of each fluid particle becomes a horrendous bookkeeping problem. Consequently, a particle description becomes unmanageable. Therefore, we find it convenient to use a different type of description. Since we shall often deal with control volume analyses, it is convenient to use the field, or Eulerian, description. This method of description focuses attention on the properties of a flow at a given point in space as a function of time. Thus, in using the Eulerian method of description, the properties of a flow field are described as functions of space coordinates and time. We shall see in Chapter 2 that this method of description is a logical outgrowth of the assumption that fluids may be treated as continuous media.

1-5 NOTE TO STUDENTS

The goal of this textbook is to provide a clear, concise introduction to the subject of fluid mechanics. It is written for you, the student. It is our strong feeling that classroom time should not be devoted to a statement of textbook material by the instructor. Instead, the time should be used to amplify the textbook material through discussion of related material and the application of basic principles to the solution of problems. The necessary conditions for accomplishing this goal are: (1) a clear, concise presentation of the fundamentals that you, the student, can read and understand, and (2) your willingness to read the text material before coming to class. We have assumed responsibility for meeting the first condition. You must assume responsibility for satisfying the second condition. There will no doubt be times when we fall short of satisfying these objectives. If so, we would appreciate hearing of these shortcomings either directly or through your instructor.

It goes without saying that an introductory text is not all-inclusive. Your instructor undoubtedly will expand on the material presented and introduce additional new material. We encourage you to refer to the many other available fluid mechanics textbooks; where another text presents a particularly good discussion of a given topic, we shall refer to it directly. We assume that you have had prior courses in thermodynamics, statics and dynamics, and differential and integral calculus. No attempt will be made to restate this subject material; however, the pertinent aspects of this previous study will be reviewed briefly when appropriate.

It is our strong belief that one learns best by *doing*. This is true whether the subject under study is fluid mechanics, thermodynamics, or golf. The fundamentals in any of these cases are few, and mastery of them comes through practice. *Thus it is extremely important, in fact essential, that you solve problems.* A large number of problems are included at the end of each chapter to provide the opportunity to gain facility in the application of the fundamentals to the solution of problems. You should avoid the temptation to adopt a "plug and chug" approach to the solution of problems. Most of the problems are of a nature that such an approach will lead to difficulty or simply will not work. In the solution of problems we strongly recommend that you proceed using the following logical steps:

1. State briefly and concisely (in your own words) the information given.
2. State the information that you have to find.
3. Draw a schematic of the system or control volume to be used in the analysis. Be sure to label the boundaries of the system or control volume and label appropriate coordinate directions.
4. Give the appropriate mathematical formulation of the *basic* laws that you consider necessary to effect a solution to the problem.
5. List the simplifying assumptions that you feel are appropriate in the problem.
6. Carry through the analysis to the point where it is appropriate to substitute numerical values.
7. Substitute numerical values (using a consistent set of units) to obtain a numerical answer. The significant figures in the answer should be consistent with the given data.
8. Check the answer and the assumptions made in effecting the solution to make sure they are reasonable.
9. Label the answer.

In your initial work this problem format may seem unnecessary. However, such an approach to the solution of problems will lead to fewer errors, save time, and permit a clearer understanding of the limitations of a particular solution. This format is used in all example problems presented in this text; answers to example problems are given to three significant figures.

1-6 DIMENSIONS AND UNITS

Engineering problems are solved to answer specific questions. It goes without saying that the answer must include units. (It makes a difference whether a pipe diameter required is 1 meter or 1 foot!) Consequently, it is appropriate at this point to present a brief review of dimensions and units. We say "review" because the topic is familiar from your earlier work in thermodynamics.[3]

The word *dimension* is used to refer to any measurable quantity; such quantities include length, time, and velocity. In terms of a particular system of dimensions all measurable quantities can be subdivided into two groups—primary quantities and secondary quantities. Primary quantities are those for which we set up arbitrary scales of measure; secondary quantities are those quantities whose dimensions are expressible in terms of the dimensions of the primary quantities.

1-6.1 SYSTEMS OF DIMENSIONS

There are three basic systems of dimensions, corresponding to the three basic ways of specifying the primary dimensions. These systems use primary dimensions of:

1. Mass $[M]$, length $[L]$, time $[t]$, temperature $[T]$.
2. Force $[F]$, length $[L]$, time $[t]$, temperature $[T]$.
3. Force $[F]$, mass $[M]$, length $[L]$, time $[t]$, temperature $[T]$.

Any valid equation that relates physical quantities must be dimensionally homogeneous; that is, each term in the equation must have the same dimensions. We recognize that Newton's second law ($\vec{F} \propto m\vec{a}$) relates the four dimensions, F, M, L, and t. Thus in system 1, force $[F]$ is a secondary dimension and the constant of proportionality in Newton's second law is dimensionless. In system 2, mass $[M]$ is a secondary dimension, and again the constant of proportionality in Newton's second law is dimensionless. In system 3, both force $[F]$ and mass $[M]$ have been selected as primary dimensions. In this case the constant of proportionality, g_c, in Newton's second law (written $\vec{F} = m\vec{a}/g_c$) is not dimensionless. The dimensions of g_c must in fact be $[ML/Ft^2]$ for the equation to be dimensionally homogeneous.

---

[3] Refer to the text you used in your beginning thermodynamics course. If it is not available, you will find a discussion of dimensions and units presented in each of these texts: W. C. Reynolds, *Thermodynamics*, 2nd ed. (New York: McGraw-Hill, 1968), pp. 15–19; G. J. Van Wylen and R. E. Sonntag, *Fundamentals of Classical Thermodynamics*, 2nd ed. (New York: John Wiley, 1973), pp. 23–29; K. Wark, *Thermodynamics*, 3rd ed. (New York: McGraw-Hill, 1977), pp. 12–18.

The numerical value of the constant of proportionality depends on the units of measure chosen for each of the primary quantities.

1-6.2   SYSTEMS OF UNITS

There is more than one way of selecting the unit of measure for each primary dimension. We shall present only the most common engineering system of units for each of the basic systems of dimensions.

a.   *MLtT*

SI, which is the official abbreviation in all languages for the Systéme International d'Unités[4], is an extension and refinement of the traditional metric system. More than 30 countries have declared it to be the only legally accepted system. The United States will most likely adopt it within the next decade.

In the SI system of units, the unit of mass is the kilogram (kg), the unit of length is the meter (m), the unit of time is the second (sec), and the unit of temperature is the Kelvin (K). Force is a secondary dimension, and its unit, the newton (N), is defined from Newton's second law as

$$1 \text{ N} \equiv \frac{1 \text{ kg} \cdot \text{m}}{\text{sec}^2}$$

In the Absolute Metric system of units, the unit of mass is the gram, the unit of length is the centimeter, the unit of time is the second, and the unit of temperature is the degree Kelvin. Since force is a secondary dimension, the unit of force, the dyne, is defined in terms of Newton's second law as

$$1 \text{ dyne} \equiv \frac{1 \text{ g} \cdot \text{cm}}{\text{sec}^2}$$

b.   *FLtT*

In the British Gravitational system of units, the unit of force is the pound (lbf), the unit of length is the foot, the unit of time is the second, and the unit of temperature is the Rankine (R). Since mass is a secondary dimension, the unit of mass, the slug, is defined in terms of Newton's second law as

$$1 \text{ slug} \equiv \frac{1 \text{ lbf} \cdot \text{sec}^2}{\text{ft}}$$

---

[4]   American Society for Testing and Materials, *Metric Practice Guide* (A Guide to the Use of SI—the International System of Units), Designation E 380-74 (American National Standard Z210.1). Philadelphia: ASTM, 1974.

c. *FMLtT*

In the English Engineering system of units, the unit of force is the pound force (lbf), the unit of mass is the pound mass (lbm), the unit of length is the foot, the unit of time is the second, and the unit of temperature is the Rankine. Since both force and mass are chosen as primary dimensions, Newton's second law is written as

$$\vec{F} = \frac{m\vec{a}}{g_c}$$

A force of one pound (1 lbf) is the force that gives a pound mass (1 lbm) an acceleration equal to the standard acceleration of gravity on Earth, 32.17 ft/sec². From Newton's second law we see that (to three significant figures)

$$1 \text{ lbf} \equiv \frac{1 \text{ lbm} \times 32.2 \text{ ft/sec}^2}{g_c}$$

or

$$g_c \equiv \frac{32.2 \text{ ft} \cdot \text{lbm}}{\text{lbf} \cdot \text{sec}^2}$$

Since a force of 1 lbf accelerates 1 lbm at 32.2 ft/sec², it would accelerate 32.2 lbm at 1 ft/sec². A slug is also accelerated at 1 ft/sec² by a force of 1 lbf. Therefore,

$$1 \text{ slug} \equiv 32.2 \text{ lbm}$$

1-6.3 PREFERRED SYSTEMS OF UNITS

In this text we shall use both the SI and the British Gravitational systems of units. In either case, the constant of proportionality in Newton's second law is dimensionless and has a value of unity. Consequently, Newton's second law is written as $\vec{F} = m\vec{a}$. In these systems, it follows that the gravitational force (the "weight"[5]) on an object of mass, m, is given by $W = mg$.

SI units and prefixes are summarized in Appendix E.

Example 1.3

The density of mercury is given as 26.3 slug/ft³. Calculate the specific weight in lbf/ft³ on Earth and on the moon (acceleration of gravity on the moon is 5.47 ft/sec²). Calculate the specific volume in m³/kg and the specific gravity of the mercury.

---

[5] Note that in the English Engineering system, the weight of an object is given by $W = mg/g_c$.

Example Problem 1.3

GIVEN:

Density of mercury, $\rho = 26.3$ slug/ft^3.
Acceleration of gravity on the moon is 5.47 ft/sec^2.

FIND:

(a) Specific weight, γ, of the mercury, in lbf/ft^3, on Earth and on the moon.
(b) Specific volume, v, in m^3/kg.
(c) Specific gravity, SG.

SOLUTION:

Definitions of specific weight, specific volume, and specific gravity are

Specific weight:
$$\gamma = \frac{\text{weight}}{\text{volume}} = \rho g$$

Specific volume:
$$v = \frac{1}{\rho}$$

Specific gravity:
$$SG = \frac{\rho}{\rho_{H_2O}} = \frac{\gamma}{\gamma_{H_2O}}$$

Properties are $\quad g_{\text{Earth}} = 32.2$ ft/sec^2, $g_{\text{moon}} = 5.47$ ft/sec^2

$$\rho_{H_2O} = 1.94 \text{ slug/ft}^3 \quad \text{(Appendix A)}$$

(a) Specific weight

$$\gamma_{\text{Earth}} = \rho g_{\text{Earth}} = \frac{26.3 \text{ slug}}{\text{ft}^3} \times \frac{32.2 \text{ ft}}{\text{sec}^2} \times \frac{\text{lbf} \cdot \text{sec}^2}{\text{slug} \cdot \text{ft}} = 847 \frac{\text{lbf}}{\text{ft}^3} \qquad \gamma_{\text{Earth}}$$

(Recall from Newton's second law that 1 lbf = 1 slug $\times$ 1 ft/sec^2.)

$$\gamma_{\text{moon}} = \rho g_{\text{moon}} = \frac{26.3 \text{ slug}}{\text{ft}^3} \times \frac{5.47 \text{ ft}}{\text{sec}^2} \times \frac{\text{lbf} \cdot \text{sec}^2}{\text{slug} \cdot \text{ft}} = 144 \frac{\text{lbf}}{\text{ft}^3} \qquad \gamma_{\text{moon}}$$

(b) Specific volume

$$v = \frac{1}{\rho} = \frac{\text{ft}^3}{26.3 \text{ slug}} \times \frac{(0.3048 \text{ m})^3}{\text{ft}^3} \times \frac{\text{slug}}{32.2 \text{ lbm}} \times \frac{\text{lbm}}{0.4536 \text{ kg}} = 7.37 \times 10^{-5} \frac{\text{m}^3}{\text{kg}} \quad v$$

(c) Specific gravity

$$SG = \frac{\rho}{\rho_{H_2O}} = \frac{26.3 \text{ slug}}{\text{ft}^3} \times \frac{\text{ft}^3}{1.94 \text{ slug}} = 13.6 \qquad SG$$

Notice that since mass is independent of the acceleration of gravity that $v_{Earth} = v_{moon}$ and $SG_{Earth} = SG_{moon}$.

The purpose of this problem was to review the definitions of density, specific weight, specific volume, and specific gravity.

summary objectives

After completing study of Chapter 1, you should be able to do the following:

1. Give operational definitions of:

fluid	Eulerian method of description
no-slip condition	dimensions
system	units
control volume	dimensional homogeneity
Lagrangian method of description	weight

2. Give examples in which fluid mechanics is important to an understanding of phenomena from everyday experience and modern technology.

3. List the five basic laws governing the motion of fluids.

4. State the three basic systems of dimensions.

5. Give typical units of physical quantities in the SI, British Gravitational, and English Engineering systems of units.

6. Solve those problems at the end of the chapter which relate to the material you have studied.

problems

1.1 A number of common substances are

Tar	Sand
"Silly Putty"	Jello
Modeling clay	Toothpaste
Wax	Shaving cream

Some of these materials exhibit characteristics of both solid and fluid behavior under different conditions. Explain and give examples.

1.2 Both gases and liquids satisfy the definition of a fluid given in Section 1–1. What characteristics distinguish these two states of matter?

1.3 Look up definitions of the term *fluid* in a dictionary and an encyclopedia. Compare these with the definition in Section 1–1. Are all three equivalent?

1.4 Give a word statement of each of the five basic conservation laws stated in Section 1–3, as they apply to a system.

1.5 Air trapped in a bicycle tire pump is compressed suddenly to $\frac{1}{5}$ of its original volume. Both heat transfer and friction may be neglected as a first approximation. Determine the final temperature of the air in the pump if the initial temperature is 20 C.

1.6 The catapult on an aircraft carrier has a piston diameter of 2 ft and a stroke of 100 ft. Initially, the cylinder contains steam at 250 psia, 600 F in a volume of 100 ft^3. Assume that heat transfer and friction are negligible as the piston moves through its stroke. Determine:
 (a) The temperature and pressure of the steam at the end of the stroke.
 (b) The work done by the steam during the expansion process.

1.7 A compressed air tank in a service station holds 0.2 m^3 of compressed air at 800 kPa (gage). Determine the amount of energy required to compress this much air isothermally from atmospheric pressure, assuming a frictionless process. (Note that the release of this much energy would be catastrophic if the tank were ruptured.)

1.8 A projectile is fired with velocity, $\vec{V}_0$, and elevation angle, θ, above the horizon. Air resistance may be neglected. Express the range of the projectile in terms of V_0 and θ. Determine the angle that gives the maximum range.

1.9 Parametric equations for the motion of a particle are

$$
\begin{aligned}
x &= A\cos\omega t \\
y &= B\sin\omega t
\end{aligned}
\qquad A > B
$$

Determine the velocity and acceleration of the particle as functions of time. Indicate the particle trajectory in the xy plane on a sketch, and locate the point(s) of maximum velocity and acceleration.

1.10 Very small particles moving in fluids experience a drag force proportional to velocity. Consider a particle of net weight, W, dropped in a fluid. The particle experiences a drag force, kV, where V is the particle speed. Determine the time required for the particle to accelerate from rest to 95 percent of its terminal speed, V_t, in terms of k, W, and g.

1.11 A sky diver with a mass of 75 kg is in free fall at an altitude of 2 km. The aerodynamic drag force acting on the sky diver is known to be $F_D = kV^2$ where $k = 0.228$ N·sec^2/m^2. Determine the maximum speed of free fall for the sky diver.

1.12 For each quantity listed, indicate dimensions using the MLtT system of dimensions, and give typical SI and English units:
 (a) Power   (b) Pressure
 (c) Modulus of elasticity   (d) Angular velocity
 (e) Energy   (f) Momentum
 (g) Shear stress   (h) Specific heat
 (i) Thermal expansion coefficient

1.13 For each quantity listed, indicate dimensions using the FLtT system of dimensions, and give typical SI and English units:

(a) Power	(b) Pressure
(c) Modulus of elasticity	(d) Angular velocity
(e) Energy	(f) Moment of a force
(g) Momentum	(h) Shear stress
(i) Strain	

1.14 Air at a pressure of 40 psia and a temperature of 70 F is moving with a speed of 100 ft/sec. Calculate:

(a) The kinetic energy per unit mass of the air.

(b) The kinetic energy per unit volume of the air.

1.15 Air at an absolute pressure of 300 kPa and a temperature of 20 C is moving at a speed of 30 m/sec. Calculate:

(a) The kinetic energy per unit mass of the air.

(b) The kinetic energy per unit volume of the air.

1.16 The stylus of a high-quality stereo system is adjusted to balance a mass of 0.75 g on a small equal arm beam balance. What force (in lbf) does the stylus exert on a record? What is the force expressed in newtons?

1.17 The unit of pressure in the SI system is the pascal. How many pounds force per square inch (psi) correspond to 1 Pa?

1.18 A container weighs 2.9 lbf when empty. When filled with water at 90 F, the mass of the container and its contents is 1.95 slug. Find the weight of water in the container, and its volume in cubic feet, using data from Appendix A.

1.19 Evaluate the change in specific weight of mercury, in lbf/ft^3, as its temperature changes from 70 to 90 F. Use the data in Appendix A.

1.20 Determine the specific gravity of air at 1 atm and 15 C, referred to water at 4 C.

1.21 Determine the specific gravity of mercury at 4 C.

1.22 Four vectors are defined by the equations

$$\vec{r}_1 = 2\hat{\imath} - \hat{\jmath} + \hat{k}$$
$$\vec{r}_2 = \hat{\imath} + 3\hat{\jmath} - 2\hat{k}$$
$$\vec{r}_3 = -2\hat{\imath} + \hat{\jmath} - 3\hat{k}$$
$$\vec{r}_4 = 3\hat{\imath} + 2\hat{\jmath} + 5\hat{k}$$

Determine scalars a, b, and c such that

$$\vec{r}_4 = a\vec{r}_1 + b\vec{r}_2 + c\vec{r}_3$$

1.23 Two vectors, $\vec{a}$ and $\vec{b}$, are given by the expressions

$$\vec{a} = xy\hat{\imath} + y^2\hat{\jmath} + 2\hat{k}$$
$$\vec{b} = x^2\hat{\imath} - xy\hat{\jmath} + z\hat{k}$$

and a scalar, ϕ, is given by

$$\phi = \frac{x^2}{2} - \frac{y^2}{2}$$

Evaluate each of the following:

(a) $\vec{a} \cdot \vec{b}$ (b) $\vec{a} \times \vec{b}$

(c) $\partial \vec{a}/\partial x$ (d) $\nabla \phi$

(e) $\nabla \cdot \vec{a}$ (f) $\nabla \times \vec{b}$

1.24 Two vectors, $\vec{c}$ and $\vec{d}$, are given by the expressions

$$\vec{c} = xyz\hat{\imath} + 2\hat{\jmath} + y^2\hat{k}$$
$$\vec{d} = x^2\hat{\imath} + y^2\hat{\jmath} + x\hat{k}$$

and a scalar, ψ, is given by

$$\psi = xy$$

Evaluate each of the following:

(a) $\vec{c} \cdot \vec{d}$ (b) $\vec{c} \times \vec{d}$

(c) $\partial \vec{c}/\partial x$ (d) $\nabla \psi$

(e) $\nabla \times \vec{c}$ (f) $\nabla \cdot \vec{d}$

1.25 Two vectors, $\vec{r}$ and $\vec{s}$, are given by the expressions

$$\vec{r} = x^2 y\hat{\imath} + z^2\hat{k}$$
$$\vec{s} = xz\hat{\imath} + xyz\hat{\jmath} + x^2 y\hat{k}$$

and a scalar, ζ, is given by

$$\zeta = \tfrac{1}{2}(x^2 - y^2)$$

Evaluate each of the following:

(a) $\vec{r} \cdot \vec{s}$ (b) $\vec{r} \times \vec{s}$

(c) $\partial \vec{s}/\partial z$ (d) $\nabla \times \vec{r}$

(e) $\nabla \cdot \vec{s}$ (f) $\nabla^2 \zeta$

Chapter 2

Fundamental Concepts

In Chapter 1 we indicated that our study of fluid mechanics will build on earlier studies in mechanics and thermodynamics. To develop a unified approach, we shall need to review some familiar topics and to introduce some new concepts and definitions. The purpose of this chapter is to develop these fundamental concepts.

2-1 FLUID AS A CONTINUUM

In our definition of a fluid, no mention was made of the molecular structure of fluids. All fluids are composed of molecules in constant motion. However, in most engineering applications we are interested in the gross or average (i.e. macroscopic) effects of many molecules. It is these macroscopic effects that we can perceive and measure. We thus treat a fluid as virtually an infinitely divisible substance (i.e. as a continuum) and do not concern ourselves with the behavior of individual molecules.

The concept of a *continuum* is the basis of classical fluid mechanics. The continuum assumption is valid in treating the behavior of fluids under normal conditions. However, it breaks down whenever the mean free path of the molecules (approximately 6.3×10^{-5} mm or 2.5×10^{-6} in. for air at STP)[1] becomes the same order of magnitude as the smallest significant characteristic dimension of the problem. In problems such as rarefied gas flow (e.g. as encountered in flights into the upper reaches of the atmosphere), we must abandon the concept of a continuum in favor of the microscopic and statistical points of view.

[1] STP (Standard Temperature and Pressure) for air are 15 C (59 F) and 101.3 kPa absolute (14.696 psia), respectively.

As a consequence of the continuum assumption, each fluid property is assumed to have a definite value at each point in space. Thus fluid properties such as density, temperature, velocity, and so on, are considered to be continuous functions of position and time.

To illustrate the concept of a property at a point, consider the manner in which we determine the density at a point. A region of fluid is shown in Fig. 2.1. We are interested in determining the density at the point C, whose coordinates are x_0, y_0, and z_0. The density is defined as mass per unit volume. Thus the mean density within the volume V would be given by $\rho = m/V$. Clearly, this will not in general be equal to the value of the density at point C. To determine the density at point C, we must select as small a volume, δV, as possible surrounding point C and determine the ratio of the mass, δm, within the volume to the volume. The question is how small can we make the volume? Let us answer this question by plotting the ratio $\delta m/\delta V$ and allowing the volume to shrink continuously in size. Assuming that the volume δV is initially relatively large (but still small compared to the volume, V) a typical plot of $\delta m/\delta V$ might appear as in Fig. 2.1b. The average density tends to approach an asymptotic value as the volume is shrunk to enclose only homogeneous fluid in the immediate neighborhood of point C. When δV becomes so small that it contains only a small number of molecules, it becomes impossible to fix a definite value for $\delta m/\delta V$; the value will vary erratically as molecules cross in and out of the volume. Thus there is a lower limiting value of δV, designated $\delta V'$ in Fig. 2.1b, allowable for use in the definition of fluid density at a point. The density at a point is then defined as

$$\rho \equiv \lim_{\delta V \to \delta V'} \frac{\delta m}{\delta V} \qquad (2.1)$$

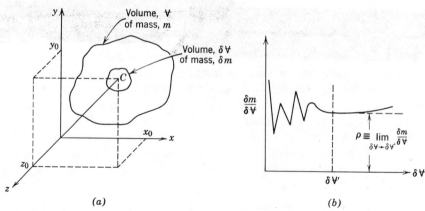

Fig. 2.1 Definition of density at a point.

Since the point C was an arbitrary point in the fluid, the density at any point could be determined in a like manner. Thus, if density determinations were made simultaneously at an infinite number of points in the fluid, we would obtain an expression for the density distribution as a function of the space coordinates, that is, $\rho = \rho(x, y, z)$, at the given instant in time. Clearly, the density at a point may vary with time as a result of work done on or by the fluid and/or heat transfer to the fluid. Thus the complete representation of density (i.e. the field representation) is given by

$$\rho = \rho(x, y, z, t) \tag{2.2}$$

Since the density is a scalar quantity, requiring only the specification of a magnitude for a complete description, the field represented by Eq. 2.2 is a scalar field.

2-2 VELOCITY FIELD

In the previous section we saw that the continuum assumption led directly to a field representation of fluid properties. We looked specifically at the density field.

In dealing with fluids in motion, we shall necessarily be concerned with the description of a velocity field. Referring again to Fig. 2.1a, we see that the fluid velocity at point C is defined as the instantaneous velocity of the center of gravity of the volume $\delta V'$ instantaneously surrounding point C. Thus, if we define a fluid particle as the small mass of fluid of fixed identity of volume $\delta V'$, we are defining the velocity at point C as the instantaneous velocity of the fluid particle which, at a given instant, is passing through the point C. The velocity at any point in the flow field is defined in a like manner. At a given instant of time the velocity field, $\vec{V}$, is a function of the space coordinates x, y, z, that is, $\vec{V} = \vec{V}(x, y, z)$. The velocity at a given point in the flow field may vary from one instant of time to another. Thus the complete representation of velocity (i.e. the velocity field) is given by

$$\vec{V} = \vec{V}(x, y, z, t) \tag{2.3}$$

If fluid properties at a point in a field do not change with time, the flow is termed a *steady flow.* Stated mathematically, the definition of steady flow is

$$\frac{\partial \eta}{\partial t} = 0$$

where η represents any fluid property. For steady flow,

$$\frac{\partial \rho}{\partial t} = 0, \quad \text{or} \quad \rho = \rho(x, y, z)$$

and

$$\frac{\partial \vec{V}}{\partial t} = 0, \qquad \text{or} \qquad \vec{V} = \vec{V}(x, y, z)$$

Thus, in steady flow, properties may vary from point to point in the field, but they must remain constant with time at any given point.

2-2.1 ONE-, TWO-, AND THREE-DIMENSIONAL FLOWS

Equation 2.3 indicates that the velocity field is a function of three space coordinates and time. Such a flow field is termed *three-dimensional* (it is also unsteady) because the velocity at any point in the flow field depends on the three coordinates required to locate the point in space.

Not all flow fields are three-dimensional. Consider, for example, the flow through a long straight pipe of constant cross section. Far from the entrance to the pipe the velocity distribution may be given by

$$u = u_{\max}\left[1 - \left(\frac{r}{R}\right)^2\right] \tag{2.4}$$

This profile is shown in Fig. 2.2, where cylindrical coordinates r, θ, and x are used to locate any point in the flow field. Since the velocity field is a function of r only, that is, it is independent of the coordinates x and θ, this is a one-dimensional flow.

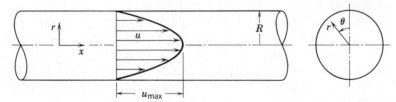

Fig. 2.2 Example of one-dimensional flow.

A flow is classified as one-, two-, or three-dimensional depending on the number of space coordinates required to specify the velocity field.[2]

An example of a two-dimensional flow is illustrated in Fig. 2.3; the velocity distribution is depicted for a flow between diverging straight walls that are imagined to be infinite in extent (in the z direction). Since the channel is considered to be infinite in the z direction, the velocity field will be identical

---

[2] Some authors choose to classify a flow as one-, two-, or three-dimensional on the basis of the number of space coordinates required to specify all fluid properties. In this text, classification of flow fields will be based on the number of space coordinates required to specify the velocity field only.

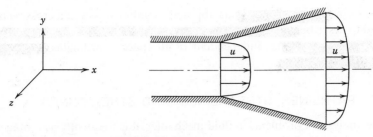

Fig. 2.3 Example of a two-dimensional flow.

in all planes perpendicular to the z axis. The velocity field is consequently a function only of the space coordinates x and y; the flow field is thus classified as two-dimensional.

As you might suspect, the complexity of analysis increases considerably with the number of dimensions of the flow field. The simplest to analyze is one-dimensional flow. For many problems encountered in engineering, a one-dimensional analysis proves adequate to provide approximate solutions of engineering accuracy.

Since all fluids satisfying the continuum assumption must have a zero relative velocity at a solid surface (to satisfy the no-slip condition), most flows are inherently two- or three-dimensional. For purposes of analysis it often is convenient to introduce the notion of uniform flow at a given cross section. In a flow that is uniform at a given cross section, the velocity is constant across any section normal to the flow. Under this assumption,[3] the two-dimensional flow of Fig. 2.3 is modeled as the flow shown in Fig. 2.4. In the flow of Fig. 2.4, the velocity field is a function of x alone, and thus the flow is one-dimensional. (Other properties, e.g. density or pressure, also may be assumed uniform at a section, if appropriate.)

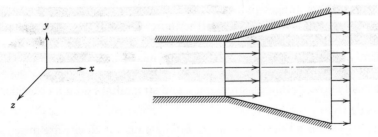

Fig. 2.4 Example of uniform flow at a section.

[3] Convenience alone does not justify this assumption; often results of acceptable accuracy are obtained. Sweeping assumptions such as uniform flow at a cross section should always be reviewed carefully to be sure they provide a reasonable analytical model of the real flow.

The term *uniform flow field* (as opposed to uniform flow at a cross section) is used to describe a flow in which the magnitude and direction of the velocity vector are constant, i.e. independent of all space coordinates, throughout the entire flow field.

2-2.2 PATHLINES, STREAKLINES, AND STREAMLINES

In the analysis of problems in fluid mechanics, it is frequently advantageous to obtain a visual representation of a flow field. Such a representation is provided by pathlines, streaklines, and streamlines.[4]

A pathline is the path or trajectory traced out by a moving fluid particle. To determine a pathline, we might identify a fluid particle at a given instant of time, for example, by the use of dye and then take a long exposure photograph of its subsequent motion. The line traced out by the particle is a pathline.

On the other hand, we may choose to focus our attention on a fixed location in space and identify, again by the use of dye, all fluid particles passing through this point. After a short period of time we then have a number of identifiable fluid particles in the flow, all of which had, at some time, passed through one fixed location in space. The line joining these fluid particles is defined as a streakline.

Streamlines are lines drawn in the flow field so that at a given instant of time they are tangent to the direction of flow at every point in the flow field. The shape of the streamlines may vary from instant to instant if the flow velocity is a function of time, that is, if the flow is unsteady. Since the streamlines are tangent to the velocity vector at every point in the flow field, there can be no flow across a streamline.

In steady flow, the velocity at each point in the flow field remains constant with time and, consequently, the streamlines do not vary from one instant to the next. This implies that a particle located on a given streamline will remain on the same streamline. Furthermore, consecutive particles passing through a fixed point in space will be on the same streamline and, subsequently, will remain on this streamline. Thus in a steady flow, pathlines, streaklines, and streamlines are identical lines in the flow field. In the case of unsteady flow, pathlines, streaklines, and streamlines do not coincide.

Example 2.1

A velocity field is given by $\vec{V} = ay\hat{\imath} + b\hat{\jmath}$; the units of velocity are m/sec and y is given in meters. $a = 2 \text{ sec}^{-1}$ and $b = 1 \text{ m/sec}$.

---

[4] Pathlines, streaklines, and streamlines are demonstrated in the film loops: S-FM047, *Pathlines, Streaklines, Streamlines, and Timelines in Steady Flow*, and S-FM048, *Pathlines, Streaklines, and Streamlines in Unsteady Flow*. These two loops are taken from the film *Flow Visualization*, S. J. Kline, principal.

(a) Is this flow field one-, two-, or three-dimensional? Why?

(b) Determine the velocity components u, v, w, at the point (1, 2, 0).

(c) Determine the slope of the streamline through the point (1, 2, 0).

Example Problem 2.1

GIVEN:

Velocity field, $V = ay\hat{\imath} + b\hat{\jmath}$, m/sec; $a = 2$ sec^{-1}, $b = 1$ m/sec, y in meters.

FIND:

(a) Dimensions of the flow field.

(b) Velocity components u, v, w at point (1, 2, 0).

(c) Slope of streamline at point (1, 2, 0).

SOLUTION:

(a) A flow is classified as one-, two-, or three-dimensional depending on the number of space coordinates required to specify the velocity field. Since the given velocity field is a function of one space coordinate, the flow field is one-dimensional.

(b) The velocity field, $\vec{V} = \hat{\imath}u + \hat{\jmath}v + \hat{k}w$. Since $\vec{V} = ay\hat{\imath} + b\hat{\jmath}$, then

$$u = ay \qquad v = b \qquad w = 0$$

At the point (1, 2, 0)

$$u = 4 \text{ m/sec} \qquad v = 1 \text{ m/sec} \qquad w = 0$$

(c) Streamlines are lines drawn in the flow field such that, at a given instant of time, they are tangent to the direction of flow at every point in the flow field. Consequently, the slope of the streamline at the point (1, 2, 0) is such that the streamline is tangent to the velocity vector at the point.

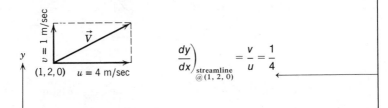

$$\left.\frac{dy}{dx}\right)_{\substack{\text{streamline} \\ @ (1,\, 2,\, 0)}} = \frac{v}{u} = \frac{1}{4}$$

$$\left\{\begin{array}{l}\text{The purpose of this problem was to illustrate:} \\ \text{(i)} \quad \text{the specification of a velocity field, and} \\ \text{(ii)} \quad \text{the definition of a streamline.}\end{array}\right\}$$

2-3   STRESS FIELD

Stresses in a medium result from forces acting on some portion of the medium. The concept of stress provides a convenient means to describe the manner in which forces acting on the boundaries of the medium are transmitted through the medium. Since force and area are both vector quantities, we might anticipate that the stress field will not be a vector field. We shall show that, in general, nine quantities are required to specify the state of stress in a fluid. (Stress is a tensor quantity of second order.)

2-3.1   SURFACE AND BODY FORCES

Surface and body forces are encountered in the study of continuum fluid mechanics. Surface forces include all forces acting on the boundaries of a medium through direct contact. Forces developed without physical contact, and distributed over the volume of the fluid, are termed *body forces.* Gravitational and electromagnetic forces are examples of body forces arising in a fluid.

The gravitational body force acting on an element of volume, $d\mathbf{V}$, is given by $\rho\vec{g}\,d\mathbf{V}$, where ρ is the density (mass per unit volume) and $\vec{g}$ is the local gravitational acceleration. Thus the gravitational body force per unit volume is $\rho\vec{g}$ and the gravitational body force per unit mass is $\vec{g}$.

Example 2.2

A body-force distribution is given as

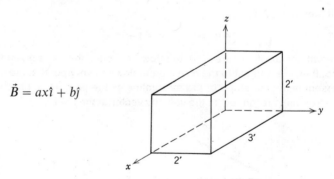

$$\vec{B} = ax\hat{\imath} + b\hat{\jmath}$$

lbf per unit mass (slug) of material acted on. The density (slug/ft^3) of the material is given by

$$\rho = cx + ez^3$$

where x, y, and z are given in feet; $a = 10$ lbf/slug$\cdot$ft, $b = 15$ lbf/slug, $c = 1$ slug/ft^4, and $e = 1$ slug/ft^6. What is the resultant body force on the material in the region shown?

Example Problem 2.2

GIVEN:

Body-force distribution per unit mass

$$\vec{B} = ax\hat{i} + b\hat{j} \quad \text{(lbf/slug)}$$

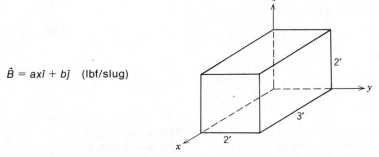

Density of material in volume shown

$$\rho = cx + ez^3 \quad \text{(slug/ft}^3\text{)}$$

where $a = 10$ lbf/slug · ft, $b = 15$ lbf/slug, $c = 1$ slug/ft^4, and $e = 1$ slug/ft^6.

FIND:

Resultant body force on material in volume shown.

SOLUTION:

Body force per unit mass is $\vec{B}$.
Multiplying by mass per unit volume (ρ) gives the force per unit volume $= \rho\vec{B}$.
Thus the body force, $d\vec{F}_B$, on an element of volume $d\forall$ is

$$d\vec{F}_B = \rho\vec{B} \, d\forall$$

and

$$\vec{F}_B = \int_{\forall} \rho\vec{B} \, d\forall = \int_{x=0}^{x_1} \int_{y=0}^{y_1} \int_{z=0}^{z_1} \rho\vec{B} \, dz \, dy \, dx, \quad \text{where } x_1 = 3 \text{ ft}, y_1 = 2 \text{ ft}, z_1 = 2 \text{ ft}$$

$$= \int_0^{x_1} \int_0^{y_1} \int_0^{z_1} (cx + ez^3)(ax\hat{i} + b\hat{j}) \, dz \, dy \, dx$$

$$\vec{F}_B = \int_0^{x_1} \int_0^{y_1} \int_0^{z_1} \{a(cx^2 + exz^3)\hat{i} + b(cx + ez^3)\hat{j}\} \, dz \, dy \, dx$$

Integrating first with respect to z and substituting limits gives

$$\vec{F}_B = \int_0^{x_1} \int_0^{y_1} \{a(cx^2 z_1 + \tfrac{1}{4}exz_1^4)\hat{i} + b(cxz_1 + \tfrac{1}{4}ez_1^4)\hat{j}\} \, dy \, dx$$

Integrating now with respect to y and substituting limits gives

$$\vec{F}_B = \int_0^{x_1} \{ay_1(cx^2z_1 + \tfrac{1}{4}exz_1^4)\hat{\imath} + by_1(cxz_1 + \tfrac{1}{4}ez_1^4)\hat{\jmath}\}\, dx$$

Finally, integrating with respect to x and substituting limits gives

$$\vec{F}_B = ay_1(\tfrac{1}{3}cx_1^3z_1 + \tfrac{1}{8}ex_1^2z_1^4)\hat{\imath} + by_1(\tfrac{1}{2}cx_1^2z_1 + \tfrac{1}{4}ex_1z_1^4)$$

Substituting numerical values gives

$$\vec{F}_B = \frac{10}{\text{slug}\cdot\text{ft}}\,\frac{\text{lbf}}{} \times 2\,\text{ft}\left\{\frac{1}{3}\times\frac{1\,\text{slug}}{\text{ft}^4}\times(3\,\text{ft})^3\times 2\,\text{ft} + \frac{1}{8}\times\frac{1\,\text{slug}}{\text{ft}^6}\times(3\,\text{ft})^2\times(2\,\text{ft})^4\right\}\hat{\imath}$$

$$+ \frac{15}{\text{slug}}\,\frac{\text{lbf}}{} \times 2\,\text{ft}\left\{\frac{1}{2}\times\frac{1\,\text{slug}}{\text{ft}^4}\times(3\,\text{ft})^2\times 2\,\text{ft} + \frac{1}{4}\times\frac{1\,\text{slug}}{\text{ft}^6}\times 3\,\text{ft}\times(2\,\text{ft})^4\right\}\hat{\jmath}$$

$$\vec{F}_B = 720\hat{\imath} + 630\hat{\jmath}\ \text{lbf} \qquad\qquad\qquad \vec{F}_B$$

Although admittedly artificial, this problem is included to:
(i) illustrate the calculation of a resultant body force, and
(ii) review the procedure for evaluating triple integrals over a volume.

2-3.2 STRESS AT A POINT

The description of the stress field is developed from the analysis of stress at a point. Consider the element of area, $\delta\vec{A}$, at the point C, acted on by the force, $\delta\vec{F}$, as shown in Fig. 2.5. (The magnitude of $\delta\vec{A}$ is the area of the element, and the direction is normal to the surface.) The stress at a point is defined as

$$\text{stress} \equiv \lim_{\delta\vec{A}\to 0} \frac{\delta\vec{F}}{\delta\vec{A}} \qquad\qquad (2.5)$$

The definition of stress requires that we evaluate the ratio of the two vectors $\delta\vec{F}$ and $\delta\vec{A}$. In order to perform this operation, let us consider what we know about vectors. The vector $\delta\vec{A}$ can be resolved into three components,

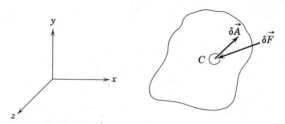

Fig. 2.5 Definition of stress.

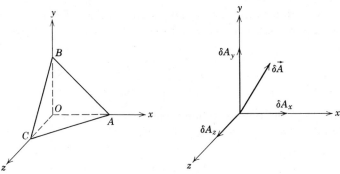

Fig. 2.6 Element of area.

that is,

$$\delta \vec{A} = \hat{\imath}\,\delta A_x + \hat{\jmath}\,\delta A_y + \hat{k}\,\delta A_z$$

A simple case is shown in Fig. 2.6 where the element of area $\delta\vec{A}$ is the element ABC.

The component δA_x is the x component of $\delta\vec{A}$, that is, the projection of $\delta\vec{A}$ on the x axis. Its magnitude is the projection of ABC on the yz plane (area OBC), that is, on the plane perpendicular to the x axis.

Similarly, δA_y is the y component of $\delta\vec{A}$, and its magnitude is the projection of ABC on the xz plane (area OCA), that is, on the plane perpendicular to the y axis.

δA_z is the z component of $\delta\vec{A}$, and its magnitude is the projection of ABC on the xy plane (area OAB), that is, on the plane perpendicular to the z axis.

The vector force, $\delta\vec{F}$, acting at the point C of Fig. 2.5 also can be resolved into three components, that is,

$$\delta\vec{F} = \hat{\imath}\,\delta F_x + \hat{\jmath}\,\delta F_y + \hat{k}\,\delta F_z$$

In defining the stress at a point, we may then consider the components δF_x, δF_y, and δF_z of the force $\delta\vec{F}$ at the point C acting on the components δA_x, δA_y, and δA_z of the area $\delta\vec{A}$ at the point C. Viewed in this manner, we are in effect defining the components of stress at point C. Thus Eq. 2.5 may be replaced by nine equations, for we have three stress components (resulting from each of the three force components δF_x, δF_y, and δF_z) acting on each of the three area components δA_x, δA_y, and δA_z.

In order to keep things straight, it is helpful to use a notation that will enable us to delineate both the plane on which the stress is acting and the direction in which the stress is acting. A double subscript notation provides such a convenient description. Thus T_{ij} denotes the stress acting on an i

plane in the j direction, where i and j each can stand for x, y, or z. Using this notation, we can write the definition of stress (Eq. 2.5) in terms of components as

$$T_{ij} \equiv \lim_{\delta A_i \to 0} \frac{\delta F_j}{\delta A_i} \qquad (2.6)$$

Equation 2.6 really represents nine scalar equations, since the subscripts i and j can take on the values x, y, z. For example,

$$T_{xy} \equiv \lim_{\delta A_x \to 0} \frac{\delta F_y}{\delta A_x}$$

defines the stress on an x plane in the y direction. The stress at a point is specified by the nine components

$$\begin{bmatrix} \sigma_{xx} & \tau_{xy} & \tau_{xz} \\ \tau_{yx} & \sigma_{yy} & \tau_{yz} \\ \tau_{zx} & \tau_{zy} & \sigma_{zz} \end{bmatrix}$$

where σ has been used to denote a normal stress and shear stresses are denoted by the symbol τ. The notation for designating stress is shown in Fig. 2.7.

Referring to the infinitesimal element shown in Fig. 2.7, we see that there are six planes (two x planes, two y planes, and two z planes) on which stresses may act. In order to designate the plane of interest, we could use terms like

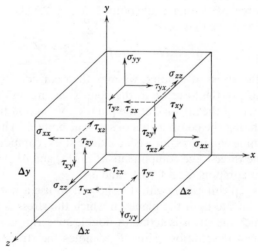

Fig. 2.7 Notation for stress.

front and back, top and bottom, or left and right. However it is more logical to name the planes in terms of the coordinate axes. The planes are named and denoted as positive or negative according to the direction of the outward drawn normal to the plane. Thus the top plane, for example, is a positive y plane and the back plane is a negative z plane.

It is also necessary to adopt a sign convention for the stress. A stress component is considered positive when the direction of the stress component and the plane on which it acts are both positive or both negative. Thus $\tau_{yx} = 5$ lbf/in.^{2} represents a shear stress on a positive y plane in the positive x direction or a shear stress on a negative y plane in the negative x direction. In Fig. 2.7 all stresses have been drawn as positive stresses. Stress components are negative when the direction of the stress component and the plane on which it acts are of opposite sign.

The specification of the stress field (a tensor field), is more involved than that of the velocity field (a vector field), or the density field (a scalar field).

2-4 NEWTONIAN FLUID: VISCOSITY

2-4.1 NEWTONIAN FLUID

We have defined a fluid as a substance that deforms continuously under the action of a shearing stress. In the absence of a shear stress, there will be no deformation. Fluids may be broadly classified according to the relation between the applied shear stress and the rate of deformation of the fluid. Fluids in which the shear stress is directly proportional to the rate of deformation are termed *Newtonian fluids.* Most common fluids such as water, air, and gasoline are closely Newtonian under normal conditions. The term *non-Newtonian* is used to classify all fluids in which the shear stress is not directly proportional to the rate of deformation.

Many common fluids exhibit non-Newtonian behavior. Two familiar examples are toothpaste and Lucite[5] paint. The latter is very "thick" when in the can, but becomes "thin" when sheared by brushing. In this way, a lot of paint is carried by the brush, to minimize the number of times it must be dipped. Toothpaste behaves as a "fluid" when squeezed from the tube. However, it does not run out by itself when the cap is removed. There is a threshold or yield stress below which toothpaste behaves as a solid. Strictly speaking, our definition of a fluid is valid only for materials that have zero yield stress. Non-Newtonian fluids will not be treated in this text.

Consider the behavior of a fluid element between the two infinite plates shown in Fig. 2.8. The upper plate moves at constant velocity, δu, under the

---

[5] Trademark, E. I. Du Pont de Nemours & Company.

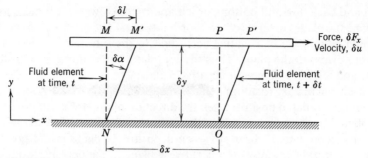

Fig. 2.8 Deformation of a fluid element.

influence of a constant applied force, δF_x. The shear stress, τ_{yx}, applied to the fluid element is given by

$$\tau_{yx} = \lim_{\delta A_y \to 0} \frac{\delta F_x}{\delta A_y} = \frac{dF_x}{dA_y}$$

where δA_y is the area of the fluid element in contact with the plate. In the time increment, δt, the fluid element is deformed from position $MNOP$ to position $M'NOP'$. The rate of deformation of the fluid is then given by

$$\text{deformation rate} = \lim_{\delta t \to 0} \frac{\delta \alpha}{\delta t} = \frac{d\alpha}{dt}$$

The fluid is Newtonian if $\tau_{yx} \propto d\alpha/dt$.

If we wish to calculate the shear stress, τ_{yx}, it is certainly desirable to formulate an expression for $d\alpha/dt$ in terms of more readily measurable quantities. This can be done easily. The distance, δl, between the points M and M' is given by

$$\delta l = \delta u \, \delta t$$

or alternatively, for small angles,

$$\delta l = \delta y \, \delta \alpha$$

Equating these two expressions for δl gives

$$\frac{\delta \alpha}{\delta t} = \frac{\delta u}{\delta y}$$

and on taking the limit of both sides of the equality we obtain

$$\frac{d\alpha}{dt} = \frac{du}{dy}$$

Thus, if the fluid of Fig. 2.8 is Newtonian,

$$\tau_{yx} \propto \frac{du}{dy} \tag{2.7}$$

The shear stress acts on a plane normal to the y axis.

2-4.2 VISCOSITY

If one considers the deformation of two different Newtonian fluids, say glycerin and water, one recognizes that they will deform at different rates under the action of the same applied shear stress. Glycerin exhibits a much larger resistance to deformation than water. Thus we say it is much more viscous. The constant of proportionality in Eq. 2.7 is the absolute (or dynamic) viscosity, μ. Thus in terms of the coordinates of Fig. 2.8, Newton's law of viscosity is given by

$$\tau_{yx} = \mu \frac{du}{dy} \tag{2.8}$$

Note that since the dimensions of τ are $[F/L^2]$ and the dimensions of du/dy are $[1/t]$, then μ has the dimensions of $[Ft/L^2]$. Since the dimensions of force, F, mass, M, length, L, and time, t, are related by Newton's second law of motion, the dimensions of μ can also be expressed as $[M/Lt]$. In the British Gravitational system, the units of viscosity are lbf·sec/ft^2 or slug/ft·sec. In the Absolute Metric system, the basic unit of viscosity is called a poise (poise $\equiv$ g/cm·sec); in the SI system the units of viscosity are kg/m·sec or Pa·sec ($=$ N·sec/m^2). The calculation of viscous shear stress is illustrated in Example 2.3.

In fluid mechanics the ratio of absolute viscosity, μ, to the density, ρ, often arises. This ratio is given the name kinematic viscosity and is represented by the symbol, ν. Since density has the dimensions $[M/L^3]$, the dimensions of ν are $[L^2/t]$. In the Absolute Metric system of units, the unit for ν is a stoke (stoke $\equiv$ cm^2/sec).

Viscosity arises in real fluids from molecular motions. Molecules from regions of high bulk velocity collide with molecules moving with lower bulk velocity, and vice versa. These collisions transport momentum from one region of fluid to another. Since the random molecular motions are affected by the temperature of the medium, viscosity is also a function of temperature. Viscosity data for a number of common Newtonian fluids are given in Appendix A.

Example 2.3

An infinite plate is moved over a second plate on a layer of liquid as shown in the diagram. For small gap width, d, we assume a linear velocity distribution in the liquid. The fluid properties are

$$\mu = 0.65 \text{ centipoise (cp)}$$
$$SG = 0.88$$

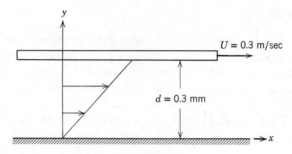

Calculate:
(a) The absolute viscosity of the fluid, in $lbf \cdot sec/ft^2$.
(b) The kinematic viscosity of the fluid, in m^2/sec.
(c) The shear stress on the upper plate, in lbf/ft^2.
(d) The shear stress on the lower plate, in Pa.
(e) Indicate the direction of each shear stress calculated in parts (c) and (d).

Example Problem 2.3

GIVEN:

Linear velocity profile in the fluid between infinite parallel plates as shown.

$$\mu = 0.65 \text{ cp } (1 \text{ poise} = 1 \text{ g/cm} \cdot \text{sec})$$
$$SG = 0.88$$

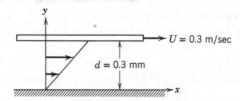

FIND:

(a) μ in units of $lbf \cdot sec/ft^2$,
(b) v in units of m^2/sec,

(c) τ on upper plate in units of lbf/ft^2,

(d) τ on lower plate in units of Pa,

(e) Direction of stress in parts (c) and (d).

SOLUTION:

Basic equation: $\tau_{yx} = \mu \dfrac{du}{dy}$ Definition: $v = \dfrac{\mu}{\rho}$

(a) $\mu = \dfrac{0.65 \text{ cp}}{} \times \dfrac{\text{poise}}{100 \text{ cp}} \times \dfrac{g}{\text{cm} \cdot \text{sec} \cdot \text{poise}} \times \dfrac{\text{lbm}}{453.6 \text{ g}} \times \dfrac{\text{slug}}{32.2 \text{ lbm}} \times \dfrac{30.48 \text{ cm}}{\text{ft}}$

$\times \dfrac{\text{lbf} \cdot \text{sec}^2}{\text{slug} \cdot \text{ft}}$

$\mu = 1.36 \times 10^{-5} \dfrac{\text{lbf} \cdot \text{sec}}{\text{ft}^2}$ 　　　　　　　　　　　　　　　　　　 μ ←

(b) $v = \dfrac{\mu}{\rho} = \dfrac{\mu}{SG\rho_{H_2O}}$

$= \dfrac{1.36 \times 10^{-5} \text{ lbf} \cdot \text{sec}}{\text{ft}^2} \times \dfrac{\text{ft}^3}{(0.88)1.94 \text{ slug}} \times \dfrac{\text{slug} \cdot \text{ft}}{\text{lbf} \cdot \text{sec}^2} \times \dfrac{(0.3048)^2 \text{ m}^2}{\text{ft}^2}$

$v = 7.40 \times 10^{-7} \dfrac{\text{m}^2}{\text{sec}}$ 　　　　　　　　　　　　　　　　　　　　 v ←

(c) $\tau_{upper} = \tau_{yx, upper} = \mu \left. \dfrac{du}{dy} \right)_{y=d}$. Since u varies linearly with y,

$\dfrac{du}{dy} = \dfrac{\Delta u}{\Delta y} = \dfrac{U - 0}{d - 0} = \dfrac{U}{d} = \dfrac{0.3 \text{ m}}{\text{sec}} \times \dfrac{1}{0.3 \text{ mm}} \times \dfrac{1000 \text{ mm}}{\text{m}} = 1000 \text{ sec}^{-1}$

$\tau_{upper} = \mu \dfrac{U}{d} = \dfrac{1.36 \times 10^{-5} \text{ lbf} \cdot \text{sec}}{\text{ft}^2} \times \dfrac{1000}{\text{sec}} = 0.0136 \dfrac{\text{lbf}}{\text{ft}^2}$ 　　　 τ_{upper} ←

(d) $\tau_{lower} = \mu \dfrac{U}{d} = \dfrac{0.0136 \text{ lbf}}{\text{ft}^2} \times \dfrac{4.448 \text{ N}}{\text{lbf}} \times \dfrac{\text{ft}^2}{(0.3048)^2 \text{ m}^2} \times \dfrac{Pa \cdot m^2}{N} = 0.651 \text{ Pa}$ 　 τ_{lower} ←

(e) Direction of shear stress on upper and lower plates.

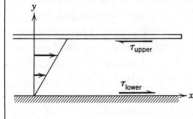

$\left\{ \begin{array}{l} \text{The upper plate is a negative } y \text{ surface, so} \\ \text{positive } \tau_{yx} \text{ acts in the negative } x \text{ direction.} \end{array} \right\}$

$\left\{ \begin{array}{l} \text{The lower plate is a positive } y \text{ surface, so} \\ \text{positive } \tau_{yx} \text{ acts in the positive } x \text{ direction.} \end{array} \right\}$

(e) ←

2-5 DESCRIPTION AND CLASSIFICATION OF FLUID MOTIONS

In Chapter 1 we listed a wide variety of typical problems encountered in fluid mechanics and outlined our method of approach to the subject. Before proceeding with our detailed study, we shall attempt a broad classification of fluid mechanics on the basis of observable physical characteristics of flow fields. Since there is much overlap in the types of flow fields encountered, there is no universally accepted classification scheme. One possible classification is shown in Fig. 2.9.

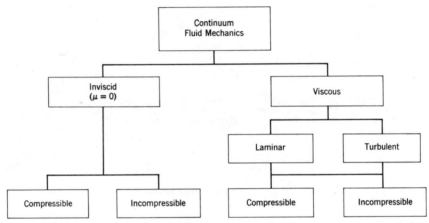

Fig. 2.9 Possible classification of continuum fluid mechanics.

2-5.1 VISCOUS AND INVISCID FLOWS

The main subdivision indicated is between inviscid and viscous flows. In an inviscid flow the fluid viscosity, μ, is assumed to be zero. Clearly, such flows do not exist; however, there are many problems wherein such an assumption will simplify the analysis and, at the same time, lead to meaningful results. (While simplification of the analysis is always desirable, the results must be reasonably accurate if the solution is to be of value.) Within the area of inviscid flow we may consider problems in two broad classes. Flows in which variations of density are small and relatively unimportant are termed incompressible. Flows in which density variations play a dominant role, as in high speed gas flows, are called compressible flows. We shall consider both of these subgroups within the area of inviscid flows.

All fluids possess viscosity and, consequently, viscous flows are of paramount importance in the study of continuum fluid mechanics. Although we shall study viscous flows in some detail later, it is instructive to consider a few examples of viscous flow phenomena.

In our discussion following the definition of a fluid (Section 1-1), we noted that in any viscous flow, the fluid in direct contact with a solid boundary has the same velocity as the boundary itself; that is, there is no slip at the boundary. For the one-dimensional viscous flow of Fig. 2.8, the shear stress was given by Eq. 2.8.

$$\tau_{yx} = \mu \frac{du}{dy} \qquad (2.8)$$

Since the fluid velocity at a stationary solid surface in a moving fluid is zero, but the bulk fluid is moving, velocity gradients and hence, shear stresses, must be present in the flow. These stresses in turn affect the motion.

As a practical case, we might look at the fluid motion around a thin wing or ship hull. Such a flow might be represented crudely by the flow over a flat plate, as shown in Fig. 2.10. The flow approaching the plate is of uniform velocity, U_∞. We are interested in providing a qualitative picture of the velocity distribution at various locations along the plate. Two such locations are denoted by x_1 and x_2. Consider first the location x_1. In order to arrive at a qualitative picture of the velocity distribution, we start by labeling the y coordinates at which the velocity is known.

From the no-slip condition, we know the velocity at point A must be zero; hence we have one point on the velocity profile. Can we locate any other points on the profile? Let us stop for a minute and ask ourselves, "What is the effect of the plate on the flow?" The plate is stationary and, therefore, exerts a retarding force on the flow; it slows the fluid in the neighborhood of the surface. At a y location sufficiently far from the plate, say point B, the flow will not be influenced by the presence of the plate. If the pressure does not vary in the x direction (as is the case for flow over a semi-infinite flat

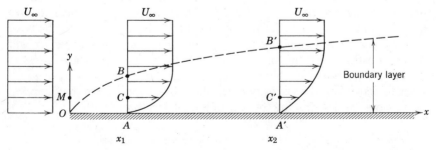

Fig. 2.10 Incompressible laminar viscous flow over a semi-infinite flat plate.

plate) the velocity at point B will be U_∞. It seems reasonable to expect the velocity to increase smoothly and monotonically from the value $u = 0$ at $y = 0$ to $u = U_\infty$ at $y = y_B$. The profile has been so drawn; thus at some point, C, intermediate between points A and B, the velocity has a value that lies between zero and U_∞, that is, for $0 \leq y \leq y_B$, then $0 \leq u \leq U_\infty$. From these characteristics of the velocity profile and our definition of the shear stress, we see that within the region $0 \leq y \leq y_B$, shear stresses are present; for $y > y_B$, the velocity gradient is zero and, hence, there are no shear stresses present.

What about the velocity profile at the location x_2? Is it exactly the same as the profile at $x = x_1$? A look at Fig. 2.10 suggests that it is not. At least it has not been drawn that way! While it is qualitatively the same, why is it not exactly the same? We might guess that the plate would influence a greater region of the flow field as we move farther down the plate. Looking again at the profile at location x_1, we see that the slower moving fluid adjacent to the plate exerts a retarding force on the faster moving fluid above it. We can see this by considering the shear stress on the plane through the point C. Since we are interested in the stress exerted on the faster moving fluid above the plane, we are looking for the direction of the shear stress on a negative y plane through the point C. From Eq. 2.8 we see that τ_{yx} on the plane through point C has a positive numerical value; consequently the shear stress must be in the negative x direction.

To establish the qualitative picture of the velocity profile at $x = x_2$, we recognize that the no-slip condition requires the velocity at the wall to be zero; this fixes the velocity at A' as zero. Since, at location x_1, the slower moving fluid exerts a retarding force on the fluid above it, we would expect the distance out to the point where the velocity is U_∞ to be increased at location x_2; that is $y_{B'} > y_B$. Furthermore, it is reasonable to expect that $u_{C'} < u_C$.

From our qualitative picture of the flow field, we see that we can divide the flow into two general regions. In the region adjacent to the boundary, shear stresses are present; this region is called the boundary layer.[6] Outside the boundary layer the velocity gradient is zero and, hence, the shear stresses are zero. In this region we may use inviscid flow theory to analyze the flow.

For a given freestream velocity, U_∞, the size of the boundary layer will depend on the properties of the particular fluid. Since the shear stress is directly proportional to the viscosity, we expect the size of the boundary layer to depend on the viscosity of the fluid. In Chapter 8, we shall develop expressions for determining the rate of boundary layer growth.

---

[6] The formation of a boundary layer is demonstrated in the film loop, S-FM006, *Boundary-Layer Formation*.

Before leaving our discussion of the viscous flow over a semi-infinite flat plate, we should stop and reflect on two points. We have implied that the shear stress on a y plane was given by the one-dimensional expression for τ, Eq. 2.8. Is the flow one-dimensional? A look at Fig. 2.10 indicates that the velocity, u, is a function of both x and y; thus the flow is two-dimensional and, hence, τ_{yx} as given by Eq. 2.8 can only be considered as approximate.[7]

In our qualitative description of the flow field, we were only concerned about the behavior of the x component of velocity, that is, the velocity, u. What about the y component of velocity, that is, the velocity, v? Is it zero throughout the flow field? To answer this question, consider the streamlines of the flow. Rather than consider all possible streamlines, let us consider the particular streamline through the point M. Recalling that a streamline is defined as a line drawn tangent to the velocity vector at every point in the flow, our first inclination might be to depict the streamline through M as a straight line parallel to the x axis. However, this would violate the requirement that there can be no flow across a streamline. Because there can be no flow across a streamline, the mass flow between adjacent streamlines (or between a streamline and a solid boundary) must be a constant. Thus in considering the incompressible viscous flow of Fig. 2.10, we recognize that the streamline through the point M cannot be a straight line parallel to the x axis.

The spacing between the streamline through the point M and the x axis must increase continuously as we move along the plate. Therefore, although small, the y component of velocity is not zero. The streamline through M crosses the dashed line we have used to denote the edge of the boundary layer. Consequently, we conclude that the edge of the boundary layer is not a streamline, and that there is flow into the boundary layer as we move down the plate. Indeed, if the boundary layer is to grow, there must be flow across the edge of the boundary layer.

We have used the incompressible flow over a semi-infinite flat plate to establish a qualitative picture of the viscous flow over a solid boundary. In that example, we had only to consider the effect of shear forces on the flow field; the pressure was constant throughout the flow field. Now let us consider a steady flow field (the incompressible flow over a cylinder) wherein both pressure forces and viscous forces are important. For steady flow, we recognize that pathlines, streaklines, and streamlines are all identical. If we were to employ some means of flow visualization, we would find the flow field to be of the general character shown in Fig. 2.11a.[8]

---

[7] As it turns out, this is a very good approximation for laminar boundary-layer flow.
[8] The details of the flow will depend on the various flow properties. For all but very low velocity flows, the qualitative picture will be as shown.

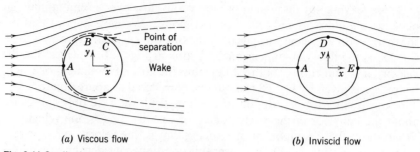

(a) Viscous flow   (b) Inviscid flow

Fig. 2.11 Qualitative picture of flow over a cylinder.

We see that the streamlines are symmetric about the x axis. The fluid along the central streamline impinges on the cylinder at point A, divides, and flows around the cylinder. Point A on the cylinder is called a stagnation point. As in flow over a flat plate, a boundary layer develops in the neighborhood of the solid surface. The velocity distribution outside the boundary layer can be determined from the spacing of the streamlines. Since there can be no flow across a streamline, we would expect the flow velocity to increase in regions where the spacing between streamlines decreases. Conversely, an increase in streamline spacing implies a decrease in flow velocity.

Consider for a moment the incompressible flow around a cylinder assuming an inviscid flow, as shown in Fig. 2.11b; this flow is symmetric about both the x and y axes. The velocity around the cylinder increases to a maximum at point D and then decreases as we move further around the cylinder. For inviscid flow, an increase in velocity is accompanied by a decrease in pressure and conversely. Thus in the case of an incompressible inviscid flow, the pressure along the surface of the cylinder decreases as we move from point A to point D and, then, increases again in going from point D to point E. Since the flow is symmetric with respect to both the x and y axes, we would also expect the pressure distribution to be symmetric with respect to these axes. This is indeed the case.

Since there are no shear stresses present in an inviscid flow, the pressure forces are the only forces we need consider in determining the net force on the cylinder. The symmetry of the pressure distribution leads to the conclusion that for an inviscid flow, there is no net force on the cylinder in either the x or y directions. The net force in the x direction is termed the *drag*. Thus for an inviscid flow over a cylinder, we are led to the conclusion that the drag is zero; this conclusion is contrary to experience, for we know that all bodies experience some drag when placed in a real flow. In treating the inviscid flow over a body we have, by the definition of inviscid flow, neglected the presence of the boundary layer. Let us go back and look again at the real flow situation.

In considering the real flow, Fig. 2.11a, we shall consider the boundary layer to be thin. Since the boundary layer is taken to be thin, it is reasonable to assume that the pressure field is qualitatively the same as in the inviscid flow case. Since the pressure decreases continuously between points A and B, a fluid element inside the boundary layer experiences a net pressure force in the direction of flow. In the region between A and B, this net pressure force is sufficient to overcome the resisting shear force and motion of the element in the flow direction is maintained.

Now consider an element of fluid inside the boundary layer on the back side of the cylinder beyond point B. Since the pressure increases in the direction of flow, the fluid element experiences a net pressure force opposite to its direction of motion. At some point around the cylinder the momentum of the fluid in the boundary layer is insufficient to carry the element further into the region of increasing pressure. The fluid layers adjacent to the solid surface will be brought to rest and the flow will separate from the surface;[9] the point at which this occurs is called the point of separation. Boundary-layer separation results in the formation of a relatively low pressure region behind a body; this region also is deficient in momentum and is called the wake. Thus, for separated flow over a body, there is a net unbalance of pressure forces in the direction of flow; this results in a pressure drag on the body. The greater the size of the wake behind a body, the greater is the pressure drag.

It is logical to ask how one might reduce the size of the wake and thus reduce the pressure drag. Since a large wake results from boundary layer separation, which in turn is related to the presence of an adverse pressure gradient (increase of pressure in the direction of flow), a reduction of the adverse pressure gradient should delay the onset of separation and, hence, reduce the drag.

The streamlining of a body reduces the magnitude of the adverse pressure gradient by spreading a given pressure rise over a larger distance. For example, if a gradually tapered section were added to the cylinder of Fig. 2.11, the flow field would appear qualitatively as shown in Fig. 2.12. Streamlining the shape of the body delays the onset of separation; although the surface area of the body and, hence, the total shear force acting on the body are increased, the drag is significantly reduced.[10]

---

[9] The flow over a variety of models, illustrating flow separation, is demonstrated in the following NCFMF film loops: S-FM012, *Flow Separation and Vortex Shedding*; S-FM004, *Separated Flows—Part I*; and S-FM005, *Separated Flows—Part II*.

[10] The effect of streamlining a body is demonstrated in the film loop, S-FM004, *Separated Flows-Part I*.

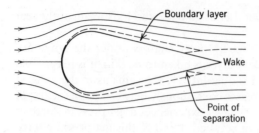

Fig. 2.12 Flow over a streamlined object.

Flow separation may also occur in internal flows (flows through ducts) as a result of rapid or abrupt changes in duct geometry.[11]

2-5.2 LAMINAR AND TURBULENT FLOWS

Viscous flows are classified into laminar or turbulent regimes on the basis of internal flow structure. In the laminar regime, flow structure is characterized by motion in laminae or layers. Flow structure in the turbulent regime is characterized by random, three-dimensional motions of fluid particles superimposed on the mean motion. In laminar flow there is no macroscopic mixing of adjacent fluid layers. A thin filament of dye injected into a laminar flow appears as a single line; there is no dispersion of dye throughout the flow, except the slow dispersion due to molecular motion. On the other hand, a dye filament injected into a turbulent flow quickly disperses throughout the flow field; the line of dye breaks up into myriad entangled threads of dye. This behavior of turbulent flow is due to small velocity fluctuations superimposed on the mean motion of a turbulent flow; the macroscopic mixing of fluid particles from adjacent layers of fluid results in rapid dispersion of the dye. The straight filament of smoke rising from a cigarette in still surroundings gives a clear picture of laminar flow. As the smoke continues to rise, it breaks up into a random, haphazard motion; this is an example of turbulent flow.[12]

Whether a flow is laminar or turbulent depends on the properties of the particular flow. In the case of flow through a pipe, the nature of the flow (laminar or turbulent) is determined by the value of a dimensionless parameter, the Reynolds number, $Re = \rho \bar{V} D/\mu$, where ρ is the density of the fluid, $\bar{V}$ the average flow velocity, D the pipe diameter, and μ the viscosity of the fluid. (The Reynolds number and other important dimensionless parameters encountered in fluid mechanics are discussed in Chapter 7.)

---

[11] Examples of separation in internal flows are shown in the film loop, S-FM015, *Incompressible Flow through Area Contractions and Expansions*.

[12] Several examples illustrating the nature of laminar and turbulent flows are shown in the film loop S-FM008, *The Occurrence of Turbulence*.

Although for $Re > 2300$ the flow is often turbulent and for $Re < 2300$ the flow is laminar, there is, in reality, no single value of the Reynolds number at which the flow changes from laminar to turbulent.[13] There is a range of values of Re over which the flow will change from laminar to turbulent. We might expect that flow conditions at the pipe inlet would affect the value of the Reynolds number at which the transition from laminar to turbulent flow would occur; this is indeed the case. Thus it is more meaningful to talk of a lower value of Reynolds number, Re_{lower}, below which the flow is always laminar and a higher value, Re_{higher}, above which the flow is always turbulent, regardless of the inlet conditions to the pipe.

Boundary layer flows may also be laminar or turbulent; the definitions of laminar and turbulent flows given earlier also apply to boundary layer flows. As we shall see in later chapters, the details of a flow field may be significantly different depending on whether the boundary layer is laminar or turbulent. Methods of analysis also differ for laminar and turbulent flow. Therefore, it is necessary to determine whether a particular flow is laminar or turbulent before beginning an analysis. We shall consider this subject in more detail in later chapters.

2-5.3 COMPRESSIBLE AND INCOMPRESSIBLE FLOWS

Flows in which variations in density are negligible are termed *incompressible*; when density variations within a flow are not negligible, the flow is called *compressible*. If one considers the two states of matter, liquid and gas, included within the definition of a fluid, one is tempted to make the general statement that all liquid flows are incompressible flows and all gas flows are compressible flows. For many practical cases the first portion of the statement is correct—that is, most liquid flows are essentially incompressible. Gas flows may also be considered incompressible provided the flow velocities are small relative to the speed of sound in the fluid; the ratio of the flow speed, V, to the speed of sound, c, in the fluid is defined as the Mach number, M, that is,

$$M = \frac{V}{c}$$

For values of $M < 0.3$, changes in density are only about 2 percent of the mean value. Thus gas flows with $M < 0.3$ can be treated as incompressible; a value of $M = 0.3$ in air at standard conditions corresponds to a speed of approximately 100 m/sec.

---

[13] The behavior of a dye stream injected into laminar and turbulent pipe flow is shown in the film loop S-FM134, *Laminar and Turbulent Pipe Flow*.

Compressible flows occur frequently in engineering applications. Common examples include compressed air systems used to power shop tools and dental drills, transmission of gases in pipelines at high pressure, and pneumatic or fluidic control and sensing systems. Compressibility effects are very important in the design of modern high speed aircraft and missiles, power plants, fans, and compressors.

Shocks, across which fluid properties such as pressure and density change abruptly, can occur in supersonic flows under some conditions. The flowrate in compressible flow can also be choked, or limited, by passage design. These, and other features of compressible flows, will be treated in detail in Chapters 9 and 10.

summary objectives

After completing study of Chapter 2, you should be able to do the following:

1. Give operational definitions of:

continuum	kinematic viscosity
property at a point	viscous flow
scalar field	inviscid flow
vector field	boundary layer
steady flow	stagnation point
uniform flow at a section	drag
pathline	point of separation
streakline	wake
streamline	laminar flow
body force	turbulent flow
surface force	compressible flow
Newtonian fluid	incompressible flow
viscosity	Mach number

2. Give examples of one-, two-, and three-dimensional flows.

3. State the convention for designating the nine components of the stress field.

4. Write Newton's law of viscosity and determine the shear stress and shear force that correspond to a given velocity profile.

5. Solve those problems at the end of the chapter that relate to the material you have studied.

problems

2.1 For the velocity fields given below, determine:

    (a) Whether the flow is one-, two-, or three-dimensional, and why

    (b) Whether the flow is steady or unsteady, and why

(The quantities a, b, and c are constants.)

(1) $\vec{V} = [ae^{-bx}]\hat{\imath}$

(2) $\vec{V} = [ax^2 e^{-bt}]\hat{\imath}$

(3) $\vec{V} = ax^2\hat{\imath} + bx\hat{\jmath}$

(4) $\vec{V} = ax\hat{\imath} + bx^2 e^{-ct}\hat{\jmath}$

(5) $\vec{V} = ax\hat{\imath} + bx^2\hat{\jmath} - cx^2\hat{k}$

(6) $\vec{V} = ax\hat{\imath} - by\hat{\jmath}$

(7) $\vec{V} = (ax + t)\hat{\imath} + by^2\hat{\jmath}$

(8) $\vec{V} = ax\hat{\imath} + by\hat{\jmath} + cx^2\hat{k}$

(9) $\vec{V} = ax\hat{\imath} + by^2\hat{\jmath} + cyt\hat{k}$

(10) $\vec{V} = ax^2\hat{\imath} + by\hat{\jmath} + cxz\hat{k}$

(11) $\vec{V} = ax^2\hat{\imath} + bxz\hat{\jmath}$

(12) $\vec{V} = ax\hat{\imath} - by\hat{\jmath} + (t - cz)\hat{k}$

(13) $\vec{V} = a(x^2 + y^2)^{1/2}(1/z^3)\hat{k}$

The air density field in the vicinity of a powerplant exhaust is approximated by

$$\rho = \rho_0 + \frac{\Delta\rho}{2}\left[\frac{r_0}{r_0 + (x^2 + y^2)^{1/2}} + e^{-kz}\right]$$

Is the field one-, two-, or three-dimensional? Is it steady or unsteady?

2.3 The density field in the exhaust pipe of a Diesel engine is approximated by

$$\rho = a[1 + be^{-cx}\cos(\omega t)]$$

Is the field one-, two-, or three-dimensional? Is it steady or unsteady?

2.4 Give an example of one-, two-, and three-dimensional flow fields. Illustrate with sketches.

2.5 A velocity field is given by

$$\vec{V} = ax\hat{\imath} + ay\hat{\jmath} + bxyt\hat{k}$$

where $a = 2 \text{ sec}^{-1}$, and $b = \text{m}^{-1} \cdot \text{sec}^{-2}$. Determine the number of dimensions of the flow field. Is it steady? Find the slope of the streamline through the point $(x, y, z) = (1, 2, 0)$ at time $t = 0$.

2.6 A velocity field is given by

$$\vec{V} = ay\hat{\imath} + bxj + c\hat{k}$$

where $a = 2 \text{ sec}^{-1}$, $b = 1 \text{ sec}^{-1}$, and $c = 2 \text{ m/sec}$. Determine the number of dimensions of the flow field. Is it steady? Determine the velocity components u, v, w, at the point $(1, 2, 0)$. Determine the slope in the xy plane of the streamline through the point $(1, 2, 0)$.

2.7 The velocity field

$$\vec{V} = ax\hat{\imath} - by\hat{\jmath}$$

can be interpreted to represent flow in a corner. Find an equation for the flow streamlines. Plot several streamlines in the first quadrant, including the one that passes through the point $(x, y) = (0, 0)$.

2.8 The velocity distribution in a certain region is given by

$$\vec{V} = 2x\hat{\imath} - ay\hat{\jmath} + (3t - bz)\hat{k}$$

Determine the number of dimensions of the flow field. Is it steady? Find the equation of the streamline through the point $(x, y, z) = (1, 1, 3)$ at $t = 0$ and $t = 1$.

2.9 Parametric equations for the position of a particle in a flow field are given as

$$x_p = c_1 e^{at}$$

and

$$y_p = c_2 e^{-bt}$$

Find the equation of the pathline for a particle located at $(x, y) = (1, 2)$ at $t = 0$. Compare with a streamline through the same point, found in Problem 2.7.

2.10 A tornado can be represented in polar coordinates by the velocity field

$$\vec{V} = -\frac{a}{r} \hat{\imath}_r + \frac{b}{r} \hat{\imath}_\theta$$

where $\hat{\imath}_r$ and $\hat{\imath}_\theta$ are unit vectors in the r and θ directions, respectively. Recall that an element of distance along the θ direction is $dl = r\,d\theta$. Show that streamlines have equations of the form

$$r = ce^{-\frac{a}{b}\theta}$$

that is, they form logarithmic spirals.

2.11 The density distribution in the fluid column shown is given by

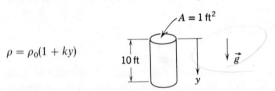

$$\rho = \rho_0(1 + ky)$$

where $\rho_0 = 0.00238$ slug/ft^3 and $k = 0.10$ ft^{-1}. Compute the body force in lbf acting on the volume.

2.12 A body force distribution is given as $\vec{B} = ax\hat{\imath} + b\hat{\jmath} + cz\hat{k}$ per unit mass of the material acted on. The density of the material is given as

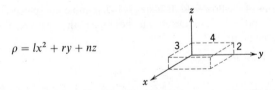

$$\rho = lx^2 + ry + nz$$

All coordinates are measured in meters. Determine the resultant body force in the region shown when: $a = 0$, $b = 0.1$ N/kg, $c = 0.5$ N/kg·m; $l = 2.0$ kg/m^5, $r = 0$, and $n = 1.0$ kg/m^4.

2.13 For the region shown in Problem 2.12, determine the resultant body force when: $a = 5.0$ N/kg·m, $b = 0$, $c = 10.0$ N/kg·m; $l = 0.1$ kg/m^5, $r = 0.5$ kg/m^4, and $n = 0$.

2.14 For the region shown in Problem 2.12, determine the resultant body force when: $a = 1.0$ N/kg·m, $b = 2.0$ N/kg, $c = 0$; $l = 0$, $r = 1.0$ kg/m^4, and $n = 2.0$ kg/m^4.

2.15 For the region shown in Problem 2.12, determine the resultant body force when: $a = 0.1$ N/kg·m, $b = 0.1$ N/kg, $c = 0.1$ N/kg·m; $l = 0$, $r = 2.0$ kg/m⁴, and $n = 4.0$ kg/m⁴.

2.16 Use double index notation to label the six shear stresses shown.

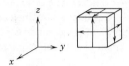

2.17 On the element shown, indicate all possible stresses represented by

$$\tau_{zx} = -10 \text{ lbf/ft}^2$$
$$\sigma_{yy} = 15 \text{ lbf/ft}^2$$

2.18 On a differential fluid element, indicate all possible stresses represented by

$$\tau_{yz} = +5 \text{ Pa}$$
$$\sigma_{xx} = -20 \text{ Pa}$$

2.19 On the element shown, indicate all possible stresses represented by

$$\tau_{rz} = -10 \text{ Pa}$$
$$\sigma_{zz} = +15 \text{ Pa}$$

2.20 The following stress levels are known to exist at a point in a continuous medium:

$$\sigma_{xx} = 200 \text{ psi}$$
$$\sigma_{yy} = -100 \text{ psi}$$
$$\tau_{xy} = \tau_{yx} = 30 \text{ psi}$$

Find the magnitudes of the normal and shear stresses acting on plane AA.

2.21 Obtain the conversion factor for converting the viscosity, μ, from units of newtons, seconds, and meters to units of pounds force, seconds, and feet. Check your answer using Appendix A.

2.22 The velocity distribution for laminar flow between parallel plates is given by

$$\frac{u}{u_{\max}} = 1 - \left(\frac{2y}{h}\right)^2$$

where h is the distance separating the plates, and the origin is placed midway between the plates. Consider a flow of water at 15 C, with $u_{max} = 0.30$ m/sec, and $h = 0.50$ mm. Calculate the shear stress on the upper plate and give its direction.

2.23 A concentric cylinder viscometer may be formed by rotating the inner member of a pair of closely fitting cylinders. The gap must be made small so that a linear velocity profile will exist. Consider a viscometer with an inner cylinder of 3 in. diameter and 6 in. height, and a clearance gap of 0.001 in. width, filled with castor oil at 90 F. Determine the torque required to turn the inner cylinder at 250 rpm.

2.24 A concentric cylinder viscometer may be formed by rotating the outer member of a pair of closely fitting cylinders. For small clearances, a linear velocity profile may be assumed in the liquid filling the clearance gap. A viscometer has an inner cylinder of 75 mm diameter and 150 mm height, with a clearance gap width of 0.02 mm. A torque of 0.021 N · m is required to turn the outer cylinder at 100 rpm. Determine the viscosity of the liquid in the clearance gap of the viscometer.

2.25 A block has a mass of 2 kg and is 0.2 m square. It slides down a smooth incline on a thin film of oil. The slope is 30° from the horizontal. The oil is SAE 30 at 20 C, the film is 0.02 mm thick, and the velocity profile may be assumed linear. Calculate the terminal speed of the block.

2.26 A block weighing 10 lbf and having dimensions 10 in. on each edge is pulled up an inclined surface on which there is a film of SAE 10 oil at 100 F. If the speed of the block is 5 ft/sec and the oil film is 0.001 in. thick, find the force required to pull the block. Assume the velocity distribution in the oil film to be linear. The surface is inclined at an angle of 15° from the horizontal.

2.27 Recording tape is to be coated on both sides with lubricant by drawing it through a narrow gap. The tape is 0.015 in. thick and 1.00 in. wide. It is centered in the gap with a clearance of 0.012 in. on each side. The lubricant, of viscosity $\mu = 0.021$ slug/ft · sec, completely fills the space between the tape and gap, for a length of 0.75 in. along the tape. If the tape can withstand a maximum tensile force of 7.5 lbf, determine the maximum speed with which it can be pulled through the gap.

2.28 Magnet wire is to be coated with varnish for insulation by drawing it through a circular die of 0.9 mm diameter. The wire diameter is 0.8 mm and it is centered in the die. The varnish ($\mu = 20$ centipoise) completely fills the space between the wire and the die for a length of 20 mm. The wire is drawn through the die at a speed of 50 m/sec. Determine the force required to pull the wire.

2.29 The air hockey game at Rocky's Rec Room has pucks of mass 30 g, with a diameter of 100 mm. The air film under a puck is 0.1 mm thick. Calculate the time required after impact for a puck to lose 10 percent of its initial speed.

2.30 The air hockey table of Problem 2.29 has been installed poorly; its surface slopes at 2° from the horizontal. A player fires a puck directly uphill at an initial speed of 10 m/sec. Determine its speed at the instant it passes through the goal, 2 m away.

2.31 From the definition of a Newtonian fluid, Eq. 2.8, the shear stress (τ_{yx}) is proportional to the shear rate (du/dy). A plot of shear stress versus shear rate is a straight line through the origin; the slope of this line equals the fluid viscosity. Now-Newtonian fluids do not exhibit such a simple relationship. Sketch a plot of shear stress versus shear rate for (a) toothpaste and (b) Silly Putty. Contrast their behavior with that of a Newtonian fluid.

2.32 The *effective viscosity* of a non-Newtonian fluid is defined as the ratio of shear stress to shear rate. Because the behavior is nonlinear for these fluids, effective viscosity varies with shear rate. Sketch a plot of shear stress versus shear rate for Lucite paint. Also sketch a plot of effective viscosity versus shear rate.

2.33 Air at standard conditions flows through a pipe of 1 in. diameter. The average velocity is 1 ft/sec. Is the flow laminar or turbulent?

2.34 Water at 15 C flows in a tube with an inside diameter of 50 mm. Determine the maximum value of average velocity for which the flow would be laminar.

2.35 Consider the incompressible laminar viscous flow over a semi-infinite flat plate shown in Fig. 2.10. Sketch the variation in shear stress, τ_{yx}, as a function of y at the locations x_1 and x_2. Also sketch the variation in shear stress along the plate surface ($y = 0$) as a function of x.

Chapter 3

Fluid Statics

By definition, a fluid must deform continuously when a shearing stress of any magnitude is applied. The absence of relative motion (and thus, angular deformation) implies the absence of shearing stress. Fluids either at rest or in "rigid-body" motion therefore are able to sustain only normal stresses. Analysis of these "flow" cases is thus appreciably simpler than for fluids undergoing angular deformation (see Section 5-4).

Mere simplicity does not justify our study of a subject. Normal forces transmitted by fluids are of importance in many practical situations. Using the principles of hydrostatics, we can compute forces on submerged objects, develop instruments for measuring pressures, and deduce properties of the atmosphere and oceans. The principles of hydrostatics may also be used to determine forces developed by hydraulic systems in applications such as presses or automobile brakes.

In a static fluid, or in a fluid undergoing rigid-body motion, a fluid particle retains its identity for all time. Since there is no relative motion within the fluid, a fluid element does not deform. We may apply Newton's second law of motion to evaluate the reaction of the particle to the applied forces.

3-1 THE BASIC EQUATION OF FLUID STATICS

Our primary objective is to obtain an equation that will enable us to determine the pressure field within the fluid. To do this, we choose a differential element of mass, dm, with sides dx, dy, and dz as shown in Fig. 3.1. The fluid

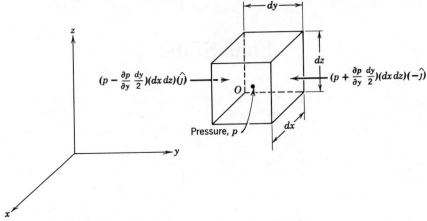

Fig. 3.1 Differential fluid element and pressure forces in *y* direction.

element is stationary relative to the stationary rectangular coordinate system shown. (Fluids in rigid-body motion will be treated in Section 3-7.)

From our previous discussion we recall that two general types of forces may be applied to a fluid. These are body forces and surface forces. The only body force that must be considered in most engineering problems is due to gravity. In some situations body forces due to electric or magnetic fields might be present, but for simplicity they will not be considered in this text.

For a differential fluid element, the body force, $d\vec{F}_B$, is

$$d\vec{F}_B = \vec{g}\,dm = \vec{g}\rho\,dV$$

where $\vec{g}$ is the local gravity vector, ρ the density, and dV is the volume of the element. In Cartesian coordinates $dV = dx\,dy\,dz$, so

$$d\vec{F}_B = \rho\vec{g}\,dx\,dy\,dz$$

In a static fluid no shear stresses can be present. Thus the only surface force is the pressure force. Pressure is a field quantity, that is, $p = p(x, y, z)$. Therefore, the pressure varies with position within the fluid. The net pressure force that results from this variation can be evaluated by summing the forces that act on the six faces of the fluid element.

Let the pressure at the center, O, of the element be p. To determine the pressure at each of the six faces of the element, we may use a Taylor series expansion of the pressure about the point O. By use of the Taylor series representation, the pressure at the left face of the differential element is

$$p_L = p + \frac{\partial p}{\partial y}(y_L - y) = p + \frac{\partial p}{\partial y}\left(-\frac{dy}{2}\right) = p - \frac{\partial p}{\partial y}\frac{dy}{2}$$

(Terms of higher order are omitted because in the limit they vanish.) The pressure on the right face of the differential element is

$$p_R = p + \frac{\partial p}{\partial y}(y_R - y) = p + \frac{\partial p}{\partial y}\frac{dy}{2}$$

The pressure *forces* acting on the two y surfaces of the differential element are shown in Fig. 3.1. Each pressure force is a product of three terms. The first is the magnitude of the pressure. The magnitude is multiplied by the area of the face to give the pressure force, and a unit vector is introduced to indicate direction. Note also in Fig. 3.1 that the pressure force on each face acts *against* the face. In other words, a positive pressure corresponds to a *compressive* stress.

Stresses and forces on the other faces of the element are obtained in the same way. Combining all such forces gives the net surface force acting on the element. Thus

$$d\vec{F}_S = \left(p - \frac{\partial p}{\partial x}\frac{dx}{2}\right)(dy\,dz)(\hat{\imath}) + \left(p + \frac{\partial p}{\partial x}\frac{dx}{2}\right)(dy\,dz)(-\hat{\imath})$$

$$+ \left(p - \frac{\partial p}{\partial y}\frac{dy}{2}\right)(dx\,dz)(\hat{\jmath}) + \left(p + \frac{\partial p}{\partial y}\frac{dy}{2}\right)(dx\,dz)(-\hat{\jmath})$$

$$+ \left(p - \frac{\partial p}{\partial z}\frac{dz}{2}\right)(dx\,dy)(\hat{k}) + \left(p + \frac{\partial p}{\partial z}\frac{dz}{2}\right)(dx\,dy)(-\hat{k})$$

Collecting and canceling terms, we obtain

$$d\vec{F}_S = \left(-\frac{\partial p}{\partial x}\hat{\imath} - \frac{\partial p}{\partial y}\hat{\jmath} - \frac{\partial p}{\partial z}\hat{k}\right)dx\,dy\,dz$$

or,

$$d\vec{F}_S = -\left(\frac{\partial p}{\partial x}\hat{\imath} + \frac{\partial p}{\partial y}\hat{\jmath} + \frac{\partial p}{\partial z}\hat{k}\right)dx\,dy\,dz \qquad (3.1a)$$

The term in parentheses is called the gradient of the pressure or simply the pressure gradient, and is designated as grad p or ∇p. In rectangular coordinates

$$\text{grad } p \equiv \nabla p \equiv \left(\hat{\imath}\frac{\partial p}{\partial x} + \hat{\jmath}\frac{\partial p}{\partial y} + \hat{k}\frac{\partial p}{\partial z}\right) \equiv \left(\hat{\imath}\frac{\partial}{\partial x} + \hat{\jmath}\frac{\partial}{\partial y} + \hat{k}\frac{\partial}{\partial z}\right)p$$

The gradient can be viewed as a vector operator; taking the gradient of a scalar field gives rise to a vector field. Using the gradient designation, Eq. 3.1a can be written as

$$d\vec{F}_S = -\operatorname{grad} p(dx\,dy\,dz) = -\nabla p\,dx\,dy\,dz \qquad (3.1b)$$

From Eq. 3.1b,

$$\operatorname{grad} p = \nabla p = -\frac{d\vec{F}_S}{dx\,dy\,dz}$$

In words, the gradient of pressure is the negative of the surface force per unit volume due to pressure. We note that the level of pressure is not important in evaluating the net pressure force. Instead, what matters is the rate at which pressure changes occur with distance, that is, the *pressure gradient*. We shall find this term very useful throughout our study of fluid mechanics.

Since no other kinds of force may be present in a static fluid, we can combine the formulations for surface and body forces that we have developed to obtain the total force acting on a fluid element. Thus

$$d\vec{F} = d\vec{F}_S + d\vec{F}_B = (-\operatorname{grad} p + \rho\vec{g})\,dx\,dy\,dz$$

or on a unit volume basis

$$\frac{d\vec{F}}{d\mathbf{V}} = \frac{d\vec{F}}{dx\,dy\,dz} = -\operatorname{grad} p + \rho\vec{g} \qquad (3.2)$$

In a static fluid, a fluid particle retains its identity as time passes. It does not deform, since there is no relative motion within the fluid. For these reasons we may apply Newton's second law of motion to evaluate the reaction of the particle to the applied forces. We may use the result also to evaluate the pressure field.

For a fluid particle, Newton's second law gives $d\vec{F} = \vec{a}\,dm = \vec{a}\rho\,d\mathbf{V}$. For a static fluid, $\vec{a} = 0$. Thus

$$\frac{d\vec{F}}{d\mathbf{V}} = \rho\vec{a} = 0$$

Substituting for $d\vec{F}/d\mathbf{V}$ from Eq. 3.2, we obtain

$$-\operatorname{grad} p + \rho\vec{g} = 0 \qquad (3.3)$$

Let us review briefly our derivation of this equation. The physical significance of each term is

$$-\operatorname{grad} p \quad + \quad \rho\vec{g} \quad = 0$$

$$\begin{Bmatrix}\text{pressure force} \\ \text{per unit volume} \\ \text{at a point}\end{Bmatrix} + \begin{Bmatrix}\text{body force per} \\ \text{unit volume} \\ \text{at a point}\end{Bmatrix} = 0$$

This is a vector equation, which means that it really consists of three component equations that must be satisfied individually. Expanding into components, we find

$$
\left.
\begin{aligned}
-\frac{\partial p}{\partial x} + \rho g_x &= 0 \qquad x \text{ direction} \\[2mm]
-\frac{\partial p}{\partial y} + \rho g_y &= 0 \qquad y \text{ direction} \\[2mm]
-\frac{\partial p}{\partial z} + \rho g_z &= 0 \qquad z \text{ direction}
\end{aligned}
\right\} \qquad (3.4)
$$

Equations 3.4 describe the pressure variation in each of the three coordinate directions in a static fluid. To simplify further, it is logical to choose a coordinate system such that the gravity vector is aligned with one of the axes. If the coordinate system is chosen such that the z axis is directed vertically, then $g_x = 0$, $g_y = 0$, and $g_z = -g$. Under these conditions, the component equations become

$$
\left.
\begin{aligned}
\frac{\partial p}{\partial x} &= 0 \\[2mm]
\frac{\partial p}{\partial y} &= 0 \\[2mm]
\frac{\partial p}{\partial z} &= -\rho g
\end{aligned}
\right\} \qquad (3.5)
$$

Equations 3.5 indicate that under the assumptions made, the pressure is independent of coordinates x and y; it depends on z alone. Since p is thus a function of a single variable, a total derivative may be used instead of a partial derivative. With these simplifications, Eqs. 3.5 finally reduce to

$$
\frac{dp}{dz} = -\rho g \equiv -\gamma \qquad (3.6)
$$

$$SG = \frac{\rho}{\rho_{H_2 O}} = \frac{\gamma}{\gamma_{H_2 O}}$$

Restrictions: 1. Static fluid
2. Gravity is the only body force
3. The z axis is vertical

This equation is the basic pressure-height relation of fluid statics. It is subject to the restrictions noted and, therefore, it must be applied only where these restrictions are reasonable for the physical situation. To determine the pressure distribution in a static fluid, Eq. 3.6 may be integrated and appropriate boundary conditions applied.

3-1.1 PRESSURE VARIATION IN A STATIC FLUID

Although ρg may be defined as $\rho g \equiv \gamma$, it has been written as ρg in Eq. 3.6 to emphasize that *both* ρ and g must be considered variables. In order to integrate Eq. 3.6 to find the pressure distribution, assumptions must be made about the variations in ρ and g.

For most fluid statics problems encountered in practical engineering situations, the variation in g will be negligible. Only for a situation such as computing very precisely the pressure change over a large elevation difference would the variation in g need to be included. For our purposes we shall assume g to be constant with elevation at any given location.

In many practical engineering problems the variation in ρ will be appreciable, and accurate results will require that it be accounted for. Several kinds of variation are easy to treat analytically. The simplest is the idealization of an incompressible fluid.

a. Incompressible Fluid

For an incompressible fluid, $\rho = \rho_0 = $ constant. Then for constant gravity,

$$\frac{dp}{dz} = -\rho_0 g = \text{constant}$$

To determine the pressure variation, we must integrate the above equation and apply appropriate boundary conditions. If the pressure at the reference level, z_0, is designated as p_0, then the pressure, p, at location z is found by integration

$$\int_{p_0}^{p} dp = -\int_{z_0}^{z} \rho_0 g\, dz$$

or

$$p - p_0 = -\rho_0 g(z - z_0) = \rho_0 g(z_0 - z)$$

For liquids, it is often convenient to take the origin of the coordinate system at the free surface (reference level) and to measure distances as positive downward from the free surface, as shown in Fig. 3.2. With h measured positive downward, then

$$z_0 - z = h$$

and

$$p = p_0 + \rho_0 g h \tag{3.7}$$

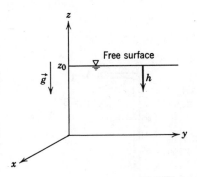

Fig. 3.2 Coordinates for determination of pressure variation in a static liquid.

This form of the basic pressure-height relation is often used to solve manometer problems. Students sometimes have trouble analyzing multiple tube manometer situations. The following rules of thumb may help:

1. Any two points at the same elevation in a continuous length of the same liquid will be at the same pressure.
2. Pressure increases as one goes *down* a liquid column (remember the pressure change on diving into a swimming pool).

Example 3.1

Water flows through pipes A and B. Oil, with specific gravity 0.8, is in the upper portion of the inverted U. Mercury (specific gravity 13.6) is in the bottom of the manometer bends. Determine the pressure difference, $p_A - p_B$, in units of $lbf/in.^2$

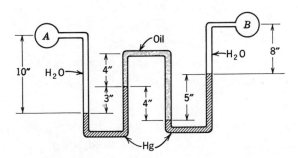

Example Problem 3.1

GIVEN:

Multiple tube manometer as shown. Specific gravity of oil is 0.8; specific gravity of mercury is 13.6.

FIND:

The pressure difference, $p_A - p_B$, in lbf/in.^{2}

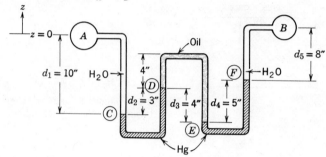

SOLUTION:

Basic equations: $\dfrac{dp}{dz} = -\rho g = -\gamma$ $\qquad SG = \dfrac{\rho}{\rho_{H_2O}} = \dfrac{\gamma}{\gamma_{H_2O}}$

$$dp = -\gamma\, dz \quad \text{and} \quad \int_{p_1}^{p_2} dp = -\int_{z_1}^{z_2} \gamma\, dz$$

For γ = constant

$$p_2 = p_1 + \gamma(z_1 - z_2)$$

Beginning at point A and applying the equation between successive points around the manometer gives

$$p_C = p_A + \gamma_{H_2O}d_1$$
$$p_D = p_C - \gamma_{Hg}d_2$$
$$p_E = p_D + \gamma_{oil}d_3$$
$$p_F = p_E - \gamma_{Hg}d_4$$
$$p_B = p_F - \gamma_{H_2O}d_5$$
$$p_A - p_B = (p_A - p_C) + (p_C - p_D) + (p_D - p_E) + (p_E - p_F) + (p_F - p_B)$$
$$= -\gamma_{H_2O}d_1 + \gamma_{Hg}d_2 - \gamma_{oil}d_3 + \gamma_{Hg}d_4 + \gamma_{H_2O}d_5$$

Substituting $\gamma = SG\gamma_{H_2O}$

$$p_A - p_B = -\gamma_{H_2O}d_1 + 13.6\gamma_{H_2O}d_2 - 0.8\gamma_{H_2O}d_3 + 13.6\gamma_{H_2O}d_4 + \gamma_{H_2O}d_5$$
$$= \gamma_{H_2O}(-d_1 + 13.6d_2 - 0.8d_3 + 13.6d_4 + d_5)$$
$$= \gamma_{H_2O}(-10 + 40.8 - 3.2 + 68 + 8)\text{ in.}$$
$$= \gamma_{H_2O} \times 103.6\text{ in.}$$
$$= \frac{62.4\text{ lbf}}{\text{ft}^3} \times \frac{103.6\text{ in.}}{} \times \frac{\text{ft}}{12\text{ in.}} \times \frac{\text{ft}^2}{144\text{ in.}^2}$$

$$p_A - p_B = 3.74\text{ lbf/in.}^2 \qquad\qquad\qquad\qquad\qquad\qquad\underline{p_A - p_B}$$

Manometers are simple and cheap, so they are used frequently for pressure measurements. Because the liquid level change is small at low pressure differential, a U-tube manometer may be difficult to read accurately. The level change can be increased by changing the manometer design or by using two liquids of slightly different density. Analysis of a typical reservoir manometer design is illustrated in Example Problem 3.2.

Example 3.2

A reservoir manometer is built with a tube diameter of 10 mm and a reservoir diameter of 30 mm. The manometer liquid is Meriam red oil with SG = 0.827. Determine the manometer deflection in millimeters per millimeter of water applied pressure differential.

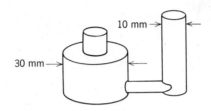

Example Problem 3.2

GIVEN:

Reservoir manometer as shown.

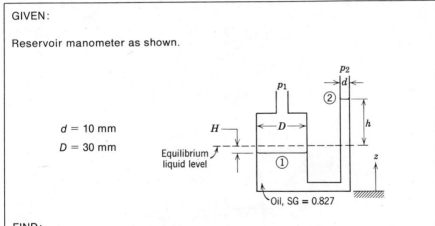

$d = 10$ mm
$D = 30$ mm

FIND:

Liquid deflection, h, in millimeters per millimeter of water applied pressure differential.

SOLUTION:

Basic equations:
$$\frac{dp}{dz} = -\rho g, \quad SG = \frac{\rho}{\rho_{H_2O}}$$

$$dp = -\rho g \, dz \quad \text{and} \quad \int_{p_1}^{p_2} dp = -\int_{z_1}^{z_2} \rho g \, dz$$

For ρ = constant

$$p_2 - p_1 = -\rho g (z_2 - z_1)$$

or

$$p_1 - p_2 = \rho g (z_2 - z_1) = \rho_{oil} g (h + H)$$

To eliminate H, note that the *volume* of manometer liquid must remain constant. Thus the volume displaced from the reservoir must be the same as that which rises into the tube,

$$\frac{\pi}{4} D^2 H = \frac{\pi}{4} d^2 h$$

or

$$H = \left(\frac{d}{D}\right)^2 h$$

Substituting gives

$$p_1 - p_2 = \rho_{oil} g h \left[1 + \left(\frac{d}{D}\right)^2\right]$$

This equation can be simplified by expressing the applied pressure differential as an equivalent water column

$$p_1 - p_2 = \rho_{H_2O} g \, \Delta h$$

and noting that $\rho_{oil} = SG_{oil} \rho_{H_2O}$. Then

$$\rho_{H_2O} g \, \Delta h = SG_{oil} \rho_{H_2O} g h \left[1 + \left(\frac{d}{D}\right)^2\right]$$

or

$$\frac{h}{\Delta h} = \frac{1}{SG_{oil}[1 + (d/D)^2]}$$

Evaluating,

$$\frac{h}{\Delta h} = \frac{1}{0.827[1 + (10/30)^2]} = 1.09 \qquad \frac{h}{\Delta h}$$

{ This problem illustrates the effect of manometer design and choice of gage liquid on sensitivity. }

b. Compressible Fluid

We have seen that pressure variation in any static fluid is described by the basic pressure-height relation

$$\frac{dp}{dz} = -\rho g \qquad (3.6)$$

Pressure variation in a compressible fluid can also be evaluated by integrating Eq. 3.6. Before this can be done, density must be expressed as a function of one of the other variables in the equation. Property information or an equation of state may be used to obtain the required relation for density.

For many liquids, density is only a weak function of temperature. Pressure and density of liquids are related by the *bulk compressibility modulus*, or modulus of elasticity,

$$E_v \equiv \frac{dp}{(d\rho/\rho)} \qquad (3.8)$$

If the bulk modulus is assumed constant, then density is only a function of pressure (i.e. the fluid is *barotropic*) and Eq. 3.8 provides the additional density relation needed to integrate the basic pressure-height relation. Bulk modulus data for some common liquids are given in Appendix A.

The density of gases generally depends on pressure and temperature. The *ideal gas* equation of state

$$p = \rho R T \qquad (3.9)$$

where R is the gas constant (see Appendix A) and T the absolute temperature, accurately models the behavior of most gases under engineering conditions. However, the use of Eq. 3.9 introduces the gas temperature as an additional variable. Therefore, an additional assumption must be made about temperature variation before Eq. 3.6 can be integrated.

Example 3.3
The maximum power output capability of an internal combustion engine decreases with altitude because the air density and hence the mass flowrate of fuel and air decrease. A truck leaves Denver (elevation 5280 ft) on a day when the local temperature and barometric pressure are 80 F and 24.8 in. of mercury, respectively. It travels through Vail Pass (elevation 10,600 ft). The temperature decreases at the rate of 3 F/1000 ft of elevation change. Determine the local barometric pressure at Vail Pass and the percent decrease in maximum power available, compared to that at Denver.

Example Problem 3.3

GIVEN:

Truck travels from Denver to Vail Pass. Engine power output is directly proportional to air density.

$$\text{Denver:} \quad z = 5{,}280 \text{ ft} \qquad \text{Vail Pass:} \quad z = 10{,}600 \text{ ft}$$

$$p = 24.8 \text{ in. Hg}$$
$$T = 80 \text{ F} \qquad\qquad \frac{dT}{dz} = -0.003 \frac{\text{F}}{\text{ft}}$$

FIND:

(a) Atmospheric pressure at Vail Pass.
(b) Percent engine power loss at Vail Pass compared to Denver.

SOLUTION:

Basic equations:
$$\frac{dp}{dz} = -\rho g \quad p = \rho R T$$

Assumptions: (1)   Static fluid
(2)   Air behaves as an ideal gas

By substituting into the basic pressure-height relation,

$$\frac{dp}{dz} = -\frac{p}{RT} g \quad \text{or} \quad \frac{dp}{p} = -\frac{g\,dz}{RT}$$

But temperature varies linearly with elevation, $dT/dz = -m$, so $T = T_0 - m(z - z_0)$

$$\frac{dp}{p} = -\frac{g\,dz}{R[T_0 - m(z - z_0)]} = \frac{g}{mR}\frac{-md(z - z_0)}{[T_0 - m(z - z_0)]}$$

By integrating from p_0 in Denver to p at Vail,

$$\ln\left(\frac{p}{p_0}\right) = \frac{g}{mR}\ln\left[\frac{T_0 - m(z - z_0)}{T_0}\right] = \frac{g}{mR}\ln\left(\frac{T}{T_0}\right)$$

or

$$\frac{p}{p_0} = \left(\frac{T}{T_0}\right)^{g/mR}$$

Evaluating gives

$$\frac{g}{mR} = \frac{32.2}{\text{sec}^2}\frac{\text{ft}}{} \times \frac{\text{ft}}{0.003 \text{ F}} \times \frac{\text{lbm} \cdot \text{R}}{53.3 \text{ ft} \cdot \text{lbf}} \times \frac{\text{slug}}{32.2 \text{ lbm}} \times \frac{\text{lbf} \cdot \text{sec}^2}{\text{slug} \cdot \text{ft}} = 6.25$$

and

$$\frac{T}{T_0} = \left[1 - \frac{0.003 \text{ F}}{\text{ft}} \times (10{,}600 - 5{,}280) \text{ ft} \times \frac{1}{(460 + 80) \text{ R}}\right] = 0.970$$

{Note that T_0 must be expressed as an absolute temperature because it came from the ideal gas equation.}

Thus

$$\frac{p}{p_0} = \left(\frac{T}{T_0}\right)^{g/mR} = (0.970)^{6.25} = 0.827$$

and

$$p = 0.827 p_0 = (0.827)24.8 \text{ in. Hg} = 20.5 \text{ in. Hg} \qquad p$$

The percentage change in power is equal to the change in density, so that

$$\frac{\Delta \mathbb{P}}{\mathbb{P}_0} = \frac{\Delta \rho}{\rho_0} = \frac{\rho - \rho_0}{\rho_0} = \frac{\rho}{\rho_0} - 1$$

By substituting from the ideal gas equation,

$$\frac{\Delta \mathbb{P}}{\mathbb{P}_0} = \frac{p}{p_0} \frac{T_0}{T} - 1 = (0.829)\left(\frac{1}{0.970}\right) - 1 = -0.145$$

or

$$\frac{\Delta \mathbb{P}}{\mathbb{P}_0} = -14.5 \text{ percent} \qquad \frac{\Delta \mathbb{P}}{\mathbb{P}_0}$$

{This problem is included to illustrate use of the ideal gas equation of state with the basic pressure-height relation to evaluate the pressure distribution in the atmosphere.}

3-2   THE STANDARD ATMOSPHERE

Several International Congresses for Aeronautics have been held so that aviation experts around the world might better be able to communicate. The outcome of one such Congress was an internationally accepted definition of the Standard Atmosphere. The sea level conditions of the U.S. Standard Atmosphere are summarized in Table 3.1.

TABLE 3.1 **Sea Level Conditions of the U.S. Standard Atmosphere**

Property	Symbol	SI	English
Temperature	T	288 K	59 F
Pressure	p	101.3 kPa (abs)	14.696 psia
Density	ρ	1.225 kg/m^3	0.002377 slug/ft^3
Specific weight	γ	—	0.07651 lbf/ft^3
Viscosity	μ	1.781×10^{-5} kg/m·sec	3.719×10^{-7} lbf·sec/ft^2

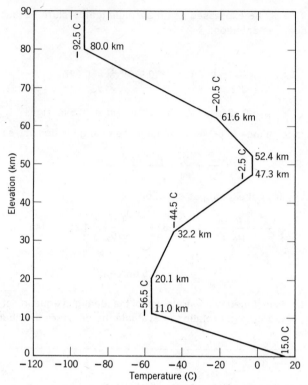

Fig. 3.3 Temperature variation with altitude in the U.S. standard atmosphere.

The temperature profile of the U.S. Standard Atmosphere is shown in Fig. 3.3. Additional property values are tabulated as functions of elevation in Appendix A.

3-3 ABSOLUTE AND GAGE PRESSURES

Pressure values must be stated with respect to a reference pressure level. If the reference level is a vacuum, pressures are termed *absolute*, as shown in Fig. 3.4.

Most pressure gages actually read a pressure *difference*—the difference between the measured pressure level and the ambient level (usually atmospheric pressure). Pressure levels measured with respect to atmospheric pressure are termed *gage* pressures.

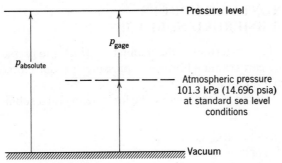

Fig. 3.4 Absolute and gage pressures, showing reference levels.

Absolute pressures must be used in all calculations with the ideal gas or other equations of state. Thus

$$p_{\text{absolute}} = p_{\text{gage}} + p_{\text{atmosphere}}$$

Atmospheric pressure may be obtained from a *barometer*, in which the height of a mercury column is measured. The measured height may be converted to engineering units using Eq. 3.7. For precise work, temperature and altitude corrections must be applied to the measured level. (Data for the specific gravity of mercury are given in Appendix A.)

**3-4   HYDRAULIC SYSTEMS

Hydraulic systems are characterized by very high pressures. As a consequence of these high system pressures, hydrostatic pressure variations may often be neglected. Automobile hydraulic brakes develop pressures up to 10 MPa (1500 psi); aircraft and machinery hydraulic actuation systems frequently are designed for pressures up to 30 MPa (4500 psi), and jacks use pressures to 70 MPa (10,000 psi). Special-purpose laboratory test equipment is commercially available for use at pressures to 1000 MPa (150,000 psi)!

Although liquids are generally considered incompressible at ordinary pressures, density changes may be appreciable at high pressures. Compressibility moduli of hydraulic fluids also may vary sharply at pressures above 50 MPa (7,000 psi). In problems involving unsteady flow, compressibility of the fluid and elasticity of the boundary structure both must be considered. The analysis of problems such as noise and vibration in hydraulic systems, actuators, and shock absorbers quickly becomes complex and is beyond the scope of this text.

** This section may be omitted without loss of continuity in the text material.

3-5 HYDROSTATIC FORCES ON SUBMERGED SURFACES

Now that we have determined the manner in which the pressure varies in a static fluid, we can examine how these pressures produce forces on surfaces submerged in a liquid.

In order to determine completely the force acting on a submerged surface, we must specify:

1. The magnitude of the force.
2. The direction of the force.
3. The line of action of the resultant force.

We shall consider both plane and curved submerged surfaces.

3-5.1 HYDROSTATIC FORCE ON PLANE SUBMERGED SURFACES

A plane submerged surface, on whose upper face we wish to determine the resultant force, is shown in Fig. 3.5. The coordinates have been chosen so that the surface lies in the xy plane.

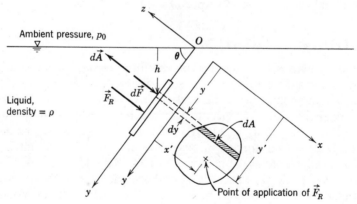

Fig. 3.5 Plane submerged surface.

Since there can be no shear stresses in a static fluid, the hydrostatic force on any element of the surface must act normal to the surface. The pressure force acting on an element of the upper surface, $d\vec{A}$, is given by

$$d\vec{F} = -p\,d\vec{A} \tag{3.10}$$

Since the positive direction of the vector $d\vec{A}$ is the outward drawn normal to the area, the negative sign in Eq. 3.10 indicates that the force, $d\vec{F}$, acts *against* the surface, that is, in a direction opposite to that of $d\vec{A}$. The *resultant* force acting on the surface is found by summing the contribution of the infinitesimal forces over the entire area. Thus

$$\vec{F}_R = \int_A -p\, d\vec{A} \qquad (3.11)$$

In order to evaluate the integral in Eq. 3.11, the pressure, p, and the element of area, $d\vec{A}$, must both be expressed in terms of the same variables. The basic pressure-height relation for a static fluid can be written as

$$\frac{dp}{dh} = \rho g$$

where h is measured positive downward from the liquid free surface. Then, if the pressure at the free surface ($h = 0$) is p_0, we may integrate the pressure-height relation to obtain an expression for the pressure, p, at any depth, h. Thus

$$\int_{p_0}^{p} dp = \int_0^h \rho g\, dh$$

and hence

$$p = p_0 + \int_0^h \rho g\, dh$$

This expression for p can then be substituted into Eq. 3.11. The geometry of the surface is expressed in terms of x and y; since the depth h is expressible in terms of y ($h = y\sin\theta$), the integration can be carried out to determine the resultant force.

The point of application of the resultant force must be such that the moment of the resultant force about any axis is equal to the moment of the distributed force about the same axis. If the position vector, from an arbitrary origin of coordinates to the point of application of the resultant force, is designated as $\vec{r}'$, then

$$\vec{r}' \times \vec{F}_R = \int \vec{r} \times d\vec{F} = -\int_A \vec{r} \times p\, d\vec{A} \qquad (3.12)$$

Referring to Fig. 3.5, we see that $\vec{r}' = \hat{i}x' + \hat{j}y'$, $\vec{r} = \hat{i}x + \hat{j}y$, and $d\vec{A} = dA\hat{k}$. Since the resultant force, $\vec{F}_R$, acts against the surface (in a direction opposite to that of $d\vec{A}$), then $\vec{F}_R = -F_R\hat{k}$. Substituting into Eq. 3.12 gives

$$(\hat{\imath}x' + \hat{\jmath}y') \times -F_R\hat{k} = \int (\hat{\imath}x + \hat{\jmath}y) \times d\vec{F} = -\int_A (\hat{\imath}x + \hat{\jmath}y) \times p\,dA\hat{k}$$

Therefore,

$$\hat{\jmath}x'F_R - \hat{\imath}y'F_R = \int_A (\hat{\jmath}xp - \hat{\imath}yp)\,dA$$

This is a vector equation, so the components must be equal. Thus

$$y'F_R = \int_A yp\,dA \qquad (3.13a)$$

and

$$x'F_R = \int_A xp\,dA \qquad (3.13b)$$

where x' and y' are the coordinates of the point of application of the resultant force. Note that Eqs. 3.11 and 3.12 can be used to determine the magnitude of the resultant force and its point of application on any plane submerged surface. They do not require that the density be a constant nor that the free surface of the liquid be at atmospheric pressure.

In considering the resultant force acting on a plane submerged surface, we have utilized vector notation to emphasize that forces and moments are vector quantities. Equations 3.11 and 3.12 are mathematical statements of basic principles familiar to you from previous courses in physics and statics:

1. The resultant force is the sum of the infinitesimal forces (Eq. 3.11).
2. The moment of the resultant force about any axis is equal to the moment of the distributed force about the same axis (Eq. 3.12).

Recognizing that the resultant force, $\vec{F}_R$, is a vector quantity, then we obtain

1. The magnitude of $\vec{F}_R$ is given by

$$F_R = |\vec{F}_R| = \int p\,dA$$

2. The direction of $\vec{F}_R$ is normal to the surface.
3. The line of action of $\vec{F}_R$ passes through the point x', y', where

$$y'F_R = \int_A yp\,dA$$
$$x'F_R = \int_A xp\,dA$$

Example 3.4
The inclined surface shown, hinged along A, is 5 m wide. Determine the resultant force, $\vec{F}_R$, of the water on the inclined surface.

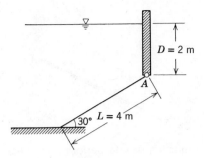

Example Problem 3.4

GIVEN:

Rectangular gate, hinged along A. $w = 5$ m

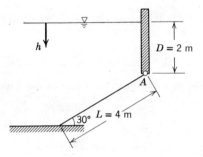

FIND:

Resultant force, $\vec{F}_R$, of the water on the gate.

SOLUTION:

In order to completely determine $\vec{F}_R$, we must specify: (a) the magnitude, (b) the direction, and (c) the line of action, of the resultant force.

Basic equations:
$$\vec{F}_R = -\int p\, d\vec{A} \qquad \frac{dp}{dh} = \rho g$$

$$\vec{r}' \times \vec{F}_R = -\int_A \vec{r} \times p\, d\vec{A}$$

Consider the gate, hinged at A, lying in the xy plane, with coordinates as shown.

$$\vec{F}_R = -\int_A p\, d\vec{A} = -\int_A pw\, dy\, \hat{k} \qquad (d\vec{A} = w\, dy\, \hat{k})$$

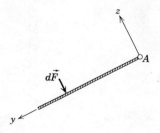

We now need p as a function of y to perform the integration. From the basic pressure-height relation,

$$\frac{dp}{dh} = \rho g$$

so

$$dp = \rho g\, dh \quad \text{and} \quad \int_{p_a}^{p} dp = \int_{0}^{h} \rho g\, dh$$

By assuming ρ = constant,

$$p = p_a + \rho g h \quad \{\text{This gives } p = p(h). \text{ We need } p = p(y).\}$$

From the diagram

$$h = D + y \sin 30° \qquad \text{where } D = 2 \text{ m}$$

Since we are interested in the force of the water on the gate, then

$$p = \rho g(D + y \sin 30°)$$

Thus

$$\vec{F}_R = -\int_A p\, d\vec{A} = -\int_0^L \rho g(D + y \sin 30°) w\, dy\, \hat{k}$$

$$= -\rho g w\left[Dy + \frac{y^2}{2} \sin 30°\right]_0^L \hat{k} = -\rho g w\left[DL + \frac{L^2}{2} \sin 30°\right]\hat{k}$$

$$= \frac{-999 \text{ kg}}{\text{m}^3} \times \frac{9.81 \text{ m}}{\text{sec}^2} \times 5 \text{ m}\left[2 \text{ m} \times 4 \text{ m} + \frac{16 \text{ m}^2}{2} \times \frac{1}{2}\right]\frac{\text{N} \cdot \text{sec}^2}{\text{kg} \cdot \text{m}}\hat{k}$$

$$\vec{F}_R = -588 \hat{k} \text{ kN} \qquad\qquad \{\text{Force acts in negative } z \text{ direction}\} \; \vec{F}_R$$

To find the line of action of the resultant force, $\vec{F}_R$, we recognize that the line of action of the resultant force must be such that the moment of the resultant force about an axis through point A equals the moment of the distributed force about the same axis. That is,

$$\vec{r}' \times \vec{F}_R = \int \vec{r} \times d\vec{F} = -\int_A \vec{r} \times p\, d\vec{A}$$

Then on substituting, we obtain

$$(\hat{i}x' + \hat{j}y') \times -F_R\hat{k} = -\int_A (\hat{i}x + \hat{j}y) \times p\, dA\hat{k}$$

and

$$\hat{j}x'F_R - \hat{i}y'F_R = \hat{j}\int_A xp\,dA - \hat{i}\int_A yp\,dA$$

Then

$$y' = \frac{1}{F_R}\int_A yp\,dA = \frac{1}{F_R}\int_0^L ypw\,dy = \frac{\rho gw}{F_R}\int_0^L y(D + y\sin 30°)\,dy$$

$$= \frac{\rho gw}{F_R}\left[\frac{D}{2}y^2 + \frac{y^3}{3}\sin 30°\right]_0^L = \frac{\rho gw}{F_R}\left[\frac{DL^2}{2} + \frac{L^3}{3}\sin 30°\right]$$

$$= \frac{999\ \text{kg}}{\text{m}^3} \times \frac{9.81\ \text{m}}{\text{sec}^2} \times \frac{5\ \text{m}}{5.88 \times 10^5\ \text{N}}\left[\frac{2\ \text{m} \times 16\ \text{m}^2}{2} + \frac{64\ \text{m}^3}{3} \times \frac{1}{2}\right]\frac{\text{N} \cdot \text{sec}^2}{\text{kg} \cdot \text{m}}$$

$$y' = 2.22\ \text{m}$$

Also

$$x' = \frac{1}{F_R}\int_A xp\,dA$$

In calculating the moment of the distributed force (right side), recall from your earlier courses in statics, that the centroid of the area element must be used for "x". Since the area element is of constant width, $x = w/2$, and

$$x' = \frac{1}{F_R}\int_A \frac{w}{2}p\,dA = \frac{w}{2F_R}\int_A p\,dA = \frac{w}{2} = 2.5\ \text{m}$$

$$\vec{r}' = 2.5\hat{i} + 2.22\hat{j}\ \text{m} \qquad \left\{\begin{array}{l}\text{Line of action of } \vec{F}_R \text{ is along} \\ \text{negative } z \text{ axis through } \vec{r}'\end{array}\right\} \qquad \vec{r}'$$

$$\left\{\begin{array}{l}\text{This problem illustrates the procedure utilized in determining the resultant} \\ \text{force, } \vec{F}_R, \text{ equivalent to a distributed force on a plane submerged surface.}\end{array}\right\}$$

Example 3.5

The rectangular gate, AB, is 5 ft wide ($w = 5$ ft) and 10 ft long ($L = 10$ ft). The gate is hinged along B. Neglecting the weight of the gate, calculate the force per unit width exerted against the stop along A.

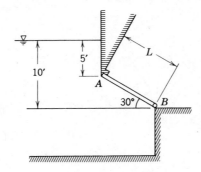

Example Problem 3.5

GIVEN:

Rectangular gate, AB, hinged along B, width, $w = 5$ ft, and length, $L = 10$ ft. Neglect weight of the gate.

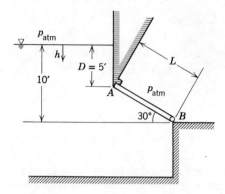

FIND:

Force per unit width against stop along A.

SOLUTION:

Basic equations: $\quad F_R = |\vec{F}_R| = \int p\,dA \qquad \dfrac{dp}{dh} = \rho g = \gamma$

$$\text{Moment,} \quad M = |\vec{M}| = Fd$$

where d is the moment arm (counterclockwise moment assumed positive).

For equilibrium,

$$\sum M = 0$$

If we consider the gate hinged along B as a free body, lying in the xy plane, with coordinates as shown, then:

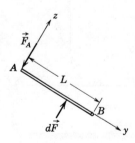

(a) $\vec{F}_A$, the force of the stop on the gate, acts along the line $y = 0$ as shown.
(b) $d\vec{F}$ is an element of the distributed force due to the water acting as shown.

Note: If in calculating the distributed force due to the water, we do not include the effect of atmospheric pressure at the free surface, then we do not need to include the force due to atmospheric pressure acting on the upper surface of the gate.

Moment about x axis through B:
(a) of $\vec{F}_A$ is $F_A L$
(b) of total distributed force is $-\int (L - y)\, dF = -\int_A (L - y)p\, dA$

Since $\sum M = 0$, then

$$F_A = \frac{1}{L} \int_A (L - y)p\, dA$$

The element of area, dA, of the gate is $dA = w\, dy$, where w is the gate width.

$$\therefore F_A = \frac{1}{L} \int_0^L (L - y)pw\, dy$$

We now need p as a function of y in order to perform the integration. From the basic pressure-height relation,

$$\frac{dp}{dh} = \gamma \qquad \therefore dp = \gamma\, dh \quad \text{and} \quad \int_{p_a}^{p} dp = \int_0^h \gamma\, dh$$

By assuming $\gamma = $ constant,

$$p = p_a + \gamma h \quad \{\text{This gives } p = p(h). \text{ We need } p = p(y).\}$$

From the diagram,

$$h = D + y \sin 30° \qquad \text{where } D = 5 \text{ ft}$$
$$\therefore p = p_a + \gamma(D + y \sin 30°)$$

When we wrote the moment equation ($\sum M_B = 0$), we did not include the moment of the force due to atmospheric pressure acting on the top of the gate. Consequently, in figuring the moment due to the water, we should not include the effect of the atmospheric pressure at the free surface. Hence the pressure due to the water alone is

$$p = \gamma(D + y \sin 30°)$$

Thus

$$F_A = \frac{1}{L} \int_0^L (L - y)pw\, dy = \frac{w}{L} \int_0^L (L - y)\gamma(D + y \sin 30°)\, dy$$

or

$$\frac{F_A}{w} = \frac{\gamma}{L} \int_0^L (DL + Ly\sin 30° - Dy - y^2 \sin 30°)\, dy$$

$$= \frac{\gamma}{L} \left[DLy + \frac{L}{2} y^2 \sin 30° - \frac{D}{2} y^2 - \frac{y^3}{3} \sin 30° \right]_0^L$$

$$= \frac{\gamma}{L} \left[DL^2 + \frac{L^3}{2} \sin 30° - \frac{DL^2}{2} - \frac{L^3}{3} \sin 30° \right]$$

$$\frac{F_A}{w} = \frac{\gamma}{L} \left[\frac{DL^2}{2} + \frac{1}{6} L^3 \sin 30° \right]$$

With $D = 5$ ft, $L = 10$ ft, and $\gamma = 62.4$ lbf/ft^3,

$$\frac{F_A}{w} = \frac{1}{10\ \text{ft}} \times \frac{62.4\ \text{lbf}}{\text{ft}^3} \left[\frac{5\ \text{ft} \times 100\ \text{ft}^2}{2} + \frac{1}{6} \times \frac{1000\ \text{ft}^3}{} \times \frac{1}{2} \right] = 2080\ \text{lbf/ft}$$

where F_A is the force of the stop on the gate. The force *on the stop* acts in the opposite direction. Therefore,

$$\frac{\vec{F}_A}{w} \text{(on stop)} = \frac{F_A}{w} \hat{k} = 2080\, \hat{k}\ \text{lbf/ft} \qquad\qquad \frac{\vec{F}_A}{w}$$

$\leftarrow$

{ This problem illustrates the direct use of the distributed moment without evaluating the resultant force and line of application separately. }

3-5.2 HYDROSTATIC FORCE ON CURVED SUBMERGED SURFACES

Although slightly more involved, determining the hydrostatic force on a curved surface is no more difficult than determining the force on a plane submerged surface. The hydrostatic force on an infinitesimal element of a curved surface, $d\vec{A}$, acts normal to the surface. However, the differential pressure force on each element of the surface acts in a different direction because of the surface curvature. Accounting for this change in direction makes the problem a little more involved.

What do we normally do when we wish to sum a series of force vectors acting in different directions? The usual procedure is to sum the components of the vectors relative to a convenient coordinate system.

Consider the curved surface shown in Fig. 3.6. The pressure force acting on the element of area, $d\vec{A}$, is given by

$$d\vec{F} = -p\, d\vec{A} \tag{3.10}$$

The resultant force is again given by

$$\vec{F}_R = -\int_A p\, d\vec{A} \tag{3.11}$$

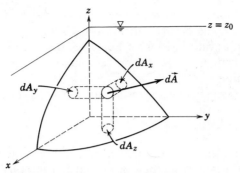

Fig. 3.6 Curved submerged surface.

We can write

$$\vec{F}_R = \hat{\imath}F_{R_x} + \hat{\jmath}F_{R_y} + \hat{k}F_{R_z} \tag{3.14}$$

where F_{R_x}, F_{R_y}, and F_{R_z} are the components of $\vec{F}_R$ in the positive x, y, and z directions, respectively.

To evaluate the components of the force in a given direction, we take the dot product of the force with the unit vector in the given direction. Then,

$$F_{R_x} = \int dF_x = \vec{F}_R \cdot \hat{\imath} = \int d\vec{F} \cdot \hat{\imath} = -\int_A p \, d\vec{A} \cdot \hat{\imath}$$

$$F_{R_x} = -\int_A p \, dA \cos \theta_x = \pm \int_{A_x} p \, dA_x \tag{3.15a}$$

Likewise,

$$F_{R_y} = -\int_A p \, dA \cos \theta_y = \pm \int_{A_y} p \, dA_y \tag{3.15b}$$

$$F_{R_z} = -\int_A p \, dA \cos \theta_z = \pm \int_{A_z} p \, dA_z \tag{3.15c}$$

where

θ_x is the angle between $d\vec{A}$ and $\hat{\imath}$

θ_y is the angle between $d\vec{A}$ and $\hat{\jmath}$

θ_z is the angle between $d\vec{A}$ and $\hat{k}$

$dA_x = dA \cos \theta_x$ is the projection of the area element dA on a plane perpendicular to the x axis

$dA_y = dA \cos \theta_y$ is the projection of the area element dA on a plane perpendicular to the y axis

$dA_z = dA \cos \theta_z$ is the projection of the area element dA on a plane perpendicular to the z axis

The sign (plus or minus) in the last equality of Eqs. 3.15 depends on the magnitude of the angle between the vector $d\vec{A}$ and the unit vectors. If a force component, as calculated from Eqs. 3.15, is negative, it simply means that, consistent with Eq. 3.14, the force component acts in the negative coordinate direction.

In general the component of the resultant force in the l direction is given by

$$F_{R_l} = \pm \int_{A_l} p \, dA_l \tag{3.16}$$

where dA_l is the projection of the area element dA on a plane perpendicular to the l direction, and the sign (plus or minus) depends on the magnitude of the angle θ_l, between $d\vec{A}$ and the unit vector in the l direction.

We see that finding the resultant force on a curved submerged surface requires the determination of each of the components of the resultant force. This necessitates application of Eq. 3.16 three times at most. As with forces on plane submerged surfaces, Eq. 3.16 can be integrated only after the pressure, p, and the element of area, dA_l, are expressed in the same variables of integration.

In considering the vertical component, F_{R_z}, of the resultant force, we note that the pressure exerted by the liquid is given by

$$p = \int_{z_s}^{z_0} \rho g \, dz$$

where z_s is the vertical coordinate of the surface, and z_0 is the vertical coordinate of the free surface. Then (refer to Fig. 3.6)

$$dF_z = -p \, dA_z = -\left(\int_{z_s}^{z_0} \rho g \, dz \right) dA_z$$

The integral represents the weight of a differential cylinder of liquid above the element of surface area, dA_z; the cylinder extends from the curved surface to the free surface. The vertical component of the resultant force is obtained by integrating over the entire surface,

$$F_z = -\int_{A_z} \int_{z_s}^{z_0} \rho g \, dz \, dA_z$$

The magnitude of the vertical component of the resultant force is equal to the total weight of the liquid directly above the surface. The minus sign indicates that a curved surface with a positive dA_z projection is subjected to a force in the negative z direction.

In working with cylindrical surfaces, that is, surfaces with a constant radius of curvature, then $dA = wR \, d\theta$, where R is the radius, and w the width, of the cylindrical surface. In these cases it is often easier to use θ as the variable of integration. Then

$$F_{R_l} = -\int_A p\,dA\cos\theta = -\int_{\theta_1}^{\theta_2} p\cos\theta\,wR\,d\theta$$

where θ is the angle between $d\vec{A}$ and the unit vector in the l direction.

To specify completely the resultant force acting on a curved submerged surface, we must specify the line of action of the force. Since we have solved for the components of the resultant force, we must then specify the line of action of each of the components. The line of action of each component of the resultant force is found by recognizing that the moment of the resultant force component about a given axis must be equal to the moment of the corresponding distributed force component about the same axis.

To find the line of action of each of the components of the resultant force on a curved surface, we would write

$$\vec{r}'_x \times \hat{i}F_{R_x} = \int \vec{r} \times dF_x\hat{i}$$

$$\vec{r}'_y \times \hat{j}F_{R_y} = \int \vec{r} \times dF_y\hat{j} \qquad\qquad (3.17)$$

$$\vec{r}'_z \times \hat{k}F_{R_z} = \int \vec{r} \times dF_z\hat{k}$$

where $\vec{r}'_x$, $\vec{r}'_y$, and $\vec{r}'_z$ are vectors to the lines of action of the components of the resultant force in the x, y, and z directions, respectively. Expressions for dF_x, dF_y, and dF_z are given by Eqs. 3.15.

To aid in visualizing the relations of Eqs. 3.15 and 3.17, consider the curved surface of constant width (in the x direction) shown in Fig. 3.7, where the element of area, $d\vec{A}$, has been enlarged.

To find the components of the force:

$$F_{R_y} = \int dF_y = -\int p\,d\vec{A}\cdot\hat{j} = -\int p\,dA\cos\theta_y = \int p\,dA\cos\beta = \int p\,dA_y$$

$$F_{R_z} = \int dF_z = -\int p\,d\vec{A}\cdot\hat{k} = -\int p\,dA\cos\theta_z = -\int p\,dA\sin\beta = -\int p\,dA_z$$

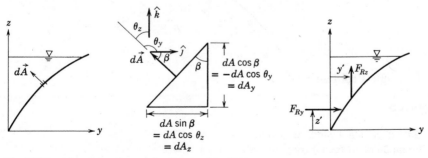

Fig. 3.7 Two-dimensional curved submerged surface.

The z component of $\vec{F}_R$ has magnitude $\int p\,dA_z$ and acts in the negative z direction. By designating the vertical component of the resultant force as F_{R_v}, then $F_{R_v} = -F_{R_z}$. To find the lines of action of the components:

$$z'\hat{k} \times F_{R_y}\hat{j} = \int z\hat{k} \times dF_y\hat{j} = \int z\hat{k} \times p\,dA_y\hat{j}$$

$$y'\hat{j} \times F_{R_z}\hat{k} = y'\hat{j} \times -F_{R_v}\hat{k} = \int y\hat{j} \times dF_z\hat{k} = \int y\hat{j} \times -p\,dA_z\hat{k}$$

From these two equations we obtain

$$z' = \frac{1}{F_{R_y}} \int_{A_y} zp\,dA_y$$

$$y' = -\frac{1}{F_{R_z}} \int_{A_z} yp\,dA_z \quad \text{or} \quad y' = \frac{1}{F_{R_v}} \int_{A_z} yp\,dA_z$$

Consequently, there is nothing new required to specify completely the resultant force on a curved submerged surface. In determining the components and their corresponding lines of action, we proceed for each component just as we did for plane submerged surfaces. Because we are dealing with a curved surface, the lines of action of the components of the resultant force will not necessarily coincide; the complete resultant may be expressed as a force plus a couple.

Example 3.6

The gate shown has a constant width, w, of 5 m. The equation of the surface is $x = y^2/a$, where $a = 4$ m. The depth of water to the right of the gate is 4 m. Find the components F_{R_x} and F_{R_y} of the resultant force due to the water and the line of action of each.

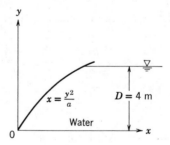

Example Problem 3.6

GIVEN:

Gate of constant width, w = 5 m.
Equation of surface in xy plane is $x = y^2/a$, where a = 4 m.
Water stands at depth, D, of 4 m to the right of the gate.

FIND:

F_{R_x}, F_{R_y} and line of action of each.

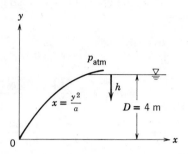

SOLUTION:

Basic equations:
$$\vec{F}_R = - \int p \, d\vec{A} \qquad \frac{dp}{dh} = \rho g.$$

Moment of a force, $\vec{M} = \vec{r} \times \vec{F}$

$$F_{R_x} = - \int_A p \, d\vec{A} \cdot \hat{\imath} = - \int_{A_x} p \, dA_x = - \int_0^D pw \, dy$$

$$F_{R_y} = - \int_A p \, d\vec{A} \cdot \hat{\jmath} = \int_{A_y} p \, dA_y = \int_0^{D^2/a} pw \, dx$$

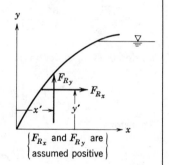

$\left\{ F_{R_x} \text{ and } F_{R_y} \text{ are} \atop \text{assumed positive} \right\}$

In order to perform the necessary integrations, we need expressions for $p(y)$ and $p(x)$ along the surface of the gate.

$$\frac{dp}{dh} = \rho g, \quad dp = \rho g \, dh \qquad \text{and} \qquad \int_{p_a}^p dp = \int_0^h \rho g \, dh$$

If we assume ρ = constant, then

$$p = p_a + \rho g h$$

Since atmospheric pressure acts on both the top of the gate and the free surface of the liquid, there is no net contribution of the atmospheric pressure force. Thus, in determining the force due to the liquid, we take $p = \rho g h$.

We now need an expression for $h = h(y)$ and $h = h(x)$ along the surface of the gate. Along the surface of the gate, $h = D - y$. Since the equation of the gate surface is $x = y^2/a$, then along the gate $y = \sqrt{a}\,x^{1/2}$ and thus h can also be written as $h = D - \sqrt{a}\,x^{1/2}$. Substituting the appropriate equations for h into the expressions for F_{R_x} and F_{R_y} gives

$$F_{R_x} = -\int_0^D pw\,dy = -\int_0^D \rho g h w\,dy = -\rho g w \int_0^D h\,dy = -\rho g w \int_0^D (D - y)\,dy$$

$$= -\rho g w \left[Dy - \frac{y^2}{2} \right]_0^D = -\rho g w \left[D^2 - \frac{D^2}{2} \right] = -\frac{\rho g w D^2}{2}$$

$$F_{R_x} = \frac{-999 \text{ kg}}{m^3} \times \frac{9.81 \text{ m}}{\sec^2} \times \frac{5 \text{ m}}{} \times \frac{(4)^2 \text{ m}^2}{2} \times \frac{N \cdot \sec^2}{kg \cdot m} = -392 \text{ kN} \qquad \underset{\longleftarrow}{F_{R_x}}$$

{The minus sign indicates F_{R_x} actually acts to the left.}

$$F_{R_y} = \int_0^{D^2/a} pw\,dx = \int_0^{D^2/a} \rho g h w\,dx = \rho g w \int_0^{D^2/a} h\,dx = \rho g w \int_0^{D^2/a} (D - \sqrt{a}\,x^{1/2})\,dx$$

$$= \rho g w \left[Dx - \frac{2}{3}\sqrt{a}\,x^{3/2} \right]_0^{D^2/a} = \rho g w \left[\frac{D^3}{a} - \frac{2}{3}\sqrt{a}\,\frac{D^3}{a^{3/2}} \right] = \frac{\rho g w D^3}{3a}$$

$$F_{R_y} = \frac{999 \text{ kg}}{m^3} \times \frac{9.81 \text{ m}}{\sec^2} \times \frac{5 \text{ m}}{} \times \frac{(4)^3 \text{ m}^3}{3} \times \frac{1}{4 \text{ m}} \times \frac{N \cdot \sec^2}{kg \cdot m} = 261 \text{ kN} \qquad \underset{\longleftarrow}{F_{R_y}}$$

To find the line of action of F_{R_x},

$$y'\hat{\jmath} \times F_{R_x}\hat{\imath} = \int y\hat{\jmath} \times dF_x\hat{\imath} = -\int y\hat{\jmath} \times p\,dA_x\hat{\imath}$$

Therefore,

$$-y'F_{R_x} = \int_{A_x} yp\,dA_x \qquad \text{and} \qquad y' = -\frac{1}{F_{R_x}} \int_{A_x} yp\,dA_x$$

$$y' = -\frac{1}{F_{R_x}} \int_0^D ypw\,dy = -\frac{1}{F_{R_x}} \int_0^D yp g h w\,dy = -\frac{w\rho g}{F_{R_x}} \int_0^D y(D - y)\,dy$$

$$= -\frac{w\rho g}{F_{R_x}} \left[\frac{D}{2}y^2 - \frac{y^3}{3} \right]_0^D$$

$$y' = -\frac{\rho g w D^3}{6F_{R_x}} = -\frac{\rho g w D^3}{6} \left[-\frac{2}{\rho g w D^2} \right] = \frac{D}{3} = \frac{4 \text{ m}}{3} = 1.33 \text{ m} \qquad \underset{\longleftarrow}{y'}$$

Similarly for x',

$$x'\hat{\imath} \times F_{R_y}\hat{\jmath} = \int x\hat{\imath} \times dF_y\hat{\jmath} = \int x\hat{\imath} \times p\,dA_y\hat{\jmath}$$

Hence

$$x'F_{R_y} = \int_{A_y} xp\,dA_y \quad \text{and} \quad x' = \frac{1}{F_{R_y}}\int_{A_y} xp\,dA_y$$

$$x' = \frac{1}{F_{R_y}}\int_0^{D^2/a} xpw\,dx = \frac{1}{F_{R_y}}\int_0^{D^2/a} x\rho ghw\,dx = \frac{w\rho g}{F_{R_y}}\int_0^{D^2/a} x(D - \sqrt{a}x^{1/2})\,dx$$

$$= \frac{w\rho g}{F_{R_y}}\left[\frac{D}{2}x^2 - \frac{2}{5}\sqrt{a}x^{5/2}\right]_0^{D^2/a} = \frac{\rho gw}{F_{R_y}}\left[\frac{D^5}{2a^2} - \frac{2}{5}\sqrt{a}\frac{D^5}{a^{5/2}}\right] = \frac{\rho gwD^5}{10F_{R_y}a^2}$$

$$= \frac{\rho gwD^5}{10a^2}\left[\frac{3a}{\rho gwD^3}\right]$$

$$x' = \frac{3D^2}{10a} = \frac{3}{10}\times(4)^2\,\text{m}^2\times\frac{1}{4\,\text{m}} = 1.2\,\text{m}$$

⟵ ──────────────── x'

{This problem illustrates the calculation of resultant force components on a curved submerged surface.}

Example 3.7

The open tank shown is filled with water to a depth of 10 ft. Determine the magnitudes and lines of action of the vertical and horizontal components of the resultant force of the water on the curved part of the tank bottom.

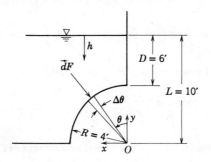

10 ft

10 ft ⟵ 12 ft ⟶ 4 ft radius

Example Problem 3.7

GIVEN:

Tank of width, w = 10 ft, filled with water to a depth of 10 ft.

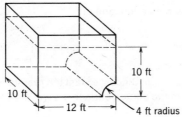

h

dF

$D = 6'$

$L = 10'$

$\Delta\theta$

θ, y

$R = 4'$

x O

FIND:

Components, F_{R_x}, F_{R_y} (and their lines of action), of resultant force due to water, on the curved portion of the tank.

SOLUTION:

Basic equations:
$$\vec{F}_R = -\int p \, d\vec{A} \qquad \frac{dp}{dh} = \rho g = \gamma$$

Moment of a force $\quad \vec{M} = \int \vec{r} \times d\vec{F} \quad$ or $\quad d\vec{M} = \vec{r} \times d\vec{F}$

$$F_{R_x} = \int dF_x = \vec{F}_R \cdot \hat{\imath} = \int d\vec{F} \cdot \hat{\imath} = -\int p \, d\vec{A} \cdot \hat{\imath} = -\int p \, dA(1) \cos(90 - \theta)$$

$$F_{R_x} = -\int p \, dA \sin \theta$$

Since for a cylindrical surface $dA = wR \, d\theta$, then θ is a logical variable of integration. Substituting limits yields

$$F_{R_x} = -\int_0^{\pi/2} pwR \, d\theta \sin \theta$$

We now need to express p as a function of θ

$$\frac{dp}{dh} = \gamma \quad \text{or} \quad dp = \gamma \, dh \quad \text{and} \quad \int_{p_{atm}}^p dp = \int_0^h \gamma \, dh$$

For $\gamma = \text{constant}$, $p = p_{\text{atm}} + \gamma h$. Since we want the resultant force due to the water (p_{atm} acts on the bottom of the curved surface, as well as on the free surface of the water), we take $p = \gamma h$. Thus

$$F_{R_x} = -\int_0^{\pi/2} \gamma h w R \sin \theta \, d\theta$$

To integrate, we need h as a function of θ. From the diagram, $h = L - y$ and along the curved surface $y = R \cos \theta$. Thus $h = L - R \cos \theta$ and

$$F_{R_x} = -\int_0^{\pi/2} \gamma (L - R \cos \theta) w R \sin \theta \, d\theta = -\gamma w R \int_0^{\pi/2} (L - R \cos \theta) \sin \theta \, d\theta$$

$$= -\gamma w R \left[-L \cos \theta + \frac{R}{2} \cos^2 \theta \right]_0^{\pi/2} = -\gamma w R \left(L - \frac{R}{2} \right)$$

$$F_{R_x} = -\frac{62.4 \text{ lbf}}{\text{ft}^3} \times 10 \text{ ft} \times 4 \text{ ft} \times \left(10 - \frac{4}{2} \right) \text{ft} = -19{,}970 \text{ lbf}$$

$$\overset{\longleftarrow}{\underset{F_{R_x}}{}}$$

{The minus sign indicates that the horizontal component of $\vec{F}_R$ acts to the right.}

$$F_{R_y} = \int dF_y = \vec{F}_R \cdot \hat{\jmath} = \int d\vec{F} \cdot \hat{\jmath} = -\int p \, d\vec{A} \cdot \hat{\jmath} = -\int_A p \, dA(1) \cos \theta$$

$$= -\int_0^{\pi/2} pwR \, d\theta \cos \theta = -\int_0^{\pi/2} \gamma hwR \cos \theta \, d\theta$$

$$= -\int_0^{\pi/2} \gamma(L - R \cos \theta)wR \cos \theta \, d\theta = -\gamma Rw \int_0^{\pi/2} (L \cos \theta - R \cos^2 \theta) \, d\theta$$

$$= -\gamma Rw \left[L \sin \theta - R \left(\frac{\theta}{2} + \frac{\sin 2\theta}{4} \right) \right]_0^{\pi/2} = -\gamma Rw \left(L - R \frac{\pi}{4} \right)$$

$$F_{R_y} = -\frac{62.4 \text{ lbf}}{\text{ft}^3} \times 4 \text{ ft} \times 10 \text{ ft} \times (10 - \pi)\text{ft} = -17{,}100 \text{ lbf} \qquad \overset{F_{R_y}}{\longleftarrow}$$

{The minus sign indicates that the vertical component of $\vec{F}_R$ acts downward.}

To find the line of action of F_{R_x}, the moment of F_{R_x} about O must be equal to the sum of moments of dF_x about O.

$$y'\hat{\jmath} \times F_{R_x}\hat{\imath} = \int y\hat{\jmath} \times dF_x\hat{\imath} = \int_A y\hat{\jmath} \times (-p \, dA \sin \theta)\hat{\imath}$$

$$-y'F_{R_x}\hat{k} = \hat{k} \int_A yp \, dA \sin \theta$$

$$y' = -\frac{1}{F_{R_x}} \int_A yp \, dA \sin \theta = -\frac{1}{F_{R_x}} \int_0^{\pi/2} y\gamma hwR \, d\theta \sin \theta$$

$$= -\frac{1}{F_{R_x}} \int_0^{\pi/2} y\gamma(L - R \cos \theta)wR \, d\theta \sin \theta$$

$$= -\frac{1}{F_{R_x}} \int_0^{\pi/2} R \cos \theta \gamma(L - R \cos \theta)wR \sin \theta \, d\theta$$

$$y' = -\frac{R^2 \gamma w}{F_{R_x}} \int_0^{\pi/2} (L \cos \theta \sin \theta - R \cos^2 \theta \sin \theta) \, d\theta$$

Integrating,

$$y' = -\frac{R^2 \gamma w}{F_{R_x}} \left[L \frac{\sin^2 \theta}{2} + \frac{R \cos^3 \theta}{3} \right]_0^{\pi/2} = -\frac{R^2 \gamma w}{F_{R_x}} \left[\frac{L}{2} - \frac{R}{3} \right]$$

$$y' = \frac{-(4)^2 \text{ ft}^2}{-19{,}970 \text{ lbf}} \times \frac{62.4 \text{ lbf}}{\text{ft}^3} \times 10 \text{ ft} \left[\frac{10}{2} - \frac{4}{3} \right] \text{ft} = 1.83 \text{ ft} \qquad \overset{y'}{\longleftarrow}$$

To find the line of action of F_{R_y}, the moment of F_{R_y} about O must be equal to the sum of moments of dF_y about O.

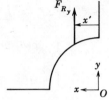

$$x' \hat{i} \times F_{R_y} \hat{j} = \int x \hat{i} \times dF_y \hat{j} = \int_A x \hat{i} \times (-p \, dA \cos \theta) \hat{j}$$

$$x' F_{R_y} \hat{k} = -\hat{k} \int_A xp \, dA \cos \theta$$

$$x' = -\frac{1}{F_{R_y}} \int_A xp \, dA \cos \theta = -\frac{1}{F_{R_y}} \int_0^{\pi/2} x\gamma(L - R\cos\theta) wR\cos\theta \, d\theta$$

$$= -\frac{1}{F_{R_y}} \int_0^{\pi/2} R\sin\theta\gamma(L - R\cos\theta) wR\cos\theta \, d\theta$$

$$= -\frac{wR^2\gamma}{F_{R_y}} \int_0^{\pi/2} (L\sin\theta\cos\theta - R\cos^2\theta\sin\theta) \, d\theta$$

$$= -\frac{wR^2\gamma}{F_{R_y}} \left[L\frac{\sin^2\theta}{2} + R\frac{\cos^3\theta}{3} \right]_0^{\pi/2} = -\frac{wR^2\gamma}{F_{R_y}} \left[\frac{L}{2} - \frac{R}{3} \right]$$

$$x' = -\frac{10 \text{ ft} \times (4)^2 \text{ ft}^2}{-17{,}100 \text{ lbf}} \times \frac{62.4 \text{ lbf}}{\text{ft}^3} \times \left(5 - \frac{4}{3}\right) \text{ft} = 2.14 \text{ ft}$$

x'

Note that since each infinitesimal force, $d\vec{F}$, acts through the center of the cylindrical surface, then the resultant force, $\vec{F}_R$, must also act through the origin, O. This fact could have been used to determine x' once y' was known, or vice versa.

$\left\{ \begin{array}{l} \text{This problem illustrates application of the basic equations for hydrostatic forces} \\ \text{to a surface with constant radius of curvature.} \end{array} \right\}$

**3-6 BUOYANCY AND STABILITY

If an object is immersed in or floating on the surface of a liquid, the force acting on it due to liquid pressure is termed *buoyancy*. Consider the object shown in Fig. 3.8, immersed in static liquid.

The vertical force on the body due to hydrostatic pressure may be found most easily by considering cylindrical volume elements similar to the one shown in Fig. 3.8. For a static fluid

$$\frac{dp}{dh} = \rho g$$

Integrating for constant ρ gives

$$p = p_0 + \rho g h$$

** This section may be omitted without loss of continuity in the text material.

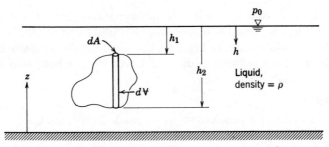

Fig. 3.8 Immersed body in static liquid.

The net vertical force on the element is therefore

$$dF_z = (p_0 + \rho g h_2) dA - (p_0 + \rho g h_1) dA = \rho g(h_2 - h_1) dA$$

But $(h_2 - h_1) dA = dV$, the volume of the element. Thus

$$F_z = \int dF_z = \int_V \rho g \, dV = \rho g V \qquad (3.18)$$

where V is the volume of the object. Thus the net vertical pressure force, or buoyancy of the object, equals the force of gravity on the liquid displaced by the object. This relation was first put to practical use by Archimedes who used it in 220 B.C. to determine the gold content in the crown of King Hiero II (Example 3.8). Consequently, it is often called "Archimedes' Principle." In more current technical applications, Eq. 3.18 is used to design displacement vessels, flotation gear, and bathyscaphes.

The line of action of the buoyancy force may be determined using the methods of Section 3-5.2. Since floating bodies are in equilibrium under body and buoyancy forces, the location of the line of action of the buoyancy force determines stability, as shown in Fig. 3.9.

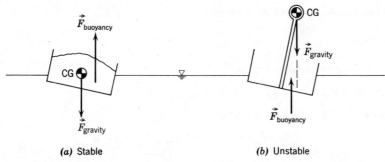

(a) Stable   *(b)* Unstable

Fig. 3.9 Stability of floating bodies.

The body force due to gravity on an object acts through its center of gravity, CG. In Fig. 3.9a, the buoyant force is offset and produces a couple that tends to right the craft. In Fig. 3.9b, the couple tends to capsize the craft. In sailing, wind loads bring additional forces onto a boat that must be considered in analyzing stability.

Example 3.8

King Hiero ordered a new crown to be made from pure gold. When he received the crown, he suspected that other metals had been used in its construction. Archimedes discovered that the crown required a force of 4.7 lbf to suspend it when immersed in water, and that it displaced 18.9 in.[3] of water. He concluded that the crown could not be pure gold. Do you agree?

Example Problem 3.8

GIVEN:

Crown volume, $\forall = 18.9$ in.[3]
Force to suspend in water, $F_{net} = 4.7$ lbf

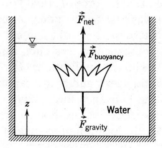

FIND:

Average density of material in crown.

SOLUTION:

Apply $\vec{F} = m\vec{a}$ to immersed crown.

Basic equations: $\qquad \sum \vec{F} = m\vec{a} \qquad F_{buoyancy} = \rho_{H_2O} g \forall \hat{k}$

Assume: $\quad \vec{a} = 0$ for crown. Then

$$\sum \vec{F} = \vec{F}_{net} + \vec{F}_{gravity} + \vec{F}_{buoyancy} = m\vec{a} = 0$$

or

$$(F_{net} - F_{gravity} + \rho_{H_2O} g \forall)\hat{k} = 0$$

But $F_{gravity} = \rho_c g V$, where ρ_c is the density of the crown material, so

$$F_{gravity} = F_{net} + \rho_{H_2O} g V = \rho_c g V$$

and

$$\rho_c = \rho_{H_2O} + \frac{F_{net}}{g V} = \frac{1.94 \text{ slug}}{ft^3} + 4.7 \text{ lbf} \times \frac{1}{18.9 \text{ in.}^3} \times \frac{sec^2}{32.2 \text{ ft}} \times \frac{slug \cdot ft}{lbf \cdot sec} \times \frac{1728 \text{ in.}^3}{ft^3}$$

$$\rho_c = 15.3 \text{ slug/ft}^3 \hspace{5cm} \rho_c$$

{This is about the same density as steel. Since gold is heavier than steel, the}
{crown could not be pure gold.}

**3-7 FLUIDS IN RIGID-BODY MOTION

A fluid in rigid-body motion moves without deformation as though it were a solid body. Since there is no deformation, there can be no shear stress. Consequently, the only surface stress on each element of fluid is that due to pressure.

A fluid particle retains its identity in rigid-body motion because the fluid does not deform. As in the case of the static fluid, we may apply Newton's second law of motion to determine the pressure field that results from a specified rigid-body motion.

In Section 3-1 we derived an expression for the total force due to pressure and gravity acting on a fluid particle of volume, dV. We obtained

$$d\vec{F} = (-\operatorname{grad} p + \rho \vec{g}) \, dV$$

or

$$\frac{d\vec{F}}{dV} = -\operatorname{grad} p + \rho \vec{g} \tag{3.2}$$

Newton's second law was written

$$d\vec{F} = \vec{a} \, dm = \vec{a} \rho \, dV$$

or

$$\frac{d\vec{F}}{dV} = \rho \vec{a}$$

Substituting from Eq. 3.2, we obtain

$$-\operatorname{grad} p + \rho \vec{g} = \rho \vec{a} \tag{3.19}$$

** This section may be omitted without loss of continuity in the text material.

The physical significance of each term in this equation is reviewed as follows:

$$-\operatorname{grad} p \qquad + \qquad \rho\vec{g} \qquad = \qquad \rho\vec{a}$$

$$\left\{\begin{array}{l}\text{pressure force} \\ \text{per unit volume} \\ \text{at a point}\end{array}\right\} + \left\{\begin{array}{l}\text{body force} \\ \text{per unit volume} \\ \text{at a point}\end{array}\right\} = \left\{\begin{array}{l}\text{mass per} \\ \text{unit} \\ \text{volume}\end{array}\right\} \times \left\{\begin{array}{l}\text{acceleration} \\ \text{of fluid} \\ \text{particle}\end{array}\right\}$$

This vector equation consists of three component equations that must be satisfied individually. In rectangular coordinates the component equations are

$$\left.\begin{array}{ll}-\dfrac{\partial p}{\partial x} + \rho g_x = \rho a_x & x \text{ direction} \\[2ex] -\dfrac{\partial p}{\partial y} + \rho g_y = \rho a_y & y \text{ direction} \\[2ex] -\dfrac{\partial p}{\partial z} + \rho g_z = \rho a_z & z \text{ direction}\end{array}\right\} \qquad (3.20)$$

Component equations for other coordinate systems can be written using the appropriate expression for grad p.

Example 3.9

As a result of a promotion, you are transferred from your present location. You must transport a fish tank in the back of your station wagon. The tank is 12 in. × 24 in. × 12 in. How much water should you leave in the tank to be reasonably sure that it will not spill over during the trip?

Example Problem 3.9

GIVEN:

Fish tank 12 in. × 24 in. × 12 in. partially filled with water to be transported in an automobile.

FIND:

Allowable depth of water for reasonable assurance that it will not spill during the trip.

SOLUTION:

The first step in effecting a solution is to formulate the problem, that is, to translate the general problem into something more specific.

We recognize that there will be motion of the water surface as a result of the car's traveling over bumps in the road, going around corners, etc. However, we shall assume that the main effect on the water surface is due to linear accelerations (and decelerations) of the car; that is, we shall neglect sloshing.

Thus we have reduced the problem to one of determining the effect of a linear acceleration on the free surface. We have not yet decided on the orientation of the tank relative to the direction of motion. Choosing the x coordinate in the direction of motion, should we align the tank with the long side parallel, or perpendicular, to the direction of motion?

If there will be no relative motion in the water, we must assume we are dealing with a constant acceleration, a_x. What is the shape of the free surface under these conditions?

Let us restate the problem to answer the original questions without making any restrictive assumptions at the outset.

GIVEN:

Tank partially filled with water (to a depth d in.) subject to constant linear acceleration, a_x. Tank height is 12 in.; length parallel to direction of motion is b in. Width perpendicular to direction of motion is c in.

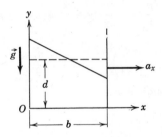

FIND:

(a) Shape of free surface under constant a_x.
(b) Allowable water height, d, to avoid spilling as a function of a_x and tank orientation.
(c) Optimum tank orientation and allowable depth.

SOLUTION:

Basic equation:
$$-\nabla p + \rho \vec{g} = \rho \vec{a}$$

$$-\left(\hat{\imath}\frac{\partial p}{\partial x} + \hat{\jmath}\frac{\partial p}{\partial y} + \hat{k}\frac{\partial p}{\partial z}\right) + \rho(\hat{\imath}g_x + \hat{\jmath}g_y + \hat{k}g_z) = \rho(\hat{\imath}a_x + \hat{\jmath}a_y + \hat{k}a_z)$$

Since p is not a function of z, $\partial p/\partial z = 0$. Also, $g_x = 0$, $g_y = -g$, $g_z = 0$ and $a_y = a_z = 0$.

$$\therefore \; -\hat{\imath}\frac{\partial p}{\partial x} - \hat{\jmath}\frac{\partial p}{\partial y} - \hat{\jmath}\rho g = \hat{\imath}\rho a_x$$

The component equations are:

$$\frac{\partial p}{\partial x} = -\rho a_x$$

$$\frac{\partial p}{\partial y} = -\rho g$$

$\left\{\begin{array}{l}\text{Recall that a partial}\\ \text{derivative means that}\\ \text{all other independent}\\ \text{variables are held constant}\\ \text{in the differentiation.}\end{array}\right.$

The problem now is to find an expression for $p = p(x, y)$. This would enable us to find the equation of the free surface. But perhaps we do not have to do that.

Since the pressure, $p = p(x, y)$, the difference in pressure between two points (x, y) and $(x + dx, y + dy)$ is

$$dp = \frac{\partial p}{\partial x}\, dx + \frac{\partial p}{\partial y}\, dy$$

Since the free surface is a line of constant pressure, then along the free surface, $dp = 0$ and

$$0 = \frac{\partial p}{\partial x}\, dx + \frac{\partial p}{\partial y}\, dy = -\rho a_x\, dx - \rho g\, dy$$

Therefore

$$\left.\frac{dy}{dx}\right)_{\text{free surface}} = -\frac{a_x}{g}$$

{The free surface is a straight line.}
In the diagram below,

d = original depth

e = height above the original depth

b = tank length parallel to direction of motion

$$e = \frac{b}{2}\tan\theta = \frac{b}{2}\left(-\frac{dy}{dx}\right)_{\text{free surface}} = \frac{b}{2}\frac{a_x}{g} \qquad \left\{\text{Only valid for } d \geq \frac{b}{2}\right\}$$

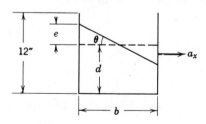

Since we want e to be smallest for a given a_x, the tank should be aligned with b as small as possible. We should align the tank with the long side perpendicular to the direction of motion, that is, choose $b = 12$ in.

With $b = 12$ in.,

$$e = 6\frac{a_x}{g} \text{ in.}$$

The maximum allowable value of $e = 12 - d$ in. Thus

$$12 - d = 6\frac{a_x}{g} \quad \text{and} \quad d_{max} = 12 - 6\frac{a_x}{g}$$

If the maximum a_x is assumed to be $\frac{2}{3}g$, then allowable $d = 8$ in.
To allow a margin of safety, perhaps we should select $d = 6$ in.

Recall that a steady acceleration has been assumed in this problem. The car would have to be driven very carefully.

This problem has been included to demonstrate:
(i) not all problems are clearly defined, nor do they have a unique answer, and
(ii) the application of the equation, $-\nabla p + \rho\vec{g} = \rho\vec{a}$.

Example 3.10

A cylindrical container, partially filled with a liquid, is rotated at a constant angular velocity, ω, about its axis as shown in the diagram. After a short period of time there is no relative motion, that is, the liquid rotates with the cylinder as if the system were a rigid body. Determine the shape of the free surface.

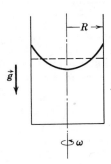

Example Problem 3.10

GIVEN:

A cylinder of liquid in solid-body rotation with angular velocity, ω, about its axis.

FIND:

The shape of the free surface.

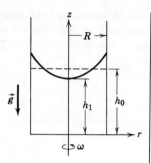

SOLUTION:

It is convenient to use a cylindrical coordinate system, r, θ, z. Since there is circumferential symmetry in this problem, the pressure p will not be a function of θ, that is, $p = p(r, z)$.

Note: The free surface is a surface of constant pressure; the problem is to find the equation of the surface.

Since $p = p(r, z)$, the differential change, dp, in pressure between two points with coordinates (r, θ, z) and $(r + dr, \theta, z + dz)$ is given by

$$dp = \frac{\partial p}{\partial r}\bigg)_z \cdot dr + \frac{\partial p}{\partial z}\bigg)_r dz$$

Consequently, we see that we need to obtain expressions for $\partial p/\partial z)_r$ and $\partial p/\partial r)_z$. This we can do by writing Newton's second law in the z and r directions, respectively, for an infinitesimal fluid element.

From Eq. 3.20 we have, for the z direction

$$-\frac{\partial p}{\partial z}\bigg)_r + \rho g_z = \rho a_z$$

Since $g_z = -g$ and $a_z = 0$, then $\partial p/\partial z)_r = -\rho g$.

To obtain an expression for $\partial p/\partial r)_z$, we apply Newton's second law in the r direction to a suitable differential element.

The pressure at the center of the element is p.

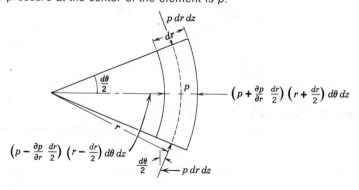

The forces acting in the $r\theta$ plane on the element are shown in the diagram. Writing Newton's second law in the r direction, we have

$$\sum dF_r = a_r\, dm = a_r \rho\, d\forall = -\omega^2 r\rho\, d\forall = -\omega^2 r\rho r\, d\theta\, dr\, dz$$

From the figure

$$\sum dF_r = \left(p - \frac{\partial p}{\partial r}\frac{dr}{2}\right)\left(r - \frac{dr}{2}\right)d\theta\, dz - \left(p + \frac{\partial p}{\partial r}\frac{dr}{2}\right)\left(r + \frac{dr}{2}\right)d\theta\, dz + 2p\, dr\, dz \sin\frac{d\theta}{2}$$

Expanding and canceling like terms, recognizing $\sin d\theta/2 = d\theta/2$ (small angles) gives

$$\sum dF_r = d\theta\, dz\left\{pr - p\frac{dr}{2} - r\frac{\partial p}{\partial r}\frac{dr}{2} + \frac{\partial p}{\partial r}\left(\frac{dr}{2}\right)^2 - pr - p\frac{dr}{2} - r\frac{\partial p}{\partial r}\frac{dr}{2} - \frac{\partial p}{\partial r}\left(\frac{dr}{2}\right)^2 + p\, dr\right\}$$

$$\sum dF_r = d\theta\, dz\left\{-r\frac{\partial p}{\partial r}\, dr\right\}$$

Then

$$-r\frac{\partial p}{\partial r}\, dr\, d\theta\, dz = -\omega^2 r\rho r\, d\theta\, dr\, dz$$

Dividing both sides by $-r\, dr\, d\theta\, dz$ results in

$$\frac{\partial p}{\partial r} = \rho\omega^2 r$$

Since

$$dp = \left.\frac{\partial p}{\partial r}\right)_z dr + \left.\frac{\partial p}{\partial z}\right)_r dz$$

Then

$$dp = \rho\omega^2 r\, dr - \rho g\, dz$$

To obtain the pressure difference between a reference point (r_1, z_1), where the pressure is p_1, and the arbitrary point (r, z), where the pressure is p, we must integrate.

$$\int_{p_1}^{p} dp = \int_{r_1}^{r} \rho\omega^2 r\, dr - \int_{z_1}^{z} \rho g\, dz$$

$$p - p_1 = \frac{\rho\omega^2}{2}(r^2 - r_1^2) - \rho g(z - z_1)$$

Taking the reference point on the cylinder axis at the free surface gives

$$p_1 = p_{atm} \qquad r_1 = 0 \qquad z_1 = h_1$$

Then

$$p - p_{atm} = \frac{\rho\omega^2 r^2}{2} - \rho g(z - h_1)$$

Since the free surface is a surface of constant pressure ($p = p_{atm}$), the equation of the free surface is given by

$$0 = \frac{\rho \omega^2 r^2}{2} - \rho g(z - h_1)$$

or

$$z = h_1 + \frac{(\omega r)^2}{2g}$$

The equation of the free surface is a parabola with vertex on the axis at $z = h_1$. We can solve for the height h_1 under conditions of rotation in terms of the original surface height, h_0, in the absence of rotation. To do this, we use the fact that the volume of fluid must remain constant. With no rotation

$$\forall = \pi R^2 h_0$$

With rotation

$$\forall = \int_0^R \int_0^z 2\pi r \, dz \, dr = \int_0^R 2\pi z r \, dr = \int_0^R 2\pi \left(h_1 + \frac{\omega^2 r^2}{2g} \right) r \, dr$$

$$\forall = 2\pi \left[h_1 \frac{r^2}{2} + \frac{\omega^2 r^4}{8g} \right]_0^R = \pi \left[h_1 R^2 + \frac{\omega^2 R^4}{4g} \right]$$

Then

$$\pi R^2 h_0 = \pi \left[h_1 R^2 + \frac{\omega^2 R^4}{4g} \right]$$

and

$$h_1 = h_0 - \frac{(\omega R)^2}{4g}$$

Finally,

$$z = h_0 - \frac{(\omega R)^2}{4g} + \frac{(\omega r)^2}{2g}$$

$$z = h_0 - \frac{(\omega R)^2}{2g} \left[\frac{1}{2} - \left(\frac{r}{R} \right)^2 \right] \qquad z(r)$$

{ This problem illustrates the application of Newton's second law to a differential element and the physical behavior of a liquid with a free surface undergoing solid body rotation. }

summary objectives

After completing study of Chapter 3, you should be able to do the following:

1. Write the basic equation of fluid statics in vector form and indicate the physical significance of each term.

2. Write the basic pressure-height relation for a static fluid and integrate it to determine the pressure variation for any given fluid property variations.

3. Define temperature and pressure conditions for the standard atmosphere.

4. State the relation between absolute and gage pressures.

5. For a plane submerged surface:
   (a) Determine the resultant force due to the fluid acting on the surface and its line of action.
   (b) Determine the external force(s) required to maintain the surface in equilibrium.

6. For a submerged surface with curvature in one plane:
   (a) Determine the components of the resultant force due to the fluid acting on the surface and their lines of action.
   (b) Determine the external force(s) required to maintain the surface in equilibrium.

**7. Determine the buoyant force on a body immersed in or floating on the surface of a liquid; determine the stability of a floating object.

**8. Apply the basic hydrostatic equation to determine the pressure field and/or free surface shape for any body of fluid in rigid-body motion.

9. Solve those problems at the end of the chapter that relate to the material you have studied.

problems

3.1 A hydraulic press for stamping sheet metal automotive parts can exert a force of 36 MN(meganewtons). The actuation stroke is 0.2 m. The press is hydraulically actuated by a system whose design pressure is 30 MPa. Neglecting friction, determine the minimum piston area needed to produce the clamping force. How much hydraulic oil must be supplied per press cycle?

3.2 A pneumatic hoist is to be designed for a service station. Shop air is available at a gage pressure of 600 kPa. The hoist must be capable of lifting automobiles up to 3000 kg. Friction in the piston-cylinder mechanism and seals causes a force of 980 N opposing the piston motion. Determine the piston diameter necessary to provide the lift force. What pressure should be maintained in the lift cylinder to lower smoothly a VW Rabbit with a mass of 895 kg?

3.3 Pipe for the Alaskan pipeline has an internal diameter of 1.22 m. Wall thicknesses of 11 and 14 mm are used. The pipe was tested hydrostatically to a pressure of 10 MPa; the maximum pressure in service is expected to be 7.79 MPa. Calculate the maximum tensile stress in the pipe wall that is to be expected in service. Will the direction of the maximum stress in the pipe wall be axial or circumferential?

3.4 Compressed nitrogen is shipped in a cylindrical tank of diameter, $D = 0.25$ m, and length $L = 1.3$ m. The gas in the tank is at an absolute pressure of 20 MPa and 20 C. Calculate the mass of gas in the tank. If the maximum allowable stress

** These objectives apply to sections that may be omitted without loss of continuity in the text material.

in the tank wall is 210 MPa, determine the theoretical minimum thickness of the cylinder wall.

3.5 On integrating the pressure-height equation for a static incompressible fluid, it was assumed that the gravitational acceleration, g, was constant. The Law of Gravitational Attraction is

$$g = g_0 \left(\frac{R}{R + h} \right)^2$$

where R is the radius of the Earth and h is altitude above the surface. Find the percent variation in g for the following two cases (take $R = 4000$ miles):
(a) $h = 6$ mile altitude
(b) $h = -4$ mile altitude

3.6 A mercury barometer is read at the same location on different days. Each time, the pressure reading is 29.5 in. of mercury, but the ambient temperatures are 70 and 95 F on the two days. Determine the actual atmospheric pressures on the two days, in lbf/ft^2, and the difference between them, in psi.

3.7 A closed container contains water at a 5 m depth. The absolute pressure above the water surface is 0.3 atm. Calculate the absolute pressure on the inside of the bottom surface of the container.

3.8 Determine the gage pressure in psig at point a in Fig. 3.10, if liquid A has a specific gravity of 0.75 and liquid B has a specific gravity of 1.20. The liquid surrounding point a is water, and the tank on the left is open to the atmosphere.

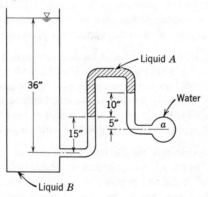

Fig. 3.10

3.9 A rectangular tank, open to the atmosphere, is filled with water to a depth of 2.5 m as shown in Fig. 3.11. A U-tube manometer is connected to the tank at a location 0.7 m above the tank bottom. If the zero level of the manometer fluid, Meriam blue (specific gravity 1.75) is 0.2 m below the connection, determine the deflection l after the manometer is connected and all the air has been removed from the connecting leg.

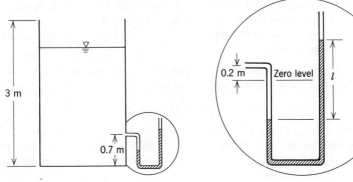

Fig. 3.11

3.10 If the tank of Problem 3.9 is sealed tightly and water drains slowly from the bottom of the tank, determine the deflection, l, after the system has attained equilibrium.

3.11 The manometer fluid of Problem 3.9 is replaced with mercury (same zero level). The tank is sealed and the air pressure is increased to a gage pressure of 0.5 atm. Determine the deflection, l.

3.12 A reservoir manometer is calibrated for use with a fluid of specific gravity 0.827. The reservoir diameter is $\frac{5}{8}$ in. and the (vertical) tube diameter is $\frac{3}{16}$ in. Calculate the required distance between marks on the vertical scale for 1 in. of water pressure difference.

3.13 The inclined manometer shown in Fig. 3.12 has reservoir diameter, D, of 90 mm and measuring tube diameter, d, of 6 mm; the manometer fluid is Meriam red oil. The length of the measuring tube is 0.6 m; $\theta = 30°$. Determine the maximum pressure, in Pa, that can be measured with the manometer.

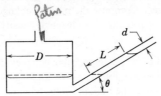

Fig. 3.12

3.14 The inclined manometer shown in Fig. 3.12 has reservoir diameter, D, of 3 in., measuring tube diameter, d, of 0.25 in., and is filled with gage oil (SG = 0.897). Compute the angle θ that will give a 5 in. oil deflection along the inclined tube for an applied pressure of 1 in. of water (gage).

3.15 The inclined manometer shown in Fig. 3.12 has reservoir diameter, D, of 96 mm and measuring tube diameter, d, of 8 mm. Determine the angle θ required to provide a 5:1 increase in liquid deflection compared to a regular U-tube manometer.

3.16 The variation of pressure with height in still air is found to be $dp/dz = -10.7$ N/m^3 at a location where the pressure is 1 atm. The coordinate, z, is measured vertically upward. Determine the temperature of the air at this location, assuming ideal gas behavior.

3.17 A model has been proposed that accounts for specific weight variations in the Earth's atmosphere according to the equation:

$$\gamma = \gamma_0 - kz$$

where $k = 1.93 \times 10^{-6}$ lbf/ft^4, $z =$ altitude above Earth's surface, and $\gamma_0 =$ specific weight of air at sea level (0.0765 lbf/ft^3). Assuming a sea level pressure of 14.7 psia, compute the pressure at an altitude of 20,000 ft. Compare with the Standard Atmosphere.

3.18 If the variation in specific weight of atmospheric air between sea level and an altitude of 3600 ft were given by $\gamma = \gamma_0 - k\sqrt{z}$, where γ_0 is the specific weight of air at sea level, z is the altitude above sea level, and $k = 0.00020$ lbf/ft$^3 \cdot$ ft$^{1/2}$, determine the pressure in psia at an altitude of 3600 ft when sea level conditions are 14.7 psia, 59 F.

3.19 As a result of changes in temperature, salinity, and pressure, the density of sea water increases with increasing depth in accordance with the equation

$$\rho = \rho_s + bh$$

where ρ_s is the density at the surface, h is depth below the surface, and b is a positive constant. Develop an algebraic equation for the pressure as a function of depth.

3.20 Water is usually assumed to be an incompressible fluid when evaluating static pressure variations. Actually, it is about 100 times more compressible than steel. The bulk modulus is defined as $E_v = dp/(d\rho/\rho)$ and it may be assumed constant (see Appendix A for values). Compute the percent change in density for water raised to a gage pressure of 100 atm if the original density at atmospheric pressure is 999 kg/m^3.

3.21 Water is usually assumed to be incompressible when evaluating static pressure variations. Actually, its compressibility can be important in the design of submersible vehicles. Assume that the bulk modulus of water is constant. Compute the pressure and water density at a depth of 4 miles in sea water. The density at the sea water surface is 64 lbm/ft^3.

3.22 Oceanographic research vessels have descended to depths approaching 10 km below sea level. At these extreme depths, the compressibility of sea water can be significant. One may model the behavior of sea water by assuming that its bulk compressibility modulus remains constant. Using this assumption, evaluate the deviations in density and pressure compared to values computed using the incompressible assumption at a depth of 10 km in sea water. Express your answers in percent.

3.23 A hypothetical dense compressible fluid is used in the system shown in Fig. 3.13. Both pistons are frictionless, and their outer faces are exposed to atmospheric pressure. The equation of state for the fluid is assumed to be

$$\gamma = Kp^{1/2} \qquad \text{where} \qquad K = 0.10 \frac{\text{lbf}^{1/2}}{\text{ft}^2}$$

Find the pressure on the inner face of the lower piston, and the force, F, required for equilibrium.

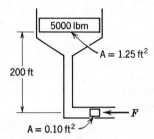

5000 lbm

A = 1.25 ft²

200 ft

A = 0.10 ft²

F

Fig. 3.13

3.24 Evaluate dp/dz for standard air on a windless day at sea level and at 5 km altitude. (The coordinate z is measured positive upward.)

3.25 Determine the change of elevation necessary to effect a 15 percent reduction in density for an isothermal atmosphere at 20 C.

3.26 At ground level in Denver, Colorado, the atmospheric pressure and temperature are 83.2 kPa and 25 C. Calculate the pressure on Pike's Peak at an elevation of 2690 m above the city assuming (a) an incompressible, and (b) an adiabatic atmosphere.

3.27 If air is assumed to be an ideal gas, knowledge of the temperature variation with altitude allows the determination of pressure at any elevation when conditions are known at a reference elevation, z_0.
(a) For $T = T_0(1 + mz)$, derive the equation for the variation of pressure as a function of altitude if the pressure at the reference elevation is p_0.
(b) Using the results of part (a), show that the variation of pressure for the isothermal case ($m \to 0$) is given by

$$\frac{p}{p_0} = e^{-(g/RT_0)(z-z_0)}$$

3.28 Because the pressure falls, water boils at a lower temperature with increasing altitude. Consequently, cake mixes and boiled eggs, among other foods, must be cooked different lengths of time. Determine the boiling temperature of water at 1000 and 2000 m elevation on a standard day, and compare with the sea level value.

3.29 Automobiles suffer a power loss with altitude as a result of the decrease in air density. If the volumetric efficiency of an engine remains constant, and the carburetion is adjusted to maintain the same air-fuel ratio, determine the percentage loss in power for an engine at 3000 m elevation, compared to sea level on a standard day.

3.30 Pressure variations that result from altitude changes can cause ear "popping" and discomfort to airplane passengers or those driving in the mountains. Each individual is affected differently, but one ear "pop"/75 m of elevation change might be a reasonable average figure. Determine the pressure change, expressed in millimeters of water, that corresponds to this elevation difference on a standard day at an altitude of 2000 m.

3.31 A device known as a deadweight tester can be used as a standard for calibration of mechanical pressure gages (the useful range is about 30 kPa to 35 MPa). Known pressures are generated by loading weights on a vertical piston-cylinder arrangement. The weighted piston is rotated to minimize frictional effects. The maximum convenient load is 100 kg. Determine an appropriate piston size to cover the pressure range given.

3.32 Many recreation facilities are currently being built using inflatable "bubble" structures. A tennis bubble to enclose four courts is shaped roughly like a circular semicylinder with a diameter of 30 m and a length of 60 m. The blowers used to inflate the structure can maintain the air pressure inside the bubble at 10 mm of water above ambient pressure. The fabric "skin" of the bubble is of uniform thickness. Determine the maximum material density, in mass per unit area, that can be used to fabricate a pressure-supported bubble.

3.33 One of the major tire companies has put into operation a tire-curing press 6.0 m in diameter. A hemispherical dome is used to contain the mold and tire while steam is introduced under pressure. The dome mass is 130 metric tons. Determine the steam pressure (gage) at which the mass of the dome is exactly balanced. What is the corresponding saturation temperature?

3.34 A mechanical pressure gage attached to the closed reservoir tank of an air compressor indicates a pressure of 827 kPa on a day when the barometer reading 750 mm of mercury. Calculate the absolute pressure in the tank. What pressure would the gage indicate if the barometer reading changed to 775 mm of mercury?

3.35 A door 1 m wide and 1.5 m high is located in a plane vertical wall of a water tank. The door is hinged along its upper edge, which is 1 m below the water surface. Atmospheric pressure acts on the outer surface of the door and at the water surface. Determine the total resultant force due to all fluids acting on the door.

3.36 If, in Problem 3.35, the water surface gage pressure is raised to 0.3 atm, determine the total resultant force from all fluids acting on the door.

3.37 A 1 ft cube is submerged as shown in Fig. 3.14. Calculate the actual force of the water on the bottom surface, and the net vertical force on the cube.

3.38 The door shown in Fig. 3.15 is 5 ft wide and 10 ft high. Find the resultant force from all fluids acting on the door.

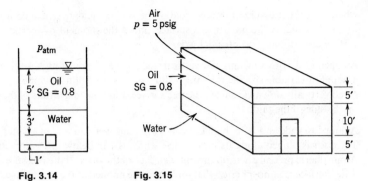

Fig. 3.14   **Fig. 3.15**

3.39 Most large dams contain flood gates that can be raised to release stored water (Fig. 3.16). The gate shown slides against a plate on each side. The gate mass is 5000 kg.
  (a)   Find the normal force on the gate due to the water.
  (b)   If μ_s (coefficient of static friction) = 0.4 between the gate and the supports, determine the magnitude of the force, R, required to start the gate in motion.

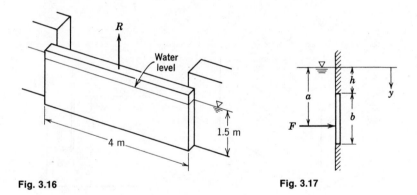

Fig. 3.16   **Fig. 3.17**

3.40 A vertical plane surface is submerged in liquid as shown in Fig. 3.17. The width of the surface is w. The density of the incompressible liquid is ρ. Find:
  (a)   A general expression for the resultant force, F
  (b)   A general expression for the vertical distance, a

3.41 Consider again the plane surface shown in Fig. 3.17. Show that the results of Problem 3.40 can be written as

$$F = p_c A \quad \text{and} \quad a = y_c + \frac{I}{y_c A}$$

where p_c is the pressure at the centroid of the surface, y_c the vertical coordinate of the centroid, A the area of the surface, and I the moment of inertia of the surface area about its centroidal axis.

3.42 An open tank with a rectangular vertical side 2 ft wide and 6 ft high is filled with a liquid of variable specific weight, γ, with $\gamma = 50 + 2y$ (lbf/ft³), where y is measured vertically downward from the free surface. Find the magnitude of the force on the side of the tank.

3.43 A door 1 m wide and 1.5 m high is located in a plane vertical wall of a water tank. The door is hinged along its upper edge, which is 1 m below the water surface. Atmospheric pressure acts on the outer surface of the door. If the pressure at the water surface is atmospheric, what force must be applied at the lower edge of the door in order to keep the door from opening?

3.44 If in Problem 3.43 the gage pressure at the water surface is 0.5 atm, what force must be applied at the lower edge of the door in order to keep the door from opening?

3.45 An aquarium at Marineland has a window located as shown in Fig. 3.18. The resultant force from sea water ($\gamma = 64$ lbf/ft³) on the window is 1280 lbf. Determine the line of application of the resultant force, in feet below the top of the window.

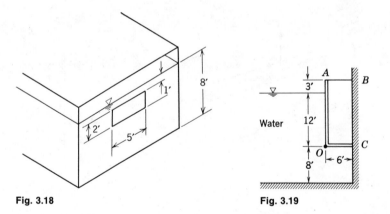

Fig. 3.18   **Fig. 3.19**

3.46 The gate AOC shown in Fig. 3.19 is 6 ft wide and is hinged at point O. Neglecting the weight of the gate, determine the force in the bar AB.

3.47 The door shown in Fig. 3.20 is hinged along its bottom. A pressure of 100 psfg is applied to the liquid free surface. Find the force, F, required to hold the door shut.

3.48 As water rises on the left side of the rectangular gate, the gate will open automatically (Fig. 3.21). At what depth above the hinge will this occur? Neglect the mass of the gate.

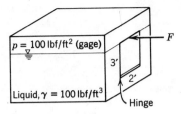

$p = 100\ \text{lbf/ft}^2$ (gage)

3'

2'

Liquid, $\gamma = 100\ \text{lbf/ft}^3$

Hinge

F

Fig. 3.20

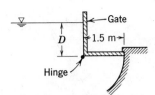

Gate

D 1.5 m

Hinge

Fig. 3.21

3.49 A tank with a center partition has a small "door" 0.5 m wide by 1 m high at the bottom. This door is hinged along the top edge. The left side has 0.6 m of water while the right side contains 1 m of nitric acid (specific gravity 1.5). What force (magnitude and direction) is required at the lower edge of the door to hold it closed?

3.50 The circular access port in the side of a water standpipe has a diameter of 0.6 m and is held in place by eight bolts evenly spaced around the circumference. If the standpipe has a diameter of 7 m and the center of the port is located 12 m below the free surface of the water, determine (a) the total force on the port and (b) the appropriate bolt diameter.

3.51 The gate shown is hinged at H (Fig. 3.22). The gate is 2 m wide normal to the plane of the diagram. Calculate the force required at A to hold the gate closed.

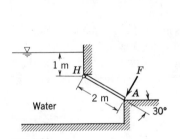

1 m H

F

2 m A

Water

30°

Fig. 3.22

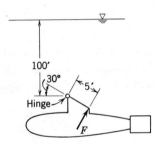

100'

30°   5'

Hinge

F

Fig. 3.23

3.52 A submarine is 100 ft below the surface as shown in Fig. 3.23. Find the net force, F, required to open the circular hatch when applied as shown. The pressure inside the submarine is equal to atmospheric pressure.

3.53 The gate shown in Fig. 3.24 is 3 m wide and for the purposes of analysis can be considered massless. For what depth of water will this rectangular gate be in equilibrium at an angle of 60° as shown?

3.54 A plane gate is held in equilibrium by the force, F, as shown in Fig. 3.25. The gate weighs 600 lbf/ft of gate width and its center of gravity is 6 ft from the hinge at O. Find the unknown force, F, when $h = 5$ ft and $\theta = 30°$.

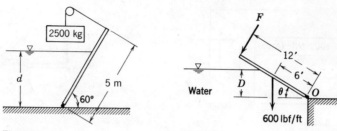

Fig. 3.24   Fig. 3.25

3.55 A gate of mass 2000 kg is mounted on a frictionless hinge along the lower edge. The length of the reservoir and gate (perpendicular to the plane of view) is 8 m. For the equilibrium conditions shown in Fig. 3.26, compute the width, b, of the gate.

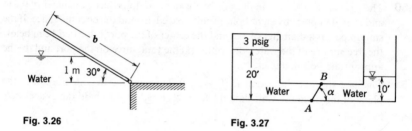

Fig. 3.26   Fig. 3.27

3.56 Gate AB is 3 ft wide and 2 ft long. It is inclined at $\alpha = 60°$ when closed. Find the moment about hinge A exerted by the water (Fig. 3.27).

3.57 The rectangular gate AB, as shown in Fig. 3.28, is 2 m wide. Find the force per unit width exerted against the stop at A. Assume the gate mass is negligible

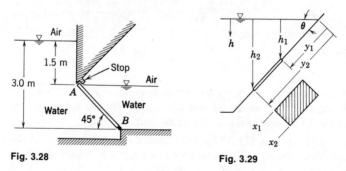

Fig. 3.28   Fig. 3.29

3.58 Due to temperature gradients in a liquid in a large tank, the density is not constant. The density variation is given by

$$\rho = \rho_0(1 + h^{1/5})$$

Find:
(a) The resultant force on the area indicated in Fig. 3.29 due only to the liquid
(b) The position y' at which the resultant force acts

3.59 The parabolic gate (Fig. 3.30) is 2 m wide. Determine the magnitude and line of action of the horizontal force on the gate due to the water; $c = 0.25 \text{ m}^{-1}$.

3.60 For the conditions of Problem 3.59, determine the magnitude and line of action of the vertical force on the gate due to the water.

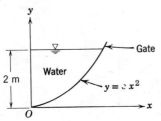

Fig. 3.30

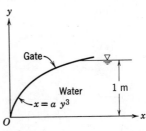

Fig. 3.31

3.61 The depth of water to the right of the gate of Problem 3.59 is increased from zero to L m. Determine the depth, L, required to reduce the moment about O to 50 percent of the value for $L = 0$.

3.62 The gate shown in Fig. 3.31 is 1.5 m wide. Determine the magnitude and moment of the vertical component of the force about O. The liquid is water; $a = 1.0 \text{ m}^{-2}$.

3.63 For the conditions of Problem 3.62, determine the magnitude and line of action of the horizontal component of the force.

3.64 If water stands at a depth of 0.5 m to the left of the gate of Problem 3.62 (Fig. 3.31), determine the total moment about O.

3.65 Determine the magnitude and line of action of the vertical force on the curved section AB (Fig. 3.32). The liquid is water and the section AB is 1 ft wide. Atmospheric pressure acts at the free surface; $k = 1.0 \text{ ft}^{-1}$.

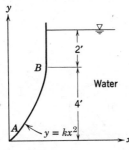

Fig. 3.32

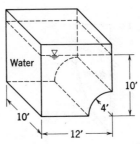

Fig. 3.33

3.66 An open tank is filled with water to the depth indicated in Fig. 3.33. Atmospheric pressure acts on all outer surfaces of the tank. Determine the magnitude and line of action of the vertical component of the force of the water on the curved part of the tank bottom.

3.67 A spillway gate formed in the shape of a circular arc is w m wide (Fig. 3.34). Find:

(a) The magnitude and direction of the vertical component of the force due to all fluids acting on the gate

(b) A point on the line of action of the total resultant force on the gate due to all fluids

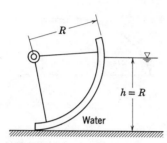

Fig. 3.34

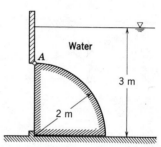

Fig. 3.35

3.68 A gate, which is in the form of a quarter-cylinder, is hinged at A and is 2 m wide normal to the paper (Fig. 3.35). The bottom of the gate is 3 m below the water surface. Determine:

(a) The magnitude of the horizontal force

(b) The line of action of the horizontal force

(c) The magnitude of the vertical force

(d) The line of action of the resultant force

3.69 The tank shown in Fig. 3.36 is 2 ft wide (i.e. 2 ft perpendicular to the xz plane). It is filled with water to a depth of 8 ft. The air between the top of the tank and the water is pressurized to 10 psig. Determine the magnitude and the line of action of the vertical force on the curved portion of the tank; $k = 0.5$ ft^{-1}.

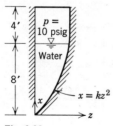

Fig. 3.36

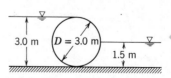

Fig. 3.37

3.70 If the water depth in the tank of Problem 3.69 is reduced to 4 ft and the air pressure is maintained at 10 psig, determine the magnitude and line of action of the vertical force on the curved portion of the tank.

3.71 A cylindrical weir has a diameter of 3 m and a length of 6 m (Fig. 3.37). Find the magnitude and direction of the resultant force acting on the weir from the water.

3.72 Find the mass of the cylinder shown in Fig. 3.38. It is 1 m long and is supported by the liquid (water). Assume no friction between the cylinder and solid wall; do not use the buoyancy force technique except to check your results.

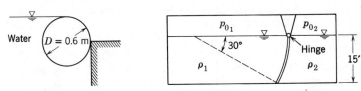

Fig. 3.38   **Fig. 3.39**

3.73 Two different liquids are separated by a weightless curved divider with a constant radius of curvature. Each has a different pressure at its free surface designated by p_{0_1} and p_{0_2} in Fig. 3.39. The hinge is frictionless. The densities, ρ_1 and ρ_2, are 33 lbm/ft^3 and 100 lbm/ft^3, respectively. Find the difference in chamber pressures that is just sufficient to keep the gate closed.

3.74 The tennis "bubble" of Problem 3.32 is subjected to a wind that blows at a speed of 50 km/hr in a direction perpendicular to the axis of the semicylindrical shape. Using polar coordinates with angle, θ, measured from the ground on the upwind side of the structure, the resulting pressure distribution may be expressed as

$$\frac{p - p_\infty}{\frac{1}{2}\rho V_w^2} = 1 - 4\sin^2\theta$$

where p is the pressure at the surface, p_∞ the atmospheric pressure, and V_w the wind speed. Determine the net vertical force exerted on the structure.

3.75 A cube is submerged in a liquid at the depth shown (Fig. 3.40). The mass of the cube is 0.5 slug and the tension in the cord is 10 lbf. Find the specific weight of the liquid.

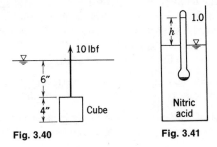

Fig. 3.40   **Fig. 3.41**

3.76 A hydrometer is a specific gravity indicator, the value being indicated by the level at which the free surface intersects the stem when floating in a liquid. The 1.0 mark is the level when in distilled water. For the unit shown in Fig. 3.41 the immersed volume in distilled water is 15 cm³. The stem is 6 mm in diameter. Find the distance, h, from the 1.0 mark to the surface when the hydrometer is placed in a nitric acid solution of specific gravity 1.5.

3.77 Completely sealed lead-acid storage batteries are a recent development on the automotive scene. These batteries feature built-in hydrometers to indicate the specific gravity of the electrolyte and thus the state of charge of the battery. When the battery is fully charged (electrolyte specific gravity between 1.28 and 1.30), the cylindrical hydrometer element of uniform diameter rises above the electrolyte surface to contact a viewing port, thus becoming visible from outside the battery. The hydrometer element is made from plastic with specific gravity 1.12. The viewing port is 10 mm above the electrolyte surface. Determine the length, L, of the hydrometer element such that it will rise to just touch the viewing port when the electrolyte reaches its fully charged specific gravity of 1.28.

3.78 A modern supertanker has a tank capacity of a half million metric tons of Arabian crude oil with SG = 0.86. The ship is essentially rectangular with a length of 400 m and a beam (width) of 65 m. The mass of the ship is approximately 230,000 metric tons. When the ship is unloaded, it is necessary to take on sea water ballast to maintain sufficient draft for stability and to keep the propeller submerged. A minimum draft of 20 m is required. Determine the maximum draft of the fully loaded tanker in sea water. Also determine what fraction of the tanks must be filled with sea water ballast when traveling unloaded.

3.79 Scientific balloons operating at pressure equilibrium with the surroundings have been used to lift instrument packages to extremely high altitudes. One such balloon, constructed of polyester with a skin thickness of 0.013 mm (0.5 mil), lifted a payload of 230 kg to an altitude of approximately 49 km. At that altitude, atmospheric conditions are 0.95 mbar and -20 C. The helium gas in the balloon was at a temperature of approximately -10 C. The specific gravity of the skin material is 1.28. Determine the diameter and mass of the balloon. Assume that the balloon is spherical.

3.80 A pressurized helium balloon is to be designed to lift a payload to an altitude of 40 km, where the atmospheric pressure and temperature are 3.0 mbar and -25C, respectively. The balloon skin is polyester (Mylar) with specific gravity of 1.28 and thickness of 0.015 mm (0.6 mil). To maintain a spherical shape, it has been decided to pressurize the helium in the balloon to a gage pressure of 0.45 mbar. Determine the maximum balloon diameter if the allowable tensile stress in the balloon skin is limited to 62 MN/m². What payload can be carried?

3.81 Find the specific weight of the sphere shown in Fig. 3.42 if its volume is 1 ft³. State all assumptions. Is the weight necessary to float the sphere?

3.82 One cubic foot of material weighing 67 lbf is allowed to sink in the water as shown in Fig. 3.43. A circular wooden rod 10 ft long and 3 in.² in cross section is

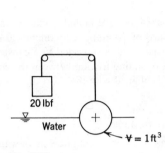

Fig. 3.42

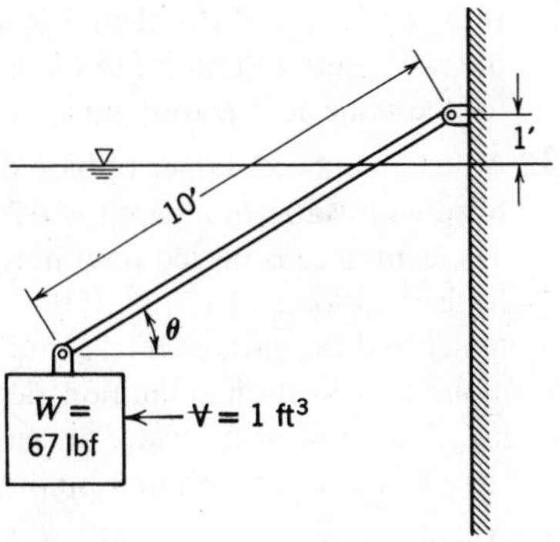

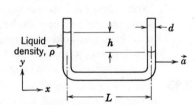

Fig. 3.43

attached to the weight and also to the wall. If the rod weighs 3 lbf, what will be the angle θ for equilibrium?

3.83 A cubical box, 1 m on a side, half filled with oil (SG = 0.80), is given a constant horizontal acceleration of 0.2 g. Determine the slope of the free surface and the pressure along the bottom of the box.

3.84 A rectangular container of water undergoes constant acceleration down an incline as shown in Fig. 3.44. Determine the slope of the free surface using the coordinate system shown.

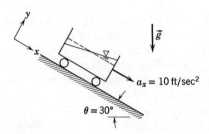

Fig. 3.44

Fig. 3.45

3.85 A crude accelerometer can be made from a liquid-filled U-tube as shown in Fig. 3.45). Derive an expression for the acceleration, $\vec{a}$, in terms of liquid level variation, h, tube geometry, and fluid properties.

3.86 A rectangular container of base dimensions 0.4 m × 0.2 m and height 0.4 m is filled with water to a depth of 0.2 m; the mass of the empty container is 10 kg. The container is placed on a plane inclined at angle of 30° to the horizontal. If the coefficient of sliding friction between the container and the plane is 0.3, determine the angle of the water surface relative to the horizontal.

3.87 If the container of Problem 3.86 slides without friction, determine the angle of the water surface relative to the horizontal. What is the slope of the free surface for the same acceleration up the plane?

3.88 A rectangular container of base dimensions 0.4 m × 0.2 m and height 0.5 m is filled with water to a depth of 0.2 m; the mass of the empty container is 10 kg. The container is placed on a horizontal surface and is subjected to a constant horizontal force of 150 N. If the coefficient of sliding friction between the container and the surface is 0.25 and the tank is aligned with the short dimension along the direction, of motion, determine:
(a) The force of the water on each end of the tank
(b) The force of the water on the bottom of the tank

3.89 A pail, 1 ft in diameter and 1 ft deep, weighs 3 lbf, and contains 8 in. of water. The pail is swung in a vertical circle of 3 ft radius at a speed of 15 ft/sec. The water may be assumed to move as a solid body. At the instant when the pail is at the top of its trajectory, compute the tension in the string and the pressure on the bottom of the pail from the water.

3.90 A sealed chamber, which contains manometer oil (SG = 0.8), rotates about its axis with angular velocity, ω. Derive an expression for the radial pressure gradient in the oil, $\partial p/\partial r$, in terms of radius, r, and angular velocity, ω.

3.91 A cylindrical container, similar to that analyzed in Example Problem 3.10, is rotated at constant angular velocity about its axis. The cylinder is 1 ft in diameter, and initially contains water that is 4 in. deep. Determine the maximum rate at which the container can be rotated before the liquid free surface just touches the bottom of the tank. Is your answer dependent on the density of the liquid? Explain.

3.92 An automobile traveling at 90 km/hr rounds a long sweeping curve of radius 250 m. The air conditioner is on and the windows are rolled up so that the air within the car may be considered to move essentially as a solid body. A child in the back seat holds the string of a balloon filled with helium. On a straight road the string was vertical, but in the curve it is not. Determine the magnitude and direction of the string angle as measured from the vertical.

3.93 Cast iron or steel molds are used in a horizontal-spindle machine for the casting of tubular castings such as liners, tubes, and so forth. A charge of molten metal is poured into the spinning mold. The radial acceleration permits uniformly thick wall sections to form. A steel liner of length $L = 2$ m, outer radius $r_0 = 0.15$ m, and inner radius $r_i = 0.10$ m is to be formed by this process; the specific gravity of steel is 7.8. To insure uniform thickness, the minimum radial acceleration should be 10 g. Determine:
(a) The required angular velocity
(b) The maximum and minimum pressures on the surface of the mold

Chapter 4

Basic Equations in Integral Form for a Control Volume

We shall begin our study of fluids in motion by developing the basic equations in integral form for application to control volumes. Why the control volume formulation rather than the system formulation? There are two basic reasons. First, since fluid media are capable of continuous distortion and deformation, it is often extremely difficult to identify and follow the same mass of fluid at all times (as must be done in applying the system formulation). Second, we are often interested, not in the motion of a given mass of fluid, but rather in the effect of the overall fluid motion on some device or structure. Thus it is more convenient to apply the basic laws to a fixed volume in space, that is, to use a control volume analysis.

There are two possible approaches to formulating the basic laws for a control volume. We could start with a word statement of the basic laws for a control volume and proceed to develop the corresponding mathematical formulation for each basic equation. For example, the conservation of mass for a control volume would be stated as: The net rate of mass efflux from a control volume (rate of outflow minus rate of inflow) plus the rate of change of mass within the control volume is equal to zero. From this word statement we could develop the mathematical formulation. We could proceed in a similar fashion for each of the other four basic laws. In doing this, we would repeat the same steps five times using the different quantities: mass, momentum, angular momentum, energy, and entropy. It seems more desirable (and certainly economical in both space and time) to develop a general formulation that is universally applicable to all five of the basic equations. It is this second approach that we shall follow. The logical question at this point is, "How shall we do it?" The answer is that we shall go back to what we already know—the basic equations for a system and the characteristics of systems and control volumes—and proceed from there.

4-1 BASIC LAWS FOR A SYSTEM

The basic laws for a system are familiar from your earlier work in mechanics and thermodynamics. We shall summarize them briefly here and for reasons that will become apparent in the next section, we shall write each of the basic equations for a system as a rate equation.

4-1.1 CONSERVATION OF MASS

Since a system is, by definition, an arbitrary collection of matter of fixed identity, a system is composed of the same quantity of matter at all times. The conservation of mass simply states that the mass, M, of the system is constant. On a rate basis, we have

$$\left.\frac{dM}{dt}\right)_{system} = 0 \qquad (4.1a)$$

where,

$$M_{system} = \int_{mass\,(system)} dm = \int_{V\,(system)} \rho\, dV \qquad (4.1b)$$

4-1.2 NEWTON'S SECOND LAW

Newton's second law states that for a system moving relative to an inertial reference frame, the sum of all external forces acting on the system is equal to the time rate of change of the linear momentum of the system, that is,

$$\vec{F} = \left.\frac{d\vec{P}}{dt}\right)_{system} \qquad (4.2a)$$

where the linear momentum, $\vec{P}$, of the system is given by

$$\vec{P}_{system} = \int_{mass\,(system)} \vec{V}\, dm = \int_{V\,(system)} \vec{V}\, \rho\, dV \qquad (4.2b)$$

**4-1.3 MOMENT OF MOMENTUM

The moment of momentum equation for a system states that the rate of change of angular momentum is equal to the sum of all torques acting on

---

** This section may be omitted without loss of continuity in the text material.

the system, that is,

$$\vec{T} = \frac{d\vec{H}}{dt}\bigg)_{\text{system}} \tag{4.3a}$$

where the angular momentum of the system is given by

$$\vec{H}_{\text{system}} = \int_{\text{mass (system)}} \vec{r} \times \vec{V} \, dm = \int_{\Psi \text{ (system)}} \vec{r} \times \vec{V} \rho \, d\Psi \tag{4.3b}$$

Torque can be produced by surface and body forces, and also by shafts that cross the system boundary, that is,

$$\vec{T} = \vec{r} \times \vec{F}_s + \int_{\text{mass (system)}} \vec{r} \times \vec{g} \, dm + \vec{T}_{\text{shaft}} \tag{4.3c}$$

4-1.4   THE FIRST LAW OF THERMODYNAMICS

The first law of thermodynamics is a statement of conservation of energy for a system, that is,

$$\delta Q + \delta W = \Delta E$$

or $\delta Q = \Delta E + \delta W$ ⟩ *use either sign convention but* **BE CONSISTENT!**

In rate form the equation can be written as

$$\dot{Q} + \dot{W} = \frac{dE}{dt}\bigg)_{\text{system}} \tag{4.4a}$$

where the total energy of the system is given by

$$E_{\text{system}} = \int_{\text{mass (system)}} e \, dm = \int_{\Psi \text{ (system)}} e\rho \, d\Psi \tag{4.4b}$$

and

$$e = u + \frac{V^2}{2} + gz \tag{4.4c}$$

In Eq. 4.4a the rate of heat transfer, $\dot{Q}$, is taken as positive when heat is added to the system from the surroundings; the rate of work, $\dot{W}$, is taken as positive when work is done on the system by its surroundings.

4-1.5   THE SECOND LAW OF THERMODYNAMICS

If an amount of heat, δQ, is transferred to a system at temperature, T, the second law of thermodynamics states that the change in entropy, dS, of the system is given by

$$dS \geq \frac{\delta Q}{T}$$

On a rate basis we can write

$$\left. \frac{dS}{dt} \right)_{system} \geq \frac{1}{T} \dot{Q} \tag{4.5a}$$

where the total entropy of the system is given by

$$S_{system} = \int_{mass\,(system)} s\,dm = \int_{\mathcal{V}\,(system)} s\rho\,d\mathcal{V} \tag{4.5b}$$

4-2 RELATION OF SYSTEM DERIVATIVES TO CONTROL VOLUME FORMULATION

In the previous section we summarized the basic equations for a system. We found that when written on a rate basis, each equation involved the time derivative of an extensive property of the system (the total mass, momentum, moment of momentum, energy, or entropy of the system). In developing the control volume formulation of each basic law from the system formulation, we shall use the symbol, N, to designate any arbitrary extensive property of the system. The corresponding intensive property (extensive property per unit mass) will be designated by η. Thus

$$N_{system} = \int_{mass\,(system)} \eta\,dm = \int_{\mathcal{V}\,(system)} \eta\rho\,d\mathcal{V} \tag{4.6}$$

By comparing Eq. 4.6 with Eqs. 4.1b, 4.2b, 4.3b, 4.4b, and 4.5b, we see that if:

$$N = M, \qquad \text{then } \eta = 1$$
$$N = \vec{P}, \qquad \text{then } \eta = \vec{V}$$
$$N = \vec{H}, \qquad \text{then } \eta = \vec{r} \times \vec{V}$$
$$N = E, \qquad \text{then } \eta = e$$
$$N = S, \qquad \text{then } \eta = s$$

The major task in going from the system to the control volume formulation of the basic laws is to express the rate of change of the arbitrary extensive property, N, for a system, in terms of time variations of this property associated with a control volume. Since mass crosses the boundaries of a control volume, time variations of the property N associated with the control volume involve the mass flux and the properties convected with it. A convenient way to account for mass flux is to use a limiting process involving a system and a control volume that coincide at a certain instant of time. Flux quantities in regions of overlap and regions surrounding the control volume are then formulated approximately, and the limiting process is applied to obtain exact results. The final equation relates the rate of change of the arbitrary extensive property, N, for a system to the time variations of this property associated with a control volume.

One final comment before beginning the derivation: the details of the derivation are involved algebraically, but the concept is simple and straightforward.

4-2.1 DERIVATION

The configuration of the system and control volume to be used in the analysis is shown in Fig. 4.1. The flow field, $\vec{V}(x, y, z, t)$, is arbitrary relative to the coordinates x, y, and z. The control volume is fixed in space; by definition, the system must always consist of the same fluid particles, and consequently it must move with the flow field. In Fig. 4.1 the boundaries of the system are shown at two different instants of time, t_0 and $t_0 + \Delta t$. At time t_0, the boundaries of the system and the control volume coincide; at time $t_0 + \Delta t$, the system occupies regions II and III. The system has been chosen so that the mass within region I enters the control volume in the time interval, Δt, and the mass in region III leaves the control volume in the same time interval.

Recall that our objective is to relate the rate of change of any arbitrary extensive property, N, of the system to the time variations of this property associated with the control volume. From the definition of a derivative, the rate of change of N_{system} is given by

$$\left. \frac{dN}{dt} \right)_{\text{system}} \equiv \lim_{\Delta t \to 0} \frac{(N_s)_{t_0 + \Delta t} - (N_s)_{t_0}}{\Delta t} \qquad (4.7)$$

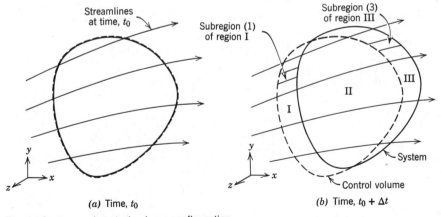

(a) Time, t_0 (b) Time, $t_0 + \Delta t$

Fig. 4.1 System and control volume configuration.

For convenience, the subscript, s, has been used to denote the system in the definition of a derivative employed in Eq. 4.7.

At time $t_0 + \Delta t$, the system occupies regions II and III; at time t_0, the system and the control volume coincide. Since

$$N_{\text{system}} = \int_{\text{mass (system)}} \eta \, dm = \int_{\mathbb{V} \text{ (sysem)}} \eta \rho \, d\mathbb{V} \qquad (4.6)$$

we can write

$$(N_s)_{t_0 + \Delta t} = (N_{\text{II}} + N_{\text{III}})_{t_0 + \Delta t} = (N_{\text{CV}} - N_{\text{I}} + N_{\text{III}})_{t_0 + \Delta t}$$

$$= \left[\int_{\text{CV}} \eta \rho \, d\mathbb{V} \right]_{t_0 + \Delta t} - \left[\int_{\text{I}} \eta \rho \, d\mathbb{V} \right]_{t_0 + \Delta t} + \left[\int_{\text{III}} \eta \rho \, d\mathbb{V} \right]_{t_0 + \Delta t}$$

and

$$N_s)_{t_0} = (N_{\text{CV}})_{t_0} = \left[\int_{\text{CV}} \eta \rho \, d\mathbb{V} \right]_{t_0}$$

Upon substituting these expressions into the definition of the system derivative, Eq. 4.7, we obtain

$$\left. \frac{dN}{dt} \right]_{\text{system}}$$

$$= \lim_{\Delta t \to 0} \frac{\left[\int_{\text{CV}} \eta \rho \, d\mathbb{V} \right]_{t_0 + \Delta t} + \left[\int_{\text{III}} \eta \rho \, d\mathbb{V} \right]_{t_0 + \Delta t} - \left[\int_{\text{I}} \eta \rho \, d\mathbb{V} \right]_{t_0 + \Delta t} - \left[\int_{\text{CV}} \eta \rho \, d\mathbb{V} \right]_{t_0}}{\Delta t}$$

$$(4.8)$$

Since the limit of a sum is equal to the sum of the limits, we can write

$$\left. \frac{dN}{dt} \right]_{\text{system}} = \lim_{\Delta t \to 0} \frac{\left[\int_{\text{CV}} \eta \rho \, d\mathbb{V} \right]_{t_0 + \Delta t} - \left[\int_{\text{CV}} \eta \rho \, d\mathbb{V} \right]_{t_0}}{\Delta t}$$

①

$$+ \lim_{\Delta t \to 0} \frac{\left[\int_{\text{III}} \eta \rho \, d\mathbb{V} \right]_{t_0 + \Delta t}}{\Delta t} - \lim_{\Delta t \to 0} \frac{\left[\int_{\text{I}} \eta \rho \, d\mathbb{V} \right]_{t_0 + \Delta t}}{\Delta t} \qquad (4.9)$$

②　　　　　　　③

Our task now is to evaluate each of the three terms in Eq. 4.9.

Term ① in Eq. 4.9

$$\lim_{\Delta t \to 0} \frac{\left[\int_{CV} \eta \rho \, d\Psi\right]_{t_0 + \Delta t} - \left[\int_{CV} \eta \rho \, d\Psi\right]_{t_0}}{\Delta t} = \lim_{\Delta t \to 0} \frac{N_{CV})_{t_0 + \Delta t} - N_{CV})_{t_0}}{\Delta t}$$

$$= \frac{\partial N_{CV}}{\partial t} = \frac{\partial}{\partial t} \int_{CV} \eta \rho \, d\Psi$$

Term ② in Eq. 4.9

$$\lim_{\Delta t \to 0} \frac{\left[\int_{III} \eta \rho \, d\Psi\right]_{t_0 + \Delta t}}{\Delta t} = \lim_{\Delta t \to 0} \frac{N_{III})_{t_0 + \Delta t}}{\Delta t}$$

To evaluate $N_{III})_{t_0 + \Delta t}$, let us look at an enlarged view of a typical subregion of region III as shown in Fig. 4.2. The vector $d\vec{A}$ has a magnitude equal to the element of area dA of the control surface; the direction of the vector $d\vec{A}$ is that of the outward drawn normal to the element of the control surface area. The angle α is the angle between $d\vec{A}$ and the velocity vector, $\vec{V}$. Since the mass in region III is that which flows *out* of the control volume during the time interval, Δt, the angle α will always be less than $\pi/2$ over the entire area of the control surface bounding region III.

For subregion (3) we can write

$$dN_3)_{t_0 + \Delta t} = (\eta \rho \, d\Psi)_{t_0 + \Delta t} = [\eta \rho (\Delta l \cos \alpha \, dA)]_{t_0 + \Delta t}$$

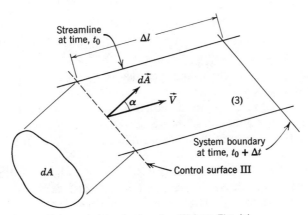

Fig. 4.2 Enlarged view of subregion (3) from Fig. 4.1.

since $d\mathbf{V} = \Delta l \cos \alpha \, dA$. Then for the entire region III,

$$N_{\text{III}})_{t_0 + \Delta t} = \left[\int_{\text{CS}_{\text{III}}} \eta \rho \, \Delta l \cos \alpha \, dA \right]_{t_0 + \Delta t}$$

where CS_{III} is the surface common to region III and the control volume. In this expression for $N_{\text{III}})_{t_0 + \Delta t}$, Δl represents the distance traveled by a particle on the system surface during the time interval, Δt, along a streamline that existed at time t_0.

Now that we have an expression for $N_{\text{III}})_{t_0 + \Delta t}$, we can evaluate term ② in Eq. 4.9:

$$\lim_{\Delta t \to 0} \frac{\left[\int_{\text{III}} \eta \rho \, d\mathbf{V} \right]_{t_0 + \Delta t}}{\Delta t} = \lim_{\Delta t \to 0} \frac{N_{\text{III}})_{t_0 + \Delta t}}{\Delta t}$$

$$= \lim_{\Delta t \to 0} \frac{\int_{\text{CS}_{\text{III}}} \eta \rho \, \Delta l \cos \alpha \, dA}{\Delta t} = \lim_{\Delta t \to 0} \int_{\text{CS}_{\text{III}}} \eta \rho \, \frac{\Delta l}{\Delta t} \cos \alpha \, dA$$

$$= \int_{\text{CS}_{\text{III}}} \eta \rho |\vec{V}| \cos \alpha |d\vec{A}|$$

The last equality follows from the fact that

$$\lim_{\Delta t \to 0} \frac{\Delta l}{\Delta t} = |\vec{V}| \qquad \text{and} \qquad dA = |d\vec{A}|$$

Term ③ in Eq. 4.9

$$- \lim_{\Delta t \to 0} \frac{\left[\int_{\text{I}} \eta \rho \, d\mathbf{V} \right]_{t_0 + \Delta t}}{\Delta t} = - \lim_{\Delta t \to 0} \frac{N_{\text{I}})_{t_0 + \Delta t}}{\Delta t}$$

To evaluate $N_{\text{I}})_{t_0 + \Delta t}$, look at an enlarged view of a typical subregion of region I as shown in Fig. 4.3.

The vector $d\vec{A}$ has a magnitude equal to the element of area dA of the control surface; the direction of the vector $d\vec{A}$ is that of the outward drawn normal from the element of the control surface area. The angle α is the angle between $d\vec{A}$ and the velocity vector, $\vec{V}$. Since the mass in region I flows *into* the control volume during the time interval, Δt, the angle α will always be greater than $\pi/2$ over the entire area of the control surface bounding region I.

For subregion (1) we can write

$$dN_1)_{t_0 + \Delta t} = (\eta \rho \, d\mathbf{V})_{t_0 + \Delta t} = \left[\eta \rho \, \Delta l(- \cos \alpha) \, dA \right]_{t_0 + \Delta t}$$

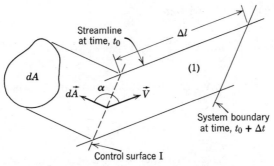

Fig. 4.3 Enlarged view of subregion (1) from Fig. 4.1.

since $d\mathbf{V} = \Delta l(-\cos\alpha)\,dA$. Why the minus sign? Recall that volume is a scalar quantity that must have a positive numerical value. Since $\alpha > \pi/2$, then $\cos\alpha$ will be negative. Hence the need for the minus sign.

Then, for the entire region I,

$$N_\mathrm{I})_{t_0+\Delta t} = \left[\int_{\mathrm{CS_I}} -\eta\rho\,\Delta l\cos\alpha\,dA\right]_{t_0+\Delta t}$$

where $\mathrm{CS_I}$ is the surface common to region I and the control volume. In this expression for $N_\mathrm{I})_{t_0+\Delta t}$, Δl represents the distance traveled by a particle on the system surface during the time interval, Δt, along a streamline that existed at time t_0.

Now that we have an expression for $N_\mathrm{I})_{t_0+\Delta t}$, we can evaluate term ③ in Eq. 4.9:

$$-\lim_{\Delta t\to 0}\frac{\left[\int_\mathrm{I}\eta\rho\,d\mathbf{V}\right]_{t_0+\Delta t}}{\Delta t} = -\lim_{\Delta t\to 0}\frac{N_\mathrm{I})_{t_0+\Delta t}}{\Delta t}$$

$$= -\lim_{\Delta t\to 0}\frac{\int_{\mathrm{CS_I}}-\eta\rho\,\Delta l\cos\alpha\,dA}{\Delta t} = \lim_{\Delta t\to 0}\int_{\mathrm{CS_I}}\eta\rho\frac{\Delta l}{\Delta t}\cos\alpha\,dA$$

$$= \int_{\mathrm{CS_I}}\eta\rho|\vec{V}|\cos\alpha\,|d\vec{A}|$$

The last equality follows from the fact that

$$\lim_{\Delta t\to 0}\frac{\Delta l}{\Delta t} = |\vec{V}| \quad\text{and}\quad dA = |d\vec{A}|$$

Now that we have obtained expressions for each of the three terms on the right side of Eq. 4.9, let us substitute these expressions back into the equation.

Equation 4.9 can then be written

$$\frac{dN}{dt}\bigg)_{\text{system}} = \frac{\partial}{\partial t} \int_{\text{CV}} \eta \rho \, d\Psi + \int_{\text{CS}_{\text{I}}} \eta \rho |\vec{V}| \cos \alpha \, |d\vec{A}| + \int_{\text{CS}_{\text{III}}} \eta \rho |\vec{V}| \cos \alpha \, |d\vec{A}|$$

Referring to Fig. 4.1, we see that the entire control surface, CS, consists of three surfaces, that is,

$$\text{CS} = \text{CS}_{\text{I}} + \text{CS}_{\text{III}} + \text{CS}_p$$

where CS_p is characterized by either $\alpha = \pi/2$ or $\vec{V} = 0$, that is, by no flow across the surface.

Consequently, we can write

$$\frac{dN}{dt}\bigg)_{\text{system}} = \frac{\partial}{\partial t} \int_{\text{CV}} \eta \rho \, d\Psi + \int_{\text{CS}} \eta \rho |\vec{V}| \cos \alpha \, |d\vec{A}| \qquad (4.10)$$

Upon recognizing that $|\vec{V}| \cos \alpha \, |d\vec{A}| = \vec{V} \cdot d\vec{A}$, Eq. 4.10 becomes

$$\frac{dN}{dt}\bigg)_{\text{system}} = \frac{\partial}{\partial t} \int_{\text{CV}} \eta \rho \, d\Psi + \int_{\text{CS}} \eta \rho \vec{V} \cdot d\vec{A} \qquad (4.11)$$

Equation 4.11 is the relation we set out to obtain, that is, a relation between $dN/dt)_{\text{system}}$ and a control volume formulation.

4-2.2 PHYSICAL MEANING

We have taken several pages to derive Eq. 4.11. Recall that our objective in deriving this equation was to obtain a general relation between the rate of change of any arbitrary extensive property, N, of a system and the time variations of this property associated with the control volume. The main reason for deriving it was to reduce the amount of algebra required to obtain the control volume formulations of the basic equations. Because the working form of each basic equation for application to control volumes is developed from Eq. 4.11, we consider the equation itself to be "basic" and rewrite it to emphasize its importance:

$$\frac{dN}{dt}\bigg)_{\text{system}} = \frac{\partial}{\partial t} \int_{\text{CV}} \eta \rho \, d\Psi + \int_{\text{CS}} \eta \rho \vec{V} \cdot d\vec{A} \qquad (4.11)$$

It is important to recall that in deriving Eq. 4.11, the limiting process (i.e. taking the limit as $\Delta t \to 0$) insured that the relation is valid at the instant when the system and the control volume coincide. In using Eq. 4.11 to go from the system formulations of the basic laws to the control volume formulations, we recognize that Eq. 4.11 relates the rate of change of any arbitrary extensive property, N, of a system to the time variations of this property associated with a control volume at the instant when the system and the control volume coincide; this is true since, as $\Delta t \to 0$, the system and the control volume occupy the same volume and have the same boundaries.

Before proceeding to use Eq. 4.11 to develop control volume formulations of the basic laws, let us make sure we understand the meaning of each of the terms and symbols in the equation:

$\left. \dfrac{dN}{dt} \right)_{\text{system}}$ is the total rate of change of any arbitrary extensive property of the system.

$\dfrac{\partial}{\partial t} \displaystyle\int_{\text{cv}} \eta \rho \, d\forall$ is the time rate of change of the arbitrary extensive property N within the control volume.

: η is the intensive property corresponding to N, that is, $\eta = N$ per unit mass.

: $\rho \, d\forall$ is an element of mass contained in the control volume.

: $\int_{\text{cv}} \eta \rho \, d\forall$ is the total amount of the extensive property N contained within the control volume.

$\displaystyle\int_{\text{cs}} \eta \rho \vec{V} \cdot d\vec{A}$ is the net rate of efflux of the extensive property N through the control surface.

: $\rho \vec{V} \cdot d\vec{A}$ is the rate of mass efflux through the element of area $d\vec{A}$ per unit time (we recognize that the dot product is a scalar product; the sign of $\rho \vec{V} \cdot d\vec{A}$ depends on the direction of the velocity vector, $\vec{V}$, relative to the area vector, $d\vec{A}$).

: $\eta \rho \vec{V} \cdot d\vec{A}$ is the rate of efflux of the extensive property N through the area $d\vec{A}$.

Two additional points about Eq. 4.11 should be made. It should be clear from the derivation that the velocity, $\vec{V}$, in the equation is measured relative to the control volume. In developing Eq. 4.11, we considered a control volume fixed relative to the reference coordinates x, y, and z. The velocity field was specified relative to the same reference coordinates. Since, in our development, the system moved in the specified velocity field, then the time rate of change of the arbitrary extensive property, N, within the control volume must be evaluated by an observer fixed in the control volume.

We shall further emphasize these points in deriving the control volume formulation of each of the basic laws. In each case we begin with the familiar system formulation and employ Eq. 4.11 to derive the control volume formulation.

4-3 CONSERVATION OF MASS

In considering the flow through a control volume, we recognize that if, during an interval of time, the amount of mass flowing into the control volume is not the same as the amount of mass flowing out of the control volume, there will be a change in the amount of mass within the control volume. If the outflow is greater than the inflow, there will be a decrease in the amount of mass in the control volume; conversely, there will be an increase in the amount of mass within the control volume if the inflow is greater than the outflow. Thus on purely physical grounds we can provide a word statement of the conservation of mass:

$$0 = \begin{bmatrix} \text{rate of mass} \\ \text{flow out of the} \\ \text{control volume} \end{bmatrix} - \begin{bmatrix} \text{rate of mass} \\ \text{flow into the} \\ \text{control volume} \end{bmatrix} + \begin{bmatrix} \text{rate of change} \\ \text{of mass inside the} \\ \text{control volume} \end{bmatrix}$$

Since the difference between the rate of mass outflow and the rate of mass inflow is equal to the net rate of efflux (outflow) of mass through the control volume, the above word statement of the conservation of mass could be written as

$$0 = \begin{bmatrix} \text{net rate of mass efflux (outflow)} \\ \text{through the control surface} \end{bmatrix} + \begin{bmatrix} \text{rate of change of mass} \\ \text{inside the control volume} \end{bmatrix}$$

Now that we have a physical formulation of conservation of mass for a control volume, let us obtain a mathematical formulation. We begin with the mathematical formulation for a system and use Eq. 4.11 to obtain the control volume formulation.

4-3.1 CONTROL VOLUME EQUATION

Recall that conservation of mass states simply that the mass of a system is constant,

$$\left. \frac{dM}{dt} \right)_{\text{system}} = 0 \tag{4.1a}$$

where

$$M_{\text{system}} = \int_{\text{mass (system)}} dm = \int_{\forall\text{ (system)}} \rho \, d\forall \tag{4.1b}$$

The system and control volume formulations are related by Eq. 4.11

$$\left. \frac{dN}{dt} \right)_{\text{system}} = \frac{\partial}{\partial t} \int_{\text{CV}} \eta \rho \, d\forall + \int_{\text{CS}} \eta \rho \vec{V} \cdot d\vec{A} \tag{4.11}$$

where

$$N_{system} = \int_{mass\,(system)} \eta\,dm = \int_{\Psi\,(system)} \eta\rho\,d\Psi \qquad (4.6)$$

To derive the control volume formulation of the conservation of mass, we set

$$N = M \quad \text{and} \quad \eta = 1$$

With this substitution, we obtain

$$\left.\frac{dM}{dt}\right)_{system} = \frac{\partial}{\partial t}\int_{CV} \rho\,d\Psi + \int_{CS} \rho\vec{V}\cdot d\vec{A} \qquad (4.12)$$

Comparing Eqs. 4.1a and 4.12, we arrive at the control volume formulation of the conservation of mass:

$$0 = \frac{\partial}{\partial t}\int_{CV} \rho\,d\Psi + \int_{CS} \rho\vec{V}\cdot d\vec{A} \qquad (4.13)$$

In Eq. 4.13 the first term represents the rate of change of mass within the control volume; the second term represents the net rate of mass efflux through the control surface. It is comforting to note that the mathematical formulation is consistent with the physical formulation that we developed at the outset.

It is worth reiterating at this point that the velocity, $\vec{V}$, in Eq. 4.13 is measured relative to the control surface. Furthermore, the dot product $\rho\vec{V}\cdot d\vec{A}$ is a scalar product. The sign depends on the direction of the velocity vector, $\vec{V}$, relative to the area vector, $d\vec{A}$. Referring back to the derivation of Eq. 4.11, we see that the dot product, $\rho\vec{V}\cdot d\vec{A}$, is positive where flow is out through the control surface, negative where flow is in through the control surface, and zero where flow is tangent to the control surface.

4-3.2 SPECIAL CASES

In special cases it is possible to simplify Eq. 4.13. Two such cases are worth noting.

Consider first the case of incompressible flow, that is, flow in which the density remains constant. Since ρ is a constant, it may be taken out from under the integral signs in Eq. 4.13. For this case we write Eq. 4.13 as

$$0 = \frac{\partial}{\partial t}\rho\int_{CV} d\Psi + \rho\int_{CS} \vec{V}\cdot d\vec{A} \qquad (4.14a)$$

The integral of $d\Psi$ over the control volume is simply the volume of the control volume. Thus we write Eq. 4.14a as

$$0 = \frac{\partial}{\partial t}[\rho\Psi] + \rho\int_{CS} \vec{V}\cdot d\vec{A} \qquad (4.14b)$$

Since ρ is a constant and the size of the control volume is fixed, the conservation of mass for incompressible flow becomes

$$0 = \int_{CS} \vec{V} \cdot d\vec{A} \tag{4.15}$$

The dimensions of the integrand in Eq. 4.15 are L^3/t. The integral of $\vec{V} \cdot d\vec{A}$ over a section of the control surface is commonly called the *volumetric flowrate or volume rate of flow*. Note that we have not assumed the flow to be steady in reducing Eq. 4.13 to the form 4.15. We have only imposed the restriction of incompressible flow. Thus Eq. 4.15 is a statement of the conservation of mass for an incompressible flow that may be steady or unsteady.

Consider now the general case of steady flow in which the flow is not incompressible. Since the flow is steady, this means that $\rho = \rho(x, y, z)$. By definition a steady flow is one in which none of the fluid properties varies with time. Consequently, the first term of Eq. 4.13 must be zero and, hence, for any steady flow the statement of conservation of mass reduces to

$$0 = \int_{CS} \rho\vec{V} \cdot d\vec{A} \tag{4.16}$$

Note that when $\rho\vec{V} \cdot d\vec{A}$ is negative, mass flows in through the control surface. Mass flows out through the control surface in regions where $\rho\vec{V} \cdot d\vec{A}$ is positive. This fact provides a quick check of the signs on the various flux terms in an analysis.

Example 4.1

Consider the steady flow of water ($\rho = 1.94$ slug/ft^3) through the device shown in the diagram. The areas are: $A_1 = 0.2$ ft^2, $A_2 = 0.5$ ft^2, and $A_3 = A_4 = 0.4$ ft^2. Mass flow out through section ③ is given as 3.88 slug/sec. The volumetric flow rate in through section ④ is given as 1 ft^3/sec, and $\vec{V}_1 = 10\hat{i}$ ft/sec. If properties are assumed uniform across all inlet and outlet flow sections, determine the flow velocity at section ②.

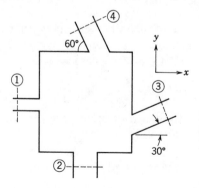

Example Problem 4.1

GIVEN:

Steady flow of water through the device. Properties uniform at all ports.

$$A_1 = 0.2 \text{ ft}^2 \qquad A_2 = 0.5 \text{ ft}^2$$
$$A_3 = A_4 = 0.4 \text{ ft}^2 \qquad \rho = 1.94 \text{ slug/ft}^3$$
$$\dot{m}_3 = 3.88 \text{ slug/sec (outflow)}$$

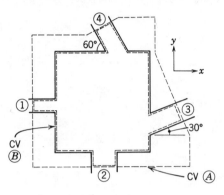

Volumetric flowrate in at ④ = 1.0 ft³/sec

$$\vec{V}_1 = 10\hat{\imath} \text{ ft/sec}$$

FIND:

Velocity at section ②.

SOLUTION:

Choose a control volume. Two possibilities are shown by dashed lines.

Basic equation:
$$0 = \frac{\partial}{\partial t} \int_{CV} \rho \, d\forall + \int_{CS} \rho \vec{V} \cdot d\vec{A}$$

For steady flow, the first term is zero by definition, so

$$0 = \int_{CS} \rho \vec{V} \cdot d\vec{A}$$

In looking at either control volume, we see that there are four sections where mass flows across the control surface. Thus we write

$$\int_{CS} \rho \vec{V} \cdot d\vec{A} = \int_{A_1} \rho \vec{V} \cdot d\vec{A} + \int_{A_2} \rho \vec{V} \cdot d\vec{A} + \int_{A_3} \rho \vec{V} \cdot d\vec{A} + \int_{A_4} \rho \vec{V} \cdot d\vec{A} = 0 \quad (1)$$

Let us look at these integrals one at a time, recognizing that properties are uniform over each area and $\rho = $ constant.

$$\int_{A_1} \rho \vec{V} \cdot d\vec{A} = -\int_{A_1} |\rho V \, dA| = -|\rho V_1 A_1| \begin{array}{c} \xrightarrow{\vec{V}_1} \\ \overleftrightarrow{\vec{A}_1} \quad \textcircled{1} \end{array} \quad \left\{ \begin{array}{l} \text{sign of } \vec{V} \cdot d\vec{A} \text{ is negative at} \\ \text{surface } \textcircled{1}. \end{array} \right\}$$

$\left\{ \begin{array}{l} \text{With the absolute value signs indicated, we have accounted for the direction of} \\ \vec{V} \text{ and } d\vec{A} \text{ in taking the dot product.} \end{array} \right\}$

Since we do not know the direction of $\vec{V}_2$, we shall leave section $\textcircled{2}$ for the moment.

$$\int_{A_3} \rho \vec{V} \cdot d\vec{A} = \int_{A_3} |\rho V \, dA| = |\rho V_3 A_3| = \dot{m}_3 \quad \left\{ \begin{array}{l} \text{sign of } \vec{V} \cdot d\vec{A} \text{ is positive} \\ \text{at surface } \textcircled{3}, \\ \text{since flow is out.} \end{array} \right\}$$

$$\int_{A_4} \rho \vec{V} \cdot d\vec{A} = -\int_{A_4} |\rho V \, dA| = -|\rho V_4 A_4| \quad \left\{ \begin{array}{l} \text{sign of } \vec{V} \cdot d\vec{A} \text{ is negative at} \\ \text{surface } \textcircled{4}. \end{array} \right\}$$

$$= -\rho |V_4 A_4| = -\rho |\dot{q}_4|$$

where $\dot{q}$ is the volumetric flowrate.
From Eq. 1 above,

$$\int_{A_2} \rho \vec{V} \cdot d\vec{A} = -\int_{A_1} \rho \vec{V} \cdot d\vec{A} - \int_{A_3} \rho \vec{V} \cdot d\vec{A} - \int_{A_4} \rho \vec{V} \cdot d\vec{A}$$

$$= +|\rho V_1 A_1| - \dot{m}_3 + \rho |\dot{q}_4|$$

$$= \left| \frac{1.94 \text{ slug}}{\text{ft}^3} \times \frac{10 \text{ ft}}{\text{sec}} \times 0.2 \text{ ft}^2 \right| - \frac{3.88 \text{ slug}}{\text{sec}} + \frac{1.94 \text{ slug}}{\text{ft}^3} \left| \frac{1.0 \text{ ft}^3}{\text{sec}} \right|$$

$$\int_{A_2} \rho \vec{V} \cdot d\vec{A} = 1.94 \text{ slug/sec}$$

Since this is positive, $\vec{V} \cdot d\vec{A}$ at section $\textcircled{2}$ is positive. Flow is out as shown in the sketch:

$$\int_{A_2} \rho \vec{V} \cdot d\vec{A} = \int_{A_2} |\rho V \, dA| = |\rho V_2 A_2| = 1.94 \text{ slug/sec} \qquad \begin{array}{c} \downarrow \vec{A}_2 \quad \textcircled{2} \\ \downarrow \vec{V}_2 \end{array}$$

$$|V_2| = \frac{1.94 \text{ slug/sec}}{\rho A_2} = \frac{1.94 \text{ slug}}{\text{sec}} \times \frac{\text{ft}^3}{1.94 \text{ slug}} \times \frac{1}{0.5 \text{ ft}^2} = 2 \text{ ft/sec}$$

Since V_2 is in the negative y direction, then

$$\vec{V}_2 = -2\hat{j} \text{ ft/sec} \qquad \qquad \xleftarrow{\hspace{2cm}} \vec{V}_2$$

$\left\{ \text{This problem illustrates the procedure for evaluating } \int_{cs} \rho \vec{V} \cdot d\vec{A}. \right\}$

Example 4.2

A tank of 0.05 m³ volume contains air at 800 kPa (absolute) and 15 C. At time, $t = 0$, air escapes from the tank through a valve with a flow area of 65 mm². The air passing through the valve has a speed of 311 m/sec and a density of 6.13 kg/m³. Properties in the rest of the tank may be assumed to be uniform. Determine the instantaneous rate of change of density in the tank at $t = 0$.

Example Problem 4.2

GIVEN:

Tank of volume $\forall = 0.05$ m³ contains air at $p = 800$ kPa (absolute), $T = 15$ C. At $t = 0$, air escapes through a valve. Air leaves with speed $V = 311$ m/sec, and density $\rho = 6.13$ kg/m³ through area $A = 65$ mm².

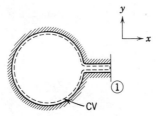

FIND:

Rate of change of air density in the tank at $t = 0$.

SOLUTION:

Choose a control volume as shown by the dashed line.

Basic equation:
$$0 = \frac{\partial}{\partial t} \int_{CV} \rho\, d\forall + \int_{CS} \rho \vec{V} \cdot d\vec{A}$$

Since properties are assumed uniform in the tank at any instant, we can take ρ out from within the integral of the first term,

$$\frac{\partial}{\partial t}\left[\rho \int_{CV} d\forall \right] + \int_{CS} \rho \vec{V} \cdot d\vec{A} = 0$$

Now, $\int_{CV} d\forall = \forall$ and hence

$$\frac{\partial}{\partial t}(\rho \forall) + \int_{CS} \rho \vec{V} \cdot d\vec{A} = 0$$

The only place where mass crosses the boundary of the control volume is at surface ①. Hence

$$\int_{CS} \rho \vec{V} \cdot d\vec{A} = \int_{A_1} \rho \vec{V} \cdot d\vec{A} \quad \text{and} \quad \frac{\partial}{\partial t}(\rho \forall) + \int_{A_1} \rho \vec{V} \cdot d\vec{A} = 0$$

At surface ① the sign of $\rho \vec{V} \cdot d\vec{A}$ is positive, so

$$\frac{\partial}{\partial t}(\rho \forall) + \int_{A_1} |\rho V \, dA| = 0$$

If it is assumed that properties are uniform over surface ①, then

$$\frac{\partial}{\partial t}(\rho \forall) + |\rho_1 V_1 A_1| = 0$$

or

$$\frac{\partial}{\partial t}(\rho \forall) = -|\rho_1 V_1 A_1|$$

Since the volume, $\forall$, of the tank is not a function of time

$$\forall \frac{\partial \rho}{\partial t} = -|\rho_1 V_1 A_1|$$

and

$$\frac{\partial \rho}{\partial t} = -\frac{|\rho_1 V_1 A_1|}{\forall}$$

At time $t = 0$,

$$\frac{\partial \rho}{\partial t} = \frac{-6.13 \text{ kg}}{m^3} \times \frac{311 \text{ m}}{\text{sec}} \times 65 \text{ mm}^2 \times \frac{1}{0.05 \text{ m}^3} \times \frac{m^2}{10^6 \text{ mm}^2}$$

$$\frac{\partial \rho}{\partial t} = -2.48 \frac{\text{kg/m}^3}{\text{sec}} \qquad \text{\{the density is decreasing\}}$$

{This problem illustrates the application of the control volume formulation of the conservation of mass to an unsteady flow.}

Example 4.3

The fluid in direct contact with a stationary solid boundary has zero velocity, that is, there is no slip at the boundary. Thus the flow over a flat plate adheres to the plate surface and forms a boundary layer, as depicted in the following figure. The flow ahead of the plate is uniform with velocity, $\vec{V} = U_0 \hat{\imath}$; $U_0 = 30$ m/sec. The velocity distribution at CD is approximated as being linear ($\vec{V}_C = U_0 \hat{\imath}$, $\vec{V}_D = 0$). The boundary-layer thickness at this location is 5 mm. The fluid is air with density $\rho = 1.24$ kg/m³. Assuming the plate width to be 0.6 m, calculate the mass flowrate across the surface BC of control volume $ABCD$.

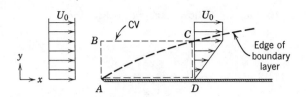

Example Problem 4.3

GIVEN:

Steady, incompressible flow over a flat plate, $\rho = 1.24$ kg/m^3. Velocity ahead of plate is uniform

$$\vec{V} = U_0\hat{\imath} \qquad U_0 = 30 \text{ m/sec}$$

At $x = x_D$: the x component of velocity varies linearly with y.

$$\text{at} \quad y = y_C = \delta, \qquad u = U_0$$
$$\text{at} \quad y = 0, \qquad u = 0$$

: boundary layer thickness, $\delta = 5$ mm
Width of plate, $w = 0.6$ m.

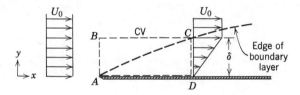

FIND:

The mass flowrate across the surface BC.

SOLUTION:

The control volume is selected as shown by the dashed lines.

Basic equation:
$$0 = \frac{\partial}{\partial t}\int_{cv} \rho \, d\forall + \int_{cs} \rho \vec{V} \cdot d\vec{A}$$

For steady flow,

$$\frac{\partial}{\partial t}\int_{cv} \rho \, d\forall = 0 \qquad \text{and, hence,} \qquad \int_{cs} \rho \vec{V} \cdot d\vec{A} = 0$$

If it is assumed that there is no flow in the z direction, then

$$0 = \int_{CS} \rho \vec{V} \cdot d\vec{A} \qquad\qquad\qquad\qquad = 0 \begin{pmatrix} \text{no flow} \\ \text{across } DA \end{pmatrix}$$

$$= \int_{A_{AB}} \rho \vec{V} \cdot d\vec{A} + \int_{A_{BC}} \rho \vec{V} \cdot d\vec{A} + \int_{A_{CD}} \rho \vec{V} \cdot d\vec{A} + \int_{A_{DA}} \rho \vec{V} \cdot d\vec{A}$$

$$\therefore \int_{A_{BC}} \rho \vec{V} \cdot d\vec{A} = -\int_{A_{AB}} \rho \vec{V} \cdot d\vec{A} - \int_{A_{CD}} \rho \vec{V} \cdot d\vec{A}$$

{We need to evaluate the integrals on the right side of the equation.}

$$\int_{A_{AB}} \rho \vec{V} \cdot d\vec{A} = -\int_{A_{AB}} |\rho u \, dA| = -\int_{y_A}^{y_B} |\rho u w \, dy| \qquad \begin{Bmatrix} \vec{V} \cdot d\vec{A} \text{ is negative} \\ dA = w \, dy \end{Bmatrix}$$

$$= -\int_0^\delta |\rho u w \, dy| = -\left| \int_0^\delta \rho U_0 w \, dy \right| \qquad \{u = U_0 \text{ over area } AB\}$$

$$= -\left| [\rho U_0 w y]_0^\delta \right| = -|\rho U_0 w \delta|$$

$$\int_{A_{CD}} \rho \vec{V} \cdot d\vec{A} = \int_{A_{CD}} |\rho u \, dA| = \int_{y_D}^{y_C} |\rho u w \, dy| \qquad \begin{Bmatrix} \vec{V} \cdot d\vec{A} \text{ is positive} \\ dA = w \, dy \end{Bmatrix}$$

$$= \int_0^\delta |\rho u w \, dy| = \left| \int_0^\delta \rho \frac{U_0 y}{\delta} w \, dy \right| \qquad \left\{ \text{across surface } CD, \ u = U_0 \frac{y}{\delta} \right\}$$

$$= \left| \left[\frac{\rho U_0 w}{\delta} \frac{y^2}{2} \right]_0^\delta \right| = \left| \frac{\rho U_0 w \delta}{2} \right|$$

$$\therefore \int_{A_{BC}} \rho \vec{V} \cdot d\vec{A} = -\int_{A_{AB}} \rho \vec{V} \cdot d\vec{A} - \int_{A_{CD}} \rho \vec{V} \cdot d\vec{A} = \rho U_0 w \delta - \frac{\rho U_0 w \delta}{2}$$

$$= \frac{\rho U_0 w \delta}{2} = \frac{1}{2} \times \frac{1.24 \text{ kg}}{\text{m}^3} \times \frac{30 \text{ m}}{\text{sec}} \times 0.6 \text{ m} \times 5 \text{ mm} \times \frac{\text{m}}{1000 \text{ mm}}$$

$$\int_{A_{BC}} \rho \vec{V} \cdot d\vec{A} = 0.0558 \text{ kg/sec} \qquad \begin{Bmatrix} \text{Positive sign indicates flow out} \\ \text{across surface } BC. \end{Bmatrix} \left. \right\} \ \dot{m}_{BC}$$

{This problem illustrates the application of the control volume formulation of the conservation of mass to the case of nonuniform flow at a section.}

4-4 MOMENTUM EQUATION FOR INERTIAL CONTROL VOLUME

We wish to develop a mathematical formulation of Newton's second law suitable for application to a control volume. In this section our derivation will be restricted to an inertial control volume, that is, a control volume not accelerating relative to a stationary frame of reference (inertial coordinate system).

In deriving the control volume formulation, the procedure is completely analogous to the procedure followed in deriving the mathematical formulation for the conservation of mass applied to a control volume. We begin with the mathematical formulation for a system and then employ Eq. 4.11 to go from the system to the control volume formulation.

4-4.1 CONTROL VOLUME EQUATION

Recall that Newton's second law for a system moving relative to an inertial coordinate system was given by Eq. 4.2a

$$\vec{F} = \left.\frac{d\vec{P}}{dt}\right)_{\text{system}} \tag{4.2a}$$

where the linear momentum, $\vec{P}$, of the system is given by

$$\vec{P}_{\text{system}} = \int_{\text{mass (system)}} \vec{V}\, dm = \int_{\Psi\,(\text{system})} \vec{V}\rho\, d\Psi \tag{4.2b}$$

and the resultant force, $\vec{F}$, includes all surface and body forces acting on the system. That is,

$$\vec{F} = \vec{F}_S + \vec{F}_B$$

The system and control volume formulations are related by Eq. 4.11

$$\left.\frac{dN}{dt}\right)_{\text{system}} = \frac{\partial}{\partial t}\int_{\text{CV}} \eta\rho\, d\Psi + \int_{\text{CS}} \eta\rho\vec{V} \cdot d\vec{A} \tag{4.11}$$

where

$$N_{\text{system}} = \int_{\text{mass (system)}} \eta\, dm = \int_{\Psi\,(\text{system})} \eta\rho\, d\Psi \tag{4.6}$$

To derive the control volume formulation of Newton's second law, we set

$$N = \vec{P} \qquad \text{and} \qquad \eta = \vec{V}$$

From Eq. 4.11, with this substitution, we obtain

$$\left. \frac{d\vec{P}}{dt} \right)_{\text{system}} = \frac{\partial}{\partial t} \int_{\text{CV}} \vec{V} \rho \, d\Psi + \int_{\text{CS}} \vec{V} \rho \vec{V} \cdot d\vec{A} \qquad (4.17)$$

From Eq. 4.2a

$$\left. \frac{d\vec{P}}{dt} \right)_{\text{system}} = \vec{F} \bigg)_{\text{on system}} \qquad (4.2a)$$

Since, in deriving Eq. 4.11, the system and the control volume coincide at time t_o,

$$\vec{F}\,]_{\text{on system}} = \vec{F}\,]_{\text{on control volume}}$$

In light of this, Eqs. 4.2a and 4.17 yield the control volume formulation of Newton's second law for a nonaccelerating control volume

$$\vec{F} = \vec{F}_S + \vec{F}_B = \frac{\partial}{\partial t} \int_{\text{CV}} \vec{V} \rho \, d\Psi + \int_{\text{CS}} \vec{V} \rho \vec{V} \cdot d\vec{A} \qquad (4.18)$$

This equation states that the sum of all forces (surface and body forces) acting on a nonaccelerating control volume is equal to the sum of the rate of change of momentum inside the control volume and the net rate of efflux of momentum through the control surface.

The derivation of the momentum equation for a control volume was straightforward. Application of this basic equation to the solution of problems will not be difficult if you exercise care in using the equation.

In using any basic equation for a control volume analysis, the first step should be to draw the boundaries of the control volume and label appropriate coordinate directions. In Eq. 4.18, the force, $\vec{F}$, represents all forces acting on the control volume. It includes both surface forces and body forces. If we denote the body force per unit mass as $\vec{B}$, then

$$\vec{F}_B = \int \vec{B} \, dm = \int_{\text{CV}} \vec{B} \rho \, d\Psi$$

The nature of the forces acting on the control volume will undoubtedly influence the choice of the control volume boundaries.

All velocities, $\vec{V}$, in Eq. 4.18 are measured relative to the control volume. The momentum flux, $\vec{V} \rho \vec{V} \cdot d\vec{A}$, through an element of the control surface area, $d\vec{A}$, is a vector. The sign of the scalar product $\rho \vec{V} \cdot d\vec{A}$ depends on the direction of the velocity vector, $\vec{V}$, relative to the area vector, $d\vec{A}$. The sign of the vector velocity, $\vec{V}$, is dependent on the choice of the coordinate system.

The momentum equation is a vector equation. As with all vector equations, it may be written in scalar component equations. Relative to an x, y, z coordinate system, the scalar components of Eq. 4.18 may be written

$$F_x = F_{S_x} + F_{B_x} = \frac{\partial}{\partial t} \int_{CV} u\rho \, d\mathcal{V} + \int_{CS} u\rho \vec{V} \cdot d\vec{A} \qquad (4.19a)$$

$$F_y = F_{S_y} + F_{B_y} = \frac{\partial}{\partial t} \int_{CV} v\rho \, d\mathcal{V} + \int_{CS} v\rho \vec{V} \cdot d\vec{A} \qquad (4.19b)$$

$$F_z = F_{S_z} + F_{B_z} = \frac{\partial}{\partial t} \int_{CV} w\rho \, d\mathcal{V} + \int_{CS} w\rho \vec{V} \cdot d\vec{A} \qquad (4.19c)$$

In using the scalar equations, it is again necessary to select a coordinate system at the outset. The positive directions of the velocity components, u, v, and w and the force components F_x, F_y, and F_z are then established relative to the selected coordinate system. As we have previously pointed out, the sign of the scalar product $\rho \vec{V} \cdot d\vec{A}$ depends on the direction of the velocity vector, $\vec{V}$, relative to the area vector, $d\vec{A}$. Thus the flux term in either Eq. 4.18 or Eqs. 4.19 is a product of two quantities, both of which have signs associated with them. It is suggested that you proceed in two steps in determining the momentum flux through any portion of a control surface. The first step should be to determine the sign of $\rho \vec{V} \cdot d\vec{A}$, that is,

$$\rho \vec{V} \cdot d\vec{A} = \rho |V \, dA| \cos \alpha = \pm |\rho V \, dA \cos \alpha|$$

Since the sign for each velocity component, u, v, and w, depends on the choice of coordinate system, the sign should be accounted for when substituting numerical values into the terms $u\rho \vec{V} \cdot d\vec{A} = u\{\pm|\rho V \, dA \cos \alpha|\}$, and so on.

Example 4.4
During a demonstration the police decide to use fire hoses to knock over a barricade built by the demonstrators. The barricade is built in the form of a flat plate as shown in the diagram. The velocity of the water leaving the nozzle is 15 m/sec relative to the nozzle; the nozzle area is 0.01 m². Assuming the water is directed normal to the plate, determine the horizontal force on the barricade.

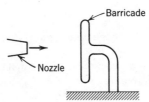

Example Problem 4.4

GIVEN:

Water from a stationary nozzle strikes a stationary flat plate as shown (water directed normal to the plate).

Jet velocity, $\vec{V} = 15\hat{\imath}$ m/sec

Nozzle area, $A_n = 0.01$ m^2

FIND:

Horizontal force on the support.

SOLUTION:

Note that we have chosen a coordinate system in defining the problem. We must now choose a suitable control volume. A number of possible choices are shown by the dashed lines.

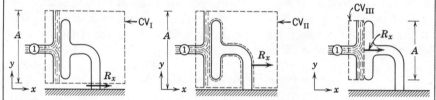

In all three cases the water from the nozzle crosses the control surface through area A_1 (assumed equal to the nozzle area) and the water is assumed to leave the control volume tangent to the plate surface, that is, in the $+y$ or $-y$ direction. Before trying to decide which is the "best" control volume to use, let us write the basic equations.

$$\vec{F} = \vec{F}_S + \vec{F}_B = \frac{\partial}{\partial t}\int_{CV} \vec{V}\rho \, d\forall + \int_{CS} \vec{V}\rho\vec{V} \cdot d\vec{A} \qquad \text{and} \qquad \frac{\partial}{\partial t}\int_{CV} \rho \, d\forall + \int_{CS} \rho\vec{V} \cdot d\vec{A} = 0$$

Regardless of our choice of control volume, the flow is steady and the basic equations become

$$\vec{F} = \vec{F}_S + \vec{F}_B = \int_{CS} \vec{V}\rho\vec{V} \cdot d\vec{A} \qquad \text{and} \qquad \int_{CS} \rho\vec{V} \cdot d\vec{A} = 0$$

Evaluating the momentum flux term will lead to the same result for all of the control volumes. We should choose the control volume that allows the most straightforward evaluation of the forces.

Remember in applying the momentum equation that the force, $\vec{F}$, represents all forces acting *on* the control volume.

Let us solve the problem using each of the three control volumes.

CV$_I$

The control volume has been selected so that the area of the left surface is equal to the area of the right surface. Denote this area by A.

The control volume cuts through the support. We denote the force of the support on the control volume as R_x and assume it to be positive. (The force of the control volume on the support is equal and opposite to R_x.)

Since we are looking for the horizontal force, we write the x component of the steady flow momentum equation

$$F_{S_x} + F_{B_x} = \int_{CS} u\rho\vec{V} \cdot d\vec{A}$$

There are no body forces in the x direction. $F_{B_x} = 0$ and

$$F_{S_x} = \int_{CS} u\rho\vec{V} \cdot d\vec{A}$$

To evaluate F_{S_x}, we must include all surface forces acting on the control volume

$$F_{S_x} = \qquad p_a A \qquad - \qquad p_a A \qquad + \qquad R_x$$

force due to atmospheric	force due to atmospheric	force of support on
pressure acts to right	pressure acts to left	control volume
(positive direction) on	(negative direction) on	(assumed positive)
left surface	right surface	

Consequently, $F_{S_x} = R_x$, and

$$R_x = \int_{CS} u\rho\vec{V} \cdot d\vec{A} = \int_{A_1} u\rho\vec{V} \cdot d\vec{A} \qquad \left\{ \begin{array}{l} \text{for mass crossing top and bottom} \\ \text{surfaces, } u = 0. \end{array} \right\}$$

$$= -\int_{A_1} u|\rho V_1\, dA| \qquad \left\{ \begin{array}{l} \text{at } ① \ \rho\vec{V} \cdot d\vec{A} = -|\rho V_1\, dA|, \text{ since direction of } \vec{V}_1 \text{ and} \\ d\vec{A}_1 \text{ are } 180° \text{ apart.} \end{array} \right\}$$

$$= -u_1|\rho V_1 A_1| \qquad \{\text{properties uniform over } A_1\}$$

$$= -\frac{15}{sec}\frac{m}{sec} \left| \frac{999}{m^3}\frac{kg}{m^3} \times \frac{15}{sec}\frac{m}{sec} \times 0.01\, m^2 \right| \frac{N \cdot sec^2}{kg \cdot m} \qquad \{u_1 = 15 \text{ m/sec}\}$$

$$R_x = -2.25 \text{ kN} \qquad \{R_x \text{ acts opposite to positive direction assumed.}\}$$

The force on the support, $K_x = -R_x = 2.25$ kN $\{\text{force on support acts to the right}\}$

CV$_{II}$

The control volume has been selected so that the area of the left surface is equal to the area of the right surface. Denote this area by A.

The control volume does not cut through the support. However, the control volume is in contact with the support over several portions of the area of the control surface. There is a force exerted by the support on the control surface. We denote the x component of this force as R_x.

Then for this control volume the x component of the momentum equation leads directly (in the same step by step manner) to the same solution as for CV$_I$ above.

CV_{III}

The control volume has been selected so that the area of the left surface and that of the right surface are equal to the area of the plate. Denote this area by A.

As in the case of CV_{II} the x component of the force of the support on the CV is denoted by R_x.

Then the x component of the momentum equation,

$$F_{S_x} = \int_{CS} u\rho\vec{V} \cdot d\vec{A}$$

yields

$$F_{Sx} = p_a A + R_x = \int_{A_1} u\rho\vec{V} \cdot d\vec{A} = -\int_{A_1} u|\rho V_1 dA| = -2.25 \text{ kN}$$

Then $R_x = -p_a A - 2.25 \text{ kN}$ and $K_x = -R_x = p_a A + 2.25 \text{ kN}$

To determine the net force on the plate, we need a free body diagram of the plate.

$$F_{net} = K_x - p_a A \quad \text{\{since atmospheric pressure acts on the back of the plate\}}$$
$$F_{net} = p_a A + 2.25 \text{ kN} - p_a A = 2.25 \text{ kN}$$

$\left\{ \begin{array}{l} \text{This problem illustrates the application of the momentum equation to an inertial} \\ \text{control volume, with emphasis on choosing a suitable control volume.} \end{array} \right\}$

Example 4.5

A metal container 2 ft high with an inside cross-sectional area of 1 ft^2 weighs 5 lbf when empty. The container is placed on a scale and water flows in through an opening in the top and out through the two equal area openings in the sides as shown in the diagram. Under steady flow conditions the height of the water in the tank is 1.9 ft. The pressure is atmospheric across all openings. Determine the reading on the scale.

$A_1 = 0.1 \text{ ft}^2$
$\vec{V}_1 = -20\hat{\jmath} \text{ ft/sec}$
$A_2 = A_3 = 0.1 \text{ ft}^2$

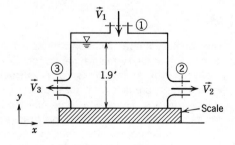

Example Problem 4.5

GIVEN:

Metal container of height 2 ft and cross-sectional area, $A = 1\ \text{ft}^2$, weighs 5 lbf when empty. Container is placed on scale. Under steady flow conditions water depth is 1.9 ft.
    Water enters vertically at section ① and leaves horizontally through sections ② and ③.

$$A_1 = 0.1\ \text{ft}^2$$
$$\vec{V}_1 = -20\hat{j}\ \text{ft/sec}$$
$$A_2 = A_3 = 0.1\ \text{ft}^2$$

The pressure is atmospheric at all openings.

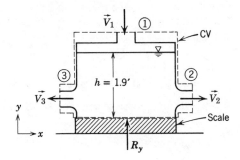

FIND:

The reading on the scale.

SOLUTION:

Choose a control volume as shown; R_y is the force of the scale on the control volume and is assumed positive.

Basic equations:

$$\vec{F}_S + \vec{F}_B = \overbrace{\frac{\partial}{\partial t}\int_{CV} \vec{V}\rho\, d\forall}^{= 0\ \text{(steady flow)}} + \int_{CS} \vec{V}\rho\vec{V}\cdot d\vec{A}$$

$$0 = \overbrace{\frac{\partial}{\partial t}\int_{CV} \rho\, d\forall}^{= 0\ \text{(sf)}} + \int_{CS} \rho\vec{V}\cdot d\vec{A}$$

We write the y component of the momentum equation

$$F_{S_y} + F_{B_y} = \int_{CS} v\rho\vec{V} \cdot d\vec{A} \tag{1}$$

$F_{S_y} = -p_{atm}A + R_y$ {Net force due to atmospheric pressure acts down.}

$F_{B_y} = -W_{tank} - W_{H_2O}$ {Both body forces act in negative y direction.}

$$W_{H_2O} = \gamma\forall = \gamma A h$$

$$\int_{CS} v\rho\vec{V} \cdot d\vec{A} = \int_{A_1} v\rho\vec{V} \cdot d\vec{A} = -\int_{A_1} v|\rho_1 V_1\, dA| \quad \begin{cases} \vec{V} \cdot d\vec{A} \text{ is negative at } ①. \\ v = 0 \text{ at sections } ② \text{ and } ③. \end{cases}$$

$$= -v_1|\rho_1 V_1 A_1| \quad \text{assuming uniform properties at } ①.$$

Substituting into Eq. 1 gives

$$-p_{atm}A + R_y - W_{tank} - \gamma A h = -v_1|\rho_1 V_1 A_1|$$

or

$$R_y = p_{atm}A + W_{tank} + \gamma A h - v_1|\rho_1 V_1 A_1|$$

Substituting numbers with $v_1 = -20$ ft/sec gives

$$R_y = p_a A + 5 \text{ lbf} + \frac{62.4 \text{ lbf}}{\text{ft}^3} \times 1 \text{ ft}^2 \times 1.9 \text{ ft}$$

$$- \left(-20 \frac{\text{ft}}{\text{sec}}\right) \left|1.94 \frac{\text{slug}}{\text{ft}^3} \left(-20 \frac{\text{ft}}{\text{sec}}\right) 0.1 \text{ ft}^2\right| \frac{\text{lbf} \cdot \text{sec}^2}{\text{slug} \cdot \text{ft}}$$

$$R_y = p_a A + 201 \text{ lbf} \quad \text{{Force of scale on CV is upward.}}$$

The force of the control volume on the scale is $K_y = -R_y = -p_a A - 201$ lbf.
   Since atmospheric pressure acts on the scale, the net force on the scale is -201 lbf.
   The minus sign indicates that the force on the scale is downward in our coordinate system.
   Therefore, the scale reading is 201 lbf.

{This problem illustrates the application of the momentum equation to an inertial control volume wherein body forces are included.}

Example 4.6

Water in an open channel flows under a sluice gate as shown in the sketch. The flow is incompressible, and is uniform at sections ① and ②. Hydrostatic

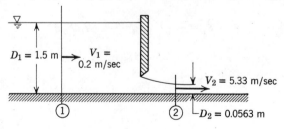

$D_1 = 1.5$ m $V_1 = 0.2$ m/sec

$V_2 = 5.33$ m/sec

①   ② $D_2 = 0.0563$ m

pressure distributions at sections ① and ② may be assumed because the flow streamlines are essentially straight there. Determine the magnitude and direction of the force exerted on the gate by the flow.

Example Problem 4.6

GIVEN:

Flow under sluice gate. Width $= w$.

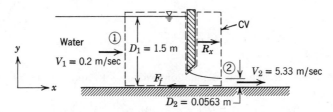

FIND:

Force exerted (per unit width) on the gate.

SOLUTION:

Choose the CV and coordinate system shown for analysis. Apply the x component of the momentum equation.

Basic equation:

$$F_{S_x} + \cancelto{0(2)}{F_{B_x}} = \cancelto{0(3)}{\frac{\partial}{\partial t}\int_{CV} u\rho\, d\forall} + \int_{CS} u\rho \vec{V} \cdot d\vec{A}$$

Assumptions: (1) F_f negligible (i.e. neglect friction on channel bottom)
(2) $F_{B_x} = 0$
(3) Steady flow
(4) Incompressible flow
(5) Uniform flow at each section
(6) Hydrostatic pressure distribution at ① and ②

Then

$$F_{S_x} = u_1\{-|\rho V_1 w D_1|\} + u_2\{|\rho V_2 w D_2|\}$$

The surface forces acting on the CV are due to pressure and the unknown force, R_x. From assumption (6),

$$\frac{dp}{dy} = -\rho g; \qquad p = p_0 + \rho g(y_0 - y) = p_{\text{atm}} + \rho g(D - y)$$

Evaluating F_{S_x} gives

$$F_{S_x} = \int_0^{D_1} p_1 \, dA_1 - \int_0^{D_2} p_2 \, dA_2 - p_{atm}(D_1 - D_2)w + R_x$$

$$= \int_0^{D_1} [p_{atm} + \rho g(D_1 - y)]w \, dy - \int_0^{D_2} [p_{atm} + \rho g(D_2 - y)]w \, dy$$

$$- p_{atm}(D_1 - D_2)w + R_x$$

$$= \cancel{p_{atm}D_1 w} + \frac{\rho g D_1^2}{2} w - \cancel{p_{atm}D_2 w} - \frac{\rho g D_2^2}{2} w - \cancel{p_{atm}D_1 w} + \cancel{p_{atm}D_2 w} + R_x$$

or

$$F_{S_x} = R_x + \frac{\rho g w}{2}(D_1^2 - D_2^2)$$

Substituting into the momentum equation, with $u_1 = V_1$ and $u_2 = V_2$, gives

$$R_x + \frac{\rho g w}{2}(D_1^2 - D_2^2) = -V_1 |\rho V_1 w D_1| + V_2 |\rho V_2 w D_2|$$

or

$$R_x = \rho w (V_2^2 D_2 - V_1^2 D_1) - \frac{\rho g w}{2}(D_1^2 - D_2^2)$$

and

$$\frac{R_x}{w} = \rho(V_2^2 D_2 - V_1^2 D_1) - \frac{\rho g}{2}(D_1^2 - D_2^2)$$

$$= \frac{999 \text{ kg}}{\text{m}^3} \left[(5.33)^2 (0.0563) - (0.2)^2(1.5) \right] \frac{\text{m}^2}{\text{sec}^2} \text{ m} \times \frac{\text{N} \cdot \text{sec}^2}{\text{kg} \cdot \text{m}}$$

$$- \frac{1}{2} \times \frac{999 \text{ kg}}{\text{m}^3} \times \frac{9.81 \text{ m}}{\text{sec}^2} \left[(1.5)^2 - (0.0563)^2 \right] \text{m}^2 \times \frac{\text{N} \cdot \text{sec}^2}{\text{kg} \cdot \text{m}}$$

$$\frac{R_x}{w} = -9.47 \text{ kN/m}$$

R_x is the unknown external force acting *on* the control volume. It is applied to the CV by the gate. Therefore, the force from all fluids *on* the gate is K_x, where $K_x = -R_x$. Thus

$$\frac{K_x}{w} = -\frac{R_x}{w} = 9.47 \text{ kN/m} \qquad \frac{K_x}{w}$$

{Thus the resulting force *on* the gate is 9.47 kN/m, applied to the right.}

Example 4.7

Water flows steadily through the 90° reducing elbow shown in the diagram. At the inlet to the elbow the absolute pressure is 221 kPa and the cross-

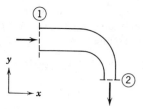

sectional area is 0.01 m². At the outlet the cross-sectional area is 0.0025 m² and the velocity is 16 m/sec. The pressure at the outlet is atmospheric. Determine the force required to hold the elbow in place.

Example Problem 4.7

GIVEN:

Steady flow of water through 90° reducing elbow

$p_1 = 221$ kPa (absolute) $A_1 = 0.01$ m²

$\vec{V}_2 = -16\hat{j}$ m/sec $A_2 = 0.0025$ m²

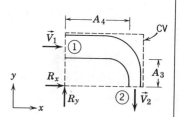

FIND:

Force required to hold elbow in place.

SOLUTION:

Choose control volume as shown by the dashed line.
R_x, R_y are components of force required to hold the elbow in place. (They are assumed positive and are forces acting on CV.)
A_3 is the area of vertical sides of CV excluding A_1, that is, $A_{\text{vertical sides}} = A_1 + A_3$.
A_4 is the area of horizontal sides of CV excluding A_2; that is, $A_{\text{horizontal sides}} = A_2 + A_4$.
Basic equations:

$$\vec{F} = \vec{F}_S + \vec{F}_B = \overset{=0\ (\text{steady flow})}{\cancel{\frac{\partial}{\partial t} \int_{\text{CV}} \vec{V} \rho\, d\forall}} + \int_{\text{CS}} \vec{V} \rho \vec{V} \cdot d\vec{A}$$

$$0 = \overset{=0\ (\text{sf})}{\cancel{\frac{\partial}{\partial t} \int_{\text{CV}} \rho\, d\forall}} + \int_{\text{CS}} \rho \vec{V} \cdot d\vec{A}$$

Assumptions: (1) Uniform flow at each section
(2) Atmospheric pressure, $p_a = 101$ kPa
(3) Incompressible flow

Writing the x component of the momentum equation results in

$$F_{S_x} = \int_{CS} u\rho \vec{V} \cdot d\vec{A} = \int_{A_1} u\rho \vec{V} \cdot d\vec{A} \qquad \{F_{B_x} = 0 \text{ and } u_2 = 0\}$$

$$p_1 A_1 + p_a A_3 - p_a(A_1 + A_3) + R_x = \int_{A_1} u\rho \vec{V} \cdot d\vec{A} \quad \left\{ \begin{array}{l} \text{Pressure over right side is} \\ p_a. \text{ Pressure over left side} \\ \text{is } p_1 \text{ on } A_1 \text{ and } p_a \text{ on } A_3. \end{array} \right\}$$

$$(p_1 - p_a)A_1 + R_x = -\int_{A_1} u|\rho V_1 \, dA| \qquad \{\vec{V} \cdot d\vec{A} \text{ is negative at } A_1\}$$

$$R_x = -p_{1_g} A_1 - u_1|\rho V_1 A_1| \qquad \{p_1 - p_a = p_{1_{\text{gage}}}\}$$

To find V_1, use the continuity equation:

$$\int_{CS} \rho \vec{V} \cdot d\vec{A} = 0 = \int_{A_1} \rho \vec{V} \cdot d\vec{A} + \int_{A_2} \rho \vec{V} \cdot d\vec{A}$$

$$\therefore 0 = -\int_{A_1} |\rho V \, dA| + \int_{A_2} |\rho V \, dA| = -|\rho V_1 A_1| + |\rho V_2 A_2|$$

and

$$|V_1| = |V_2| \frac{A_2}{A_1} = \frac{16}{\sec} \frac{m}{} \times \frac{0.0025}{0.01} = 4 \text{ m/sec} \qquad \therefore \vec{V}_1 = 4\hat{\imath} \text{ m/sec}$$

$$R_x = -p_{1_g} A_1 - u_1|\rho V_1 A_1|$$

$$= -\frac{1.20 \times 10^5 \text{ N}}{m^2} \times 0.01 \text{ m}^2 - \frac{4 \text{ m}}{\sec} \left| \frac{999 \text{ kg}}{m^3} \times \frac{4 \text{ m}}{\sec} \times 0.01 \text{ m}^2 \times \frac{N \cdot \sec^2}{kg \cdot m} \right|$$

$$R_x = -1.36 \text{ kN} \qquad \{R_x \text{ to hold elbow acts to left.}\} \hspace{2cm} \underline{R_x}$$

Writing the y component of the momentum equation gives

$$F_{S_y} + F_{B_y} = \int_{CS} v\rho \vec{V} \cdot d\vec{A} = \int_{A_2} v\rho \vec{V} \cdot d\vec{A} \qquad \{v_1 = 0\}$$

$$p_a A_4 + p_a A_2 - p_a A_4 - p_a A_2 + F_{B_y} + R_y = \int_{A_2} v|\rho V \, dA| \quad \left\{ \begin{array}{l} \text{Pressure is } p_a \text{ over} \\ \text{top and bottom of CV.} \\ \vec{V} \cdot d\vec{A} \text{ is positive at } ②. \end{array} \right\}$$

$$F_{B_y} + R_y = v_2|\rho V_2 A_2|$$

$$R_y = -F_{B_y} + v_2|\rho V_2 A_2| \quad \left\{ \begin{array}{l} \text{Since we do not know the volume of} \\ \text{the elbow, we cannot evaluate } F_{B_y}. \end{array} \right\}$$

Substituting numbers, recognizing $\vec{V}_2 = -16\hat{\jmath}$ m/sec, so $v_2 = -16$ m/sec

$$R_y = -F_{B_y} + \left(\frac{-16 \text{ m}}{\sec} \right) \left| \frac{999 \text{ kg}}{m^3} \left(\frac{-16 \text{ m}}{\sec} \right) 0.0025 \text{ m}^2 \times \frac{N \cdot \sec^2}{kg \cdot m} \right|$$

$$R_y = -F_{B_y} - 639 \text{ N}$$

Neglecting F_{B_y} gives

$$R_y = -639 \text{ N} \quad \{R_y \text{ to hold elbow acts down.}\} \qquad\qquad R_y$$

$\left\{\begin{array}{l}\text{This problem illustrates the application of the momentum equation to an inertial}\\\text{control volume in which the pressure is not atmospheric across the entire}\\\text{control surface.}\end{array}\right\}$

Example 4.8

A horizontal conveyor belt moving at 3 ft/sec receives sand from a hopper. The sand falls vertically from the hopper to the belt at a speed of 5 ft/sec and a flowrate of 500 lbm/sec (the density of sand is approximately 2700 lbm/cubic yard). The conveyor belt is initially empty but begins to fill with sand. If friction in the drive system and rollers is negligible, find the tension required to pull the belt while the conveyor is filling.

Example Problem 4.8

GIVEN:

Conveyor and hopper shown in sketch.

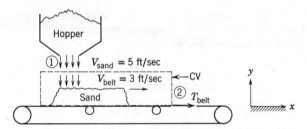

FIND:

T_{belt} at the instant shown.

SOLUTION:

Use control volume and coordinates shown. Apply x component of momentum equation.

Basic equations:

$$F_{S_x} + \cancel{F_{B_x}}^{=\,0(2)} = \frac{\partial}{\partial t}\int_{CV} u\rho\, d\forall + \int_{CS} u\rho\vec{V}\cdot d\vec{A}; \qquad 0 = \frac{\partial}{\partial t}\int_{CV}\rho\, d\forall + \int_{CS}\rho\vec{V}\cdot d\vec{A}$$

Assumptions: (1) $F_{S_x} = T_{belt} = T$
(2) $F_{B_x} = 0$
(3) Uniform flow at section ①
(4) All sand on belt moves with $V_{belt} = V_b$

Then

$$T = \frac{\partial}{\partial t} \int_{CV} u\rho \, d\forall + u_1\{-|\rho V_1 A_1|\} + u_2\{|\rho V_2 A_2|\}$$

Since $u_1 = 0$, and there is no flow at section ②, then

$$T = \frac{\partial}{\partial t}[V_b M_s] = M_s \frac{\partial V_b}{\partial t}^{\,0} + V_b \frac{\partial M_s}{\partial t}$$

where M_s is the mass of sand on the belt. From the continuity equation,

$$\frac{\partial}{\partial t} \int_{CV} \rho \, d\forall = \frac{\partial}{\partial t} M_s = -\int_{CS} \rho \vec{V} \cdot d\vec{A} = \dot{m}_s = 500 \text{ lbm/sec}$$

Then

$$T = V_b \dot{m}_s = \frac{3}{\text{sec}} \frac{\text{ft}}{} \times \frac{500 \text{ lbm}}{\text{sec}} \times \frac{\text{slug}}{32.2 \text{ lbm}} \times \frac{\text{lbf} \cdot \text{sec}^2}{\text{slug} \cdot \text{ft}}$$

$$T = 46.6 \text{ lbf} \hspace{6cm} T$$

$\left\{\begin{array}{l}\text{The purpose of this problem is to illustrate an application of the momentum}\\ \text{equation to a problem in which the rate of change of momentum within the}\\ \text{control volume is not equal to zero.}\end{array}\right\}$

**4-4.2 DIFFERENTIAL CONTROL VOLUME ANALYSIS

The control volume chosen for application of the basic equations need not be finite in size. If detailed information is desired, a suitable control volume of differential size sometimes can be used for analysis.

To illustrate the use of differential control volumes, let us apply the continuity and momentum equations to a steady incompressible flow without friction, as shown in Fig. 4.4. The control volume chosen is fixed in space and bounded by flow streamlines, and is thus an element of a stream tube. The length of the control volume is ds.

Because the control volume is bounded by streamlines, the only flow across the bounding surfaces occurs at the end sections. These are located at coordinates s and $s + ds$, measured along the central streamline.

Properties at the inlet section are assigned arbitrary symbolic values. Properties at the outlet section are assumed to differ by a differential amount.

** This section may be omitted without loss of continuity in the text material.

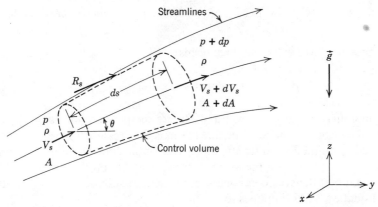

Fig. 4.4 Differential control volume for momentum analysis of flow along a streamline.

Thus at $s + ds$, the flow velocity is assumed to be $V_s + dV_s$, and so forth. The differential changes, dp, dV_s, and dA, are all assumed to be positive in formulating the problem. (As in a free body analysis in statics or dynamics, the actual algebraic sign of each differential change will be determined from the results of the analysis.)

Now let us apply the continuity equation and the s component of the momentum equation to the control volume of Fig. 4.4.

a. Continuity Equation

Basic equation:
$$0 = \overset{=0(1)}{\cancel{\frac{\partial}{\partial t} \int_{CV} \rho \, d\mathbf{V}}} + \int_{CS} \rho \vec{V} \cdot d\vec{A} \tag{4.13}$$

Assumptions: (1) Steady flow
(2) No flow across bounding streamlines
(3) Incompressible flow, $\rho = $ constant

Then
$$0 = \{-|\rho V_s A|\} + \{|\rho(V_s + dV_s)(A + dA)|\}$$
or
$$\rho V_s A = \rho(V_s + dV_s)(A + dA) \tag{4.20}$$

b. s Component of the Momentum Equation

Momentum equation:
$$F_{S_s} + F_{B_s} = \overset{=0(1)}{\cancel{\frac{\partial}{\partial t} \int_{CV} u_s \rho \, d\mathbf{V}}} + \int_{CS} u_s \rho \vec{V} \cdot d\vec{A} \tag{4.21}$$

Assumption: (4) No friction: $R_s = 0$ and F_{S_s} is due to pressure forces only.

The pressure force will have three terms:

$$F_{S_s} = pA - (p + dp)(A + dA) + \left(p + \frac{dp}{2}\right)(dA) \qquad (4.22a)$$

The first and second terms in Eq. 4.22a are the pressure forces on the end faces of the control surface. The third term is the pressure force acting in the s direction on the bounding stream surface of the control volume. Its magnitude is the product of the average pressure acting on the stream surface, $p + \frac{1}{2}dp$, times the area component of the stream surface in the s direction, dA. Equation 4.22a simplifies to

$$F_{S_s} = -A\,dp - \tfrac{1}{2}\,dp\,dA \qquad (4.22b)$$

The body force component in the s direction is

$$F_{B_s} = \rho g_s\,d\mathcal{V} = \rho(-g \sin\theta)\left(A + \frac{dA}{2}\right)ds$$

But $\sin\theta\,ds = dz$, so that

$$F_{B_s} = -\rho g\left(A + \frac{dA}{2}\right)dz \qquad (4.22c)$$

The momentum flux will be

$$\int_{CS} u_s\rho\vec{V}\cdot d\vec{A} = V_s\{-|\rho V_s A|\} + (V_s + dV_s)\{|\rho(V_s + dV_s)(A + dA)|\}$$

since there is no mass flux across the stream surfaces. The terms in braces are equal from continuity, so

$$\int_{CS} u_s\rho\vec{V}\cdot d\vec{A} = V_s(-\rho V_s A) + (V_s + dV_s)(\rho V_s A)$$

$$= (\rho V_s A)\,dV_s \qquad (4.23)$$

Substitution of Eqs. 4.22b, 4.22c, and 4.23 into the momentum equation gives

$$-A\,dp - \tfrac{1}{2}\,dp\,dA - \rho gA\,dz - \tfrac{1}{2}\rho g\,dA\,dz = \rho V_s A\,dV_s$$

Dividing by ρA and noting that products of differentials are negligible compared to the remaining terms, we obtain the result

$$-\frac{dp}{\rho} - g\,dz = V_s\,dV_s = d\left(\frac{V_s^2}{2}\right)$$

or

$$\frac{dp}{\rho} + d\left(\frac{\overline{V_s^2}}{2}\right) + g\,dz = 0 \qquad (4.24)$$

For incompressible flow, this equation may be integrated to obtain

$$\frac{p}{\rho} + \frac{V_s^2}{2} + gz = \text{constant}$$

or dropping the subscript, s,

$$\frac{p}{\rho} + \frac{V^2}{2} + gz = \text{constant} \qquad (4.25)$$

This equation is subject to the restrictions:

1. Steady flow.
2. No friction.
3. Flow along a streamline.
4. Incompressible flow.

By applying the momentum equation to an infinitesimal streamtube control volume, for steady incompressible flow without friction, we have derived a relation among the pressure, velocity, and elevation. This relationship is very powerful and useful. For example, it could have been used to evaluate the pressure at the inlet of the reducing elbow analyzed in Example Problem 4.7 or to determine the velocity of water leaving the sluice gate of Example Problem 4.6. In both of these flow situations the restrictions required to derive Eq. 4.25 are reasonable idealizations of the actual flow behavior. The restrictions must be emphasized heavily because they do not always form a realistic model for flow behavior; consequently, they must be justified carefully each time Eq. 4.25 is applied.

Equation 4.25 is a form of the Bernoulli equation. It will be derived again in detail in Chapter 6, because it is such a useful tool for flow analysis and because an alternative derivation will give added insight into the need for care in applying the equation.

4-4.3 CONTROL VOLUME MOVING WITH CONSTANT VELOCITY

In the preceding problems, which illustrate applications of the momentum equation to inertial control volumes, we have considered only stationary control volumes. It should be emphasized that a control volume (fixed relative to reference frame xyz) moving with a constant velocity, $\vec{V}_{rf}$, relative to a fixed (inertial) reference frame XYZ, is also inertial, since it has no acceleration with respect to XYZ.

Equation 4.11, which expresses system derivatives in terms of control volume variables, is valid for any motion of the coordinate system xyz (fixed to the control volume), provided that:

1. All velocities are measured relative to the control volume.
2. All time derivatives are measured relative to the control volume.

To emphasize this point, we rewrite Eq. 4.11 as

$$\frac{dN}{dt}\bigg)_{\text{system}} = \frac{\partial}{\partial t} \int_{\text{CV}} \eta \rho \, d\mathbf{V} + \int_{\text{CS}} \eta \rho \vec{V}_{xyz} \cdot d\vec{A} \qquad (4.26)$$

Since all time derivatives must be measured relative to the control volume, in using this equation to obtain the momentum equation for an inertial control volume from the system formulation, we really set

$$N = \vec{P}_{xyz} \qquad \text{and} \qquad \eta = \vec{V}_{xyz}$$

to obtain the equation

$$\vec{F} = \vec{F}_S + \vec{F}_B = \frac{\partial}{\partial t} \int_{\text{CV}} \vec{V}_{xyz} \rho \, d\mathbf{V} + \int_{\text{CS}} \vec{V}_{xyz} \rho \vec{V}_{xyz} \cdot d\vec{A} \qquad (4.27)$$

Equation 4.27 is the formulation of Newton's second law applied to any inertial control volume (stationary or moving with a constant velocity). It is identical to Eq. 4.18 except that we have included the subscript xyz to emphasize that quantities must be measured relative to the control volume. The momentum equation is applied to an inertial control volume moving with a constant velocity in Example 4.9.

Example 4.9
The sketch shows a vane with a turning angle of 60°. The vane moves at constant speed, $U = 10 \, \text{m/sec}$, and receives a jet of water that leaves a stationary nozzle with speed, $V = 30 \, \text{m/sec}$. The nozzle has an exit area of 0.003 m². Determine the force of the water on the moving vane.

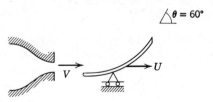

Example Problem 4.9

GIVEN:

Vane having a turning angle, $\theta = 60°$, moves with constant velocity, $\vec{U} = 10\hat{\imath}$ m/sec. Water from a constant area nozzle, $A = 0.003$ m^2, with velocity $\vec{V} = 30\hat{\imath}$ m/sec flows over the vane as shown.

FIND:

The force of the water on the vane.

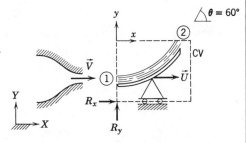

SOLUTION:

Select a control volume moving with the vane at velocity, $\vec{U}$, as shown by the dashed line. R_x and R_y are the components of the force required to maintain the velocity of the control volume at $10\hat{\imath}$ m/sec.
The control volume is inertial, since it is not accelerating ($U = $ constant). Remember that all velocities must be measured relative to the control volume in applying the basic equations.

Basic equations:
$$\vec{F}_S + \vec{F}_B = \frac{\partial}{\partial t} \int_{CV} \vec{V}_{xyz} \rho \, d\forall + \int_{CS} \vec{V}_{xyz} \rho \vec{V}_{xyz} \cdot d\vec{A}$$

$$0 = \frac{\partial}{\partial t} \int_{CV} \rho \, d\forall + \int_{CS} \rho \vec{V}_{xyz} \cdot d\vec{A}$$

Assumptions:
(1) Flow is steady relative to vane
(2) Magnitude of relative velocity along the vane is constant, that is, $|\vec{V}_1| = |\vec{V}_2| = V - U$
(3) Properties are uniform at sections ① and ②
(4) $F_{B_x} = F_{B_y} = 0$
(5) Incompressible flow

The x component of the momentum equation is

$$F_{S_x} + \overbrace{F_{B_x}}^{= 0(4)} = \overbrace{\frac{\partial}{\partial t} \int_{CV} u\rho \, d\forall}^{= 0(1)} + \int_{CS} u\rho \vec{V} \cdot d\vec{A}$$

There is no net pressure force, since p_{atm} acts on all sides of the CV. Thus

$$R_x = \int_{A_1} u\{-|\rho V \, dA|\} + \int_{A_2} u\{|\rho V \, dA|\} = -u_1|\rho V_1 A_1| + u_2|\rho V_2 A_2|$$

From the continuity equation

$$0 = \int_{A_1} \{-|\rho V\, dA|\} + \int_{A_2} |\rho V\, dA| = -|\rho V_1 A_1| + |\rho V_2 A_2|$$

or

$$|\rho V_1 A_1| = |\rho V_2 A_2|$$

Therefore,

$$R_x = (u_2 - u_1)|\rho V_1 A_1|$$

All velocities must be measured relative to the CV, so we note that

$$u_1 = V - U \qquad u_2 = (V - U)\cos\theta$$
$$V_1 = V - U \qquad V_2 = V - U$$

Substituting,

$$R_x = [(V - U)\cos\theta - (V - U)]|\rho(V - U)A_1| = (V - U)(\cos\theta - 1)|\rho(V - U)A_1|$$

$$= \frac{(30 - 10)\ \text{m}}{\text{sec}}\ (0.50 - 1)\left|\frac{999\ \text{kg}}{\text{m}^3}\ \frac{(30 - 10)\ \text{m}}{\text{sec}} \times 0.003\ \text{m}^2\left|\frac{\text{N}\cdot\text{sec}^2}{\text{kg}\cdot\text{m}}\right.\right.$$

$$R_x = -599\ \text{N}$$

R_x is the force required to maintain the vane velocity of $\vec{U}$. The force of the water on the vane is $K_x = -R_x = 599\ \text{N}$ {to right} $\qquad\qquad K_x$

Writing the y component of the momentum equation,

$$F_{S_y} + \overset{= 0(4)}{\cancel{F_{B_y}}} = \overset{= 0(1)}{\cancel{\frac{\partial}{\partial t}\int_{\text{CV}} v\rho\, d\forall}} + \int_{\text{CS}} v\rho\vec{V}\cdot d\vec{A}$$

$$R_y = \int_{\text{CS}} v\rho\vec{V}\cdot d\vec{A} = \int_{A_2} v\rho\vec{V}\cdot d\vec{A} \qquad \{v_1 = 0\}$$

$$= \int_{A_2} v|\rho V\, dA| = v_2|\rho V_2 A_2| = v_2|\rho V_1 A_1| \qquad \{\text{Recall}\ |\rho V_2 A_2| = |\rho V_1 A_1|\}$$

$$= (V - U)\sin\theta|\rho(V - U)A_1|$$

$$= \frac{(30 - 10)\ \text{m}}{\text{sec}}\ (0.866)\left|\frac{999\ \text{kg}}{\text{m}^3}\ \frac{(30 - 10)\ \text{m}}{\text{sec}} \times 0.003\ \text{m}^2\left|\frac{\text{N}\cdot\text{sec}^2}{\text{kg}\cdot\text{m}}\right.\right.$$

$$R_y = 1.04\ \text{kN} \qquad \{\text{upward}\}$$

The force of the water on the vane is

$$K_y = -R_y = -1.04\ \text{kN} \qquad \{\text{downward}\} \qquad\qquad K_y$$

This problem illustrates the point that in applying the momentum equation to an inertial control volume all velocities are measured relative to the control volume.

4-5 MOMENTUM EQUATION FOR CONTROL VOLUME WITH RECTILINEAR ACCELERATION

For an inertial control volume, that is, a control volume not accelerating relative to a stationary frame of reference, we developed Eq. 4.27 (same as Eq. 4.18) as the control volume formulation of Newton's second law.

$$\vec{F} = \vec{F}_S + \vec{F}_B = \frac{\partial}{\partial t} \int_{CV} \vec{V}_{xyz} \rho \, d\Psi + \int_{CS} \vec{V}_{xyz} \rho \vec{V}_{xyz} \cdot d\vec{A} \qquad (4.27)$$

Since we are interested in analyzing control volumes that may accelerate relative to an inertial coordinate system (a rocket must accelerate if it is to get off the ground), it is logical to ask whether Eq. 4.27 can be used for accelerating control volumes. To answer this question we should look at the items we employed in deriving Eq. 4.27. In the derivation of Eq. 4.27 we used the relation Eq. 4.11 and the system equation (Eq. 4.2).

In deriving the control volume formulation (Section 4-2.1), the control volume was fixed relative to xyz; the flow field, $\vec{V}(x, y, z, t)$, was specified relative to the coordinates x, y, and z. No restriction was placed on the motion of the xyz reference frame. Therefore, Eq. 4.11 (or Eq. 4.26) is valid at any instant for any arbitrary motion of the coordinates x, y, and z provided that all time derivatives and velocities in the equation are measured relative to the control volume.

The system equation

$$\vec{F} = \frac{d\vec{P}}{dt} \bigg)_{system} \qquad (4.2a)$$

where the linear momentum, $\vec{P}$, of the system is given by

$$\vec{P}_{system} = \int_{mass \, (system)} \vec{V} \, dm = \int_{\Psi \, (system)} \vec{V} \rho \, d\Psi \qquad (4.2b)$$

is valid only for velocities measured relative to an inertial reference frame. Thus, if we denote the inertial reference frame by XYZ, then Newton's second law states that

$$\vec{F} = \frac{d\vec{P}_{XYZ}}{dt} \bigg)_{system} \qquad (4.28)$$

Consider a pinball machine on a train. Here the inertial coordinate system XYZ can be taken fixed to the ground. The motion of a pinball is described most easily relative to a coordinate system xyz fixed on the train. Clearly, the description of the motion of the ball relative to the train (coordinates x, y, and z) is different from the description relative to the ground (coordinates X, Y, and Z) if the train is in motion.

In order to obtain the momentum equation for a linearly accelerating control volume, that is, one which is undergoing translation without rotation relative to an inertial reference frame (XYZ), we shall employ Eq. 4.11 and the appropriate system equation, Eq. 4.28. In Eq. 4.11, all time derivatives and velocities are measured relative to the control volume; thus it is necessary to relate $\vec{P}_{XYZ}$ of the system to $\vec{P}_{xyz}$ of the system.

4-5.1   CONTROL VOLUME EQUATION

Let us write Newton's second law for a system, remembering that the acceleration must be measured relative to an inertial reference frame that we have designated XYZ. We write

$$\vec{F} = \frac{d\vec{P}_{XYZ}}{dt}\bigg]_{\text{system}} = \frac{d}{dt}\int_{\text{mass (system)}} \vec{V}_{XYZ}\,dm = \int_{\text{mass (system)}} \frac{d\vec{V}_{XYZ}}{dt}\,dm$$

$$\vec{F} = \int_{\text{mass (system)}} \vec{a}_{XYZ}\,dm \tag{4.29}$$

The only problem now is to obtain a suitable expression for $\vec{a}_{XYZ}$, for the special case in which the coordinates x, y, and z are undergoing pure translation, without rotation, relative to the inertial frame XYZ.

Since the motion of xyz is pure translation, without rotation, relative to an inertial reference frame XYZ, then

$$\vec{a}_{XYZ} = \vec{a}_{xyz} + \vec{a}_{rf} \tag{4.30}$$

where

$\vec{a}_{XYZ}$ is the rectilinear acceleration of the system relative to the inertial reference frame XYZ,

$\vec{a}_{xyz}$ is the rectilinear acceleration of the system relative to the noninertial reference frame xyz, and

$\vec{a}_{rf}$ is the rectilinear acceleration of the noninertial reference frame xyz relative to the inertial frame XYZ.

For this case we write the system equation as

$$\vec{F} = \int_{\text{mass (system)}} \vec{a}_{XYZ}\,dm = \int_{\text{mass (system)}} (\vec{a}_{xyz} + \vec{a}_{rf})\,dm$$

Alternately,

$$\vec{F} - \int_{\text{mass (system)}} \vec{a}_{rf}\,dm = \int_{\text{mass (system)}} \vec{a}_{xyz}\,dm$$

Since

$$\vec{a}_{xyz} = \frac{d\vec{V}_{xyz}}{dt}$$

then

$$\vec{F} - \int_{\text{mass (system)}} \vec{a}_{rf}\, dm = \int_{\text{mass (system)}} \frac{d\vec{V}_{xyz}}{dt}\, dm$$

$$= \frac{d}{dt} \int_{\text{mass (system)}} \vec{V}_{xyz}\, dm = \left. \frac{d\vec{P}_{xyz}}{dt} \right)_{\text{system}}$$

Since $dm = \rho\, d\mathbf{V}$, the system equation becomes

$$\vec{F} - \int_{\mathbf{V}\text{ (system)}} \vec{a}_{rf}\rho\, d\mathbf{V} = \left. \frac{d\vec{P}_{xyz}}{dt} \right)_{\text{system}} \qquad (4.31a)$$

where the linear momentum, $\vec{P}_{xyz}$, of the system is given by

$$\vec{P}_{xyz_{\text{system}}} = \int_{\text{mass (system)}} \vec{V}_{xyz}\, dm = \int_{\mathbf{V}\text{ (system)}} \vec{V}_{xyz}\rho\, d\mathbf{V} \qquad (4.31b)$$

and the force, $\vec{F}$, includes all surface and body forces acting on the system.

The system formulation is related to the formulation for a moving control volume by Eq. 4.26

$$\left. \frac{dN}{dt} \right)_{\text{system}} = \frac{\partial}{\partial t} \int_{\text{CV}} \eta \rho\, d\mathbf{V} + \int_{\text{CS}} \eta \rho \vec{V}_{xyz} \cdot d\vec{A} \qquad (4.26)$$

where

$$N_{\text{system}} = \int_{\text{mass (system)}} \eta\, dm = \int_{\mathbf{V}\text{ (system)}} \eta \rho\, d\mathbf{V} \qquad (4.6)$$

To derive the control volume formulation of Newton's second law, we set

$$N = \vec{P}_{xyz} \qquad \text{and} \qquad \eta = \vec{V}_{xyz}$$

From Eq. 4.26, with this substitution, we obtain

$$\left. \frac{d\vec{P}_{xyz}}{dt} \right)_{\text{system}} = \frac{\partial}{\partial t} \int_{\text{CV}} \vec{V}_{xyz}\rho\, d\mathbf{V} + \int_{\text{CS}} \vec{V}_{xyz}\rho \vec{V}_{xyz} \cdot d\vec{A} \qquad (4.32)$$

From the system equation,

$$\left. \frac{d\vec{P}_{xyz}}{dt} \right)_{\text{system}} = \vec{F}_{\text{on system}} - \int_{\mathbf{V}\text{ (system)}} \vec{a}_{rf}\rho\, d\mathbf{V} \qquad (4.31a)$$

Since the system and the control volume coincided at time t_0, then,

$$\vec{F}_{\text{on system}} - \int_{\mathbf{V}\text{ (system)}} \vec{a}_{rf}\rho\, d\mathbf{V} = \vec{F}_{\text{on CV}} - \int_{\text{CV}} \vec{a}_{rf}\rho\, d\mathbf{V}$$

In light of this, Eqs. 4.31a and 4.32 may be combined to yield the formulation of Newton's second law for a control volume accelerating, without rotation, relative to an inertial reference frame:

$$\vec{F} - \int_{CV} \vec{a}_{rf}\rho \, dV = \frac{\partial}{\partial t} \int_{CV} \vec{V}_{xyz}\rho \, dV + \int_{CS} \vec{V}_{xyz}\rho \vec{V}_{xyz} \cdot d\vec{A} \qquad (4.33)$$

Since $\vec{F} = \vec{F}_S + \vec{F}_B$ and the body force, $\vec{F}_B$, can be written

$$\vec{F}_B = \int_{CV} \vec{B}\rho \, dV$$

then Eq. 4.33 becomes

$$\vec{F}_S + \int_{CV} \vec{B}\rho \, dV - \int_{CV} \vec{a}_{rf}\rho \, dV = \frac{\partial}{\partial t} \int_{CV} \vec{V}_{xyz}\rho \, dV + \int_{CS} \vec{V}_{xyz}\rho \vec{V}_{xyz} \cdot d\vec{A} \quad (4.34)$$

In comparing the momentum equation for a control volume with rectilinear acceleration, Eq. 4.34, to that for a nonaccelerating control volume, Eq. 4.27, we see that the only difference is the presence of one additional term in Eq. 4.34. When the control volume is not accelerating relative to an inertial reference frame, XYZ, that is, when $\vec{a}_{rf} = 0$, then Eq. 4.34 reduces to Eq. 4.27.

The precautions concerning the use of Eq. 4.27 also apply to the use of Eq. 4.34. Before attempting to apply either equation, one should draw the boundaries of the control volume and label appropriate coordinate directions. For an accelerating control volume, one must label a coordinate system (xyz) on the control volume and an inertial reference frame (XYZ).

In Eq. 4.34, $\vec{F}_S$ represents all surface forces acting on the control volume. One should recognize that both the remaining terms on the left side of the equation may be functions of time, since the mass within the control volume may vary with time. Furthermore, the acceleration, $\vec{a}_{rf}$, of the reference frame xyz relative to an inertial frame in general will be a function of time.

All velocities in Eq. 4.34 are measured relative to the control volume. The momentum flux, $\vec{V}_{xyz}\rho\vec{V}_{xyz} \cdot d\vec{A}$, through an element of the control surface area, $d\vec{A}$, is a vector. The sign of the scalar product $\rho\vec{V}_{xyz} \cdot d\vec{A}$ depends on the direction of the velocity vector, $\vec{V}_{xyz}$, relative to the area vector, $d\vec{A}$. The sign of the vector velocity, $\vec{V}_{xyz}$, is dependent on the choice of the coordinate system.

The momentum equation is a vector equation. As with all vector equations, it may be written in scalar component equations. The scalar components of Eq. 4.34 are

$$F_{S_x} + \int_{CV} B_x\rho \, dV - \int_{CV} a_{rf_x}\rho \, dV = \frac{\partial}{\partial t} \int_{CV} u_{xyz}\rho \, dV + \int_{CS} u_{xyz}\rho \vec{V}_{xyz} \cdot d\vec{A}$$

$$(4.35a)$$

$$F_{S_y} + \int_{CV} B_y \rho \, d\Psi - \int_{CV} a_{rf_y} \rho \, d\Psi = \frac{\partial}{\partial t} \int_{CV} v_{xyz} \rho \, d\Psi + \int_{CS} v_{xyz} \rho \vec{V}_{xyz} \cdot d\vec{A}$$

(4.35b)

$$F_{S_z} + \int_{CV} B_z \rho \, d\Psi - \int_{CV} a_{rf_z} \rho \, d\Psi = \frac{\partial}{\partial t} \int_{CV} w_{xyz} \rho \, d\Psi + \int_{CS} w_{xyz} \rho \vec{V}_{xyz} \cdot d\vec{A}$$

(4.35c)

Example 4.10

A vane with turning angle $\theta = 60°$ is attached to a cart. The cart and vane of mass 75 kg roll on a level track. Friction and air resistance may be neglected. The vane receives a jet of water, which leaves a stationary nozzle horizontally at 35 m/sec. The nozzle exit area is 0.003 m². Determine the velocity of the cart as a function of time and plot the results.

Example Problem 4.10

GIVEN:

Vane and cart as sketched.

FIND:

(a) $U(t)$.
(b) Plot results.

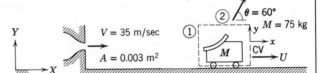

SOLUTION:

Choose the control volume and coordinate systems shown for the analysis. Note that XY is a fixed frame, while frame xy moves with the cart. Apply the x component of the momentum equation.

Basic equation:
$$\overset{=\,0(1)}{\cancel{F_{S_x}}} + \overset{=\,0(2)}{\cancel{F_{B_x}}} - \int_{CV} a_{rf_x} \rho \, d\Psi = \overset{\simeq\,0(3)}{\frac{\partial}{\partial t} \int_{CV} u_{xyz} \rho \, d\Psi} + \int_{CS} u_{xyz} \rho \vec{V}_{xyz} \cdot d\vec{A}$$

Assumptions: (1) $F_{S_x} = 0$, since no resistance is present
              (2) $F_{B_x} = 0$
              (3) Neglect the rate of change of u_{xyz} for water in contact with the vane;

$$\frac{\partial}{\partial t} \int_{CV} u_{xyz} \rho \, d\Psi \simeq 0$$

(4) Uniform flow at sections ① and ②, $\rho = 999 \ kg/m^3$

(5) $|\vec{V}_{xyz_1}| = |\vec{V}_{xyz_2}|$; that is, water stream is not slowed by friction on the vane

(6) $A_2 = A_1 = A$

Then

$$-\int_{CV} a_{rf_x}\rho d\forall = u_{xyz_1}\{-|\rho V_{xyz_1}A_1|\} + u_{xyz_2}\{|\rho V_{xyz_2}A_2|\}$$

where all quantities are measured relative to the xyz frame. Dropping subscripts rf and xyz,

$$-\int_{CV} a_x\rho \, d\forall = u_1\{-|\rho V_1 A_1|\} + u_2\{|\rho V_2 A_2|\} \qquad (1)$$

Let us evaluate these terms separately:

$$-\int_{CV} a_x\rho \, d\forall = -a_x M_{CV} = -a_x M = -\frac{dU}{dt} M$$

$$u_1\{-|\rho V_1 A_1|\} = (V - U)\{-|\rho(V - U)A|\} = -\rho(V - U)^2 A$$

$$u_2\{|\rho V_2 A_2|\} = (V - U)\cos\theta\{|\rho(V - U)A|\} = \rho(V - U)^2 A \cos\theta$$

Since $V \geq U$, the absolute value signs have been dropped from the flux terms. Substitution in Eq. 1 gives

$$-M\frac{dU}{dt} = -\rho(V - U)^2 A + \rho(V - U)^2 A \cos\theta$$

$$-M\frac{dU}{dt} = (\cos\theta - 1)\rho(V - U)^2 A$$

Separating variables,

$$\frac{dU}{(V - U)^2} = \frac{(1 - \cos\theta)\rho A}{M} dt = b \, dt$$

Integrating between limits $U = 0$ at $t = 0$, and $U = U$ at $t = t$,

$$\int_0^U \frac{dU}{(V - U)^2} = \frac{1}{(V - U)}\Big]_0^U = \int_0^t b \, dt = bt$$

or

$$\frac{1}{(V - U)} - \frac{1}{V} = bt$$

which becomes

$$\frac{U}{V(V - U)} = bt$$

Solving for U, we obtain

$$\frac{U}{V} = \frac{Vbt}{1 + Vbt}$$

Evaluating the term, Vb, gives

$$Vb = V \frac{(1 - \cos\theta)\rho A}{M}$$

$$Vb = \frac{35}{\sec} \frac{m}{\sec} \times \frac{(1 - 0.5)}{75 \text{ kg}} \times \frac{999 \text{ kg}}{m^3} \times \frac{0.003 \text{ m}^2}{} = 0.699 \text{ sec}^{-1}$$

Thus

$$\frac{U}{V} = \frac{0.699t}{1 + 0.699t} \qquad (t \text{ in sec})$$

$U(t)$

Plot:

$\frac{U}{V}$

Example 4.11

A small rocket, with an initial mass of 400 kg is to be launched vertically. Upon ignition the rocket consumes fuel at the rate of 5 kg/sec and ejects gas at atmospheric pressure with a speed of 1500 m/sec relative to the rocket. Determine the initial acceleration of the rocket, and the rocket velocity after 10 sec if air resistance is neglected.

Example Problem 4.11

GIVEN:

Small rocket accelerates vertically from rest.
Air resistance may be neglected.
Rate of fuel consumption, $\dot{m}_{out} = 5$ kg/sec
Exhaust velocity, $V_e = 1500$ m/sec, relative to rocket

FIND:

(a) Initial acceleration of the rocket.
(b) Rocket velocity after 10 sec.

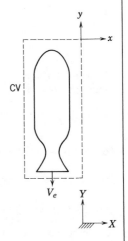

SOLUTION:

Choose a control volume as shown by dashed lines. Because the control volume is accelerating, we need to define an inertial coordinate system XY. The coordinate system xy is attached to the CV. Apply the y component of the momentum equation.

Basic equation:

$$F_{S_y} + \int_{CV} B_y \rho \, d\forall - \int_{CV} a_{rf_y} \rho \, d\forall = \frac{\partial}{\partial t} \int_{CV} v_{xyz} \rho \, d\forall + \int_{CS} v_{xyz} \rho \vec{V}_{xyz} \cdot d\vec{A}$$

Assumptions: (1) Atmospheric pressure acts on all surfaces of the CV. Since air resistance is neglected, then $F_{S_y} = 0$

            (2) Only body force is that due to gravity and hence $B_y = -g$

            (3) Flow leaving the rocket is uniform and V_e is constant

Under these assumptions the momentum equation reduces to

$$-\int_{CV} g \rho \, d\forall - \int_{CV} a_{rf_y} \rho \, d\forall = \frac{\partial}{\partial t} \int_{CV} v_{xyz} \rho \, d\forall + \int_{CS} v_{xyz} \rho \vec{V}_{xyz} \cdot d\vec{A} \qquad (1)$$

      Ⓐ   Ⓑ   Ⓒ   Ⓓ

Let us look at the equation term by term:

Ⓐ $-\int_{CV} g \rho \, d\forall = -g \int_{CV} \rho \, d\forall = -g M_{CV},$ since g is constant

We recognize that the mass of the CV will be a function of time because mass is leaving the CV at a rate $\dot{m}$.

To determine M_{CV} as a function of time, we use the conservation of mass equation

$$\frac{\partial}{\partial t} \int_{CV} \rho \, d\forall + \int_{CS} \rho \vec{V} \cdot d\vec{A} = 0$$

Then

$$\frac{\partial}{\partial t} \int_{CV} \rho \, d\forall = -\int_{CS} \rho \vec{V} \cdot d\vec{A} = -\int_{A_e} \rho \vec{V} \cdot d\vec{A} = -\int_{A_e} |\rho V \, dA| = -|\dot{m}|$$

The minus sign indicates that the mass of the CV is decreasing with time. Since the mass of the CV is only a function of time, we can write

$$\frac{dM_{CV}}{dt} = -|\dot{m}|$$

To find the mass of the CV at any time, t, we integrate

$$\int_{M_0}^{M} dM_{CV} = -\int_{0}^{t} |\dot{m}| \, dt \qquad \text{where at } t = 0, M_{CV} = M_0, \text{ and at } t = t, M_{CV} = M$$

Then $M - M_0 = -|\dot{m}|t$ or $M = M_0 - \dot{m}t$

{Since $\dot{m}$ is positive, we have dropped the absolute value sign.}

    Substituting the expression for M into term Ⓐ, we obtain

$$-\int_{CV} g \rho \, d\forall = -g M_{CV} = -g(M_0 - \dot{m}t)$$

Ⓑ $-\int_{CV} a_{rf_y} \rho \, d\forall$

The acceleration, a_{rf_y}, of the CV is that seen by an observer in the XY coordinate system. Thus a_{rf_y} is not a function of the coordinates xyz, and

$$- \int_{CV} a_{rf_y} \rho \, d\forall = - a_{rf_y} \int_{CV} \rho \, d\forall = - a_{rf_y} M_{CV} = - a_{rf_y} (M_0 - \dot{m}t)$$

© $\quad \dfrac{\partial}{\partial t} \int_{CV} v_{xyz} \rho \, d\forall$

is the time rate of change of the y momentum of the fluid in the control volume measured relative to the control volume.

While the y momentum of the fluid inside the CV, measured relative to the CV, is a large number, it does not change appreciably with time. To see this, we must recognize that:

(1) The unburned fuel and the rocket structure have zero momentum relative to the rocket.
(2) The velocity of the gas at the nozzle exit remains constant with time as does the velocity at various points in the nozzle.

Consequently, it is reasonable to assume that

$$\frac{\partial}{\partial t} \int_{CV} v_{xyz} \rho \, d\forall \approx 0$$

Ⓓ $\quad \displaystyle\int_{CS} v_{xyz} \rho \vec{V}_{xyz} \cdot d\vec{A} = \int_{A_e} v_{xyz} |\rho V_{xyz} \, dA| = v_{xyz} |\dot{m}|$

Since $\vec{V}_e = - V_e \hat{j}$, then

$$v_{xyz} |\dot{m}| = - V_e |\dot{m}| = - V_e \dot{m}$$

Substituting terms Ⓐ through Ⓓ into Eq. 1, we obtain

$$- g(M_0 - \dot{m}t) - a_{rf_y}(M_0 - \dot{m}t) = - V_e \dot{m}$$

or

$$a_{rf_y} = \frac{V_e \dot{m}}{M_0 - \dot{m}t} - g \tag{2}$$

At time, $t = 0$

$$\left. a_{rf_y} \right)_{t=0} = \frac{V_e \dot{m}}{M_0} - g = \frac{1500 \text{ m}}{\text{sec}} \times \frac{5 \text{ kg}}{\text{sec}} \times \frac{1}{400 \text{ kg}} - \frac{9.81 \text{ m}}{\text{sec}^2}$$

$$\left. a_{rf_y} \right)_{t=0} = 8.94 \text{ m/sec}^2 \qquad\qquad\qquad\qquad\qquad\quad \overleftarrow{\hspace{2cm}} \quad a_{rf_y})_{t=0}$$

The acceleration of the CV is by definition

$$a_{rf_y} = \frac{dV_{CV}}{dt}$$

Substituting from Eq. 2,

$$\frac{dV_{CV}}{dt} = \frac{V_e \dot{m}}{M_0 - \dot{m}t} - g$$

Separating variables and integrating gives

$$V_{CV} = \int_0^{V_{CV}} dV_{CV} = \int_0^t \frac{V_e \dot{m}\, dt}{M_0 - \dot{m}t} - \int_0^t g\, dt = -V_e \ln\left[\frac{M_0 - \dot{m}t}{M_0}\right] - gt$$

At $t = 10$ sec

$$V_{CV} = -\; 1500\; \frac{m}{sec} \times \ln\left[\frac{350\; kg}{400\; kg}\right] - 9.81\; \frac{m}{sec^2} \times 10\; sec$$

$$V_{CV} = 102\; m/sec \qquad\qquad V_{CV})_{t = 10\,sec}$$

{ This problem illustrates the application of the momentum equation to a linearly accelerating control volume. }

**4-6 MOMENTUM EQUATION FOR CONTROL VOLUME WITH ARBITRARY ACCELERATION

In Section 4-5 we formulated the momentum equation for a control volume with rectilinear acceleration. The purpose of this section is to extend the formulation to include rotation and angular acceleration of the control volume in addition to translation and rectilinear acceleration.

First, we develop an expression for Newton's second law in an arbitrary, noninertial coordinate system. Then we use Eq. 4.26 to complete the formulation for a control volume.

4-6.1 CONTROL VOLUME EQUATION

Newton's second law for a system moving relative to an inertial coordinate system is given by

$$\vec{F} = \frac{d\vec{P}_{XYZ}}{dt}\bigg)_{system}$$

Since

$$\vec{P}_{XYZ}\bigg)_{system} = \int_{M\,(system)} \vec{V}_{XYZ}\, dm,$$

and M (system) is constant, then

$$\vec{F} = \frac{d}{dt}\int_{M\,(system)} \vec{V}_{XYZ}\, dm = \int_{M\,(system)} \frac{d\vec{V}_{XYZ}}{dt}\, dm$$

** This section may be omitted without loss of continuity in the text material.

or

$$\vec{F} = \int_{M \text{ (system)}} \vec{a}_{XYZ} \, dm \tag{4.36}$$

The basic problem is to relate $\vec{a}_{XYZ}$ to the acceleration, $\vec{a}_{xyz}$, measured relative to a noninertial coordinate system and other variables. For this purpose, consider the noninertial reference frame, xyz, shown in Fig. 4.5.

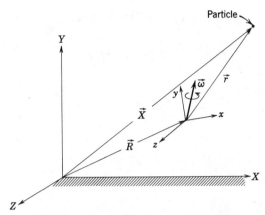

Fig. 4.5 Location of a particle in inertial (XYZ) and noninertial (xyz) reference frames.

The noninertial frame, xyz, is located by position vector $\vec{R}$ relative to the fixed frame. The noninertial frame rotates with angular velocity $\vec{\omega}$.[1] The position of a particle is located relative to the moving frame by position vector $\vec{r} = \hat{\imath}x + \hat{\jmath}y + \hat{k}z$. Relative to the inertial reference frame, XYZ, the position of the particle is denoted by position vector $\vec{X}$. From the geometry of the figure, $\vec{X} = \vec{R} + \vec{r}$.

The velocity of the particle relative to an observer in the XYZ system is

$$\vec{V}_{XYZ} = \frac{d\vec{X}}{dt} = \frac{d\vec{R}}{dt} + \frac{d\vec{r}}{dt} = \vec{V}_{rf} + \frac{d\vec{r}}{dt} \tag{4.37}$$

We must be careful in evaluating $d\vec{r}/dt$, because both the magnitude $|\vec{r}|$ and the orientation of the unit vectors, $\hat{\imath}, \hat{\jmath}$, and $\hat{k}$ are functions of time. Thus

$$\frac{d\vec{r}}{dt} = \frac{d}{dt}(x\hat{\imath} + y\hat{\jmath} + z\hat{k}) = \hat{\imath}\frac{dx}{dt} + x\frac{d\hat{\imath}}{dt} + \hat{\jmath}\frac{dy}{dt} + y\frac{d\hat{\jmath}}{dt} + \hat{k}\frac{dz}{dt} + z\frac{d\hat{k}}{dt} \tag{4.38a}$$

The terms dx/dt, dy/dt, and dz/dt are the velocity components of the particle

---

[1] Note that any arbitrary motion can be decomposed into a translation plus a rotation.

relative to xyz. Thus

$$\vec{V}_{xyz} = \hat{\imath}\frac{dx}{dt} + \hat{\jmath}\frac{dy}{dt} + \hat{k}\frac{dz}{dt} \tag{4.38b}$$

You may recall from dynamics that for a rotating coordinate system,

$$\vec{\omega} \times \vec{r} = x\frac{d\hat{\imath}}{dt} + y\frac{d\hat{\jmath}}{dt} + z\frac{d\hat{k}}{dt} \tag{4.38c}$$

(This result is also derived in Example Problem 4.12.) Combining Eqs. 4.38a, b, and c, we obtain

$$\frac{d\vec{r}}{dt} = \vec{V}_{xyz} + \vec{\omega} \times \vec{r} \tag{4.38d}$$

Substituting into Eq. 4.37,

$$\vec{V}_{XYZ} = \vec{V}_{rf} + \vec{V}_{xyz} + \vec{\omega} \times \vec{r} \tag{4.39}$$

The acceleration of the particle relative to an observer in the XYZ system is

$$\vec{a}_{XYZ} = \frac{d\vec{V}_{XYZ}}{dt} = \frac{d\vec{V}_{rf}}{dt} + \frac{d\vec{V}_{xyz}}{dt} + \frac{d}{dt}(\vec{\omega} \times \vec{r})$$

or

$$\vec{a}_{XYZ} = \vec{a}_{rf} + \frac{d\vec{V}_{xyz}}{dt} + \frac{d}{dt}(\vec{\omega} \times \vec{r}) \tag{4.40}$$

Both $\vec{V}_{xyz}$ and $\vec{r}$ are measured relative to xyz, so the same caution observed in developing Eq. 4.38d applies. Thus

$$\frac{d\vec{V}_{xyz}}{dt} = \vec{a}_{xyz} + \vec{\omega} \times \vec{V}_{xyz} \tag{4.41a}$$

and

$$\frac{d}{dt}(\vec{\omega} \times \vec{r}) = \frac{d\vec{\omega}}{dt} \times \vec{r} + \vec{\omega} \times \frac{d\vec{r}}{dt}$$

$$= \dot{\vec{\omega}} \times \vec{r} + \vec{\omega} \times (\vec{V}_{xyz} + \vec{\omega} \times \vec{r})$$

or

$$\frac{d}{dt}(\vec{\omega} \times \vec{r}) = \dot{\vec{\omega}} \times \vec{r} + \vec{\omega} \times \vec{V}_{xyz} + \vec{\omega} \times (\vec{\omega} \times \vec{r}) \tag{4.41b}$$

Substituting Eqs. 4.41a and b into Eq. 4.40, we obtain

$$\vec{a}_{XYZ} = \vec{a}_{rf} + \vec{a}_{xyz} + 2\vec{\omega} \times \vec{V}_{xyz} + \vec{\omega} \times (\vec{\omega} \times \vec{r}) + \dot{\vec{\omega}} \times \vec{r} \tag{4.42}$$

The physical meaning of each term in Eq. 4.42 is

$\vec{a}_{XYZ}$:   Absolute rectilinear acceleration of a particle relative to a fixed reference frame

$\vec{a}_{rf}$: Absolute rectilinear acceleration of a moving reference frame relative to a fixed frame

$\vec{a}_{xyz}$: Rectilinear acceleration of a particle *relative* to a moving reference frame (this acceleration would be that seen by an observer on the moving frame)

$2\vec{\omega} \times \vec{V}_{xyz}$: Coriolis acceleration due to motion of the particle *within* the moving frame

$\vec{\omega} \times (\vec{\omega} \times \vec{r})$: Centripetal acceleration due to rotation of the moving frame

$\dot{\vec{\omega}} \times \vec{r}$: Tangential acceleration due to angular acceleration of the moving reference frame

Substituting $\vec{a}_{XYZ}$, as given by Eq. 4.42, into Eq. 4.36, we obtain

$$\vec{F}_{\text{system}} = \int_{M\,(\text{system})} \left[\vec{a}_{rf} + \vec{a}_{xyz} + 2\vec{\omega} \times \vec{V}_{xyz} + \vec{\omega} \times (\vec{\omega} \times \vec{r}) + \dot{\vec{\omega}} \times \vec{r}\right] dm$$

or

$$\int_{M\,(\text{system})} \vec{a}_{xyz}\, dm = \vec{F} - \int_{M\,(\text{system})} \left[\vec{a}_{rf} + 2\vec{\omega} \times \vec{V}_{xyz} + \vec{\omega} \times (\vec{\omega} \times \vec{r}) + \dot{\vec{\omega}} \times \vec{r}\right] dm$$

(4.43a)

But

$$\int_{M\,(\text{system})} \vec{a}_{xyz}\, dm = \int_{M\,(\text{system})} \frac{d\vec{V}_{xyz}}{dt}\, dm = \frac{d}{dt} \int_{M\,(\text{system})} \vec{V}_{xyz}\, dm = \left.\frac{d\vec{P}_{xyz}}{dt}\right)_{\text{system}}$$

(4.43b)

where all time derivatives are those seen by an observer fixed in the non-inertial frame, *xyz*. Combining Eqs. 4.43a and 4.43b, we obtain

$$\left.\frac{d\vec{P}_{xyz}}{dt}\right)_{\text{system}} = \vec{F} - \int_{M\,(\text{system})} \left[\vec{a}_{rf} + 2\vec{\omega} \times \vec{V}_{xyz} + \vec{\omega} \times (\vec{\omega} \times \vec{r}) + \dot{\vec{\omega}} \times \vec{r}\right] dm$$

or

$$\left.\frac{d\vec{P}_{xyz}}{dt}\right)_{\text{system}} = \vec{F}_S + \vec{F}_B - \int_{V\,(\text{system})} \left[\vec{a}_{rf} + 2\vec{\omega} \times \vec{V}_{xyz} \right.$$
$$\left. + \vec{\omega} \times (\vec{\omega} \times \vec{r}) + \dot{\vec{\omega}} \times \vec{r}\right]\rho\, dV$$

(4.44)

Equation 4.44 is a statement of Newton's second law for a system. It is an expression for the rate of change of momentum, $\vec{P}_{xyz}$, measured relative to *xyz*, as seen by an observer in *xyz*. The system and control volume formulations are related by Eq. 4.26,

$$\left.\frac{dN}{dt}\right)_{\text{system}} = \frac{\partial}{\partial t} \int_{\text{CV}} \eta\rho\, dV + \int_{\text{CS}} \eta\rho\vec{V}_{xyz} \cdot d\vec{A}$$

(4.26)

To obtain the control volume formulation, we set $N = \vec{P}_{xyz}$ and $\eta = \vec{V}_{xyz}$. Then Eqs. 4.26 and 4.44 may be combined to give

$$\vec{F}_S + \vec{F}_B - \int_{\text{CV}} [\vec{a}_{rf} + 2\vec{\omega} \times \vec{V}_{xyz} + \vec{\omega} \times (\vec{\omega} \times \vec{r}) + \dot{\vec{\omega}} \times \vec{r}]\rho\, d\Psi$$

$$= \frac{\partial}{\partial t} \int_{\text{CV}} \vec{V}_{xyz}\rho\, d\Psi + \int_{\text{CS}} \vec{V}_{xyz}\rho \vec{V}_{xyz} \cdot d\vec{A} \quad (4.45)$$

Equation 4.45 is the most general control volume formulation of Newton's second law. Comparing the momentum equation for a control volume moving with arbitrary acceleration, Eq. 4.45, to that for a control volume moving with rectilinear acceleration, Eq. 4.34, we see that the only difference is the presence of three additional terms on the left side of Eq. 4.45. These terms result from the angular motion of the noninertial reference frame, xyz. Note that Eq. 4.45 reduces to Eq. 4.34 when the angular terms are zero, and to Eq. 4.27 for an inertial control volume.

The precautions concerning the use of Eqs. 4.27 and 4.34 also apply to the use of Eq. 4.45. Before attempting to apply this equation, one should draw the boundaries of the control volume and label appropriate coordinate directions. For a control volume moving with arbitrary acceleration, one must label a coordinate system (xyz) on the control volume and an inertial reference frame (XYZ).

Example 4.12

A reference frame, xyz, moves arbitrarily with respect to a fixed frame, XYZ. A particle moves with velocity, $\vec{V}_{xyz} = (dx/dt)\hat{\imath} + (dy/dt)\hat{\jmath} + (dz/dt)\hat{k}$, relative to frame xyz. Show that the absolute velocity of the particle is given by

$$\vec{V}_{XYZ} = \vec{V}_{rf} + \vec{V}_{xyz} + \vec{\omega} \times \vec{r}$$

Example Problem 4.12

GIVEN:

Fixed and noninertial frames as shown.

FIND:

$\vec{V}_{XYZ}$ in terms of $\vec{V}_{xyz}$, $\vec{\omega}$, $\vec{r}$, and $\vec{V}_{rf}$.

SOLUTION:

From the geometry of the sketch, $\vec{X} = \vec{R} + \vec{r}$, so

$$\vec{V}_{XYZ} = \frac{d\vec{X}}{dt} = \frac{d\vec{R}}{dt} + \frac{d\vec{r}}{dt} = \vec{V}_{rf} + \frac{d\vec{r}}{dt}$$

Since

$$\vec{r} = x\hat{\imath} + y\hat{\jmath} + z\hat{k}$$

then

$$\frac{d\vec{r}}{dt} = \frac{dx}{dt}\hat{\imath} + \frac{dy}{dt}\hat{\jmath} + \frac{dz}{dt}\hat{k} + x\frac{d\hat{\imath}}{dt} + y\frac{d\hat{\jmath}}{dt} + z\frac{d\hat{k}}{dt}$$

or

$$\frac{d\vec{r}}{dt} = \vec{V}_{xyz} + x\frac{d\hat{\imath}}{dt} + y\frac{d\hat{\jmath}}{dt} + z\frac{d\hat{k}}{dt}$$

The problem now is to evaluate $d\hat{\imath}/dt$, $d\hat{\jmath}/dt$, and $d\hat{k}/dt$ due to the angular motion of frame xyz. To evaluate these derivatives, we must consider the rotation of each unit vector due to the three components of the angular velocity, $\vec{\omega}$, of frame xyz.

Consider the unit vector $\hat{\imath}$. It will rotate in the xy plane due to ω_z, as follows:

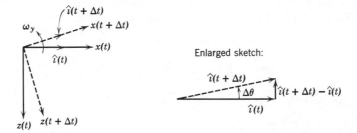

Now

$$\frac{d\hat{\imath}}{dt}\bigg]_{\text{due to } \omega_z} = \lim_{\Delta t \to 0}\left[\frac{\hat{\imath}(t + \Delta t) - \hat{\imath}(t)}{\Delta t}\right] = \lim_{\Delta t \to 0}\left[\frac{(1)\,\Delta\theta\hat{\jmath}}{\Delta t}\right] = \lim_{\Delta t \to 0}\left[\frac{(1)\omega_z\Delta t\hat{\jmath}}{\Delta t}\right]$$

$$\frac{d\hat{\imath}}{dt}\bigg]_{\text{due to } \omega_z} = \hat{\jmath}\omega_z$$

Similarly, $\hat{\imath}$ will rotate in the xz plane due to ω_y

Then

$$\frac{d\hat{\imath}}{dt}\bigg]_{\text{due to }\omega_y} = \lim_{\Delta t \to 0} \left[\frac{\hat{\imath}(t + \Delta t) - \hat{\imath}(t)}{\Delta t}\right] = \lim_{\Delta t \to 0} \left[\frac{(1)\,\Delta\theta(-\hat{k})}{\Delta t}\right] = \lim_{\Delta t \to 0} \left[\frac{(1)\omega_y\,\Delta t(-\hat{k})}{\Delta t}\right]$$

$$\frac{d\hat{\imath}}{dt}\bigg]_{\text{due to }\omega_y} = -\hat{k}\omega_y$$

Rotation in the yz plane due to ω_x does not affect $\hat{\imath}$. Combining terms,

$$\frac{d\hat{\imath}}{dt} = \omega_z\hat{\jmath} - \omega_y\hat{k}$$

By similar reasoning,

$$\frac{d\hat{\jmath}}{dt} = \omega_x\hat{k} - \omega_z\hat{\imath}$$

and

$$\frac{d\hat{k}}{dt} = \omega_y\hat{\imath} - \omega_x\hat{\jmath}$$

Thus

$$x\frac{d\hat{\imath}}{dt} + y\frac{d\hat{\jmath}}{dt} + z\frac{d\hat{k}}{dt} = (z\omega_y - y\omega_z)\hat{\imath} + (x\omega_z - z\omega_x)\hat{\jmath} + (y\omega_x - x\omega_y)\hat{k}$$

But

$$\vec{\omega} \times \vec{r} = \begin{vmatrix} \hat{\imath} & \hat{\jmath} & \hat{k} \\ \omega_x & \omega_y & \omega_z \\ x & y & z \end{vmatrix} = (z\omega_y - y\omega_z)\hat{\imath} + (x\omega_z - z\omega_x)\hat{\jmath} + (y\omega_x - x\omega_y)\hat{k}$$

Combining these results, we obtain

$$\vec{V}_{XYZ} = \vec{V}_{rf} + \vec{V}_{xyz} + \vec{\omega} \times \vec{r} \qquad\qquad \vec{V}_{XYZ}$$

Example 4.13

A small "pod" is propelled by a jet. The pod is mounted on a strut and rotates about a vertical axis at a radius of 1 m from a fixed center. The pod has a mass of 2 kg; the jet velocity, density, and area are 200 m/sec, 1.5 kg/m³ and 75 mm², respectively. At a particular instant, the pod moves at a speed of 50 m/sec, and the air drag is 10 N. Friction and the mass of the strut may be neglected. Determine the angular acceleration of the pod at the given instant.

Example Problem 4.13

GIVEN:

System shown in sketch.

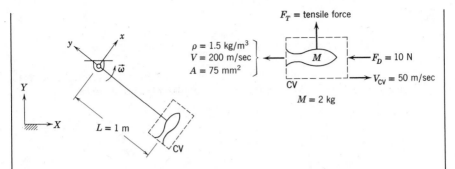

FIND:

$\dot{\vec{\omega}}$

SOLUTION:

Basic equation:
$$\vec{F}_S + \vec{F}_B - \int_{CV} [\overset{=\,0(3)}{\cancel{\vec{a}_{rf}}} + 2\vec{\omega} \times \overset{\simeq\,0(4)}{\cancel{\vec{V}_{xyz}}} + \vec{\omega} \times (\vec{\omega} \times \vec{r}) + \dot{\vec{\omega}} \times \vec{r}]\rho\,d\forall$$

$$= \frac{\partial}{\partial t} \int_{CV} \overset{\simeq\,0(4)}{\cancel{\vec{V}_{xyz}}}\rho\,d\forall + \int_{CS} \vec{V}_{xyz}\rho\vec{V}_{xyz} \cdot d\vec{A}$$

Assumptions: (1) $\vec{F}_S = -F_D\hat{\imath} + F_T\hat{\jmath} + F_V\hat{k}$
(2) $\vec{F}_B = -Mg\hat{k}$
(3) $\vec{a}_{rf} = 0$
(4) Neglect $\vec{V}_{xyz}$ within CV
(5) Uniform flow from CV

Then

$$-F_D\hat{\imath} + F_T\hat{\jmath} + F_V\hat{k} - Mg\hat{k} - \int_{CV} [\vec{\omega} \times (\vec{\omega} \times \vec{r}) + \dot{\vec{\omega}} \times \vec{r}]\rho\,d\forall = -V\hat{\imath}\{|\rho VA|\}$$

At the instant shown,

$$\vec{\omega} = \frac{V_{CV}}{r}\hat{k} = \frac{50 \text{ m}}{\text{sec}} \times \frac{1}{1 \text{ m}}\hat{k} = 50 \text{ sec}^{-1}\,\hat{k}$$

$$\vec{r} = -1 \text{ m }\hat{\jmath}$$

$$\vec{\omega} \times \vec{r} = \frac{50}{\text{sec}}\hat{k} \times (-1 \text{ m }\hat{\jmath}) = 50\hat{\imath} \text{ m/sec}$$

$$\vec{\omega} \times (\vec{\omega} \times \vec{r}) = \frac{50}{\text{sec}}\hat{k} \times \frac{50 \text{ m}}{\text{sec}}\hat{\imath} = 2500\hat{\jmath} \text{ m/sec}^2$$

$$\dot{\vec{\omega}} \times \vec{r} = \dot{\omega}\hat{k} \times (-L\hat{\jmath}) = \dot{\omega}L\hat{\imath}$$

Substituting,

$$-F_D\hat{i} + F_T\hat{j} + F_V\hat{k} - Mg\hat{k} - \left(2500 \, \frac{m}{sec^2}\hat{j} + \dot\omega L\hat{i}\right)M = -\rho V^2 A\hat{i}$$

This vector equation can be written as three scalar component equations. We can solve explicitly for $\dot\omega$ by considering the coefficients of unit vector $\hat{i}$. Thus

$$-F_D - \dot\omega LM = -\rho V^2 A$$

or

$$\dot\omega = \frac{\rho V^2 A}{LM} - \frac{F_D}{LM}$$

$$= \frac{1.5 \, kg}{m^3} \times \frac{(200)^2 \, m^2}{sec^2} \times 75 \, mm^2 \times \frac{1}{1 \, m} \times \frac{1}{2 \, kg} \times \frac{m^2}{10^6 \, mm^2}$$

$$-10 \, N \times \frac{1}{1 \, m} \times \frac{1}{2 \, kg} \times \frac{kg \cdot m}{N \cdot sec^2}$$

$$\dot\omega = -2.75\hat{k} \, rad/sec^2$$

Thus $\dot{\vec\omega} = \dot\omega\hat{k} = -2.75 \, \hat{k} \, rad/sec^2$ $\qquad\qquad\qquad\qquad\qquad\qquad\qquad \underset{\dot{\vec\omega}}{\longleftarrow}$

The pod is decelerating, since $\dot\omega < 0$.

The tensile force can be obtained by considering the coefficients of unit vector $\hat{j}$. Thus

$$F_T - \frac{2500 \, m}{sec^2} \times M = 0$$

or

$$F_T = \frac{2500 \, m}{sec^2} \times 2 \, kg \times \frac{N \cdot sec^2}{kg \cdot m} = 5.00 \, kN$$

The remaining component equation gives

$$F_V = Mg = 2 \, kg \times \frac{9.81 \, m}{sec^2} \times \frac{N \cdot sec^2}{kg \cdot m} = 19.6 \, N$$

**4-7 MOMENT OF MOMENTUM

Next we wish to develop the moment of momentum equation for application to a control volume. We shall begin with the mathematical statement for a system and then use Eq. 4.26 to obtain equations for use with fixed and rotating control volumes.

** This section may be omitted without loss of continuity in the text material.

4-7.1 EQUATION FOR FIXED CONTROL VOLUME

The moment of momentum equation for a system was given by

$$\vec{T} = \frac{d\vec{H}}{dt}\Bigg)_{\text{system}} \tag{4.3a}$$

where $\vec{T}$ = total torque exerted on the system by its surroundings, and
$\vec{H}$ = angular momentum of the system.

$$\vec{H} = \int_{M \text{ (system)}} \vec{r} \times \vec{V} \, dm = \int_{\Psi \text{ (system)}} \vec{r} \times \vec{V} \rho \, d\Psi \tag{4.3b}$$

As for all system equations, all quantities in Eq. 4.3b must be formulated with respect to an inertial reference frame. Reference frames at rest or moving with constant velocity are inertial, and Eq. 4.3b can be used directly to develop the control volume form of the moment of momentum equation. (Rotating reference frames are noninertial and will be treated in Section 4-7.3.)

The position vector, $\vec{r}$, locates each mass or volume element of the system with respect to the coordinate system. The torque, $\vec{T}$, applied to a system may be written

$$\vec{T} = \vec{r} \times \vec{F}_s + \int_{M \text{ (system)}} \vec{r} \times \vec{g} \, dm + \vec{T}_{\text{shaft}} \tag{4.3c}$$

The relation between the system and control volume formulations is given by

$$\frac{dN}{dt}\Bigg)_{\text{system}} = \frac{\partial}{\partial t} \int_{CV} \eta \rho \, d\Psi + \int_{CS} \eta \rho \vec{V} \cdot d\vec{A} \tag{4.11}$$

where

$$N_{\text{system}} = \int_{M \text{ (system)}} \eta \, dm$$

If we set $N = \vec{H}$ and $\eta = \vec{r} \times \vec{V}$, then

$$\frac{d\vec{H}}{dt}\Bigg)_{\text{system}} = \frac{\partial}{\partial t} \int_{CV} \vec{r} \times \vec{V} \rho \, d\Psi + \int_{CS} \vec{r} \times \vec{V} \rho \vec{V} \cdot d\vec{A} \tag{4.46}$$

Combining Eqs. 4.3a, 4.3c, and 4.46, we obtain

$$\vec{r} \times \vec{F}_s + \int_{M \text{ (system)}} \vec{r} \times \vec{g} \, dm + \vec{T}_{\text{shaft}} = \frac{\partial}{\partial t} \int_{CV} \vec{r} \times \vec{V} \rho \, d\Psi + \int_{CS} \vec{r} \times \vec{V} \rho \vec{V} \cdot d\vec{A}$$

Since the system and control volume coincide at time t_0,

$$\vec{T}_{\text{system}} = \vec{T}_{CV}$$

and

$$\vec{r} \times \vec{F}_s + \int_{CV} \vec{r} \times \vec{g}\rho\, d\mathbf{V} + \vec{T}_{\text{shaft}} = \frac{\partial}{\partial t} \int_{CV} \vec{r} \times \vec{V}\rho\, d\mathbf{V} + \int_{CS} \vec{r} \times \vec{V}\rho\vec{V} \cdot d\vec{A}$$

$$(4.47)$$

Equation 4.47 is a general vector equation for the moment of momentum for an inertial control volume. The left side of the equation is an expression for all the torques that act on the control volume. The right side consists of terms that express the rate of change of angular momentum within the control volume and the rate of efflux of angular momentum from the control volume. All velocities, $\vec{V}$, in Eq. 4.47 are measured relative to the fixed control volume.

For analysis of rotating machinery, Eq. 4.47 is often used in scalar form by considering only the component of the equation that is directed along the axis of rotation. This application is illustrated in the next section.

4-7.2 APPLICATION TO TURBOMACHINERY

Fluid handling devices that direct the flow with blades or vanes attached to a rotating member are termed *turbomachines*. In contrast to positive displacement machinery, the fluid in a turbomachine is never confined. All work interactions between a fluid and a turbomachine rotor result from the dynamic effects of the rotor on the fluid stream.

Turbines extract energy from a fluid stream. The assembly of blades attached to the turbine shaft is called the *wheel* or *runner*. The two most general classifications of turbines are *impulse* and *reaction* turbines. Impulse turbines are driven by one or more high speed free jets. Each jet is accelerated to high speed in a nozzle external to the turbine wheel. If friction and gravity are neglected, neither the fluid pressure nor its speed relative to the runner change during its passage over the turbine vanes. Thus for an impulse turbine the fluid expansion from high to low pressure takes place in nozzles external to the blades, and the runner does not flow full.

In reaction turbines, part of the fluid expansion takes place externally and part within the moving blades. The external acceleration takes place and the flow is turned to enter the runner in the proper direction as it passes through a set of stationary blades called *guide vanes*. Because additional fluid acceleration relative to the rotor occurs within the moving blades, both the relative velocity and pressure of the stream change across the runner. The combination of a stationary blade row and a moving blade row is called a turbine *stage.* Reaction turbine runners may be designed to flow full of fluid; as a consequence, reaction turbines can produce more power for a given volume than impulse turbines.

Turbines range from simple devices such as the windmill to complex steam and gas turbines with many stages of carefully designed blading. All of these devices can be analyzed in idealized form by applying Eq. 4.47.

Prime movers, which add energy to a fluid stream, are called pumps when the flow is liquid (or slurry), and fans, blowers, or compressors for gas and vapor handling units, depending on pressure rise. (Flow through fans is essentially incompressible; blowers have a small pressure rise; compressors are units designed for a larger pressure rise.) Pumps and turboblowers can also be analyzed in idealized form by applying Eq. 4.47.

Flow through a turbomachine may be nearly axial, nearly radial, or a combination of the two, referred to as *mixed flow.* The basic design for a given application is usually chosen on the basis of efficiency. In general, axial machines handle the largest flows at high efficiency, followed by mixed flow and radial flow units. For pumps and blowers, radial flow units usually are designed for high pressure rise, followed by mixed flow and axial units for lower pressure rise applications.

For turbomachinery analysis it is convenient in general to choose a fixed control volume enclosing the rotating member, the rotor or impeller, for analysis of torque reactions. Torques due to surface forces may be ignored as a first approximation. The body force contribution may be neglected by symmetry. Then for steady flow, Eq. 4.47 becomes

$$\vec{T}_{shaft} = \int_{CS} \vec{r} \times \vec{V} \rho \vec{V}_{xyz} \cdot d\vec{A} \tag{4.48a}$$

Let us now write this equation in scalar form and then illustrate its application to axial and radial flow machines.

The most practical coordinate system is one chosen with the z axis aligned with the axis of rotation of the machine. The term on the right side of Eq. 4.48a is the product of $\vec{r} \times \vec{V}$ with the mass flowrate at each section. For uniform flow into the rotor at section ① and out of the rotor at section ②, Eq. 4.48a becomes

$$T_{shaft}\hat{k} = (r_2 V_{t_2} - r_1 V_{t_1})\dot{m}\hat{k} \tag{4.48b}$$

or finally in scalar form,

$$T_{shaft} = (r_2 V_{t_2} - r_1 V_{t_1})\dot{m} \tag{4.48c}$$

The velocities that appear in Eq. 4.48c are the tangential components of the absolute velocity of the fluid crossing the control surface. Equation 4.48c may be used to calculate the torque that must be applied to the control volume to cause the change in angular momentum of the flowing fluid. Equation 4.48c is the basic relationship between torque and moment of momentum for all forms of turbomachines, including turbines and prime movers. It is often called the *Euler turbine equation.*

The rate of work done on a turbomachine rotor is given by the dot product of the rotor angular velocity, $\vec{\omega}$, and the applied torque, $\vec{T}_{\text{shaft}}$. Using Eq. 4.48b,

$$\dot{W}_{\text{in}} = \vec{\omega} \cdot \vec{T}_{\text{shaft}} = \omega \hat{k} \cdot T_{\text{shaft}} \hat{k}$$

or

$$\dot{W}_{\text{in}} = \omega T_{\text{shaft}} = \omega (r_2 V_{t_2} - r_1 V_{t_1}) \dot{m} \qquad (4.49a)$$

According to Eq. 4.49a, the moment of momentum of the fluid is increased by addition of shaft work (as in a prime mover). For a turbine, $\dot{W}_{\text{in}} < 0$ and the moment of momentum of the fluid must decrease.

Equation 4.49a may be written in two other useful forms. Introducing $U = r\omega$, the linear speed of the rotor at radius r, then

$$\dot{W}_{\text{in}} = (U_2 V_{t_2} - U_1 V_{t_1}) \dot{m} \qquad (4.49b)$$

Dividing Eq. 4.49b by $\dot{m}g$, we obtain a quantity with dimensions of length, often termed the *head* added to the flow,

$$\Delta h = \frac{\dot{W}_{\text{in}}}{\dot{m}g} = \frac{1}{g}(U_2 V_{t_2} - U_1 V_{t_1}) \qquad (4.49c)$$

Equations 4.48 and 4.49 are simplified forms of the moment of momentum equation for a control volume. They are all written for a fixed control volume under the assumptions of steady flow and uniform flow at each section. The equations show that only the difference in the product rV_t or UV_t between the outlet and inlet sections is important in determining the torque applied to the rotor or the energy transfer to the fluid. No restriction has been made on geometry; the fluid may enter and leave at different radii.

The equations that we have derived also suggest the importance of clearly defining the velocity components of the fluid and rotor at inlet and outlet sections. For this purpose it is useful to develop *velocity polygons* for the inlet and outlet flows. Figure 4.6 shows two such velocity polygons and introduces the notation to be used for blade and flow angles.

Each point on the blade of a turbomachine follows a circular path. Velocity polygons are plotted in a plane formed by unwrapping this circular blade path onto a plane surface. In the idealized situation at the design point, the flow relative to the rotor is assumed to enter and leave tangent to the blade profile at each section. This idealized inlet condition is sometimes called *shockless* entry flow. Blade angles, β, are measured relative to the circumferential direction, as shown in Fig. 4.6a. Knowledge of the blade angle, β_1, fixes the direction of the relative inlet velocity at design conditions.

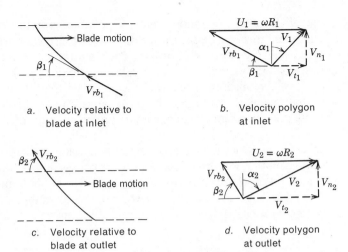

a. Velocity relative to blade at inlet

b. Velocity polygon at inlet

c. Velocity relative to blade at outlet

d. Velocity polygon at outlet

Fig. 4.6 Geometry and notation used to develop velocity polygons for a turbomachine.

The runner speed at the inlet is $U_1 = \omega R_1$, and is therefore specified by the impeller geometry and operating speed. The absolute fluid velocity is the vector sum of the impeller velocity and the velocity of the flow relative to the blade. The absolute velocity may be determined graphically as shown in Fig. 4.6b. The angle of the absolute fluid velocity, α_1, is measured from the normal as shown. The tangential component of the absolute velocity, V_{t_1}, and the component normal to the flow area, V_{n_1}, are also shown in Fig. 4.6b. Note from the geometry of Fig. 4.6b that at each section the normal components of the *absolute* velocity, V, and the velocity relative to the blade, V_{rb}, are equal.

Velocity polygons at the outlet section are constructed in a similar fashion. The runner speed at the outlet is $U_2 = \omega R_2$, which again is known from the geometry and operating conditions of the turbomachine. The relative flow is again assumed to leave the impeller tangent to the blades, as shown in Fig. 4.6c. This idealizing assumption fixes its direction.

For an impulse turbine, there is no acceleration of the flow relative to the blade. Therefore, the magnitudes of its outlet and inlet velocities are equal. For a reaction turbine or pump, the velocity relative to the blade in general changes in magnitude. The continuity equation must be applied, together with the impeller geometry, to determine the normal component of the velocity. This value, together with the outlet blade angle, is sufficient to establish the velocity relative to the blade at the impeller outlet. The velocity polygon is completed by adding the velocity relative to the blade and the wheel velocity vectorially as shown in Fig. 4.6d.

Knowledge of the inlet and outlet velocity polygons provides all the information required to calculate the torque or power absorbed or delivered by the impeller using any of Eqs. 4.48 or 4.49. The resulting values represent the performance of a turbomachine under idealized conditions at the design operating point of the machine, since we have assumed that all flows are uniform and that they enter and leave the rotor tangent to the blades. The idealized results thus represent the upper limits to performance of a turbomachine.

Performance of an actual machine is determined using the same basic approach but accounting for variations in flow properties across the blade span at inlet and outlet sections, and for deviations between blade and flow directions. Such detailed calculations are beyond the scope of this text. The idealized results that we have developed are applied next to axial and radial flow machines.

Example 4.14

An axial flow fan operates at 1200 rpm. The blade tip diameter is 1.1 m and the hub diameter 0.8 m. The blade inlet and exit angles are 30 and 60°, respectively. Inlet guide vanes give the absolute flow entering the first stage an angle of 30°. The fluid is air at standard conditions, and the flow may be considered as incompressible. There is no change in the axial component of velocity across the rotor. The relative flow may be assumed to enter and leave the rotor at the geometric blade angles, and properties at the mean blade diameter may be used for calculations. For these idealized conditions, draw the inlet velocity polygon, determine the volumetric flow rate of the fan, and sketch the rotor blade shapes. Using the data so obtained, draw the outlet velocity polygon, and calculate the torque and power required to drive the fan.

Example Problem 4.14

GIVEN:

Flow through rotor of axial flow fan.

Tip diameter: 1.1 m
Hub diameter: 0.8 m
Operating speed: 1200 rpm
Absolute inlet angle: 30°
Blade inlet angle: 30°
Blade outlet angle: 60°

Fluid is air at standard conditions.
Use properties at mean diameter of blades.

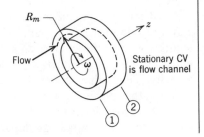

FIND:

(a) Inlet velocity polygon.
(b) Volumetric flowrate.
(c) Rotor blade shape.
(d) Outlet velocity polygon.
(e) Rotor torque.
(f) Power required.

SOLUTION:

Apply the moment of momentum equation to a fixed control volume.

Computing equations:
$$\vec{T}_{shaft} = \int_{CS} \vec{r} \times \vec{V} \rho \vec{V} \cdot d\vec{A} \tag{4.48a}$$

$$0 = \overbrace{\frac{\partial}{\partial t} \int_{CV} \rho \, d\forall}^{= 0(2)} + \int_{CS} \rho \vec{V} \cdot d\vec{A}$$

Assumptions: (1) Neglect torques due to body or surface forces
(2) Steady flow
(3) Uniform flow at inlet and outlet sections
(4) Incompressible flow
(5) No change in axial flow area
(6) Use mean radius of rotor blades

The inlet velocity polygon is

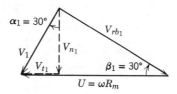

The blade shapes are

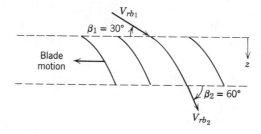

From continuity

$$0 = \{-|\rho V_{n_1} A_1|\} + \{|\rho V_{n_2} A_2|\}$$

or

$$Q = V_{n_1} A_1 = V_{n_2} A_2$$

Since $A_1 = A_2$, then $V_{n_1} = V_{n_2}$, and the outlet velocity polygon is as shown in the following figure:

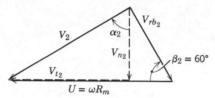

At the mean blade radius

$$U = R_m \omega = \frac{D_m}{2} \omega$$

$$U = \frac{\frac{1}{2}(1.1 + 0.8)\ m}{2} \times \frac{1200\ rev}{min} \times \frac{2\pi\ rad}{rev} \times \frac{min}{60\ sec} = 59.7\ m/sec$$

From the geometry of the inlet velocity polygon,

$$U = V_{n_1}(\tan \alpha_1 + \cot \beta_1)$$

so that

$$V_{n_1} = \frac{U}{\tan \alpha_1 + \cot \beta_1} = \frac{59.7}{sec}\frac{m}{} \times \frac{1}{\tan 30° + \cot 30°} = 25.9\ m/sec$$

Consequently,

$$V_1 = \frac{V_{n_1}}{\cos \alpha_1} = \frac{25.9}{sec}\frac{m}{} \times \frac{1}{\cos 30°} = 29.9\ m/sec$$

$$V_{t_1} = V_1 \sin \alpha_1 = \frac{29.9}{sec}\frac{m}{} \times \sin 30° = 15.0\ m/sec$$

and

$$V_{rb_1} = \frac{V_{n_1}}{\sin \beta_1} = \frac{25.9}{sec}\frac{m}{} \times \frac{1}{\sin 30°} = 51.8\ m/sec$$

The volumetric flowrate is

$$Q = V_{n_1} A_1 = \frac{\pi}{4} V_{n_1}(D_t^2 - D_h^2) = \frac{\pi}{4} \times \frac{25.9}{sec}\frac{m}{} [(1.1)^2 - (0.8)^2]\ m^2$$

$$Q = 11.6\ m^3/sec \qquad\qquad\qquad\qquad\qquad\qquad Q$$

From the geometry of the outlet velocity polygon,

$$\tan \alpha_2 = \frac{V_{t_2}}{V_{n_2}} = \frac{U - V_{n_2} \cot \beta_2}{V_{n_2}} = \frac{U - V_{n_1} \cot \beta_2}{V_{n_1}}$$

or

$$\alpha_2 = \tan^{-1} \left[\frac{59.7 \dfrac{m}{sec} - 25.9 \dfrac{m}{sec} \times \cot 60°}{25.9 \dfrac{m}{sec}} \right] = 60.0°$$

and

$$V_2 = \frac{V_{n_2}}{\cos \alpha_2} = \frac{V_{n_1}}{\cos \alpha_2} = 25.9 \frac{m}{sec} \times \frac{1}{\cos 60.0°} = 51.8 \text{ m/sec}$$

Finally,

$$V_{t_2} = V_2 \sin \alpha_2 = 51.8 \frac{m}{sec} \times \sin 60.0° = 44.8 \text{ m/sec}$$

The moment of momentum equation becomes

$$\vec{T} = T_z \hat{k} = \int_{CS} \vec{R}_m \times \vec{V} \rho \vec{V} \cdot d\vec{A} = \hat{k} \int_{CS} R_m V_t \rho \vec{V} \cdot d\vec{A}$$

so that for uniform flow

$$T_z = R_m V_{t_1}\{-|\rho V_{n_1} A_1|\} + R_m V_{t_2}\{|\rho V_{n_2} A_2|\} = \rho Q R_m (V_{t_2} - V_{t_1})$$

$$= \frac{1.23 \text{ kg}}{m^3} \times \frac{11.6 \text{ m}^3}{sec} \times \frac{0.95 \text{ m} (44.8 - 15.0)}{2} \frac{m}{sec} \times \frac{N \cdot sec^2}{kg \cdot m}$$

$$T_z = 202 \text{ N} \cdot \text{m} \overset{\longleftarrow}{} \qquad\qquad T_z$$

Thus the torque *on* the CV is in the same sense as $\vec{\omega}$. The power required is

$$\dot{W}_{in} = \vec{\omega} \cdot \vec{T} = \omega_z T_z = \frac{1200 \text{ rev}}{min} \times \frac{2\pi \text{ rad}}{rev} \times \frac{min}{60 \text{ sec}} \times \frac{202 \text{ N} \cdot \text{m}}{} \times \frac{W \cdot sec}{N \cdot m}$$

$$\dot{W}_{in} = 25.4 \text{ kW} \overset{\longleftarrow}{} \qquad\qquad \dot{W}_{in}$$

$\left\{\text{This problem illustrates the construction of velocity polygons, and the application of the moment of momentum equation for a fixed control volume to an axial flow machine under idealized conditions.}\right\}$

Example 4.15

Water at 150 gal/min enters a mixed flow pump impeller axially. The inlet velocity is axial and uniform. The outlet diameter of the impeller is 4 in. Flow leaves the impeller at a velocity of 10 ft/sec relative to the radial blades. The impeller speed is 3450 rpm. Determine the impeller exit width, *b*, the torque input to the impeller, and the horsepower supplied.

Example Problem 4.15

GIVEN:

Flow as shown in the following figure.

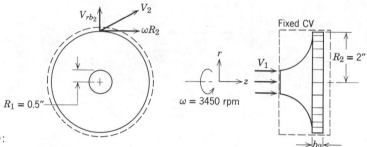

FIND:

(a) b_2.
(b) T_{shaft}.
(c) $\dot{W}_{in}$.

SOLUTION:

Apply the moment of momentum equation to a fixed control volume.

Computing equations:

$$\vec{T}_{\text{shaft}} = \int_{CS} \vec{r} \times \vec{V} \rho \vec{V} \cdot d\vec{A} \qquad (4.48a)$$

$$0 = \overbrace{\frac{\partial}{\partial t} \int_{CV} \rho \, d\forall}^{= 0(2)} + \int_{CS} \rho \vec{V} \cdot d\vec{A}$$

Assumptions:
(1) Neglect torques due to body or surface forces
(2) Steady flow
(3) Uniform flow at inlet and outlet sections
(4) Incompressible flow

Then, from continuity,

$$0 = \left\{ -\left| \rho V_1 \pi R_1^2 \right| \right\} + \left\{ \left| \rho V_{rb_2} 2\pi R_2 b_2 \right| \right\}$$

or

$$\rho Q = \rho V_{rb_2} 2\pi R_2 b_2$$

so that

$$b_2 = \frac{Q}{2\pi R_2 V_{rb_2}} = \frac{1}{2\pi} \times \frac{150 \text{ gal}}{\text{min}} \times \frac{1}{2 \text{ in.}} \times \frac{\text{sec}}{10 \text{ ft}} \times \frac{\text{ft}^3}{7.48 \text{ gal}} \times \frac{\text{min}}{60 \text{ sec}} \times \frac{12 \text{ in.}}{\text{ft}}$$

$$b_2 = 0.0319 \text{ ft or } 0.383 \text{ in.} \qquad\qquad\qquad b_2$$

The axial inlet flow has no z component of moment of momentum.

From the moment of momentum equation with uniform exit flow,

$$\hat{i}_z T_{\text{shaft}} = \vec{r}_2 \times \vec{V}_2\{|\rho Q|\}$$

At section ②,

$$\vec{r}_2 = R_2 \hat{i}_r$$
$$\vec{V}_2 = V_{rb_2} \hat{i}_r + \omega R_2 \hat{i}_\theta$$

so

$$\vec{r}_2 \times \vec{V}_2 = R_2(\omega R_2)\hat{i}_z = \omega R_2^2 \hat{i}_z$$

Thus

$$T_{\text{shaft}} = \omega R_2^2 \rho Q = \frac{3450 \text{ rev}}{\text{min}} \times \frac{(2)^2 \text{ in.}^2}{} \times \frac{1.94 \text{ slug}}{\text{ft}^3} \times \frac{150 \text{ gal}}{\text{min}}$$

$$\times \frac{2\pi \text{ rad}}{\text{rev}} \times \frac{\text{min}^2}{3600 \text{ sec}^2} \times \frac{\text{ft}^3}{7.48 \text{ gal}} \times \frac{\text{ft}^2}{144 \text{ in.}^2} \times \frac{\text{lbf} \cdot \text{sec}^2}{\text{slug} \cdot \text{ft}}$$

$$\underline{T_{\text{shaft}} = 6.51 \text{ ft} \cdot \text{lbf}} \qquad\qquad\qquad\qquad\qquad T_{\text{shaft}}$$

and

$$\dot{W}_{in} = \omega T_{\text{shaft}} = \frac{3450 \text{ rev}}{\text{min}} \times \frac{6.51 \text{ ft} \cdot \text{lbf}}{} \times \frac{2\pi \text{ rad}}{\text{rev}} \times \frac{\text{min}}{60 \text{ sec}} \times \frac{\text{hp} \cdot \text{sec}}{550 \text{ ft} \cdot \text{lbf}}$$

$$\underline{\dot{W}_{in} = 4.28 \text{ hp}} \qquad\qquad\qquad\qquad\qquad\qquad\qquad \dot{W}_{in}$$

4-7.3 EQUATION FOR ROTATING CONTROL VOLUME

In some problems, such as those involving impulse turbines or rotating spray systems, it is most convenient to express all fluid velocities relative to the rotating component. The most convenient control volume to use is a non-inertial one that rotates with the component. Our objective in this section is to develop such a formulation.

Inertial and noninertial reference frames were related in Section 4-6. Figure 4.5, p. 163, showed the notation used. For a system,

$$\vec{T}_{\text{system}} = \frac{d\vec{H}}{dt}\bigg)_{\text{system}} \tag{4.3a}$$

The angular momentum of a system in general motion must be specified relative to an inertial reference frame, so using the notation of Fig. 4.5,

$$\vec{H}_{\text{system}} = \int_{M \text{ (system)}} (\vec{R} + \vec{r}) \times \vec{V}_{XYZ} \, dm = \int_{\Psi \text{ (system)}} (\vec{R} + \vec{r}) \times \vec{V}_{XYZ} \rho \, d\Psi$$

With $\vec{R} = 0$, then the xyz frame is restricted to rotation about XYZ, and the equation becomes

$$\vec{H}_{\text{system}} = \int_{M \text{ (system)}} \vec{r} \times \vec{V}_{XYZ} \, dm = \int_{\Psi \text{ (system)}} \vec{r} \times \vec{V}_{XYZ} \rho \, d\Psi$$

so that

$$\vec{T}_{\text{system}} = \frac{d}{dt} \int_{M \text{ (system)}} \vec{r} \times \vec{V}_{XYZ} \, dm$$

Since the mass of a system is constant,

$$\vec{T}_{\text{system}} = \int_{M \text{ (system)}} \frac{d}{dt} (\vec{r} \times \vec{V}_{XYZ}) \, dm$$

or

$$\vec{T}_{\text{system}} = \int_{M \text{ (system)}} \left(\frac{d\vec{r}}{dt} \times \vec{V}_{XYZ} + \vec{r} \times \frac{d\vec{V}_{XYZ}}{dt} \right) dm \tag{4.50}$$

From the analysis of Section 4-6,

$$\vec{V}_{XYZ} = \vec{V}_{rf} + \frac{d\vec{r}}{dt} \tag{4.37}$$

With xyz restricted to pure rotation, $\vec{V}_{rf} = 0$. The first term under the integral on the right side of Eq. 4.50 is then

$$\frac{d\vec{r}}{dt} \times \frac{d\vec{r}}{dt} = 0$$

Thus Eq. 4.50 reduces to

$$\vec{T}_{\text{system}} = \int_{M \text{ (system)}} \vec{r} \times \frac{d\vec{V}_{XYZ}}{dt} \, dm = \int_{M \text{ (system)}} \vec{r} \times \vec{a}_{XYZ} \, dm \tag{4.51}$$

From Eq. 4.42 with $\vec{a}_{rf} = 0$ (since xyz does not translate),

$$\vec{a}_{XYZ} = \vec{a}_{xyz} + 2\vec{\omega} \times \vec{V}_{xyz} + \vec{\omega} \times (\vec{\omega} \times \vec{r}) + \dot{\vec{\omega}} \times \vec{r}$$

Substituting into Eq. 4.51,

$$\vec{T}_{\text{system}} = \int_{M \text{ (system)}} \vec{r} \times [\vec{a}_{xyz} + 2\vec{\omega} \times \vec{V}_{xyz} + \vec{\omega} \times (\vec{\omega} \times \vec{r}) + \dot{\vec{\omega}} \times \vec{r}] \, dm$$

or

$$\vec{T}_{\text{system}} - \int_{M \text{ (system)}} \vec{r} \times [2\vec{\omega} \times \vec{V}_{xyz} + \vec{\omega} \times (\vec{\omega} \times \vec{r}) + \dot{\vec{\omega}} \times \vec{r}] \, dm$$

$$= \int_{M \text{ (system)}} \vec{r} \times \vec{a}_{xyz} \, dm = \int_{M \text{ (system)}} \vec{r} \times \frac{d\vec{V}_{xyz}}{dt} \, dm \tag{4.52}$$

Now, using the time rate of change as observed from the system, we can write the last term as

$$\int_{M\,(system)} \vec{r} \times \frac{d\vec{V}_{xyz}}{dt}\, dm = \frac{d}{dt}\int_{M\,(system)} \vec{r} \times \vec{V}_{xyz}\, dm = \frac{d\vec{H}_{xyz}}{dt}\bigg)_{system} \qquad (4.53)$$

The torque on the system is given by

$$\vec{T}_{system} = \vec{r} \times \vec{F}_S + \int_{M\,(system)} \vec{r} \times \vec{g}\, dm + \vec{T}_{shaft} \qquad (4.3c)$$

The relation between the system and control volume formulations is

$$\frac{dN}{dt}\bigg)_{system} = \frac{\partial}{\partial t}\int_{CV} \eta\rho\, d\mathbb{V} + \int_{CS} \eta\rho\vec{V}_{xyz} \cdot d\vec{A} \qquad (4.26)$$

where

$$N_{system} = \int_{M\,(system)} \eta\, dm$$

Setting N equal to $\vec{H}_{xyz})_{system}$ and $\eta = \vec{r} \times \vec{V}_{xyz}$, then

$$\frac{d\vec{H}_{xyz}}{dt}\bigg)_{system} = \frac{\partial}{\partial t}\int_{CV} \vec{r} \times \vec{V}_{xyz}\rho\, d\mathbb{V} + \int_{CS} \vec{r} \times \vec{V}_{xyz}\rho\vec{V}_{xyz} \cdot d\vec{A} \qquad (4.54)$$

Combining Eqs. 4.52, 4.53 and 4.54, we obtain

$$\vec{r} \times \vec{F}_S + \int_{M\,(system)} \vec{r} \times \vec{g}\, dm + \vec{T}_{shaft}$$

$$- \int_{M\,(system)} \vec{r} \times [2\vec{\omega} \times \vec{V}_{xyz} + \vec{\omega} \times (\vec{\omega} \times \vec{r}) + \dot{\vec{\omega}} \times \vec{r}]\, dm$$

$$= \frac{\partial}{\partial t}\int_{CV} \vec{r} \times \vec{V}_{xyz}\rho\, d\mathbb{V} + \int_{CS} \vec{r} \times \vec{V}_{xyz}\rho\vec{V}_{xyz} \cdot d\vec{A}$$

Since the system and control volume coincide at time t_0,

$$\vec{r} \times \vec{F}_S + \int_{CV} \vec{r} \times \vec{g}\rho\, d\mathbb{V} + \vec{T}_{shaft}$$

$$- \int_{CV} \vec{r} \times [2\vec{\omega} \times \vec{V}_{xyz} + \vec{\omega} \times (\vec{\omega} \times \vec{r}) + \dot{\vec{\omega}} \times \vec{r}]\rho\, d\mathbb{V}$$

$$= \frac{\partial}{\partial t}\int_{CV} \vec{r} \times \vec{V}_{xyz}\rho\, d\mathbb{V} + \int_{CS} \vec{r} \times \vec{V}_{xyz}\rho\vec{V}_{xyz} \cdot d\vec{A} \qquad (4.55)$$

Equation 4.55 is the formulation of moment of momentum for a (noninertial) control volume rotating about an axis fixed in space. All fluid velocities and rates of change in Eq. 4.55 are evaluated relative to the control volume. Application of the equation to a rotating nozzle is illustrated in Example 4.16.

Example 4.16

Water flows at the rate of 0.15 m³/sec through the rotating nozzle assembly shown in the sketch. The assembly is rotated at 30 rpm. The mass of the arm and nozzle are negligible compared to that of the water inside. Determine the torque required to drive the device and the reaction torques at the flange.

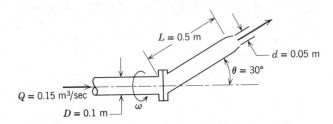

Example Problem 4.16

GIVEN:

Rotating nozzle shown in sketch. Neglect mass of arm and nozzle.

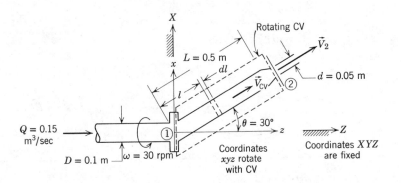

FIND:

Torque required to drive assembly and reaction torques at the flange.

SOLUTION:

Apply the moment of momentum equation to the (noninertial) rotating control volume shown.

Basic equations:

$$0 = \underbrace{\frac{\partial}{\partial t} \int_{CV} \rho \, d\forall}_{= \, 0(1)} + \int_{CS} \rho \vec{V}_{xyz} \cdot d\vec{A}$$

$$\vec{T}_{CV} - \int_{CV} \vec{r} \times [2\vec{\omega} \times \vec{V}_{xyz} + \vec{\omega} \times (\vec{\omega} \times \vec{r}) + \underbrace{\dot{\vec{\omega}} \times \vec{r}}_{= \, 0(3)}] \, \rho \, d\forall$$

$$= \underbrace{\frac{\partial}{\partial t} \int_{CV} \vec{r} \times \vec{V}_{xyz} \rho \, d\forall}_{= \, 0(1)} + \int_{CS} \vec{r} \times \vec{V}_{xyz} \rho \vec{V}_{xyz} \cdot d\vec{A}$$

where $\vec{T}_{CV}$ represents all external torques on the control volume.

Assumptions: (1) Steady flow
(2) Uniform flow at each section
(3) $\dot{\vec{\omega}}$ = constant
(4) Neglect length of nozzle section compared to L
(5) Neglect mass of arm and nozzle assembly compared to that of water inside
(6) Neglect torque due to body force

From continuity

$$V_{CV} = \frac{Q}{A_{CV}} = \frac{4}{\pi} \frac{Q}{D^2} = \frac{4}{\pi} \times \frac{0.15 \text{ m}^3}{\text{sec}} \times \frac{1}{(0.1)^2 \text{ m}^2} = 19.1 \text{ m/sec}$$

$$V_2 = \frac{Q}{A_2} = \frac{4}{\pi} \frac{Q}{d^2} = \frac{4}{\pi} \times \frac{0.15 \text{ m}^3}{\text{sec}} \times \frac{1}{(0.05)^2 \text{ m}^2} = 76.4 \text{ m/sec}$$

Thus, since the CV is fixed in the *xyz* system,

$$\vec{V}_{CV} = V_{CV}(\sin \theta \hat{\imath} + \cos \theta \hat{k})$$
$$\vec{V}_2 = V_2(\sin \theta \hat{\imath} + \cos \theta \hat{k})$$

Since the moment of momentum at section ① is zero by symmetry, the basic equation reduces to

$$\vec{T}_{CV} - \int_{CV} \vec{r} \times [2\vec{\omega} \times \vec{V}_{CV} + \vec{\omega} \times (\vec{\omega} \times \vec{r})] \rho \, d\forall = \vec{r}_2 \times \vec{V}_2 \{|\rho Q|\} \qquad (1)$$

Let us evaluate these terms separately. By assumption (4),

$$\vec{r}_2 = L(\sin \theta \hat{\imath} + \cos \theta \hat{k})$$

and

$$\vec{r}_2 \times \vec{V}_2 = L(\sin \theta \hat{\imath} + \cos \theta \hat{k}) \times V_{CV}(\sin \theta \hat{\imath} + \cos \theta \hat{k})$$
$$= LV_{CV}[\sin \theta \cos \theta(-\hat{\jmath}) + \cos \theta \sin \theta(+\hat{\jmath})] = 0$$

Thus there is no contribution from the control surface integral.

Within the control volume $\vec{r} = \ell(\sin\theta\hat{\imath} + \cos\theta\hat{\jmath})$. Since $\vec{\omega} = \omega\hat{k}$, then

$$\vec{\omega} \times \vec{r} = \omega\hat{k} \times \ell(\sin\theta\hat{\imath} + \cos\theta\hat{k}) = \omega\ell\sin\theta\hat{\jmath}$$
$$\vec{\omega} \times (\vec{\omega} \times \vec{r}) = \omega\hat{k} \times \omega\ell\sin\theta\hat{\jmath} = \omega^2\ell\sin\theta(-\hat{\imath})$$

and

$$\vec{r} \times [\vec{\omega} \times (\vec{\omega} \times \vec{r})] = \ell(\sin\theta\hat{\imath} + \cos\theta\hat{k}) \times \omega^2\ell\sin\theta(-\hat{\imath})$$
$$= \omega^2\ell^2\sin\theta\cos\theta(-\hat{\jmath}) \tag{2}$$

Also within the control volume, $\vec{V}_{CV} = V_{CV}(\sin\theta\hat{\imath} + \cos\theta\hat{k})$, so that

$$2\vec{\omega} \times \vec{V}_{CV} = 2\omega\hat{k} \times V_{CV}(\sin\theta\hat{\imath} + \cos\theta\hat{k}) = 2\omega V_{CV}\sin\theta\hat{\jmath}$$

and

$$\vec{r} \times [2\vec{\omega} \times \vec{V}_{CV}] = \ell(\sin\theta\hat{\imath} + \cos\theta\hat{k}) \times 2\omega V_{CV}\sin\theta\hat{\jmath}$$
$$= 2\omega\ell V_{CV}\sin^2\theta\hat{k} + 2\omega\ell V_{CV}\sin\theta\cos\theta(-\hat{\imath}) \tag{3}$$

Substituting Eqs. 2 and 3 into Eq. 1, and using $d\forall = A_{CV}\,d\ell$ gives

$$\vec{T}_{CV} = \int_0^L [-2\omega\ell V_{CV}\sin\theta\cos\theta\hat{\imath} - \omega^2\ell^2\sin\theta\cos\theta\hat{\jmath} + 2\omega\ell V_{CV}\sin^2\theta\hat{k}]\rho A_{CV}\,d\ell$$

$$\vec{T}_{CV} = [-\omega L^2 V_{CV}\sin\theta\cos\theta\hat{\imath} - \tfrac{1}{3}\omega^2 L^3\sin\theta\cos\theta\hat{\jmath} + \omega L^2 V_{CV}\sin^2\theta\hat{k}]\rho A_{CV}$$

The torque required to maintain steady rotation of the nozzle assembly is

$$T_{shaft} = T_{CV_z} = \omega L^2 \rho V_{CV} A_{CV}\sin^2\theta = \omega L^2 \rho Q \sin^2\theta$$

$$= \frac{30\text{ rev}}{\text{min}} \times \frac{(0.5)^2\text{ m}^2}{} \times \frac{999\text{ kg}}{\text{m}^3} \times \frac{0.15\text{ m}^3}{\text{sec}}\left(\frac{1}{2}\right)^2 \times \frac{2\pi\text{ rad}}{\text{rev}} \times \frac{\text{min}}{60\text{ sec}} \times \frac{\text{N}\cdot\text{sec}^2}{\text{kg}\cdot\text{m}}$$

$$T_{shaft} = 29.4\text{ N}\cdot\text{m} \qquad\qquad\qquad T_{shaft}$$

The reaction moments at the flange are

$$M_x = -T_{CV_x} = \omega L^2 \rho Q \sin\theta\cos\theta$$

$$= \frac{30\text{ rev}}{\text{min}} \times \frac{(0.5)^2\text{ m}^2}{} \times \frac{999\text{ kg}}{\text{m}^3} \times \frac{0.15\text{ m}^3}{\text{sec}}\left(\frac{1}{2}\right)\left(\frac{\sqrt{3}}{2}\right)$$

$$\times \frac{2\pi\text{ rad}}{\text{rev}} \times \frac{\text{min}}{60\text{ sec}} \times \frac{\text{N}\cdot\text{sec}^2}{\text{kg}\cdot\text{m}}$$

$$M_x = 51.0\text{ N}\cdot\text{m} \qquad\text{(applied }to\text{ flange }by\text{ CV)} \qquad\qquad M_x$$

$$M_y = -T_{CV_y} = \tfrac{1}{3}\omega^2 L^3 \rho A_{CV}\sin\theta\cos\theta$$

$$= \frac{1}{3} \times \frac{(30)^2\text{ rev}^2}{\text{min}^2} \times \frac{(0.5)^3\text{ m}^3}{} \times \frac{999\text{ kg}}{\text{m}^3} \times \frac{\pi(0.10)^2\text{ m}^2}{4}\left(\frac{1}{2}\right)\left(\frac{\sqrt{3}}{2}\right)\frac{(2\pi)^2\text{ rad}^2}{\text{rev}^2}$$

$$\times \frac{\text{min}^2}{3600\text{ sec}^2} \times \frac{\text{N}\cdot\text{sec}^2}{\text{kg}\cdot\text{m}}$$

$$M_y = 1.40\text{ N}\cdot\text{m} \qquad\text{(applied }to\text{ flange }by\text{ CV)} \qquad\qquad M_y$$

This problem has been included to illustrate use of the moment of momentum equation for a (noninertial) rotating control volume. Note from the results that torques about the x and z axes are required to produce the Coriolis accelerations of the fluid, and that the reaction about the y axis results from the centripetal acceleration. (The body force would also contribute to the reaction about the y axis at the instant shown.)

4-8 THE FIRST LAW OF THERMODYNAMICS

Development of the control volume formulation of the first law of thermodynamics should not prove difficult for us at this point. We have used Eq. 4.11 often enough to make the pattern of developing the control volume equation from the corresponding system equation reasonably clear. Furthermore, the steady state form of the first law is familiar to you from your earlier work in thermodynamics.

Recall that the system formulation of the first law was given by

$$\dot{Q} + \dot{W} = \frac{dE}{dt}\bigg)_{\text{system}} \tag{4.4a}$$

where the total energy of the system is given by

$$E_{\text{system}} = \int_{M\,(\text{system})} e\,dm = \int_{\Psi\,(\text{system})} e\rho\,d\Psi \tag{4.4b}$$

and

$$e = u + \frac{V^2}{2} + gz$$

In Eq. 4.4a the rate of heat transfer, $\dot{Q}$, is taken as positive when heat is added to the system from the surroundings; the rate of work, $\dot{W}$, is taken as positive when work is done on the system by its surroundings.

The system and control volume formulations are related by

$$\frac{dN}{dt}\bigg)_{\text{system}} = \frac{\partial}{\partial t}\int_{\text{CV}} \eta\rho\,d\Psi + \int_{\text{CS}} \eta\rho\vec{V}\cdot d\vec{A} \tag{4.11}$$

where

$$N_{\text{system}} = \int_{M\,(\text{system})} \eta\,dm = \int_{\Psi\,(\text{system})} \eta\rho\,d\Psi \tag{4.6}$$

Then, to derive the control volume formulation of the first law of thermodynamics, we set

$$N = E \quad \text{and} \quad \eta = e$$

From Eq. 4.11, with this substitution, we obtain

$$\frac{dE}{dt}\bigg)_{\text{system}} = \frac{\partial}{\partial t}\int_{\text{CV}} e\rho\,d\Psi + \int_{\text{CS}} e\rho\vec{V}\cdot d\vec{A} \tag{4.56}$$

In deriving Eq. 4.11, the system and the control volume coincided at time t_0, so

$$[\dot{Q} + \dot{W}]_{\text{system}} = [\dot{Q} + \dot{W}]_{\text{control volume}}$$

In light of this, Eqs. 4.4a and 4.56 yield the control volume formulation of the first law of thermodynamics,

$$\dot{Q} + \dot{W} = \frac{\partial}{\partial t} \int_{\text{CV}} e\rho \, d\mathbf{V} + \int_{\text{CS}} e\rho \vec{V} \cdot d\vec{A} \qquad (4.57)$$

where

$$e = u + \frac{V^2}{2} + gz$$

Note that for steady flow the first term on the right side of Eq. 4.57 is zero. Is Eq. 4.57 the form of the first law you used in your earlier thermodynamics course? Even for steady flow, Eq. 4.57 is not quite the same form that you used previously in applying the first law to control volume problems. To obtain a formulation suitable and convenient for problem solutions, let us take a closer look at the work term, $\dot{W}$.

4-8.1   RATE OF WORK FOR A CONTROL VOLUME

The term $\dot{W}$ in Eq. 4.57 has a positive numerical value when work is done on the control volume by the surroundings. This interaction can take place only at the boundaries of the control volume.

The rate of work done on the control volume is conveniently subdivided into four classifications.

$$\dot{W} = \dot{W}_s + \dot{W}_{\text{normal}} + \dot{W}_{\text{shear}} + \dot{W}_{\text{other}}$$

Let us consider these separately:

1.   Shaft Work

We shall designate shaft work by W_s and, hence, the rate of work transferred in through the control surface by shaft work is designated by $\dot{W}_s$.

2.   Work Done on the Control Surface by Normal Stresses

Recall that work requires a force to be moved through a distance. Thus, in moving a force, $\vec{F}$, through an infinitesimal distance, $d\vec{s}$, the work done is given by

$$\delta W = \vec{F} \cdot d\vec{s}$$

To obtain the rate at which work is done by the force, divide by the time increment Δt and take the limit as $\Delta t \to 0$. Thus, the rate of work done by the force, $\vec{F}$, is given by

$$\dot{W} = \lim_{\Delta t \to 0} \frac{\delta W}{\Delta t} = \lim_{\Delta t \to 0} \frac{\vec{F} \cdot d\vec{s}}{\Delta t}$$

that is,

$$\dot{W} = \vec{F} \cdot \vec{V}$$

The rate of work done on an element of area, $d\vec{A}$, of the control surface by normal stresses is given by

$$d\vec{F} \cdot \vec{V} = \sigma_{nn} \, d\vec{A} \cdot \vec{V}$$

The total rate of work done on the enitre control surface by normal stresses is given by

$$\dot{W}_{\text{normal}} = \int_{\text{CS}} \sigma_{nn} \, d\vec{A} \cdot \vec{V} = \int_{\text{CS}} \sigma_{nn} \vec{V} \cdot d\vec{A}$$

3. Work Done on the Control Surface by Shear Stresses

Just as there is work done by the normal stresses at the boundaries of the control volume, so may there be work done by the shear stresses.

The shear force acting on an element of area of the control surface is given by

$$d\vec{F} = \vec{\tau} \, dA$$

where the shear stress vector, $\vec{\tau}$, is the shear stress acting in the plane of dA.

The rate of work done on the entire control surface by shear stresses is given by

$$\dot{W}_{\text{shear}} = \int_{\text{CS}} \vec{\tau} \, dA \cdot \vec{V} = \int_{\text{CS}} \vec{\tau} \cdot \vec{V} \, dA$$

This integral is better expressed as three terms

$$\dot{W}_{\text{shear}} = \int_{\text{CS}} \vec{\tau} \cdot \vec{V} \, dA$$

$$= \int_{A \text{ (shafts)}} \vec{\tau} \cdot \vec{V} \, dA + \int_{A \text{ (solid surface)}} \vec{\tau} \cdot \vec{V} \, dA + \int_{A \text{ (ports)}} \vec{\tau} \cdot \vec{V} \, dA$$

We have already accounted for the first term, since we included $\dot{W}_{\text{shaft}}$ previously. At solid surfaces, $\vec{V} = 0$, so the second term is zero (for a fixed control volume). Thus

$$\dot{W}_{\text{shear}} = \int_{A \text{ (ports)}} \vec{\tau} \cdot \vec{V} \, dA$$

This last term can be made zero by proper choice of control surfaces. If we choose a control surface that cuts across each port perpendicular to the flow, then $d\vec{A}$ is parallel to $\vec{V}$. Since $\vec{\tau}$ is in the plane of dA, then $\vec{\tau}$ is perpendicular to $\vec{V}$. Thus

$$\vec{\tau} \cdot \vec{V} = 0 \qquad \text{and} \qquad \dot{W}_{\text{shear}} = 0$$

4. Other Work

Electrical energy could be added to the control volume. Also, electromagnetic energy, for example, in radar or laser beams, could be absorbed. In most problems, such contributions will be absent, but we should note them in our general formulation.

With all of the terms in $\dot{W}$ evaluated, we obtain

$$\dot{W} = \dot{W}_s + \int_{\text{CS}} \sigma_{nn} \vec{V} \cdot d\vec{A} + \dot{W}_{\text{shear}} + \dot{W}_{\text{other}} \qquad (4.58)$$

4-8.2 CONTROL VOLUME EQUATION

Substituting the expression for $\dot{W}$ from Eq. 4.58 into Eq. 4.57 gives

$$\dot{Q} + \dot{W}_s + \int_{\text{CS}} \sigma_{nn} \vec{V} \cdot d\vec{A} + \dot{W}_{\text{shear}} + \dot{W}_{\text{other}} = \frac{\partial}{\partial t} \int_{\text{CV}} e\rho \, d\Psi + \int_{\text{CS}} e\rho \vec{V} \cdot d\vec{A}$$

Rearranging this equation, we obtain

$$\dot{Q} + \dot{W}_s + \dot{W}_{\text{shear}} + \dot{W}_{\text{other}} = \frac{\partial}{\partial t} \int_{\text{CV}} e\rho \, d\Psi + \int_{\text{CS}} e\rho \vec{V} \cdot d\vec{A} - \int_{\text{CS}} \sigma_{nn} \vec{V} \cdot d\vec{A}$$

Since $\rho = 1/v$, then

$$\int_{\text{CS}} \sigma_{nn} \vec{V} \cdot d\vec{A} = \int_{\text{CS}} \sigma_{nn} v\rho \vec{V} \cdot d\vec{A}$$

Hence

$$\dot{Q} + \dot{W}_s + \dot{W}_{\text{shear}} + \dot{W}_{\text{other}} = \frac{\partial}{\partial t} \int_{\text{CV}} e\rho \, d\Psi + \int_{\text{CS}} (e - \sigma_{nn} v)\rho \vec{V} \cdot d\vec{A}$$

Viscous effects can make the normal stress, σ_{nn}, different from the negative of the thermodynamic pressure, $-p$. However, for most flows of common engineering interest, $\sigma_{nn} \simeq -p$. Then

$$\dot{Q} + \dot{W}_s + \dot{W}_{\text{shear}} + \dot{W}_{\text{other}} = \frac{\partial}{\partial t} \int_{\text{CV}} e\rho \, d\Psi + \int_{\text{CS}} (e + pv)\rho \vec{V} \cdot d\vec{A}$$

Finally, substituting $e = u + V^2/2 + gz$ into the last term, we obtain the familiar form of the first law formulation for a control volume,

$$\dot{Q} + \dot{W}_s + \dot{W}_{\text{shear}} + \dot{W}_{\text{other}} = \frac{\partial}{\partial t} \int_{\text{CV}} e\rho \, d\mathcal{V} + \int_{\text{CS}} \left(u + pv + \frac{V^2}{2} + gz \right) \rho \vec{V} \cdot d\vec{A}$$

(4.59)

Each work term in Eq. 4.59 represents the rate of work done on the control volume.

Example 4.17
Air at 14.7 psia, 70 F, enters a machine with negligible velocity, and is discharged at 50 psia, 100 F through a pipe of area 1 ft². The flowrate is 20 lbm/sec. The power input to the machine is 600 hp. Determine the rate of heat transfer.

Example Problem 4.17

GIVEN:

Air enters a compressor at ① with conditions

$$p_1 = 14.7 \text{ psia} \qquad T_1 = 70 \text{ F} \qquad V_1 \approx 0$$

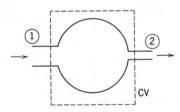

and is discharged at ② with conditions

$$p_2 = 50 \text{ psia} \qquad T_2 = 100 \text{ F} \qquad A_2 = 1 \text{ ft}^2$$

The air flowrate is 20 lbm/sec and the power input to the machine is 600 hp.

FIND:

The rate of heat transfer.

SOLUTION:

Basic equations:

$$\dot{Q} + \dot{W}_s + \overset{=0(4)}{\cancel{\dot{W}_{shear}}} = \overset{=0(\,)}{\cancel{\frac{\partial}{\partial t} \int_{CV} e\rho \, d\forall}} + \int_{CS} \left(u + pv + \frac{V^2}{2} + gz \right) \rho \vec{V} \cdot d\vec{A}$$

$$0 = \overset{=0(1)}{\cancel{\frac{\partial}{\partial t} \int_{CV} \rho \, d\forall}} + \int_{CS} \rho \vec{V} \cdot d\vec{A}$$

Assumptions:
(1) Steady flow
(2) Properties uniform over inlet and outlet sections
(3) Treat air as an ideal gas, $p = \rho RT$
(4) Area of CV at ① and ② perpendicular to velocity, thus $\dot{W}_{shear} = 0$
(5) $z_1 = z_2$
(6) Inlet kinetic energy is negligible

Under the assumptions listed, the first law becomes

$$\dot{Q} + \dot{W}_s = \int_{CS} \left(u + pv + \frac{V^2}{2} + gz \right) \rho \vec{V} \cdot d\vec{A}$$

$$\dot{Q} + \dot{W}_s = \int_{CS} \left(h + \frac{V^2}{2} + gz \right) \rho \vec{V} \cdot d\vec{A} \qquad \{h = u + pv\}$$

or

$$\dot{Q} = -\dot{W}_s + \int_{CS} \left(h + \frac{V^2}{2} + gz \right) \rho \vec{V} \cdot d\vec{A}$$

For uniform properties (assumption 2) we can write

$$\dot{Q} = -\dot{W}_s + \left(h_1 + \overset{\approx 0(6)}{\cancel{\frac{V_1^2}{2}}} + gz_1 \right) \{-|\rho_1 V_1 A_1|\} + \left(h_2 + \frac{V_2^2}{2} + gz_2 \right) \{|\rho_2 V_2 A_2|\}$$

For steady flow, from conservation of mass,

$$\int_{CS} \rho \vec{V} \cdot d\vec{A} = 0$$

Therefore, $-|\rho_1 V_1 A_1| + |\rho_2 V_2 A_2| = 0$, that is $|\rho_1 V_1 A_1| = |\rho_2 V_2 A_2| = \dot{m}$. Hence, we can write

$$\dot{Q} = -\dot{W}_s + \dot{m} \left[(h_2 - h_1) + \frac{V_2^2}{2} + g(z_2 \overset{=0(5)}{\cancel{-z_1}}) \right]$$

Assume that air behaves as an ideal gas with constant c_p. Then $h_2 - h_1 = c_p(T_2 - T_1)$, and

$$\dot{Q} = -\dot{W}_s + \dot{m}\left[c_p(T_2 - T_1) + \frac{V_2^2}{2}\right]$$

From continuity $|V_2| = \dot{m}/\rho_2 A_2$. Since $p_2 = \rho_2 R T_2$, then

$$|V_2| = \frac{\dot{m}}{A_2}\frac{R T_2}{p_2} = \frac{20\text{ lbm}}{\text{sec}} \times \frac{1}{1\text{ ft}^2} \times \frac{53.3\text{ ft}\cdot\text{lbf}}{\text{lbm}\cdot\text{R}} \times \frac{560\text{ R}}{50\text{ lbf}} \times \frac{\text{in.}^2}{144\text{ in.}^2} \times \frac{\text{ft}^2}{144\text{ in.}^2}$$

$$|V_2| = 82.9\text{ ft/sec}$$

$$\dot{Q} = -\dot{W}_s + \dot{m}c_p(T_2 - T_1) + \dot{m}\frac{V_2^2}{2}$$

$$= -\frac{600\text{ hp}}{} \times \frac{550\text{ ft}\cdot\text{lbf}}{\text{hp}\cdot\text{sec}} \times \frac{\text{Btu}}{778\text{ ft}\cdot\text{lbf}} + \frac{20\text{ lbm}}{\text{sec}} \times \frac{0.24\text{ Btu}}{\text{lbm}\cdot\text{R}} \times \frac{30\text{ R}}{}$$

$$+ \frac{20\text{ lbm}}{\text{sec}} \times \frac{(82.9)^2}{2}\frac{\text{ft}^2}{\text{sec}^2} \times \frac{\text{slug}}{32.2\text{ lbm}} \times \frac{\text{Btu}}{778\text{ ft}\cdot\text{lbf}} \times \frac{\text{lbf}\cdot\text{sec}^2}{\text{slug}\cdot\text{ft}}$$

$$\dot{Q} = -277\text{ Btu/sec} \quad \{\text{heat rejection}\} \qquad\qquad \dot{Q}$$

$$\left\{\begin{array}{l}\text{In addition to demonstrating a straightforward application of the first law,}\\ \text{this problem illustrates the need for keeping units straight.}\end{array}\right\}$$

Example 4.18

A tank of volume 0.1 m³ is connected to a high pressure air line; both line and tank are initially at a uniform temperature of 20 C. The initial tank gage pressure is 100 kPa. The absolute line pressure is 2.0 MPa; the line is large enough so that its temperature and pressure may be assumed to remain constant. The tank temperature is monitored by a fast-response thermocouple. At the instant after the valve is opened, the tank temperature rises at the rate of 0.05 C/sec. Determine the instantaneous flowrate of air into the tank. (Heat transfer is negligible.)

Example Problem 4.18

GIVEN:

Air supply pipe and tank as shown.

At $t = 0^+$, $\dfrac{\partial T}{\partial t} = 0.05\,\dfrac{\text{C}}{\text{sec}}$

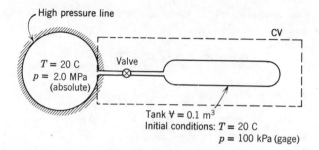

High pressure line

$T = 20\ C$
$p = 2.0\ MPa$
(absolute)

Valve

CV

Tank $\Psi = 0.1\ m^3$
Initial conditions: $T = 20\ C$
$p = 100\ kPa$ (gage)

FIND:

$\dot{m}$ at $t = 0^+$.

SOLUTION:

Choose CV shown, apply energy equation.

$$= 0(1) \quad = 0(2) \quad = 0(3) \quad = 0(4)$$

Basic equation: $\cancel{\dot{Q}} + \cancel{\dot{W}_s} + \cancel{\dot{W}_{shear}} + \cancel{\dot{W}_{other}} = \dfrac{\partial}{\partial t}\displaystyle\int_{CV} e\rho\, d\Psi + \int_{CS} (e + pv)\rho\vec{V}\cdot d\vec{A};$

$$\simeq 0(5) \quad \simeq 0(6)$$

$$e = u + \cancel{\frac{V^2}{2}} + \cancel{gz}$$

Assumptions:
(1) $\dot{Q} = 0$ (given)
(2) $\dot{W}_s = 0$
(3) $\dot{W}_{shear} = 0$
(4) $\dot{W}_{other} = 0$
(5) Velocities in line and tank are small
(6) Neglect potential energy
(7) Uniform flow at tank inlet
(8) Properties uniform in tank
(9) Ideal gas, $p = \rho RT$, $du = c_v\, dT$

Then

$$0 = \frac{\partial}{\partial t}\int_{CV} u_{tank}\rho\, d\Psi + (u_{line} + pv)\{-|\rho VA|\}$$

But initially, T is uniform, so $u_{tank} = u_{line} = u$, and

$$0 = \frac{\partial}{\partial t}\int_{CV} u\rho\, d\Psi + (u + pv)\{-|\rho VA|\}$$

Since tank properties are uniform, $\partial/\partial t$ may be replaced by d/dt, and

$$0 = \frac{d}{dt}[uM] - (u + pv)\dot{m}$$

or

$$0 = u\frac{dM}{dt} + M\frac{du}{dt} - u\dot{m} - pv\dot{m} \tag{1}$$

The term dM/dt may be evaluated from continuity:

Basic equation: $\quad 0 = \dfrac{\partial}{\partial t}\displaystyle\int_{CV} \rho \, d\forall + \displaystyle\int_{CS} \rho\vec{V}\cdot d\vec{A}$

$$0 = \frac{dM}{dt} + \{-|\rho VA|\} \quad \text{or} \quad \frac{dM}{dt} = \dot{m}$$

Substituting into Eq. 1 gives

$$0 = u\dot{m} + M\frac{du}{dt} - u\dot{m} - pv\dot{m} = Mc_v\frac{dT}{dt} - pv\dot{m}$$

or

$$\dot{m} = \frac{Mc_v(dT/dt)}{pv} = \frac{\rho\forall c_v(dT/dt)}{pv} = \frac{\rho\forall c_v(dT/dt)}{RT} \tag{2}$$

But at $t = 0$,

$$\rho = \rho_{\text{tank}} = \frac{p_{\text{tank}}}{RT} = \frac{(1.00 + 1.01)10^5 \text{ N}}{\text{m}^2} \times \frac{\text{kg}\cdot\text{K}}{287 \text{ N}\cdot\text{m}} \times \frac{1}{293 \text{ K}} = 2.39 \text{ kg/m}^3$$

Substituting into Eq. 2,

$$\dot{m} = \frac{2.39 \text{ kg}}{\text{m}^3} \times 0.1 \text{ m}^3 \times \frac{717 \text{ N}\cdot\text{m}}{\text{kg}\cdot\text{K}} \times \frac{0.05 \text{ K}}{\text{sec}} \times \frac{\text{kg}\cdot\text{K}}{287 \text{ N}\cdot\text{m}} \times \frac{1}{293 \text{ K}}$$

$$\dot{m} = 0.102 \text{ g/sec} \qquad\qquad \dot{m}$$

{ This problem illustrates the application of the energy equation to an unsteady flow situation.

4-9 THE SECOND LAW OF THERMODYNAMICS

Recall that the system formulation of the second law is

$$\left.\frac{dS}{dt}\right)_{\text{system}} \geq \frac{1}{T}\dot{Q} \tag{4.5a}$$

where the total entropy of the system is given by

$$S_{\text{system}} = \int_{\text{mass (system)}} s \, dm = \int_{\forall \text{ (system)}} s\rho \, d\forall \tag{4.5b}$$

The relation between system and control volume formulations is

$$\left.\frac{dN}{dt}\right)_{\text{system}} = \frac{\partial}{\partial t}\int_{CV} \eta\rho \, d\forall + \int_{CS} \eta\rho\vec{V}\cdot d\vec{A} \tag{4.11}$$

where

$$N_{\text{system}} = \int_{\text{mass (system)}} \eta \, dm = \int_{\forall \text{ (system)}} \eta \rho \, d\forall \qquad (4.6)$$

To derive the control volume formulation of the second law of thermodynamics, we set

$$N = S \quad \text{and} \quad \eta = s$$

From Eq. 4.11, with this substitution, we obtain

$$\left. \frac{dS}{dt} \right)_{\text{system}} = \frac{\partial}{\partial t} \int_{\text{CV}} s\rho \, d\forall + \int_{\text{CS}} s\rho \vec{V} \cdot d\vec{A} \qquad (4.59)$$

From Eq. 4.5a

$$\left. \frac{dS}{dt} \right)_{\text{system}} \geq \frac{1}{T} \dot{Q}$$

The system and the control volume coincided at time t_0; thus

$$\left. \frac{1}{T} \dot{Q} \right)_{\text{system}} = \left. \frac{1}{T} \dot{Q} \right)_{\text{CV}} = \int_{\text{CS}} \frac{1}{T} \left(\frac{\dot{Q}}{A} \right) dA$$

In light of this, Eqs. 4.5a and 4.59 yield the control volume formulation of the second law of thermodynamics

$$\frac{\partial}{\partial t} \int_{\text{CV}} s\rho \, d\forall + \int_{\text{CS}} s\rho \vec{V} \cdot d\vec{A} \geq \int_{\text{CS}} \frac{1}{T} \left(\frac{\dot{Q}}{A} \right) dA \qquad (4.60)$$

In Eq. 4.60, the term $(\dot{Q}/A)$ represents the heat flux per unit area into the control volume through the area element, dA. To evaluate the term

$$\int_{\text{CS}} \frac{1}{T} \left(\frac{\dot{Q}}{A} \right) dA$$

both the local heat flux, $(\dot{Q}/A)$, and local temperature, T, must be known for each area element of the control surface.

Example 4.19

Air to be used in a climate control system is first cooled to remove water vapor and then heated to a comfortable temperature as it flows through a 0.3 m square duct. The air flowrate through the duct is 3.5 kg/sec. Heat is removed at the rate of 70 kJ/sec in a section of duct 1.5 m long, where the air and duct temperature are reduced from 30 C to 10 C. Heat is supplied at 35 kJ/sec in a second duct section 1 m long, where the temperature of the duct and air are increased from 10 C to 20 C. Determine the change in specific entropy in J/kg · K.

Example Problem 4.19

GIVEN:

Air flow in a duct as shown.

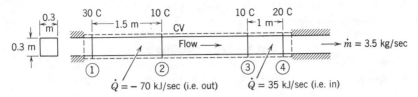

FIND:

$s_4 - s_1$, in J/kg · K

SOLUTION:

Apply the second law of thermodynamics

$$= 0(1)$$

Basic equation: $\quad \dfrac{\overset{\downarrow}{\partial}}{\partial t} \displaystyle\int_{CV} s\rho\, d\forall + \int_{CS} s\rho \vec{V} \cdot d\vec{A} \geq \int_{CS} \dfrac{1}{T}\left(\dfrac{\dot{Q}}{A}\right) dA$

Assumptions: (1)   Steady flow
                    (2)   Uniform flow at each cross section
                    (3)   Heat transfer takes place at mean temperature in each section of duct
                    (4)   Reversible processes, that is, equality holds in second law

Then

$$s_1\{-|\rho_1 V_1 A_1|\} + s_4\{|\rho_4 V_4 A_4|\} = \int_{CS} \dfrac{1}{T}\dfrac{\dot{Q}}{A}\, dA$$

or

$$(s_4 - s_1)\dot{m} = \sum\left[\dfrac{1}{T}\left(\dfrac{\dot{Q}}{A}\right)A\right] = \sum\left[\dfrac{\dot{Q}}{T}\right]$$

Thus

$$s_4 - s_1 = \dfrac{1}{\dot{m}}\sum\left[\dfrac{\dot{Q}}{T}\right]$$

$$= \dfrac{\text{sec}}{3.5\ \text{kg}}\left[-\dfrac{70\ \text{kJ}}{\text{sec}} \times \dfrac{1}{(273+20)\text{K}} + \dfrac{35\ \text{kJ}}{\text{sec}} \times \dfrac{1}{(273+15)\text{K}}\right]$$

$$s_4 - s_1 = -32.5\ \text{J/kg} \cdot \text{K} \qquad\qquad\qquad\qquad s_4 - s_1$$

> This problem was included to illustrate the application of the second law of thermodynamics to a control volume. There are two artificial features of this problem:
> (i) reversible heat transfer—in real problems, heat transfer is almost always irreversible.
> (ii) use of average temperature for cooling and heating.

summary objectives

After completing study of Chapter 4, you should be able to do the following:

1. Write each of the five basic laws (conservation of mass, Newton's second law, moment of momentum**, the first law of thermodynamics, and the second law of thermodynamics) for a system as a rate equation.

2. If the extensive property in the rate equations of Summary Objective 1 is designated N, define the corresponding intensive property, designated η, in each of the basic equations.

3. Write the equation that relates the rate of change of any arbitrary extensive property, N, of a system, to the time variations of the property associated with a control volume. Give the physical significance of each quantity in the equation.

4. Write the control volume formulation of the conservation of mass and state the physical meaning of each term in the equation. Apply the equation to the solution of flow problems.

5. Write the control volume formulation of the momentum equation for an inertial control volume and state the physical meaning of each term in the equation. Apply the equation to the solution of flow problems.

**6. State the relationship among fluid properties (the Bernoulli equation) that results from applying the momentum equation to a differential control volume. List the restrictions on the use of the Bernoulli equation.

7. Write the control volume formulation of the momentum equation for a control volume with rectilinear acceleration and state the physical meaning of each term in the equation. Apply the equation to the solution of flow problems.

**8. Write the control volume formulation of the momentum equation for a control volume with arbitrary acceleration and state the physical meaning of each term in the equation. Apply the equation to the solution of flow problems.

**9. Write the control volume formulation of the moment of momentum equation for (a) a fixed and (b) a rotating control volume and state the physical meaning of each term in the equation. Apply the equation to the solution of flow problems.

10. Write the control volume formulation of the first law of thermodynamics and state the physical meaning of each term in the equation. Apply the equation to the solution of flow problems.

** These objectives apply to sections that may be omitted without loss of continuity in the text material.

11. Write the control volume formulation of the second law of thermodynamics and state the physical meaning of each term in the equation. Apply the equation to the solution of flow problems.

12. Solve those problems at the end of the chapter that relate to the material you have studied.

problems

4.1 A police investigation of tire marks showed that a car traveling along a straight level street had skidded for a total distance of 50 m after the brakes were applied. The coefficient of friction between the tires and the pavement is estimated to be $\mu = 0.6$. What was the probable speed of the car when the brakes were applied?

4.2 A body of mass M is at rest on a horizontal plane. At time $t = 0$, the body is acted upon by a constant force, F_1, of 10 lbf. The force is applied for a period of 5 sec. Immediately upon removal of this force, the body is acted upon by a constant force, F_2, of 3 lbf in the opposite direction. There is no friction. Determine the length of time of action by F_2 required to bring the body to rest.

4.3 A flywheel consisting of a steel disk 20 mm thick and 0.6 m in diameter rotates at 20,000 rpm. The flywheel is mounted in a subway car with its axis of rotation placed lengthwise. The car, traveling at a speed of 15 m/sec, rounds a curve of radius 80 m. Determine the reaction torque exerted by the flywheel assembly on the subway car. The specific gravity of steel is 7.8.

4.4 A fluid mechanics laboratory experiment consists of a cylindrical tank containing water and mounted on a turntable, as in Example 3.11. The cylinder diameter is 0.2 m; the initial water depth, h_0, is 40 mm. Measurements show that when the turntable is switched off from a speed of 78 rpm, all motion of the liquid ceases 180 sec later. Determine the average torque applied to the water during this interval.

4.5 A weight is dropped from a height of 10 ft above the end of a spring. When the motion of the weight is stopped by the spring, the spring has been compressed 2 ft. If the spring constant is 1 lbf/in., calculate the magnitude of the weight.

4.6 Calculate the amount of work required to accelerate a 1500 kg automobile from rest to a speed of 25 m/sec on a level highway if air resistance and mechanical losses are neglected.

4.7 The average rate of heat loss from a person to the surroundings when not actively working is about 600 Btu/hr. Suppose that in an auditorium containing 6000 people the ventilation system fails.
(a) How much does the internal energy of the air in the auditorium increase during the first 15 min after the ventilation system fails?
(b) Considering the auditorium and all the people as a system, and assuming no heat transfer to the surroundings, how much does the internal energy of the system change? How do you account for the fact that the temperature of the air increases?

4.8   Air is expanded isothermally in a nonflow process from initial conditions of 170 C, 276 kPa (absolute) and 0.05 m³ to a final volume of 0.1 m³. Determine the heat transferred, in joules.

4.9   Air at 20 C and an absolute pressure of 1 atm is compressed adiabatically, without friction, to an absolute pressure of 3 atm. Determine the internal energy change in J/kg.

4.10   A cylinder fitted with a piston contains 5 lbm of wet steam at a pressure of 200 psia and a quality of 80 percent. The piston is restrained by a spring arranged so that for zero volume in the cylinder the spring is fully extended. The spring force is proportional to the spring displacement. The weight of the piston may be neglected so that the force on the spring is exactly balanced by the pressure forces on the steam

Heat is transferred to the steam until its volume is 150 percent of the initial volume.
(a)   What is the final pressure?
(b)   What is the quality (if saturated) or temperature (if superheated) in the final state?
(c)   Draw a pV diagram and determine the work delivered to the spring.
(d)   Determine the heat transferred.

4.11   In order to cool a six-pack as quickly as possible, it is placed in a freezer for a period of 2 hr. If the room temperature is 25 C and the cooled beverage is at a final temperature of 5 C, determine the change in the specific entropy of the beverage.

4.12   The velocity field in the region shown in Fig. 4.7 is given by

$$\vec{V} = ay\hat{j} + b\hat{k}$$

where $a = 10\,\text{sec}^{-1}$ and $b = 5\,\text{m/sec}$. For a depth, w, perpendicular to the diagram, an element of area ① may be represented by $w\,dz(-\hat{j})$ and an element of area ② by $w\,dy(-\hat{k})$. (Note that both are drawn *outward* from the control volume, hence the minus sign.)

(a)   Find an expression for $\vec{V} \cdot d\vec{A}_①$.

(b)   Evaluate $\int_{A_①} \vec{V} \cdot d\vec{A}_①$.

(c)   Find an expression for $\vec{V} \cdot d\vec{A}_②$.
(d)   Find an expression for $\vec{V}(\vec{V} \cdot d\vec{A}_②)$.

(e)   Evaluate $\int_{A_②} \vec{V}(\vec{V} \cdot d\vec{A}_②)$.

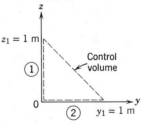

Fig. 4.7

4.13   A flow field is given by $\vec{V} = ay\hat{i} + b\hat{k}$, where $a = 2\,\text{sec}^{-1}$ and $b = 1\,\text{ft/sec}$. The volumetric flowrate, Q, through a surface is given by

$$Q = \int_A \vec{V} \cdot d\vec{A}$$

Evaluate the volumetric flowrate through the shaded surface in Fig. 4.8. All dimensions are in feet.

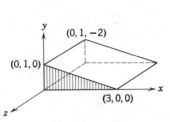

Fig. 4.8

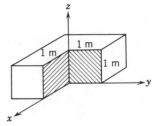

Fig. 4.9

4.14 The shaded area shown in Fig. 4.9 is in a flow where the velocity field is given by

$$\vec{V} = ax\hat{\imath} + by\hat{\jmath}$$

where $a = b = 1 \ \text{sec}^{-1}$.
Evaluate the following integrals over the shaded area

(a) $\int \vec{V} \cdot d\vec{A}$ (b) $\int \vec{V}(\vec{V} \cdot d\vec{A})$

4.15 The area shown shaded in Fig. 4.10 is in a flow where the velocity field is given by

$$\vec{V} = -ax\hat{\imath} + by\hat{\jmath} + c\hat{k}$$

where $a = b = 1 \ \text{sec}^{-1}$ and $c = 1 \ \text{m/sec}$.
Write a vector expression for an element of the shaded area. Evaluate the following integrals over the shaded area.

(a) $\int \vec{V} \cdot d\vec{A}$

(b) $\int \vec{V}(\vec{V} \cdot d\vec{A})$

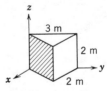

Fig. 4.10

4.16 The velocity distribution for laminar flow in a long circular tube is given by the one-dimensional expression

$$\vec{V} = u\hat{\imath} = u_{\text{max}}\left[1 - \left(\frac{r}{R}\right)^2\right]\hat{\imath}$$

For this profile, use the area vector $d\vec{A} = dA\hat{\imath} = 2\pi r \ dr\hat{\imath}$ to evaluate

(a) $\int \vec{V} \cdot d\vec{A}$ (b) $\int \vec{V}(\vec{V} \cdot d\vec{A})$

for the tube cross section.

4.17 Consider the flow of an incompressible fluid with vector velocity field:

$$\vec{V} = (ax + bt)\hat{\imath} - cy\hat{\jmath}$$

where $a = 1 \sec^{-1}$, $b = 2\,\text{m/sec}^2$ and $c = 1 \sec^{-1}$. For the control volume shown (a 1 m cube, Fig. 4.11), evaluate the rate of change of momentum within the control volume.

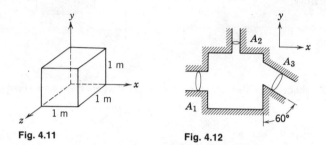

Fig. 4.11   Fig. 4.12

4.18 Fluid with a density of 1050 kg/m³ is flowing steadily through the rectangular box shown in Fig. 4.12. If

$$A_1 = 0.05\,\text{m}^2 \qquad A_2 = 0.01\,\text{m}^2 \qquad A_3 = 0.06\,\text{m}^2$$
$$\vec{V}_1 = 4\hat{\imath}\,\text{m/sec} \qquad \vec{V}_2 = -8\hat{\jmath}\,\text{m/sec}$$

determine the velocity $\vec{V}_3$.

4.19 Consider steady, incompressible flow through the device shown in Fig. 4.13. Determine the volumetric flowrate through port 3.

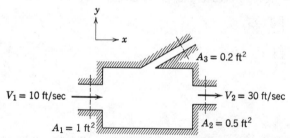

Fig. 4.13

4.20 Air at standard atmospheric conditions enters a compressor at a rate of 20 m³/min. The air is discharged at 800 kPa (absolute) and 60 C. If the velocity in the discharge line must be limited to 20 m/sec, calculate the required diameter of the line.

4.21 In the incompressible flow through the device shown in Fig. 4.14, velocities may be considered uniform over the inlet and outlet sections. If the fluid flowing is water, determine an expression for the mass flowrate at section ③. The following conditions are known:

$$A_1 = 0.1 \text{ m}^2 \qquad A_2 = 0.2 \text{ m}^2 \qquad A_3 = 0.15 \text{ m}^2$$
$$V_1 = 5 \text{ m/sec} \qquad V_2 = 10 + 5\cos(4\pi t) \text{ m/sec}$$

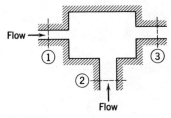

Flow →

① ③

②

Flow

Fig. 4.14

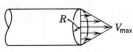

R

V_{max}

Fig. 4.15

4.22 Water flowing out of a circular pipe is assumed to have a linearly varying velocity distribution as shown in Fig. 4.15. What is the average velocity of the outward flow in terms of V_{max}?

4.23 Water is flowing steadily through a pipe of length L and radius $R = 3$ in. (Fig. 4.16). Calculate the value of the uniform inlet velocity, U, if the velocity distribution across the outlet is given by

$$u = 10\left[1 - \frac{r^2}{R^2}\right] \text{ ft/sec}$$

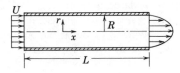

U

r

R

x

L

Fig. 4.16

4.24 Water enters a wide, flat channel of height, $2h$, with a uniform velocity of 5 m/sec. At the outlet of the channel the velocity distribution is given by

$$\frac{u}{U_{max}} = 1 - \left(\frac{y}{h}\right)^2$$

The coordinate y is measured from the centerline of the channel. Determine the exit centerline velocity, U_{max}.

4.25 A tank of 0.5 m³ volume contains compressed air. A valve is opened and air escapes with a velocity of 300 m/sec through an opening of 130 mm² area. Air temperature passing through the opening is -15 C and the absolute pressure is 350 kPa. Find the rate of change of density of the air in the tank at this moment.

4.26 Air enters a tank through an area of 0.2 ft² with a velocity of 15 ft/sec and a density of 0.03 slug/ft³. Air leaves with a velocity of 5 ft/sec and a density equal to that in the tank. The initial density of the air in the tank is 0.02 slug/ft³. The total tank volume is 20 ft³. Exit area is 0.4 ft². Find the initial rate of change of density in the tank.

4.27   The vessel shown in Fig. 4.17 is 6 ft long in the direction normal to the diagram. There is no flow into the vessel, but a chemical reaction that occurs in the vessel generates gas that leaves through the four openings (each 1 ft by 6 ft in cross-sectional area) as shown. The velocity relative to the vessel and the density of the gas leaving vary with radius as

$$V = \frac{10}{r} \quad \text{and} \quad \rho = 0.0020 + 0.001r$$

where V is in ft/sec, ρ is in slug/ft^3, and r is in feet. Determine the rate of change of mass in the vessel in slug/sec.

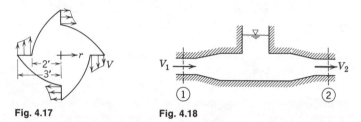

Fig. 4.17   **Fig. 4.18**

4.28   A section of pipe carrying water contains an expansion chamber with a free surface whose area is 2 m^2 (Fig. 4.18). The inlet and outlet pipes are both 1 m^2 in area. At a given instant, the velocity at section ① is 3 m/sec into the chamber. Water flows out at section ② at 4 m^3/sec. Both flows are uniform. Find the rate of change of free surface level at the given instant, in m/sec. Indicate whether the level rises or falls.

4.29   A cylindrical tank, 0.3 m in diameter, drains through a hole in its bottom. At the instant when the water depth is 0.6 m, the flowrate from the tank is observed to be 4 kg/sec. Determine the rate of change of water level at this instant.

4.30   A cylindrical tank of diameter $D = 50$ mm, drains through an opening, $d = 5$ mm, in the bottom of the tank. The speed of the liquid leaving the tank may be approximated by $V = \sqrt{2gy}$, where y is the height from the tank bottom to the free surface. If the tank is initially filled with water to a depth $y_0 = 0.4$ m, determine the water depth at time, $t = 12$ sec.

4.31   For the conditions of Problem 4.30, determine the time required to drain the tank completely.

4.32   A funnel of half angle θ, drains through a hole of area A, at the vertex. The speed of the liquid leaving the funnel is approximated by $V = \sqrt{2gy}$, where y is the height of the liquid free surface above the hole. The tank is initially filled to height y_0. Obtain an expression for the time, t, required to drain the funnel. Express the result in terms of the initial volume, V_0, of liquid in the funnel and the initial volumetric flow rate, $Q_0 = A\sqrt{2gy_0} = AV_0$.

4.33   Solve Problem 4.32 if the funnel is inverted so that liquid leaves through a hole of area A in the large end.

4.34 In ancient Egypt, circular vessels filled with water sometimes were used as crude clocks. The vessels were shaped in such a way that, as water drained from the bottom, the surface level dropped at a constant rate, s. Assume that water drains from a hole of area, A. The water speed leaving the vessel is approximated by $V = \sqrt{2gh}$, where h is the height of the liquid free surface above the jet exit. Find an expression for the radius of the vessel, r, as a function of height, h. Determine the volume of water needed so that the clock will operate for n hours.

4.35 Motion of a hydraulic cylinder is cushioned at the end of its stroke by a piston that enters a hole as shown in Fig. 4.19. The cavity and cylinder are filled with hydraulic fluid of uniform density, ρ. Obtain an expression for the velocity, V_1, at which hydraulic fluid escapes from the cylindrical hole as a function of piston displacement, x.

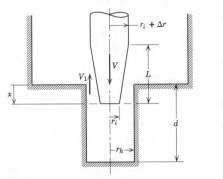

Fig. 4.19

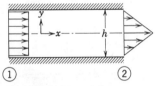

Fig. 4.20

4.36 Consider the steady flow of water between two parallel plates a distance h feet apart (Fig. 4.20). At section ① the velocity is uniform across the width; the velocity distribution is assumed to be linear at section ②. The flow is identical in all planes parallel to the plane of the diagram. Calculate the ratio of the x-direction momentum flux at section ② to that at section ① for the assumed velocity distributions.

4.37 Evaluate the net rate of efflux of momentum through the control surface of Problem 4.18.

4.38 For the conditions of Problem 4.23, evaluate the ratio of the x-direction momentum flux at the pipe outlet to that at the inlet.

4.39 For the conditions of Problem 4.24, evaluate the ratio of the x-direction momentum flux at the channel outlet to that at the inlet.

4.40 Oil is being loaded into a barge through a 150 mm diameter pipe (Fig. 4.21). The oil, $\rho = 900$ kg/m³, leaves the pipe with a uniform speed of 5 m/sec. Determine the force of the rope on the barge.

4.41 A jet of water issuing from a stationary nozzle with a speed of 15 m/sec (jet area = 0.05 m²) strikes a turning vane mounted on a cart as shown in Fig. 4.22.

The vane turns the jet through an angle $\theta = 50°$. Determine the value of M required to hold the cart stationary.

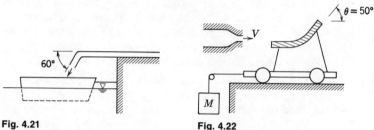

Fig. 4.21   Fig. 4.22

4.42   If the vane angle, θ, of Problem 4.41 is adjustable, plot the mass, M, needed to hold the cart stationary as a function of θ for $0 \le \theta \le 180°$.

4.43   A large tank is affixed to a cart as shown in Fig. 4.23. Water issues from the tank through a 600 mm^2 nozzle at a speed of 10 m/sec. The water level in the tank is maintained constant by adding water through a vertical pipe. Determine the tension in the wire holding the cart stationary.

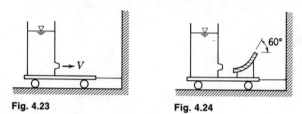

Fig. 4.23   Fig. 4.24

4.44   A turning vane (Fig. 4.24), which deflects the water through an angle of 60°, is attached to the cart under the conditions of Problem 4.43. Determine the tension in the wire holding the cart stationary, and the force of the vane on the cart.

4.45   If the turning vane of Problem 4.44 deflects the water through an angle of 90°, determine the tension in the wire holding the cart stationary.

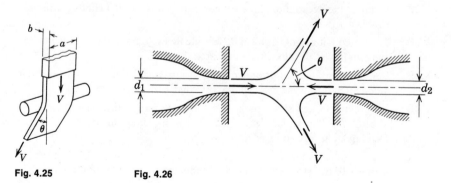

Fig. 4.25   Fig. 4.26

4.46 A circular cylinder inserted across a stream of flowing water deflects the stream through an angle θ as shown in Fig. 4.25. (This is termed the "Coanda effect.") For $a = 0.5$ in., $b = 0.1$ in., $V = 10$ ft/sec, and $\theta = 20°$, determine the horizontal component of the force on the cylinder due to the flowing water.

4.47 In a fluidic device, two circular coaxial jets of incompressible liquid with speed, V, interact as shown in Fig. 4.26. The interaction region is open to atmosphere. Liquid leaves the interaction region in a conical sheet. The angle, θ, of the resulting flow must be known to predict the operating characteristics of the device. Obtain an expression for the cone angle, θ, in terms of d_2/d_1.

4.48 Dry ready-mix concrete is packaged in 90 lb sacks; the density of the dry mix is 100 lbm/ft³. The sacks are supported by pallets as they move through the automatic bagging machine. If the filling time is 4 sec and the vertical velocity of the mix is 10 ft/sec, determine the maximum vertical force on the pallet.

4.49 A farmer purchases 675 kg of bulk grain from the local co-op. The grain is loaded into his pickup truck from a hopper with an outlet diameter of 0.3 m. The loading operator determines the payload by observing the indicated gross mass of the truck as a function of time. The grain flow from the hopper ($\dot{m} = 40$ kg/sec) is terminated when the indicated scale reading reaches the desired gross mass. If the grain density is 600 kg/m³, determine the true payload.

4.50 A large "weigh tank" is to be used in the calibration of a flowmeter. Measurements of weight as a function of time are to be made. Water enters the tank vertically from the flow-metering system at a speed of 20 ft/sec through a 1.5 in. diameter pipe. If the weight of the empty tank is 50 lbf, determine the scale reading at $t = 10$ sec. What is the true weight of the water in the tank at $t = 10$ sec?

4.51 When full, the tank of Problem 4.50 is drained by opening a valve in the drain elbow. The exit plane of the drain pipe is at the same elevation as the bottom of the tank; the drain diameter is 1 in. The tank of diameter 2.5 ft is filled to a depth of 3 ft. Determine the percent error in the scale reading 2 sec after the valve is opened. The vertical speed of efflux from the drain may be approximated by $V = \sqrt{2gh}$, where h is the height of the free surface above the exit plane of the drain pipe.

4.52 Shown in Fig. 4.27 is an adjustable fire-fighting nozzle assembly. In the box marked M is the connection that makes it possible to turn and elevate the nozzle. For the position shown, calculate the horizontal force exerted by this assembly on the supply pipe when the supply gage pressure is 500 kPa and the jet speed of water leaving the nozzle, as a free jet at atmospheric pressure, is 35 m/sec. The nozzle diameter is 25 mm.

4.53 Water flows steadily through the double nozzle assembly shown (Fig. 4.28) in a horizontal plane. The cross-sectional area of the supply line is 0.01 m², and the outlet cross-sectional area of each of the two nozzles is 0.001 m². The gage pressure of the water entering at section ① is 130 kPa. The water speed at each nozzle outlet is 15 m/sec. Calculate the net horizontal force of the nozzle assembly on the flange of the supply pipe at section ①.

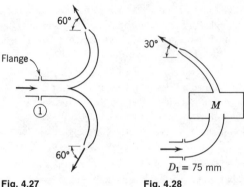

Fig. 4.27   **Fig. 4.28**

4.54 Water flows steadily through a fire hose and nozzle. The hose is 75 mm inside diameter, and the nozzle tip is 25 mm i.d.; water gage pressure in the hose is 510 kPa, and the stream leaving the nozzle is uniform. The exit speed and pressure are 32 m/sec, and atmospheric, respectively. Find the force transmitted by the coupling between the nozzle and hose. Indicate whether the coupling is in tension or compression.

4.55 Consider the steady adiabatic flow of air through a long straight pipe with a cross-sectional area of 0.5 ft^2. At the inlet, the air is at 30 psia, 140 F, and has a velocity of 500 ft/sec. At the exit, the air is at 11.3 psia and has a velocity of 985 ft/sec. Calculate the axial force of the air on the pipe. (Be sure to make the direction clear.)

4.56 A gas flows steadily through a heated porous pipe of constant 0.2 m^2 cross-sectional area. At the pipe inlet, the absolute pressure is 340 kPa, the density is 5.1 kg/m^3, and the mean velocity is 152 m/sec. The fluid passing through the porous wall leaves in a direction normal to the pipe axis, and the total flowrate through the porous wall is 29.2 kg/sec. At the pipe outlet, the absolute pressure is 280 kPa and the density is 2.6 kg/m^3. Determine the force in the axial direction of the fluid on the pipe.

4.57 Water is flowing steadily through the 180° elbow shown in Fig. 4.29. At the inlet to the elbow the gage pressure, is 96 kPa. The water discharges to atmospheric pressure. Properties are assumed to be uniform over the inlet and outlet areas; $A_1 = 2600 \text{ mm}^2$, $A_2 = 650 \text{ mm}^2$, $V_1 = 3.05$ m/sec. Find the horizontal component of the force required to hold the elbow in place.

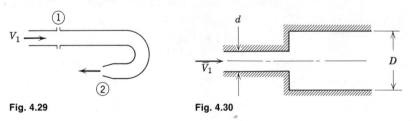

Fig. 4.29   **Fig. 4.30**

4.58 Consider flow through the sudden expansion shown in Fig. 4.30. If the flow is incompressible and friction is neglected, show that the pressure rise, $\Delta p = p_2 - p_1$, is given by

$$\frac{\Delta p}{\frac{1}{2}\rho \overline{V}_1^2} = 2\left(\frac{d}{D}\right)^2\left[1-\left(\frac{d}{D}\right)^2\right]$$

4.59 The General Electric CF6-50A jet engine has produced a thrust force of 52,000 lbf on the test stand. A typical installation is shown in Fig. 4.31, together with some test data. Fuel enters the top of the engine vertically at a rate equal to 2 percent of the mass flowrate of the inlet air. For the given conditions, compute the air flowrate through the engine in lbm/sec.

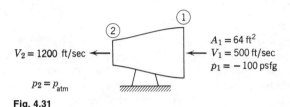

$A_1 = 64$ ft^2
$V_1 = 500$ ft/sec
$p_1 = -100$ psfg

$V_2 = 1200$ ft/sec

$p_2 = P_{atm}$

Fig. 4.31

4.60 Consider the incompressible flow of fluid in a boundary layer as depicted in Example 4.3. Show that the drag force of the fluid on the surface is given by

$$D = \int_0^\delta \rho u(U - u)w\,dy$$

Evaluate the drag force for the conditions of Example 4.3.

4.61 Experimental measurements are made in a low speed wind tunnel to determine the drag force on a circular cylinder. Velocity measurements at two sections where the pressure is uniform and equal give the results shown in Fig. 4.32. Evaluate the drag force on the cylinder, per unit width.

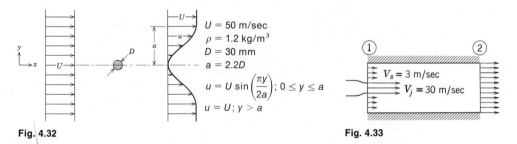

$U = 50$ m/sec
$\rho = 1.2$ kg/m^3
$D = 30$ mm
$a = 2.2D$

$u = U \sin\left(\dfrac{\pi y}{2a}\right); 0 \leq y \leq a$

$u = U; y > a$

$V_s = 3$ m/sec
$V_j = 30$ m/sec

Fig. 4.32   **Fig. 4.33**

4.62 A water jet pump has jet area, 0.01 m^2, and jet speed, 30 m/sec (Fig. 4.33). The jet is within a secondary stream of water having a speed of 3 m/sec. The total area of the duct (the sum of the jet and secondary stream areas) is 0.075 m^2. The water is thoroughly mixed and leaves the jet pump in a uniform stream.

The pressures of the jet and secondary stream are the same at the pump inlet. Determine the speed at the pump exit, and the pressure rise, $p_2 - p_1$.

4.63 A fluid of constant density, ρ, enters a pipe of radius R with a uniform velocity, V. At a downstream section the velocity varies with radius r according to the equation

$$u = 2V\left(1 - \frac{r^2}{R^2}\right)$$

The pressure at sections ① (inlet) and ② (downstream) are p_1 and p_2, respectively. Show that the frictional force, F, of the pipe walls on the fluid between sections ① and ② is

$$F = \pi R^2\left[-(p_1 - p_2) + \tfrac{1}{3}\rho V^2\right]$$

in a direction opposing the flow.

4.64 Liquid flows through a tee, as shown in Fig. 4.34. A portion of the flow is drawn off vertically through the branch ③; the remainder continues through the run and leaves at ②. Obtain an expression for the pressure change, $\Delta p = p_2 - p_1$, across the tee. Express your results in terms of inlet flow properties and the ratio Q_3/Q_1. Plot $\Delta p/\rho\bar{V}_1^2$ as a function of Q_3/Q_1.

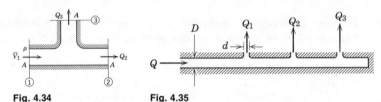

Fig. 4.34   Fig. 4.35

4.65 Water flows through a circular spray tube containing three circular jets, as shown in Fig. 4.35. The flow areas of the three jet orifices are equal, with $d = 5$ mm. The jet speed leaving each orifice may be approximated by $V = \sqrt{2p/\rho}$, where p is the static gage pressure at each orifice. The tube diameter is $D = 15$ mm, and the flowrate from orifice ③ is $Q_3 = 1.37 \times 10^{-4}$ m³/sec. Determine the pressure and flowrate at orifices ① and ②, and the pressure and flowrate that must be supplied to the spray tube.

4.66 A jet of water is directed against a vane (see Fig. 4.36), which could be a blade in a turbine or in any other piece of hydraulic machinery. The water leaves the

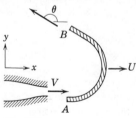

Fig. 4.36

stationary 50 mm diameter nozzle with a speed of 20 m/sec and enters the vane tangent to the surface at A. The inside surface of the vane at B makes an angle, $\theta = 150°$ with the x direction. Compute the resultant force of the fluid on the vane when the vane is moving with a constant speed, $U = 5$ m/sec, in the positive x direction.

4.67 Consider a single vane with turning angle, θ, moving horizontally at constant speed, U, under the influence of an impinging jet as shown in Fig. 4.36. The absolute speed of the jet is V. Obtain general expressions for the resultant force and power that the vane could produce. Show that the power is maximized when $U = V/3$.

4.68 Water from a stationary nozzle impinges on a moving vane with a turning angle of 120°. The vane moves away from the nozzle with constant speed, $U = 30$ ft/sec, and receives a jet that leaves the nozzle with speed, $V = 100$ ft/sec. The nozzle has an exit area of 0.04 ft². Find the force of the fluid on the moving vane.

4.69 A water jet, issuing from a stationary nozzle, encounters a vane curved through an angle of 90° that is moving away from the nozzle at a constant speed of 15 m/sec. The jet has a cross-sectional area of 600 mm² and a speed of 30 m/sec. Determine the force on the vane from the jet.

4.70 A jet of oil (SG = 0.8) strikes a curved blade that turns the fluid through an angle of 180°. The jet area is 1200 mm² and its speed relative to the stationary nozzle is 20 m/sec. The blade moves toward the nozzle at 10 m/sec. Determine the force on the blade from the jet.

4.71 A jet of lye solution (SG = 1.10) 50 mm in diameter has an absolute speed of 15 m/sec. It strikes a single flat plate that is moving away from the nozzle with an absolute speed of 5 m/sec. The plate makes an angle of 60° with the horizontal. Calculate the force on the plate from the jet. Assume no friction force along the plate surface.

4.72 A boat propelled by thrust from a water jet is to achieve a speed of 30 mph. Towing tests have shown that a force of 200 lbf is required to move the boat at this speed. The water for the pump-jet system may be assumed to enter the inlet at a speed (relative to the boat) equal to the boat speed. Inlet and outlet pressures may be assumed equal to that of undisturbed water at the same level. The pump flowrate is 900 gal/min. Determine the exit jet speed relative to the boat required to sustain the desired boat speed.

4.73 The vane/cart assembly of mass, $M = 30$ kg, shown in Fig. 4.37 is driven by a water jet. The water leaves the stationary nozzle of area, $A = 0.02$ m² with a

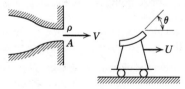

Fig. 4.37

speed of 20 m/sec. The coefficient of kinetic friction between the assembly and the surface is 0.10. Plot the terminal speed of the assembly as a function of vane turning angle θ, for $0 \le \theta \le \pi/2$. At what angle, θ, does the assembly begin to move if the coefficient of static friction is 0.15?

4.74 A sled, of mass 15 kg, is moving along a horizontal track; the coefficient of rolling friction between sled and track is 0.02. At the instant when the sled speed is 30 m/sec, a small air jet affixed to the sled is activated in an effort to reduce the sled speed. The air jet discharges to the atmosphere through a nozzle whose area is 600 mm^2. The air discharges with a speed of 150 m/sec relative to the sled and a density of 1.1 kg/m^3. Calculate the sled acceleration at the instant following activation of the jet.

4.75 A rocket sled weighing 10,000 lbf, traveling 600 mph, is to be braked by lowering a scoop into a water trough (Fig. 4.38). The scoop is 6 in. wide. Determine the acceleration when the scoop is immersed to a depth of 3 in. below the water surface.

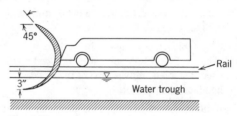

Fig. 4.38

4.76 For the conditions of Problem 4.75 determine the time required (after lowering the scoop into the water) to bring the sled to a speed of 20 mph.

4.77 A rocket sled is to be slowed from an initial speed of 300 m/sec by lowering a scoop into a water trough. The scoop is 0.3 m wide; it deflects the water through 150°. The trough is 800 m in length. The mass of the sled is 8000 kg. At the initial speed it experiences an aerodynamic drag force of 90 kN. The aerodynamic force is proportional to the square of the sled speed. It is desired to slow the sled to 100 m/sec. Determine the depth to which the scoop must be lowered into the water.

4.78 Consider the vehicle shown in Fig. 4.39. Starting from rest, it is propelled by a hydraulic catapult (liquid jet). The jet strikes the curved surface and makes a 180° turn, leaving horizontally. Air and rolling resistance may be neglected. Using the notation shown:
(a) Obtain an equation for the acceleration of the vehicle at any time, t.
(b) Determine the time required for the vehicle to reach $U = V/2$.

4.79 The mass of the cart in Problem 4.78 is 100 kg. The jet of water leaves the nozzle (area of 0.001 m^2) with a speed of 30 m/sec. Determine the speed of the cart 5 sec after the jet is directed against the cart.

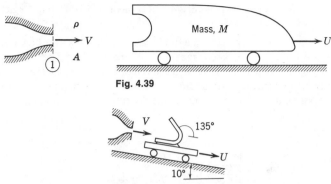

Fig. 4.39

Fig. 4.40

4.80   A jet, $\rho = 990$ kg/m³, issues from a stationary nozzle; the jet is deflected by a vane through an angle of 135°. The vane is attached to the cart as shown in Fig. 4.40. Assume that the cart moves without friction down the 10° plane. The cart mass is 100 kg. The jet leaves the nozzle at atmospheric pressure with speed $V = 65$ m/sec relative to the nozzle. If the area of the jet is $A = 500$ mm², determine the speed of the cart 5 sec after the jet is directed against the vane.

4.81   The acceleration of the vane/cart assembly shown in Fig. 4.37 is to be controlled as it accelerates from rest by changing the vane angle, θ. A constant acceleration, $a = 1.5$ m/sec², is desired. The water jet leaves the nozzle of area, $A = 0.025$ m², with a speed of 15 m/sec. The vane/cart assembly has a mass of 55 kg; neglect friction. Determine θ at $t = 5$ sec.

4.82   The vane angle, θ, of the vane/cart assembly in Problem 4.73 is set at 60°. Determine the cart speed at time, $t = 1$ sec.

4.83   A rocket sled with initial mass of 3 metric tons, including 1 ton of fuel, rests on a level section of ground. At time $t = 0$, the solid fuel of the rocket is ignited and the rocket burns fuel at the rate of 75 kg/sec. The exit speed of the exhaust gas relative to the rocket is 1000 m/sec. Neglecting friction and air resistance, calculate the accleration of the sled at time $t = 10$ sec.

4.84   A vehicle is moving at a speed of 50 ft/sec along level ground under the action of a constant force, $F = 100$ lbf. At time $t = 0$, mass begins leaving the vehicle through a hole in the bottom. Assuming the mass leaves the vehicle vertically at a rate of 10 lbm/sec and the vehicle continues to move under the action of the constant force, F, determine the speed of the vehicle after 20 sec. The initial mass of the vehicle is 2000 lbm.

4.85   A box car is moving to the left with a speed of 15 ft/sec and is accelerating in the same direction at 0.5 ft/sec². A stream of grain falls into the car from a vertical stationary pipe. The velocity of the grain is 24 ft/sec in a downward direction relative to the pipe. The mass flowrate of grain leaving the pipe is 50 lbm/sec. Determine the force required to move the car as indicated at the instant when the total mass of car and contents is 15,000 lbm. Neglect friction between the car and the track.

4.86 A manned space vehicle with a total mass of 1600 kg is traveling in level flight (Fig. 4.41). A retro-rocket on the vehicle is fired for 10 sec. The rocket discharges 8 kg/sec to ambient pressure with a speed of 2800 m/sec relative to the vehicle. Neglect any frictional drag forces. Calculate the deceleration of the vehicle at the end of the rocket firing, that is, at $t = 10$ sec.

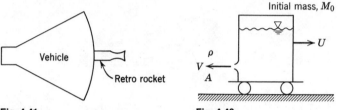

Fig. 4.41   **Fig. 4.42**

4.87 A cart is propelled by a liquid jet issuing horizontally from a tank as shown in Fig. 4.42. The track is horizontal; resistance to motion may be neglected. The tank is pressurized so that the jet speed may be considered constant. Obtain a general expression for the speed of the cart as it accelerates from rest.

4.88 A rocket sled has an initial mass of 4 metric tons including 1 ton of fuel. The motion resistance in the track on which the sled rides and that of the air total kU, where k is 75 N/m/sec, and U is the speed of the sled in m/sec. The exit speed of the exhaust gas relative to the rocket is 1500 m/sec, and it burns fuel at the rate of 90 kg/sec. Compute the speed of the sled after 10 sec.

4.89 A rocket sled of initial mass, M_0, is fired at time, $t = 0$, along a horizontal track. Fuel is burned at the rate, $\dot{m}$; the speed of the exhuast gas relative to the rocket is V. If the total motion resistance is given by kU^2 where U is the sled speed, determine the speed, U, of the rocket, as a function of time.

4.90 A rocket cart that is initially at rest and weighs 1610 lbf is to be fired and is to have a constant acceleration of 20 ft/sec² (Fig. 4.43). To accomplish this, the exhaust gases will be deflected through an angle θ that varies as a function of time. This will account for a frictional force that is proportional to the speed squared, $F = 0.002U^2$, where F is in lbf and U is in ft/sec. The rocket exhausts gases at the rate of 1 slug/sec at constant speed, $V_e = 1000$ ft/sec, relative to the vehicle.

(a) Find an expression for $\cos \theta$ as a function of time, t.
(b) Determine t at which $\cos \theta$ is a minimum.
(c) Find θ_{max}.

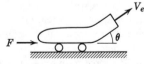

Fig. 4.43

4.91 A missile is to be launched from a straight track, inclined 30° above the horizontal. The mass of the missile is 250 kg. It burns propellant at the rate, $\dot{m} = 2$ kg/sec; the exhaust gas speed is 2750 m/sec relative to the missile. Friction between the missile and launcher is negligible. Determine the initial acceleration of the missile.

4.92 Neglecting air resistance, what speed would a vertically directed rocket attain in 10 sec if it starts from rest, has initial mass of 200 kg, burns 10 kg/sec, and ejects gas at 1200 m/sec relative to the rocket?

4.93 A "home-made" solid propellant rocket has an initial mass of 20 lbm; 15 lbm of this is fuel. The rocket is directed vertically upward from rest, burns fuel at a constant rate of 0.5 lbm/sec, and ejects exhaust gas at a speed of 3500 ft/sec relative to the rocket. Assume that the pressure at the exit is atmospheric and that air resistance may be neglected. Calculate the rocket speed after 20 sec, and the distance traveled by the rocket in 20 sec.

4.94 A small, solid fuel rocket motor is fired on a test stand. The combustion chamber is circular, with a diameter of 100 mm. Fuel, of density 1660 kg/m³, burns uniformly at the rate of 12.7 mm/sec. Measurements show that the exhaust gases leave the rocket at ambient pressure, at a speed of 2750 m/sec. The absolute pressure and temperature in the combustion chamber are 7.0 MPa and 3610 K, respectively. The combustion products may be treated as an ideal gas with a molecular mass of 25.8. Evaluate the rate of change of mass and of linear momentum within the rocket motor. Express the rate of change of linear momentum within the motor as a percentage of the motor thrust.

4.95 The moving tank shown in Fig. 4.44 is to be slowed by lowering a scoop to pick up water from a trough. The initial mass and speed of the cart and its contents are M_0 and U_0, respectively. Neglect external forces due to pressure or friction and assume that the track is horizontal. Apply the continuity and momentum equations to show that at any instant, $U = U_0 M_0 / M$. Obtain a general expression for U/U_0 as a function of time.

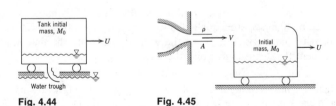

Fig. 4.44   **Fig. 4.45**

4.96 The tank shown in Fig. 4.45 rolls with negligible resistance along a horizontal track. It is to be accelerated from rest by a liquid jet that strikes the vane and is deflected into the tank. The initial mass of the tank is M_0. Use the continuity and momentum equations to show that at any instant the mass of the vehicle and liquid contents is $M = M_0 V/(V - U)$. Obtain a general expression for U/V as a function of time.

4.97   The large irrigation sprinkler (Fig. 4.46) is fabricated such that a "nozzle section" of length, l, can be mounted to the boom arm of length, L; both the boom and nozzle section have inside diameter, D; the opening at the exit of the nozzle section is of diameter, d. The boom, inclined at angle, θ, to the horizontal, rotates with angular velocity, ω, about the vertical axis. Determine the force of the nozzle section on the boom given that the weight of the nozzle section is $W = 50$ lbf and the following operating conditions:

$$D = 4 \text{ in.} \qquad d = 1 \text{ in.} \qquad \theta = 30°$$
$$L = 10 \text{ ft} \qquad l = 3 \text{ ft} \qquad \omega = 3 \text{ rpm}$$
$$V_2 = 120 \text{ ft/sec} \qquad p_1 = 100 \text{ psig}$$

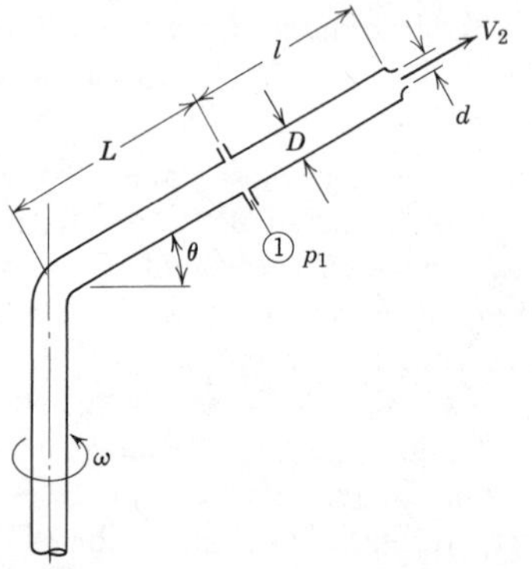

Fig. 4.46 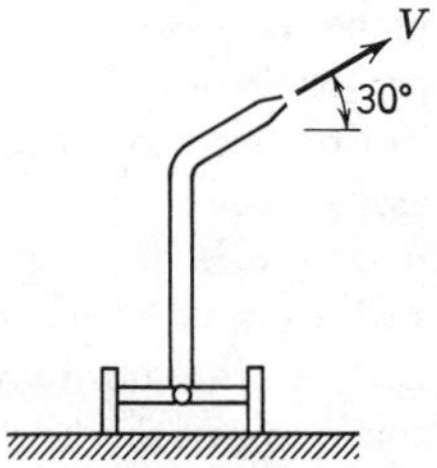   Fig. 4.47

4.98   The circular spray tube of Problem 4.65 is rotated about a vertical axis, with angular velocity, $\bar{\omega} = 10\,\hat{k}$ rpm. The spacing between the jets is $l = 0.5$ m; jet ① is located a distance, $l = 0.5$ m, from the axis of rotation. Determine the flowrate at orifices ① and ②, and the pressure and flowrate that must be supplied to the spray tube. Plot the pressure distribution along the tube centerline as a function of radial position.

4.99   A large irrigation sprinkler unit, mounted on a cart, discharges water with a speed of 40 m/sec at an angle of 30° to the horizontal (Fig. 4.47). The 50 mm diameter nozzle is 3 m above the ground. Calculate the magnitude of the moment that tends to overturn the cart.

4.100 Crude oil (SG = 0.95) from a tanker dock flows through a pipe of 0.4 m diameter in the configuration shown in Fig. 4.48. The flowrate is $0.58 \text{ m}^3/\text{sec}$, and the gage pressures are shown in the diagram. Determine the force and torque that are exerted by the pipe assembly on its supports.

4.101 Liquid in a thin sheet of width, w, and thickness, h, flows from a slot and strikes a stationary inclined flat plate, as shown in Fig. 4.49. Experiments show that the resultant force of the liquid jet on the plate does not act through point O, where the jet centerline intersects the plate. Determine the magnitude and line of application of the resultant force as functions of θ. Evaluate the equilibrium angle of the plate if a force is applied at point O. Neglect any viscous effects.

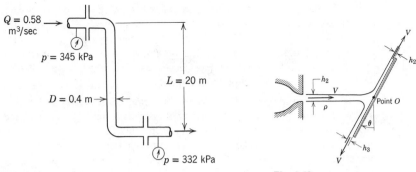

Fig. 4.48   **Fig. 4.49**

4.102 A Pelton wheel is a form of water turbine well adapted to situations of high head and low flowrate. The wheel consists of a series of vanes mounted on a rotor, as shown in Fig. 4.50. One or more jets are arranged to strike the buckets tangentially. In practice it is possible to deflect the jet stream through angle, θ, of up to 165°. Consider the Pelton wheel and single jet arrangement shown in Fig. 4.50. Obtain an expression for the torque exerted by the water stream on the wheel and the corresponding power output. (Let the bucket speed be $U = \omega R$.) Determine the value of U/V required to maximize the power produced by the wheel.

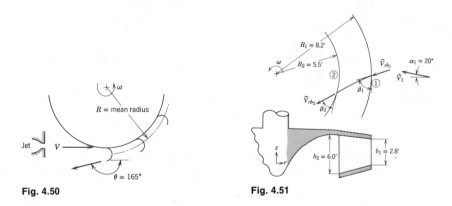

Fig. 4.50   **Fig. 4.51**

4.103 The hydraulic turbines at Grand Coulee Dam on the Columbia River are of the Francis, or radial inflow, type. [Francis turbines generally are preferred for locations with medium available head (80 to 600 ft) and in large installations can produce 90 to 95 percent of the theoretically possible power output.] The simplified geometry of the installation is shown schematically in Fig. 4.51. Design conditions for this installation call for a turbine speed of 120 rpm at a head of 330 ft. The absolute inlet speed of the water is given by $V_1 = 0.95\sqrt{2gH}$, where H is the design head. Assume that flow is steady and uniform at each section. Assume also that the flow enters and leaves the runner tangent to the blade tips, and that the absolute velocity at the turbine outlet is purely radial. Using the conditions shown in Fig. 4.51, develop the velocity polygons for the flow entering and leaving the runner. Determine the runner blade angles, β_1 and β_2, and the torque and power produced by the turbine.

4.104 A centrifugal pump operates at 1750 rpm. The impeller geometry is shown in Fig. 4.52. Water enters without swirl at volumetric flowrate, Q. The inlet and outlet radii and impeller passage height are shown on the diagram. The flow may be assumed to be steady and uniform at each section, and to enter and leave the impeller tangent to the blades. Construct the velocity polygons for the entering and leaving flows. Determine the design flowrate, in gallons per minute, for $\beta_1 = 30^0$. Evaluate the torque and power required to drive the pump, as functions of flowrate, for $\beta_2 = 75, 90,$ and 105°.

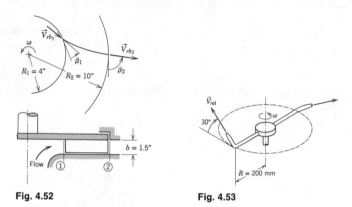

Fig. 4.52   **Fig. 4.53**

4.105 A small lawn sprinkler is shown in Fig. 4.53. The sprinkler operates at an inlet gage pressure of 140 kPa. The total flowrate of water through the sprinkler is 4.0 liters/min. Each jet discharges at a speed of 17 m/sec relative to the sprinkler arm, in a direction inclined 30° above the horizontal. The sprinkler rotates about a vertical axis. Friction in the bearing causes a torque of 0.18 N·m opposing rotation. Determine the steady speed of rotation of the sprinkler and the approximate area covered by the spray.

4.106 Air enters a compressor at standard conditions with a speed of 75 m/sec and leaves at an absolute pressure and temperature of 200 kPa and 345 K, with a speed of 125 m/sec. The flowrate is 1 kg/sec. The cooling water circulating around the compressor casing removes 18 kJ/kg of air. Determine the power required by the compressor.

4.107 Air enters a compressor at 14 psia, 80 F with negligible velocity and is discharged at 70 psia, 500 F with a velocity of 500 ft/sec. If the power input is 3200 hp and the flowrate is 20 lbm/sec, determine the heat transferred in Btu/lbm of air.

4.108 Air is drawn from the atmosphere into a turbomachine. At the exit, conditions are gage pressure of 500 kPa and 130 C. The exit speed is 100 m/sec, and the mass flowrate is 0.8 kg/sec. Flow is steady and there is no heat transfer. Compute the shaft work interaction with the surroundings, in kW.

4.109 A turbine is supplied by 0.6 m³/sec of water from a 0.3 m pipe; the discharge pipe has a 0.4 m diameter. Determine the pressure drop across the turbine if it delivers 60 kW.

4.110 A pump draws water from a reservoir through a 150 mm diameter suction pipe and delivers it to a 75 mm diameter discharge pipe. The end of the suction pipe is 2 m below the free surface of the reservoir. The pressure gage on the discharge pipe (2 m above the reservoir surface) reads 170 kPa. The average speed in the discharge pipe is 3 m/sec. If the pump has an efficiency of 75 percent, determine the power required to drive the pump.

4.111 A pump system in a dishwasher circulates water at 55 gal/min. The total head, $p/\rho g + V^2/2g$, leaving the pump is 120 in. of water. The head at the pump inlet may be neglected. If the power supplied to the pump is 0.4 hp, determine its efficiency.

4.112 Sand, at the rate of 50 kg/sec, falls from a hopper onto a moving conveyor belt as shown in Fig. 4.54. Any friction in the drive pulleys and support rollers may be neglected, as they are supported on air bearings. End effects at the discharge end may also be neglected. If $\theta = 30°$, $L = 175$ m, $R = 1$ m, and $V = 3$ m/sec, determine the power required to drive the conveyor.

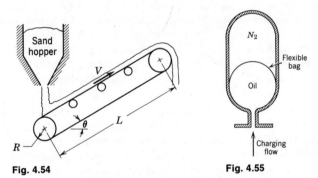

Fig. 4.54   **Fig. 4.55**

4.113 A hydraulic accumulator consists of a pressure vessel with a flexible bag inside, as shown in Fig. 4.55. The space surrounding the bag is precharged with nitrogen gas to a gage pressure of 15 MPa. The bag is to be filled rapidly with hydraulic oil to a maximum allowable gage pressure of 30 MPa. The gas may be assumed to undergo an isentropic compression process, such that $pv^k =$ constant, where k is the ratio of specific heats. The accumulator is to store 0.005 m³ of hydraulic oil. Determine the initial volume of the charge gas and the energy stored in the accumulator when it is fully charged.

4.114 The fluidic device of Problem 4.47 is built with input jets of equal diameter. Flow control is accomplished by variations in jet speed. Obtain an expression for the cone angle, θ, in terms of V_2/V_1. Assume that the kinetic energy of the exit stream is k times the total kinetic energy of the inlet streams, where $0 < k \le 1$.

4.115 The propulsive efficiency, η, of a jet-propelled device is defined as the ratio of the useful work output to the mechanical energy input to the fluid. Determine the propulsive efficiency for the jet boat of Problem 4.72.

4.116 A jet-propelled aircraft traveling at 200 m/sec takes in 40 kg/sec of air and discharges it at 500 m/sec relative to the aircraft. Determine the propulsive efficiency (defined as the ratio of the useful work output to the mechanical energy input to the fluid) of the aircraft.

4.117 A fire truck draws water from an open constant level reservoir through a hose of 0.2 ft² cross section. The water is discharged from the pump through a 0.2 ft² hose to a surge tank pressurized to 120 psig with air. The pump adds 970 ft·lbf/slug of energy to the fluid; the water level in the surge tank is 3 ft above the pump centerline. From the surge tank the water flows steadily through a hose of 0.04 ft² area to a nozzle of 0.01 ft² area. The nozzle is held at the same level as the water free surface in the surge tank. Neglect all frictional losses and sketch the system. Find:
(a)   Velocity of the water leaving the nozzle.
(b)   Elevation of the pump relative to the reservoir.
(c)   Power required to drive the pump if it has an efficiency of 80 percent.
(d)   Maximum height above the reservoir that the stream of water can reach.

4.118 All major harbors are equipped with "fire boats" for combating ship fires. A 75 mm diameter hose is attached to the discharge of a 10 kW pump on such a boat. The nozzle attached to the end of the hose has a diameter of 25 mm. If the nozzle discharge is held 3 m above the surface of the water determine the volume flowrate through the nozzle, the maximum height to which the water will rise, and the force on the boat if the water jet is directed horizontally over the stern.

4.119 The total mass of the helicopter-type craft shown in Fig. 4.56 is 1500 kg. The pressure of the air is atmospheric at both inlet and outlet. Assume that the flow is steady and one-dimensional. Treat the air as being incompressible at standard conditions and calculate for a hovering position the speed of air leaving the craft, and the minimum power that must be delivered to the air by the propeller.

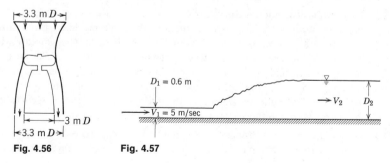

Fig. 4.56   **Fig. 4.57**

4.120 Liquid flowing at high speed in a wide, horizontal open channel under some conditions can undergo a hydraulic jump, as shown in Fig. 4.57. For a suitably chosen control volume, the flows entering and leaving the jump may be considered uniform with hydrostatic pressure distribution (see Example Problem 4.6). Consider a channel of width, w, with water flow at $D_1 = 0.6$ m and $V_1 = 5$ m/sec. Show that in general, $D_2 = D_1[\sqrt{1 + 8V_1^2/gD_1} - 1]/2$. Evaluate the change in mechanical energy through the hydraulic jump. If heat transfer to the surroundings is negligible, determine the change in water temperature through the jump.

Chapter 5

Introduction to Differential
Analysis of Fluid Motion

In Chapter 4, we developed the basic equations in integral form for a control volume. The integral equations are particularly useful when we are interested in the gross behavior of a flow field and its effect on various devices. However, the integral approach does not provide a detailed (i.e. point by point) knowledge of the flow field.

To obtain this detailed knowledge, we must apply the equations of fluid motion in differential form. In this chapter we shall develop differential equations for the conservation of mass and Newton's second law of motion. Since we are interested in formulating differential equations, our analysis will be in terms of infinitesimal systems and control volumes.

5-1 REVIEW OF THE FIELD CONCEPT

In Chapter 2, we found that the continuum assumption—that is, the assumption that a fluid could be treated as a continuous distribution of matter—led directly to a field representation of fluid properties. The property fields are defined by continuous functions of the space coordinates and time. We may speak of a scalar field $[\rho = \rho(x, y, z, t)]$, a vector field $[\vec{V} = \vec{V}(x, y, z, t)]$ or a tensor field (the stress field).

5-2 THE CONTINUITY EQUATION

We shall derive the differential equation for the conservation of mass in both rectangular and cylindrical coordinates. In both cases the derivation is carried out by applying conservation of mass to a differential control volume.

5-2.1 RECTANGULAR COORDINATE SYSTEM

In rectangular coordinates, the control volume chosen is an infinitesimal cube with sides of length dx, dy, dz as shown in Fig. 5.1. The density at the center, O, of the control volume is ρ and the velocity there is $\vec{V} = \hat{\imath}u + \hat{\jmath}v + \hat{k}w$.

To determine the values of the properties at each of the six faces of the control surface, we must use a Taylor series expansion of the properties about the point O. For example, at the right face,

$$\left.\rho\right)_{x+dx/2} = \rho + \left(\frac{\partial \rho}{\partial x}\right)\frac{dx}{2} + \left(\frac{\partial^2 \rho}{\partial x^2}\right)\frac{1}{2!}\left(\frac{dx}{2}\right)^2 + \cdots$$

Neglecting higher order terms, we can write

$$\left.\rho\right)_{x+dx/2} = \rho + \left(\frac{\partial \rho}{\partial x}\right)\frac{dx}{2}$$

Similarly,

$$\left.u\right)_{x+dx/2} = u + \left(\frac{\partial u}{\partial x}\right)\frac{dx}{2}$$

The corresponding terms at the left face are

$$\left.\rho\right)_{x-dx/2} = \rho + \left(\frac{\partial \rho}{\partial x}\right)\left(-\frac{dx}{2}\right) = \rho - \left(\frac{\partial \rho}{\partial x}\right)\frac{dx}{2}$$

$$\left.u\right)_{x-dx/2} = u + \left(\frac{\partial u}{\partial x}\right)\left(-\frac{dx}{2}\right) = u - \left(\frac{\partial u}{\partial x}\right)\frac{dx}{2}$$

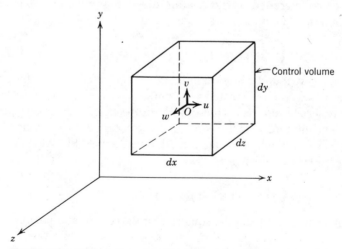

Fig. 5.1 Differential control volume in rectangular coordinates.

A word statement of the conservation of mass is

$$\begin{bmatrix} \text{net rate of mass efflux} \\ \text{through the control surface} \end{bmatrix} + \begin{bmatrix} \text{rate of change of mass} \\ \text{inside the control volume} \end{bmatrix} = 0$$

To evaluate the first term in this equation, we must consider the mass flux through each of the six surfaces of the control surface, that is, we must evaluate $\int_{CS} \rho \vec{V} \cdot d\vec{A}$. The details of this evaluation are shown in Table 5.1.

We see that the net rate of mass efflux through the control surface is given by

$$\left[\frac{\partial \rho u}{\partial x} + \frac{\partial \rho v}{\partial y} + \frac{\partial \rho w}{\partial z} \right] dx\, dy\, dz$$

The mass inside the control volume at any instant is the product of the mass per unit volume, ρ, and the volume, $dx\, dy\, dz$. Thus the rate of change of mass

TABLE 5.1 **Mass Flux through the Control Surface of a Rectangular Differential Control Volume**

Surface	$\int \rho \vec{V} \cdot d\vec{A}$
Left $(-x)$	$= -\left[\rho - \left(\frac{\partial \rho}{\partial x}\right)\frac{dx}{2}\right]\left[u - \left(\frac{\partial u}{\partial x}\right)\frac{dx}{2}\right]dy\,dz = -\rho u\,dy\,dz + \frac{1}{2}\left[u\left(\frac{\partial \rho}{\partial x}\right) + \rho\left(\frac{\partial u}{\partial x}\right)\right]dx\,dy\,dz$
Right $(+x)$	$= \left[\rho + \left(\frac{\partial \rho}{\partial x}\right)\frac{dx}{2}\right]\left[u + \left(\frac{\partial u}{\partial x}\right)\frac{dx}{2}\right]dy\,dz = \rho u\,dy\,dz + \frac{1}{2}\left[u\left(\frac{\partial \rho}{\partial x}\right) + \rho\left(\frac{\partial u}{\partial x}\right)\right]dx\,dy\,dz$
Bottom $(-y)$	$= -\left[\rho - \left(\frac{\partial \rho}{\partial y}\right)\frac{dy}{2}\right]\left[v - \left(\frac{\partial v}{\partial y}\right)\frac{dy}{2}\right]dx\,dz = -\rho v\,dx\,dz + \frac{1}{2}\left[v\left(\frac{\partial \rho}{\partial y}\right) + \rho\left(\frac{\partial v}{\partial y}\right)\right]dx\,dy\,dz$
Top $(+y)$	$= \left[\rho + \left(\frac{\partial \rho}{\partial y}\right)\frac{dy}{2}\right]\left[v + \left(\frac{\partial v}{\partial y}\right)\frac{dy}{2}\right]dx\,dz = \rho v\,dx\,dz + \frac{1}{2}\left[v\left(\frac{\partial \rho}{\partial y}\right) + \rho\left(\frac{\partial v}{\partial y}\right)\right]dx\,dy\,dz$
Back $(-z)$	$= -\left[\rho - \left(\frac{\partial \rho}{\partial z}\right)\frac{dz}{2}\right]\left[w - \left(\frac{\partial w}{\partial z}\right)\frac{dz}{2}\right]dx\,dy = -\rho w\,dx\,dy + \frac{1}{2}\left[w\left(\frac{\partial \rho}{\partial z\cdot}\right) + \rho\left(\frac{\partial w}{\partial z}\right)\right]dx\,dy\,dz$
Front $(+z)$	$= \left[\rho + \left(\frac{\partial \rho}{\partial z}\right)\frac{dz}{2}\right]\left[w + \left(\frac{\partial w}{\partial z}\right)\frac{dz}{2}\right]dx\,dy = \rho w\,dx\,dy + \frac{1}{2}\left[w\left(\frac{\partial \rho}{\partial z}\right) + \rho\left(\frac{\partial w}{\partial z}\right)\right]dx\,dy\,dz$

Then,

$$\int_{CS} \rho \vec{V} \cdot d\vec{A} = \left[\left\{u\left(\frac{\partial \rho}{\partial x}\right) + \rho\left(\frac{\partial u}{\partial x}\right)\right\} + \left\{v\left(\frac{\partial \rho}{\partial y}\right) + \rho\left(\frac{\partial v}{\partial y}\right)\right\} + \left\{w\left(\frac{\partial \rho}{\partial z}\right) + \rho\left(\frac{\partial w}{\partial z}\right)\right\}\right] dx\,dy\,dz$$

or

$$\int_{CS} \rho \vec{V} \cdot d\vec{A} = \left[\frac{\partial \rho u}{\partial x} + \frac{\partial \rho v}{\partial y} + \frac{\partial \rho w}{\partial z}\right] dx\,dy\,dz$$

inside the control volume is given by

$$\frac{\partial \rho}{\partial t} \, dx \, dy \, dz$$

In rectangular coordinates the differential equation for the conservation of mass is then

$$\frac{\partial \rho u}{\partial x} + \frac{\partial \rho v}{\partial y} + \frac{\partial \rho w}{\partial z} + \frac{\partial \rho}{\partial t} = 0 \qquad (5.1a)$$

Equation 5.1a may be written more compactly in vector notation, since

$$\frac{\partial \rho u}{\partial x} + \frac{\partial \rho v}{\partial y} + \frac{\partial \rho w}{\partial z} = \nabla \cdot \rho \vec{V}$$

where ∇ in rectangular coordinates is given by

$$\nabla = \hat{\imath} \frac{\partial}{\partial x} + \hat{\jmath} \frac{\partial}{\partial y} + \hat{k} \frac{\partial}{\partial z}$$

Then the conservation of mass may be written as

$$\nabla \cdot \rho \vec{V} + \frac{\partial \rho}{\partial t} = 0 \qquad (5.1b)$$

Two flow cases for which the differential continuity equation may be simplified are worthy of note. For incompressible flow, $\rho = $ constant; that is, the density is neither a function of space coordinates, nor time. For incompressible flow, the continuity equation simplifies to

$$\frac{\partial u}{\partial x} + \frac{\partial v}{\partial y} + \frac{\partial w}{\partial z} = 0 \qquad \text{\emph{Incompressible}}$$

or

$$\nabla \cdot \vec{V} = 0$$

Thus the velocity field, $\vec{V}(x, y, z, t)$, for incompressible flow must satisfy $\nabla \cdot \vec{V} = 0$.

For steady flow, all fluid properties are, by definition, independent of time. Thus at most $\rho = \rho(x, y, z)$, and for steady flow, the continuity equation can be written as

$$\frac{\partial \rho u}{\partial x} + \frac{\partial \rho v}{\partial y} + \frac{\partial \rho w}{\partial z} = 0 \qquad \text{\emph{steady flow}}$$

or

$$\nabla \cdot \rho \vec{V} = 0$$

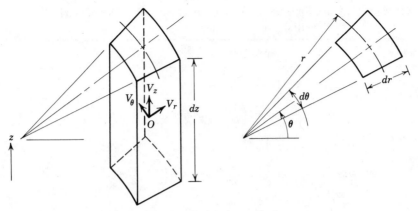

Fig. 5.2 Differential control volume in cylindrical coordinates.

5-2.2 CYLINDRICAL COORDINATE SYSTEM

In cylindrical coordinates, a suitable differential control volume is shown in Fig. 5.2. The density at the center, O, of the control volume is ρ and the velocity there is $\vec{V} = \hat{i}_r V_r + \hat{i}_\theta V_\theta + \hat{i}_z V_z$, where $\hat{i}_r$, $\hat{i}_\theta$, and $\hat{i}_z$ are unit vectors in the r, θ, and z directions, respectively, and V_r, V_θ, and V_z are the velocity components in the r, θ, and z directions, respectively.

The conservation of mass states that:

$$\begin{bmatrix} \text{net rate of mass efflux} \\ \text{through the control surface} \end{bmatrix} + \begin{bmatrix} \text{rate of change of mass} \\ \text{inside the control volume} \end{bmatrix} = 0$$

To evaluate the first term in this equation, we must consider the mass flux through each of the six faces of the control surface; that is, we must evaluate $\int_{cs} \rho \vec{V} \cdot d\vec{A}$. The properties at each of the six faces of the control surface are obtained from a Taylor series expansion of the properties about the point O. The details of the mass flux evaluation are shown in Table 5.2.

We see that the net rate of mass efflux through the control surface is given by

$$\left[\rho V_r + r \frac{\partial \rho V_r}{\partial r} + \frac{\partial \rho V_\theta}{\partial \theta} + r \frac{\partial \rho V_z}{\partial z} \right] dr\, d\theta\, dz$$

The mass inside the control volume at any instant is the product of the mass per unit volume, ρ, and the volume, $r\, d\theta\, dr\, dz$. Thus the rate of change of mass inside the control volume is given by

$$\frac{\partial \rho}{\partial t} r\, d\theta\, dr\, dz$$

TABLE 5.2 Mass Flux through the Control Surface of a Cylindrical Differential Control Volume

Surface	$\int \rho \vec{V} \cdot d\vec{A}$
Inside $(-r)$	$= -\left[\rho - \left(\dfrac{\partial \rho}{\partial r}\right)\dfrac{dr}{2}\right]\left[V_r - \left(\dfrac{\partial V_r}{\partial r}\right)\dfrac{dr}{2}\right]\left(r - \dfrac{dr}{2}\right)d\theta\,dz$
	$= -\rho V_r r\,d\theta\,dz + \rho V_r \dfrac{dr}{2}d\theta\,dz + \rho\left(\dfrac{\partial V_r}{\partial r}\right)r\dfrac{dr}{2}d\theta\,dz + V_r\left(\dfrac{\partial \rho}{\partial r}\right)r\dfrac{dr}{2}d\theta\,dz$
Outside $(+r)$	$= \left[\rho + \left(\dfrac{\partial \rho}{\partial r}\right)\dfrac{dr}{2}\right]\left[V_r + \left(\dfrac{\partial V_r}{\partial r}\right)\dfrac{dr}{2}\right]\left(r + \dfrac{dr}{2}\right)d\theta\,dz$
	$= \rho V_r r\,d\theta\,dz + \rho V_r \dfrac{dr}{2}d\theta\,dz + \rho\left(\dfrac{\partial V_x}{\partial r}\right)r\dfrac{dr}{2}d\theta\,dz + V_r\left(\dfrac{\partial \rho}{\partial r}\right)r\dfrac{dr}{2}d\theta\,dz$
Front $(-\theta)$	$= -\left[\rho - \left(\dfrac{\partial \rho}{\partial \theta}\right)\dfrac{d\theta}{2}\right]\left[V_\theta - \left(\dfrac{\partial V_\theta}{\partial \theta}\right)\dfrac{d\theta}{2}\right]dr\,dz$
	$= -\rho V_\theta\,dr\,dz + \rho\left(\dfrac{\partial V_\theta}{\partial \theta}\right)\dfrac{d\theta}{2}dr\,dz + V_\theta\left(\dfrac{\partial \rho}{\partial \theta}\right)\dfrac{d\theta}{2}dr\,dz$
Back $(+\theta)$	$= \left[\rho + \left(\dfrac{\partial \rho}{\partial \theta}\right)\dfrac{d\theta}{2}\right]\left[V_\theta + \left(\dfrac{\partial V_\theta}{\partial \theta}\right)\dfrac{d\theta}{2}\right]dr\,dz$
	$= \rho V_\theta\,dr\,dz + \rho\left(\dfrac{\partial V_\theta}{\partial \theta}\right)\dfrac{d\theta}{2}dr\,dz + V_\theta\left(\dfrac{\partial \rho}{\partial \theta}\right)\dfrac{d\theta}{2}dr\,dz$
Bottom $(-z)$	$= -\left[\rho - \left(\dfrac{\partial \rho}{\partial z}\right)\dfrac{dz}{2}\right]\left[V_z - \left(\dfrac{\partial V_z}{\partial z}\right)\dfrac{dz}{2}\right]r\,d\theta\,dr$
	$= -\rho V_z r\,d\theta\,dr + \rho\left(\dfrac{\partial V_z}{\partial z}\right)\dfrac{dz}{2}r\,d\theta\,dr + V_z\left(\dfrac{\partial \rho}{\partial z}\right)\dfrac{dz}{2}r\,d\theta\,dr$
Top $(+z)$	$= \left[\rho + \left(\dfrac{\partial \rho}{\partial z}\right)\dfrac{dz}{2}\right]\left[V_z + \left(\dfrac{\partial V_z}{\partial z}\right)\dfrac{dz}{2}\right]r\,d\theta\,dr$
	$= \rho V_z r\,d\theta\,dr + \rho\left(\dfrac{\partial V_z}{\partial z}\right)\dfrac{dz}{2}r\,d\theta\,dr + V_z\left(\dfrac{\partial \rho}{\partial z}\right)\dfrac{dz}{2}r\,d\theta\,dr$

Then,

$$\int_{cs} \rho \vec{V} \cdot d\vec{A} = \left[\rho V_r + r\left\{\rho\left(\dfrac{\partial V_r}{\partial r}\right) + V_r\left(\dfrac{\partial \rho}{\partial r}\right)\right\} + \left\{\rho\left(\dfrac{\partial V_\theta}{\partial \theta}\right) + V_\theta\left(\dfrac{\partial \rho}{\partial \theta}\right)\right\} + r\left\{\rho\left(\dfrac{\partial V_z}{\partial z}\right) + V_z\left(\dfrac{\partial \rho}{\partial z}\right)\right\}\right]dr\,d\theta\,dz$$

or

$$\int_{cs} \rho \vec{V} \cdot d\vec{A} = \left[\rho V_r + r\dfrac{\partial \rho V_r}{\partial r} + \dfrac{\partial \rho V_\theta}{\partial \theta} + r\dfrac{\partial \rho V_z}{\partial z}\right]dr\,d\theta\,dz$$

In cylindrical coordinates the differential equation for the conservation of mass is then

$$\rho V_r + r\frac{\partial \rho V_r}{\partial r} + \frac{\partial \rho V_\theta}{\partial \theta} + r\frac{\partial \rho V_z}{\partial z} + r\frac{\partial \rho}{\partial t} = 0$$

Dividing by r gives

$$\frac{\rho V_r}{r} + \frac{\partial \rho V_r}{\partial r} + \frac{1}{r}\frac{\partial \rho V_\theta}{\partial \theta} + \frac{\partial \rho V_z}{\partial z} + \frac{\partial \rho}{\partial t} = 0$$

or

$$\frac{1}{r}\frac{\partial r\rho V_r}{\partial r} + \frac{1}{r}\frac{\partial \rho V_\theta}{\partial \theta} + \frac{\partial \rho V_z}{\partial z} + \frac{\partial \rho}{\partial t} = 0 \qquad (5.2)$$

In cylindrical coordinates the vector operator, ∇, is given by

$$\nabla = \hat{\imath}_r \frac{\partial}{\partial r} + \hat{\imath}_\theta \frac{1}{r}\frac{\partial}{\partial \theta} + \hat{\imath}_z \frac{\partial}{\partial z}$$

Thus in vector notation the conservation of mass may be written[1]

$$\nabla \cdot \rho \vec{V} + \frac{\partial \rho}{\partial t} = 0$$

For incompressible flow, ρ = constant, and Eq. 5.2 reduces to

$$\frac{1}{r}\frac{\partial r V_r}{\partial r} + \frac{1}{r}\frac{\partial V_\theta}{\partial \theta} + \frac{\partial V_z}{\partial z} = 0$$

For steady flow, Eq. 5.2 reduces to

$$\frac{1}{r}\frac{\partial r\rho V_r}{\partial r} + \frac{1}{r}\frac{\partial \rho V_\theta}{\partial \theta} + \frac{\partial \rho V_z}{\partial z} = 0$$

Example 5.1
For a two-dimensional flow, the x component of velocity is given by $u = ax^2 - bx + by$. Determine a possible y component for steady, incompressible flow. How many possible y components are there?

Example Problem 5.1

GIVEN:

Two-dimensional flow in which $u = ax^2 - bx + by$.

FIND:

Possible v for steady, two-dimensional, incompressible flow.

---

[1] In carrying out the divergence operation in cylindrical coordinates we must remember that

$$\frac{\partial \hat{\imath}_r}{\partial \theta} = \hat{\imath}_\theta \quad \text{and} \quad \frac{\partial \hat{\imath}_\theta}{\partial \theta} = -\hat{\imath}_r$$

(This may be shown using the method of Example Problem 4.12.)

SOLUTION:

Basic equation: $\nabla \cdot \rho\vec{V} + \dfrac{\partial \rho}{\partial t} = 0$

For incompressible flow, ρ = constant, and we can write $\nabla \cdot \vec{V} = 0$. In rectangular coordinates

$$\frac{\partial u}{\partial x} + \frac{\partial v}{\partial y} + \frac{\partial w}{\partial z} = 0$$

Assume two-dimensional flow in the xy plane. Then partial derivatives with respect to z are zero, and

$$\frac{\partial u}{\partial x} + \frac{\partial v}{\partial y} = 0$$

Then
$$\frac{\partial v}{\partial y} = -\frac{\partial u}{\partial x} = -2ax + b$$

{This gives an expression for the rate of change of v holding x constant.}

This equation can be integrated to obtain an expression for v. On integrating, we obtain for steady flow

$v = -2axy + by + f(x)$ {The function of x, that is, $f(x)$, appears because we had the partial derivative of v with respect to y.}

Since any function $f(x)$ is allowable, any number of expressions for v could satisfy the differential continuity equation under the given conditions. The simplest expression for v would be obtained by setting $f(x) = 0$, that is,

$$v = -2axy + by$$

{This "artificial" problem illustrates the use of the differential continuity equation and points out the difference between a partial and a total derivative.}

Example 5.2

A compressible flow field is described by

$$\rho\vec{V} = [ax\hat{\imath} - bxy\hat{\jmath}]e^{-kt}$$

where x and y are coordinates in meters, t is time in seconds, and a, b, and k are constants with appropriate units so that ρ and $\vec{V}$ are in kg/m^3 and m/sec, respectively. Calculate the rate of change of density per unit time at the point $x = 3$ m, $y = 2$ m, $z = 2$ m, for $t = 0$.

Example Problem 5.2

GIVEN:

A compressible flow field, $\rho \vec{V} = [ax\hat{\imath} - bxy\hat{\jmath}]e^{-kt}$

x,y = coordinates in m

t = time in sec

a = constant with units of kg/m$^3 \cdot$ sec

b = constant with units of kg/m$^4 \cdot$ sec

k = constant with units of sec^{-1}

FIND:

$\partial \rho / \partial t$ at point (3, 2, 2) at time $t = 0$.

SOLUTION:

Basic equation: $\nabla \cdot \rho \vec{V} + \dfrac{\partial \rho}{\partial t} = 0$

$$\frac{\partial \rho}{\partial t} = -\nabla \cdot \rho \vec{V} = -\left[\hat{\imath} \frac{\partial}{\partial x} + \hat{\jmath} \frac{\partial}{\partial y} + \hat{k} \frac{\partial}{\partial z} \right] \cdot [ax\hat{\imath} - bxy\hat{\jmath}]e^{-kt}$$

$$\frac{\partial \rho}{\partial t} = -[a - bx]e^{-kt} = (bx - a)e^{-kt}$$

For $t = 0$, at the point (3, 2, 2)

$$\frac{\partial \rho}{\partial t} = \left(3 \text{ m} \times \frac{b}{\text{m}^4 \cdot \text{sec}} \frac{\text{kg}}{\text{m}^4 \cdot \text{sec}} - \frac{a}{\text{m}^3 \cdot \text{sec}} \frac{\text{kg}}{\text{m}^3 \cdot \text{sec}} \right) e^{-k(0)} = (3b - a) \text{ kg/m}^3 \cdot \text{sec} \qquad \underset{\longleftarrow}{\frac{\partial \rho}{\partial t}}$$

$\left\{ \begin{array}{l} \text{This "artificial" problem illustrates the use of the differential continuity equation} \\ \text{in vector form.} \end{array} \right\}$

**5-3 STREAM FUNCTION FOR TWO-DIMENSIONAL INCOMPRESSIBLE FLOW

It is convenient to have a means of concisely describing the form of any particular pattern of flow. An adequate description should portray the notion of the shape of the streamlines (including the boundaries) and the scale of the velocity at representative points in the flow. A mathematical device that

** This section may be omitted without loss of continuity in the text material.

serves this purpose is the stream function, ψ. The stream function is formulated as a relation between the streamlines and the statement of the conservation of mass.

For a two-dimensional incompressible flow in the xy plane, the conservation of mass, Eq. 5.1a, can be written

$$\frac{\partial u}{\partial x} + \frac{\partial v}{\partial y} = 0 \qquad (5.3)$$

If a continuous function, $\psi(x, y, t)$, called the stream function, is defined such that

---

$$u \equiv \frac{\partial \psi}{\partial y} \qquad \text{and} \qquad v \equiv -\frac{\partial \psi}{\partial x} \qquad (5.4)$$

---

then the continuity equation, Eq. 5.3, is satisfied exactly, since

$$\frac{\partial u}{\partial x} + \frac{\partial v}{\partial y} = \frac{\partial^2 \psi}{\partial x \, \partial y} - \frac{\partial^2 \psi}{\partial y \, \partial x} = 0$$

Recall that streamlines are lines drawn in the flow field, such that, at a given instant of time, they are tangent to the direction of flow at every point in the flow field. Thus if $d\vec{r}$ is an element of length along a streamline, the equation of the streamline is given by

$$\vec{V} \times d\vec{r} = 0 = (\hat{\imath}u + \hat{\jmath}v) \times (\hat{\imath}\,dx + \hat{\jmath}\,dy)$$
$$= \hat{k}(u\,dy - v\,dx)$$

That is, the equation of a streamline in a two-dimensional flow is

$$u\,dy - v\,dx = 0$$

Substituting for the velocity components u and v in terms of the stream function, ψ, from Eq. 5.4, then along a streamline

$$\frac{\partial \psi}{\partial x}\,dx + \frac{\partial \psi}{\partial y}\,dy = 0 \qquad (5.5)$$

Since $\psi = \psi(x, y, t)$, then at an instant, t_0, $\psi = \psi(x, y, t_0)$; at this instant, a change in ψ may be evaluated as though $\psi = \psi(x, y)$. Thus at any instant

$$d\psi = \frac{\partial \psi}{\partial x}\,dx + \frac{\partial \psi}{\partial y}\,dy \qquad (5.6)$$

Comparing Eqs. 5.5 and 5.6, we see that along an instantaneous streamline, $d\psi = 0$; that is, ψ is a constant along a streamline. Since the differential of

ψ is exact, the integral of $d\psi$ between any two points in a flow field, that is, $\psi_2 - \psi_1$, is dependent only on the end points of integration.

From the definition of a streamline, we recognize that there can be no flow across a streamline. Thus, if the streamlines in a two-dimensional, incompressible flow field, at a given instant of time, are as shown in Fig. 5.3, the rate of flow between streamlines ψ_1 and ψ_2 across the lines AB, BC, DE, and DF must be equal.

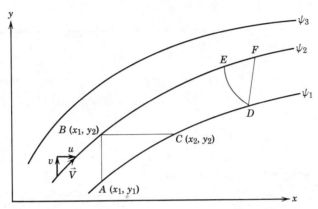

Fig. 5.3 Instantaneous streamlines in a two-dimensional flow.

The volumetric flowrate, Q, between the streamlines ψ_1 and ψ_2 can be evaluated by considering the flow across AB or across BC. For a unit depth, the flowrate across AB is

$$Q = \int_{y_1}^{y_2} u \, dy = \int_{y_1}^{y_2} \frac{\partial \psi}{\partial y} \, dy$$

Along AB, $x = $ constant, and $d\psi = \partial \psi / \partial y \, dy$. Therefore,

$$Q = \int_{y_1}^{y_2} \frac{\partial \psi}{\partial y} \, dy = \int_{\psi_1}^{\psi_2} d\psi = \psi_2 - \psi_1$$

For a unit depth, the flowrate across BC is

$$Q = \int_{x_1}^{x_2} v \, dx = -\int_{x_1}^{x_2} \frac{\partial \psi}{\partial x} \, dx$$

Along BC, $y = $ constant, and $d\psi = \partial \psi / \partial x \, dx$. Therefore,

$$Q = -\int_{x_1}^{x_2} \frac{\partial \psi}{\partial x} \, dx = -\int_{\psi_2}^{\psi_1} d\psi = \psi_2 - \psi_1$$

stream function for two-dimensional incompressible flow/5-3 *233*

Thus the volumetric rate of flow (per unit depth) between any two stream-lines can be written as the difference between the constant values of ψ defining the two streamlines.

For a two-dimensional, incompressible flow in the $r\theta$ plane, the conservation of mass, Eq. 5.2, can be written as

$$\frac{\partial r V_r}{\partial r} + \frac{\partial V_\theta}{\partial \theta} = 0 \tag{5.7}$$

The stream function $\psi(r, \theta, t)$ is then defined such that

$$V_r \equiv \frac{1}{r}\frac{\partial \psi}{\partial \theta} \qquad \text{and} \qquad V_\theta \equiv -\frac{\partial \psi}{\partial r} \tag{5.8}$$

With ψ defined according to Eq. 5.8, the continuity equation, Eq. 5.7, is satisfied exactly.

Example 5.3
Given the velocity field, $\vec{V} = -ay\hat{\imath} + ax\hat{\jmath}$, determine the family of ψ functions that will yield this velocity field.

Example Problem 5.3

GIVEN:

Velocity field, $\vec{V} = -ay\hat{\imath} + ax\hat{\jmath}$

FIND:

The stream function, ψ.

SOLUTION:

From Eqs. 5.4, $u = \partial\psi/\partial y$ and $v = -\partial\psi/\partial x$. For the given velocity field,

$$u = -ay = \frac{\partial\psi}{\partial y}$$

Integrating with respect to y,

$$\psi = \int \frac{\partial\psi}{\partial y}\, dy = \int -ay\, dy = -\frac{a}{2}y^2 + f(x)$$

where $f(x)$ is arbitrary. However, $f(x)$ may be evaluated using the equation for v.

Thus, since $v = ax$,

$$v = -\frac{\partial \psi}{\partial x} = -\frac{\partial f(x)}{\partial x} = -\frac{df(x)}{dx} = ax$$

Then

$$f(x) = \int \frac{df}{dx}\,dx = \int -ax\,dx = -\frac{a}{2}x^2 + c$$

Substituting gives

$$\psi = -\frac{a}{2}x^2 - \frac{a}{2}y^2 + c$$

ψ

Check:

$$u = \frac{\partial \psi}{\partial y} = \frac{\partial}{\partial y}\left(-\frac{a}{2}x^2 - \frac{a}{2}y^2 + c\right) = -ay$$

$$v = -\frac{\partial \psi}{\partial x} = -\frac{\partial}{\partial x}\left(-\frac{a}{2}x^2 - \frac{a}{2}y^2 + c\right) = ax$$

$\left\{\text{This problem illustrates the relation between the velocity field and the stream}\atop\text{function.}\right\}$

5-4   MOTION OF A FLUID ELEMENT (KINEMATICS)

Before formulating the effects of forces on fluid motion (dynamics), let us consider, first, only the motion (kinematics) of a fluid element in a flow field. For convenience, we follow an infinitesimal element of fixed identity (mass), as shown in Fig. 5.4.

As the infinitesimal element of mass dm moves in a flow field, several things may happen to it. Perhaps the most obvious of these is that the element translates; that is, it undergoes a linear displacement from a location x, y, z to a different location x_1, y_1, z_1. The element may also rotate; that is, the orientation of the element as shown in Fig. 5.4, wherein the sides of

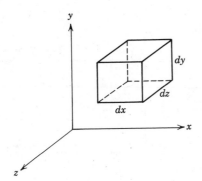

Fig. 5.4 Infinitesimal element of fluid.

the element are parallel to the coordinate axes x, y, z, may change as a result of pure rotation about any one (or all three) of the coordinate axes. In addition the element may deform. The deformation may be subdivided into two parts—linear and angular deformation. Linear deformation involves a change in shape of the element without a change in the orientation of the element: a deformation in which planes of the element that were originally perpendicular (for example, the top and side of the element) remain perpendicular. Angular deformation involves a distortion of the element in which planes that were originally perpendicular are no longer perpendicular. In general a fluid element may undergo a combination of translation, rotation, and linear and angular deformation during the course of its motion.

These four components of fluid motion are illustrated pictorially in Fig. 5.5 for motion in the xy plane. For a general three-dimensional flow, similar motions of the particle would be depicted in the yz and xz planes. For pure translation or rotation, the fluid element retains its shape; that is, there is no deformation. Thus shear stresses do not arise as a result of pure translation or rotation (recall from Chapter 2 that in a Newtonian fluid the shear stress is directly proportional to the rate of angular deformation). We postpone further discussion of deformation and consider first the question of fluid rotation.

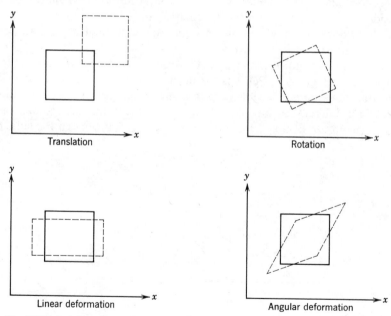

Fig. 5.5 Pictorial representation of the components of fluid motion.

**5-5 FLUID ROTATION

The rotation, $\vec{\omega}$, of a fluid particle is defined as the average angular velocity of any two mutually perpendicular line elements of the particle. Rotation is a vector quantity. A particle moving in a general three-dimensional flow field may rotate about all three coordinate axes. Thus, in general,

$$\vec{\omega} = \hat{\imath}\omega_x + \hat{\jmath}\omega_y + \hat{k}\omega_z$$

where ω_x is the rotation about the x axis, ω_y is the rotation about the y axis, and ω_z is the rotation about the z axis. The positive sense of rotation is given by the right-hand rule.

To obtain a mathematical expression for fluid rotation, consider the motion of a fluid element in the xy plane. The components of the velocity at every point in the flow field are given by $u(x, y)$ and $v(x, y)$. The rotation of a fluid element in such a flow field is illustrated in Fig. 5.6. The two mutually perpendicular lines oa and ob will rotate to the position shown in the time interval Δt only if the velocities at the points a and b are different from the velocity at o.

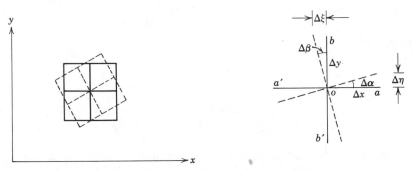

Fig. 5.6 Rotation of a fluid element in a two-dimensional flow field.

Consider first the rotation of the line oa of length Δx. Rotation of this line is due to variations of the y component of velocity, that is, variations in v. If the y component of velocity at the point o is taken as v_o, then the y component of velocity at point a can be written, using a Taylor series expansion, as

$$v = v_o + \frac{\partial v}{\partial x}\Delta x$$

** This section may be omitted without loss of continuity in the text material.

The angular velocity of the line oa is given by

$$\omega_{oa} = \lim_{\Delta t \to 0} \frac{\Delta \alpha}{\Delta t} = \lim_{\Delta t \to 0} \frac{\Delta \eta / \Delta x}{\Delta t}$$

Since

$$\Delta \eta = \frac{\partial v}{\partial x} \Delta x \, \Delta t$$

$$\omega_{oa} = \lim_{\Delta t \to 0} \frac{(\partial v / \partial x) \Delta x \, \Delta t / \Delta x}{\Delta t} = \frac{\partial v}{\partial x}$$

Rotation of the line ob, of length Δy, results from variations of the x component of velocity, that is, variations in u. If the x component of velocity at the point o is taken as u_o, then the x component of velocity at point b can be written, using a Taylor series expansion, as

$$u = u_o + \frac{\partial u}{\partial y} \Delta y$$

The angular velocity of the line ob is given by

$$\omega_{ob} = \lim_{\Delta t \to 0} \frac{\Delta \beta}{\Delta t} = \lim_{\Delta t \to 0} \frac{\Delta \xi / \Delta y}{\Delta t}$$

Since

$$\Delta \xi = - \frac{\partial u}{\partial y} \Delta y \, \Delta t$$

$$\omega_{ob} = \lim_{\Delta t \to 0} \frac{-(\partial u / \partial y) \Delta y \, \Delta t / \Delta y}{\Delta t} = - \frac{\partial u}{\partial y}$$

(The negative sign is introduced to give a positive value of ω_{ob}. According to our sign convention, counterclockwise rotation is positive.)

The rotation of the fluid element about the z axis is the average angular velocity of the two mutually perpendicular line elements, oa and ob, in the xy plane.

$$\omega_z = \frac{1}{2} \left(\frac{\partial v}{\partial x} - \frac{\partial u}{\partial y} \right)$$

By considering the rotation of two mutually perpendicular lines in the yz and xz planes, one can show that

$$\omega_x = \frac{1}{2} \left(\frac{\partial w}{\partial y} - \frac{\partial v}{\partial z} \right)$$

and

$$\omega_y = \frac{1}{2}\left(\frac{\partial u}{\partial z} - \frac{\partial w}{\partial x}\right)$$

Then

$$\vec{\omega} = \hat{\imath}\omega_x + \hat{\jmath}\omega_y + \hat{k}\omega_z = \frac{1}{2}\left[\hat{\imath}\left(\frac{\partial w}{\partial y} - \frac{\partial v}{\partial z}\right) + \hat{\jmath}\left(\frac{\partial u}{\partial z} - \frac{\partial w}{\partial x}\right) + \hat{k}\left(\frac{\partial v}{\partial x} - \frac{\partial u}{\partial y}\right)\right] \quad (5.9)$$

We recognize the term in the square brackets as

$$\text{curl } \vec{V} = \nabla \times \vec{V}$$

Then, in vector notation, we can write

$$\vec{\omega} = \tfrac{1}{2}\nabla \times \vec{V} \quad (5.10)$$

The factor of $\frac{1}{2}$ can be eliminated in Eq. 5.10 by defining a quantity called the vorticity, $\vec{\zeta}$, to be twice the rotation.

$$\vec{\zeta} \equiv 2\vec{\omega} = \nabla \times \vec{V} \quad (5.11)$$

In cylindrical coordinates,

$$\vec{V} = \hat{\imath}_r V_r + \hat{\imath}_\theta V_\theta + \hat{\imath}_z V_z$$

and

$$\nabla = \hat{\imath}_r \frac{\partial}{\partial r} + \hat{\imath}_\theta \frac{1}{r}\frac{\partial}{\partial \theta} + \hat{\imath}_z \frac{\partial}{\partial z}$$

The vorticity, in cylindrical coordinates, is then[2]

$$\nabla \times \vec{V} = \hat{\imath}_r \left(\frac{1}{r}\frac{\partial V_z}{\partial \theta} - \frac{\partial V_\theta}{\partial z}\right) + \hat{\imath}_\theta \left(\frac{\partial V_r}{\partial z} - \frac{\partial V_z}{\partial r}\right) + \hat{\imath}_z \left(\frac{1}{r}\frac{\partial r V_\theta}{\partial r} - \frac{1}{r}\frac{\partial V_r}{\partial \theta}\right) \quad (5.12)$$

**5-6 IRROTATIONAL FLOW

An *irrotational flow* is one in which there is no rotation; that is, fluid elements moving in the flow field do not undergo any rotation. For $\vec{\omega} = 0$, $\nabla \times \vec{V} = 0$ and from Eq. 5.9,

$$\frac{\partial w}{\partial y} - \frac{\partial v}{\partial z} = \frac{\partial u}{\partial z} - \frac{\partial w}{\partial x} = \frac{\partial v}{\partial x} - \frac{\partial u}{\partial y} = 0 \quad (5.13)$$

---

[2] In carrying out the curl operation, recall that $\hat{\imath}_r$ and $\hat{\imath}_\theta$ are functions of θ (see footnote[1] on p. 229).
** This section may be omitted without loss of continuity in the text material. (Note that Sections 5-3 and 5-5 contain background material needed for study of this section.)

In cylindrical coordinates, the irrotationality condition requires that

$$\frac{1}{r}\frac{\partial V_z}{\partial \theta} - \frac{\partial V_\theta}{\partial z} = \frac{\partial V_r}{\partial z} - \frac{\partial V_z}{\partial r} = \frac{1}{r}\frac{\partial r V_\theta}{\partial r} - \frac{1}{r}\frac{\partial V_r}{\partial \theta} = 0 \tag{5.14}$$

5-6.1 VELOCITY POTENTIAL

In Section 5–3 we formulated the stream function, ψ, as a relation between the streamlines and the statement of the conservation of mass for two-dimensional, incompressible flow.

We can also formulate a relation called the potential function, ϕ, for a velocity field subject to the irrotationality condition. To do so, we must make use of the fundamental vector identity[3]

$$\text{curl}(\text{grad } \phi) = \nabla \times \nabla \phi = 0 \tag{5.15}$$

which holds if ϕ is any scalar function (of the space coordinates and time) having continuous first and second derivatives.

Then, for an irrotational flow in which $\nabla \times \vec{V} = 0$, a scalar function, ϕ, must exist such that the gradient of ϕ is equal to the velocity vector, $\vec{V}$. In order that the positive direction of flow be in the direction of decreasing ϕ (analogous to the positive direction of heat transfer being defined in the direction of decreasing temperature), we define ϕ such that

$$\vec{V} \equiv -\nabla \phi \tag{5.16}$$

Thus

$$u = -\frac{\partial \phi}{\partial x} \qquad v = -\frac{\partial \phi}{\partial y} \qquad w = -\frac{\partial \phi}{\partial z} \tag{5.17}$$

With the potential function defined in this way, the irrotationality condition, Eq. 5.13, is satisfied identically.

In cylindrical coordinates,

$$\nabla = \hat{\imath}_r \frac{\partial}{\partial r} + \hat{\imath}_\theta \frac{1}{r}\frac{\partial}{\partial \theta} + \hat{\imath}_z \frac{\partial}{\partial z}$$

From Eq. 5.16, then, in cylindrical coordinates

$$V_r = -\frac{\partial \phi}{\partial r} \qquad V_\theta = -\frac{1}{r}\frac{\partial \phi}{\partial \theta} \qquad V_z = -\frac{\partial \phi}{\partial z} \tag{5.18}$$

[3] The proof of this identity may be found in any book treating vector analysis, or it may be proved by expanding Eq. 5.15 into components (Appendix D).

5-6.2 STREAM FUNCTION AND VELOCITY POTENTIAL FOR IRROTATIONAL, TWO-DIMENSIONAL, INCOMPRESSIBLE FLOW

For a two-dimensional, incompressible, irrotational flow we have expressions for the velocity components u and v in terms of both the stream function, ψ, and the velocity potential, ϕ.

$$u = \frac{\partial \psi}{\partial y} \qquad v = -\frac{\partial \psi}{\partial x} \tag{5.4}$$

$$u = -\frac{\partial \phi}{\partial x} \qquad v = -\frac{\partial \phi}{\partial y} \tag{5.17}$$

Substituting for u and v from Eq. 5.4 into the irrotationality condition,

$$\frac{\partial v}{\partial x} - \frac{\partial u}{\partial y} = 0 \tag{5.13}$$

we obtain

$$\frac{\partial^2 \psi}{\partial x^2} + \frac{\partial^2 \psi}{\partial y^2} = 0 \tag{5.19a}$$

Substituting for u and v from Eq. 5.17 into the continuity equation,

$$\frac{\partial u}{\partial x} + \frac{\partial v}{\partial y} = 0 \tag{5.3}$$

we obtain

$$\frac{\partial^2 \phi}{\partial x^2} + \frac{\partial^2 \phi}{\partial y^2} = 0 \tag{5.19b}$$

Equations 5.19a and b are forms of Laplace's equation—an equation that arises in many areas of the physical sciences and engineering. Any function ψ or ϕ that satisfies Laplace's equation represents a possible incompressible, irrotational flow field.

In Section 5-3 we showed that the stream function, ψ, is constant along a streamline. For $\psi = $ constant, $d\psi = (\partial\psi/\partial x)\,dx + (\partial\psi/\partial y)\,dy = 0$. The slope of a streamline, that is, the slope of a line of constant ψ, is given by

$$\left.\frac{dy}{dx}\right)_\psi = -\frac{\partial\psi/\partial x}{\partial\psi/\partial y} = -\frac{-v}{u} = \frac{v}{u} \tag{5.20}$$

Along a line of constant ϕ, $d\phi = 0$, and hence

$$d\phi = \frac{\partial \phi}{\partial x}\,dx + \frac{\partial \phi}{\partial y}\,dy = 0$$

Consequently,

$$\frac{dy}{dx}\Big)_\phi = -\frac{\partial\phi/\partial x}{\partial\phi/\partial y} = -\frac{u}{v} \qquad (5.21)$$

The last equality of Eq. 5.21 follows from use of Eq. 5.17. Comparing Eqs. 5.20 and 5.21, we see that the slope of a constant ψ line at any point is the negative reciprocal of the slope of the constant ϕ line at that point. Consequently, lines of constant ψ and constant ϕ are orthogonal.

5-6.3   IRROTATIONAL FLOW AND VISCOSITY

Before leaving the subject of irrotational flow, we should make a few additional comments. The velocity potential, ϕ, only exists for an irrotational flow. The stream function, ψ, is defined to satisfy the continuity equation; the stream function is not subject to the restriction of irrotational flow.

Under what conditions might we expect to have an irrotational flow? A fluid particle moving, without rotation, in a flow field cannot develop a rotation under the action of a body force or normal surface (i.e. pressure) forces; the development of rotation in a fluid particle, initially without rotation, requires the action of a shear stress on the surface of the particle. Since shear stress is proportional to the rate of angular deformation, then a particle that is initially without rotation will not develop a rotation without a simultaneous angular deformation. Does it then follow that a flow field in which shear stresses are present will be a rotational flow field? We can obtain a qualitative answer to this question by considering the angular deformation of a fluid element in the xy plane as shown in Fig. 5.7. The rate of angular deformation of the fluid element is the rate of change of the angle between two mutually perpendicular lines in the fluid. Referring to

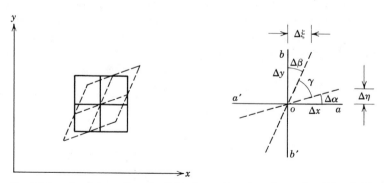

Fig. 5.7 Angular deformation of a fluid element in a two-dimensional flow field.

Fig. 5.7, the rate of angular deformation of the fluid element is the rate of change of the angle between lines *oa* and *ob*; that is, the rate of angular deformation is given by

$$\frac{d\gamma}{dt} = \frac{d\alpha}{dt} + \frac{d\beta}{dt}$$

Now,

$$\frac{d\alpha}{dt} = \lim_{\Delta t \to 0} \frac{\Delta\alpha}{\Delta t} = \lim_{\Delta t \to 0} \frac{\Delta\eta/\Delta x}{\Delta t} = \lim_{\Delta t \to 0} \frac{(\partial v/\partial x)\,\Delta x\,\Delta t/\Delta x}{\Delta t} = \frac{\partial v}{\partial x}$$

and

$$\frac{d\beta}{dt} = \lim_{\Delta t \to 0} \frac{\Delta\beta}{\Delta t} = \lim_{\Delta t \to 0} \frac{\Delta\xi/\Delta y}{\Delta t} = \lim_{\Delta t \to 0} \frac{(\partial u/\partial y)\,\Delta y\,\Delta t/\Delta y}{\Delta t} = \frac{\partial u}{\partial y}$$

Consequently, the rate of angular deformation is

$$\frac{d\gamma}{dt} = \frac{\partial v}{\partial x} + \frac{\partial u}{\partial y} \tag{5.22}$$

The shear stress is related to the rate of angular deformation through the viscosity. In a viscous flow (wherein velocity gradients are present) it is highly unlikely that $\partial v/\partial x$ will be equal and opposite to $\partial u/\partial y$ throughout the flow field (for example, consider the boundary-layer flow of Fig. 2.10 and the flow over a cylinder, shown in Fig. 2.11). The presence of viscous forces means the flow is rotational.[4]

The condition of irrotationality may be a valid assumption for those regions of a flow in which viscous forces are negligible.[5] (For example, such a region exists outside the boundary layer in the flow over a solid surface.) The theory for irrotational flow is developed in terms of an imaginary fluid called an ideal fluid whose viscosity is identically zero. Since, in an irrotational flow, the velocity field may be defined by the potential function, ϕ, the theory is often referred to as potential flow theory.[6]

All real fluids possess viscosity, but there are many situations in which the assumption of inviscid flow leads to a considerable simplification in the analysis and, at the same time, gives meaningful results.

[4] A rigorous proof employing the complete equations of motion for a fluid particle is given in W. H. Li and S. H. Lam, *Principles of Fluid Mechanics* (Reading, Mass.: Addison-Wesley, 1964), pp. 142–145.
[5] Examples of rotational and irrotational motion are shown in the film loops: S-FM014A, *Visualization of Vorticity with Vorticity Meter—Part I*; S-FM014B, *Visualization of Vorticity with Vorticity Meter—Part II*.
[6] Anyone interested in a detailed study of potential flow theory may find the following books of interest: V. L. Streeter, *Fluid Dynamics* (New York: McGraw-Hill, 1948); H. R. Valentine, *Applied Hydrodynamics* (London: Butterworths, 1959); J. M. Robertson, *Hydrodynamics in Theory and Application* (Englewood Cliffs, N.J.: Prentice-Hall, 1965).

Example 5.4

An incompressible flow field is characterized by the stream function

$$\psi = 3ax^2y - ay^3$$

where $a = 1/\text{m} \cdot \text{sec}$.

(a) Show that this flow is irrotational.
(b) Show that the magnitude of the velocity at any point in the flow field depends only on the distance of the point from the origin of coordinates.
(c) Sketch a few streamlines for the flow in the quadrant $x > 0$, $y > 0$.

Example Problem 5.4

GIVEN:

The stream function for an incompressible flow field is given by $\psi = 3ax^2y - ay^3$, where $a = 1/\text{m} \cdot \text{sec}$.

SHOW:

(a) That flow is irrotational.
(b) That magnitude of the velocity at any point depends only on the distance of the point from the origin.
(c) Sketch streamlines.

SOLUTION:

Since $\psi = \psi(x, y)$, the flow field is two-dimensional. If the flow is to be irrotational, then ω_z must be zero.

Since $\quad \omega_z = \dfrac{1}{2}\left[\dfrac{\partial v}{\partial x} - \dfrac{\partial u}{\partial y}\right] \quad$ and $\quad u = \dfrac{\partial \psi}{\partial y}, \quad v = -\dfrac{\partial \psi}{\partial x},$

then for irrotational flow

$$\frac{\partial v}{\partial x} - \frac{\partial u}{\partial y} = -\frac{\partial^2 \psi}{\partial x^2} - \frac{\partial^2 \psi}{\partial y^2} = -\nabla^2\psi = 0$$

Evaluating $\nabla^2\psi$, we obtain

$$\nabla^2\psi = \frac{\partial^2}{\partial x^2}(3ax^2y - ay^3) + \frac{\partial^2}{\partial y^2}(3ax^2y - ay^3) = 6ay - 6ay = 0$$

Since $\nabla^2\psi = 0$, the flow is irrotational.

The magnitude of the velocity at a point is given by

$$|\vec{V}| = \sqrt{u^2 + v^2}$$

$$u = \frac{\partial \psi}{\partial y} = \frac{\partial}{\partial y}(3ax^2y - ay^3) = 3ax^2 - 3ay^2$$

$$v = -\frac{\partial \psi}{\partial x} = -\frac{\partial}{\partial x}(3ax^2y - ay^3) = -6axy$$

$$|\vec{V}| = \sqrt{u^2 + v^2} = \sqrt{(3ax^2 - 3ay^2)^2 + (-6axy)^2}$$
$$= \sqrt{9a^2(x^4 - 2x^2y^2 + y^4) + 36a^2x^2y^2}$$
$$= \sqrt{9a^2(x^4 + 2x^2y^2 + y^4)} = 3a\sqrt{(x^2 + y^2)^2}$$

$$|\vec{V}| = 3a(x^2 + y^2) = 3ar^2 \qquad\qquad |\vec{V}|$$

where $r = |\vec{r}| = |x\hat{\imath} + y\hat{\jmath}|$.

A few streamlines, corresponding to lines along which ψ is constant, are shown in the following figure.

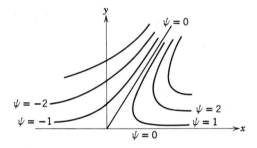

Since there is no flow across streamlines, and the flow is irrotational (i.e. without viscous effects), any streamline may be imagined to represent a solid surface. If we let the lines $\psi = 0$ represent solid surfaces, we obtain a picture of flow in a corner with either an obtuse or acute angle.

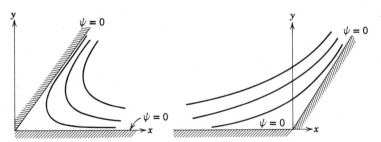

⎰ This problem illustrates the relations among the stream function, the velocity ⎱
⎱ field, and the irrotationality condition. ⎰

Example 5.5

Consider the flow field given by $\psi = ax^2 - ay^2$, where $a = 1 \text{ sec}^{-1}$.
(a) Show that the flow is irrotational.
(b) Determine the velocity potential for this flow.

Example Problem 5.5

GIVEN:

Incompressible flow field with $\psi = ax^2 - ay^2$, where $a = 1 \text{ sec}^{-1}$.

REQUIRED:

(a) Show that the flow is irrotational.
(b) Determine the velocity potential for this flow.

SOLUTION:

If the flow is irrotational, then $\omega_z = 0$. Since

$$\omega_z = \frac{\partial v}{\partial x} - \frac{\partial u}{\partial y} \quad \text{and} \quad u = \frac{\partial \psi}{\partial y}, \quad v = -\frac{\partial \psi}{\partial x},$$

then

$$u = \frac{\partial}{\partial y}(ax^2 - ay^2) = -2ay \quad \text{and} \quad v = -\frac{\partial}{\partial x}(ax^2 - ay^2) = -2ax$$

Thus

$$\omega_z = \frac{\partial v}{\partial x} - \frac{\partial u}{\partial y} = \frac{\partial}{\partial x}(-2ax) - \frac{\partial}{\partial y}(-2ay) = -2a + 2a = 0$$

Therefore, the flow is irrotational.

The velocity components can be written in terms of the velocity potential as

$$u = -\frac{\partial \phi}{\partial x} \quad \text{and} \quad v = -\frac{\partial \phi}{\partial y}$$

Consequently, $\quad u = -\dfrac{\partial \phi}{\partial x} = -2ay \quad$ and $\quad \dfrac{\partial \phi}{\partial x} = 2ay$

Integrating with respect to x gives $\phi = 2axy + f(y)$, where $f(y)$ is an arbitrary function of y. Then

$$v = -2ax = -\frac{\partial \phi}{\partial y} = -\frac{\partial}{\partial y}[2axy + f(y)]$$

Therefore, $-2ax = -2ax - \dfrac{\partial f(y)}{\partial y} = -2ax - \dfrac{df}{dy}$, so $\dfrac{df}{dy} = 0$ and $f = \text{constant}$.

Therefore, $\phi = 2axy + \text{constant}$ $\hspace{4cm} \phi$

We can also show that lines of constant ψ and ϕ are orthogonal.

$$\psi = ax^2 - ay^2$$

For $\psi = \text{constant}$, $d\psi = 0 = 2ax\,dx - 2ay\,dy$, hence $\left.\dfrac{dy}{dx}\right)_{\psi=c} = \dfrac{x}{y}$

For $\phi = \text{constant}$, $d\phi = 0 = 2ay\,dx + 2ax\,dy$, hence $\left.\dfrac{dy}{dx}\right)_{\phi=c} = -\dfrac{y}{x}$

$\left\{ \text{This problem illustrates the relation among the stream function, the velocity} \atop \text{field, and the velocity potential.} \right\}$

5-7 MOMENTUM EQUATION

A dynamic equation describing fluid motion may be obtained by applying Newton's second law to a particle. To derive the differential form of the momentum equation, we shall apply Newton's second law to an infinitesimal system of mass, dm.

Recall that Newton's second law for a finite system is given by

$$\vec{F} = \left.\dfrac{d\vec{P}}{dt}\right)_{\text{system}} \tag{4.2a}$$

where the linear momentum, $\vec{P}$, of the system is given by

$$\vec{P}_{\text{system}} = \int_{\text{mass (system)}} \vec{V}\,dm \tag{4.2b}$$

Then, for an infinitesimal system of mass, dm, Newton's second law can be written

$$d\vec{F} = dm\,\left.\dfrac{d\vec{V}}{dt}\right)_{\text{system}} \tag{5.23}$$

We recognize that the force term, $d\vec{F}$, can be written in terms of the surface forces and body forces acting on the element of fixed mass, dm. Indeed, we will eventually have to formulate $d\vec{F}$. For the moment we shall delay that and concentrate our efforts on determining a suitable expression for $d\vec{V}/dt]_{\text{system}}$.

5-7.1 ACCELERATION OF A FLUID PARTICLE

Let us remember first that we are dealing with an element of fixed mass, dm. As discussed in Section 1-4.3, one may obtain the equation of motion for a particle by applying Newton's second law to that particle. The disadvantage of this approach is that a separate equation is required for each particle. Thus the bookkeeping for many particles becomes a problem.

5-7.2 ACCELERATION OF A FLUID PARTICLE IN A VELOCITY FIELD

A more general description of acceleration can be obtained by considering a a particle moving in a velocity field. The basic hypothesis of continuum fluid mechanics has led us to a field description of fluid flow in which the properties of a flow field are defined by continuous functions of the space coordinates and time. In particular, the velocity field is given by $\vec{V} = \vec{V}(x, y, z, t)$. The field description is very powerful, since information for the entire flow is given by one equation.

The problem, then, is to retain the field description for fluid properties and obtain an expression for the acceleration of a fluid particle as it translates in a flow field. Stated simply, the problem is:

Given the velocity field, $\vec{V} = \vec{V}(x, y, z, t)$, find the acceleration of a fluid particle, $\vec{a}_p$.

Consider a particle moving in a velocity field. At time, t, the particle is at the position x, y, z and has a velocity corresponding to the velocity at that point in space at time t, that is,

$$\vec{V}_p]_t = \vec{V}(x, y, z, t)$$

At time, $t + dt$, the particle has moved to a new position, with coordinates $x + dx$, $y + dy$, $z + dz$, and has a velocity given by

$$\vec{V}_p]_{t+dt} = \vec{V}(x + dx, y + dy, z + dz, t + dt)$$

This is shown pictorially in Fig. 5.8.

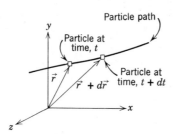

Fig. 5.8 Motion of a particle in a flow field.

The particle velocity at time t (position $\vec{r}$) is given by $\vec{V}_p = \vec{V}(x, y, z, t)$. Then $d\vec{V}_p$, the change in the velocity of the particle, in moving from location $\vec{r}$ to $d\vec{r}$ is given by

$$d\vec{V}_p = \frac{\partial \vec{V}}{\partial x} dx_p + \frac{\partial \vec{V}}{\partial y} dy_p + \frac{\partial \vec{V}}{\partial z} dz_p + \frac{\partial \vec{V}}{\partial t} dt$$

The total acceleration of the particle is given by

$$\vec{a}_p = \frac{d\vec{V}_p}{dt} = \frac{\partial \vec{V}}{\partial x} \frac{dx_p}{dt} + \frac{\partial \vec{V}}{\partial y} \frac{dy_p}{dt} + \frac{\partial \vec{V}}{\partial z} \frac{dz_p}{dt} + \frac{\partial \vec{V}}{\partial t}$$

Since

$$\frac{dx_p}{dt} = u \qquad \frac{dy_p}{dt} = v \qquad \text{and} \qquad \frac{dz_p}{dt} = w$$

then

$$\vec{a}_p = \frac{d\vec{V}_p}{dt} = u\frac{\partial \vec{V}}{\partial x} + v\frac{\partial \vec{V}}{\partial y} + w\frac{\partial \vec{V}}{\partial z} + \frac{\partial \vec{V}}{\partial t}$$

To remind us that calculation of the acceleration of a fluid particle in a velocity field requires a special derivative, it is given the symbol $D\vec{V}/Dt$. Thus

$$\frac{D\vec{V}}{Dt} \equiv \vec{a}_p = u\frac{\partial \vec{V}}{\partial x} + v\frac{\partial \vec{V}}{\partial y} + w\frac{\partial \vec{V}}{\partial z} + \frac{\partial \vec{V}}{\partial t} \qquad (5.24)$$

$\leftarrow$ *acceleratic.*
of a
point

The derivative, $D\vec{V}/Dt$, defined by Eq. 5.24, is commonly called the *substantial derivative* to remind us that it is computed following a particle of "substance." It often is called the material or particle derivative.

From Eq. 5.24 we recognize that a fluid particle moving in a flow field may undergo an acceleration for either of two reasons. It may be accelerated because it is convected into a region of higher (or lower) velocity. For example, in the steady flow through a nozzle, in which, by definition, the velocity field is not a function of time, a fluid particle will accelerate as it moves through the nozzle. The particle is convected into a region of higher velocity. If a flow field is unsteady, a fluid particle will undergo an acceleration, a "local" acceleration, because the velocity field is a function of time.

The physical significance of the terms in Eq. 5.24 is

$$\vec{a}_p = \underset{\substack{\text{total} \\ \text{acceleration} \\ \text{of a particle}}}{\frac{D\vec{V}}{Dt}} = \underset{\substack{\text{convective} \\ \text{acceleration}}}{u\frac{\partial \vec{V}}{\partial x} + v\frac{\partial \vec{V}}{\partial y} + w\frac{\partial \vec{V}}{\partial z}} + \underset{\substack{\text{local} \\ \text{acceleration}}}{\frac{\partial \vec{V}}{\partial t}}$$

For a two-dimensional flow, say $\vec{V} = \vec{V}(x, y, t)$, Eq. 5.24 reduces to

$$\frac{D\vec{V}}{Dt} = u\frac{\partial \vec{V}}{\partial x} + v\frac{\partial \vec{V}}{\partial y} + \frac{\partial \vec{V}}{\partial t}$$

For a one-dimensional flow, say $\vec{V} = \vec{V}(x, t)$, Eq. 5.24 becomes

$$\frac{D\vec{V}}{Dt} = u\frac{\partial \vec{V}}{\partial x} + \frac{\partial \vec{V}}{\partial t}$$

Finally, for a steady flow in three dimensions, Eq. 5.24 becomes

$$\frac{D\vec{V}}{Dt} = u\frac{\partial \vec{V}}{\partial x} + v\frac{\partial \vec{V}}{\partial y} + w\frac{\partial \vec{V}}{\partial z}$$

which is not necessarily zero. Thus a fluid particle can undergo a convective acceleration due to its motion, even in a steady velocity field.

Equation 5.24 is a vector equation. As with all vector equations, it may be written in scalar component equations. Relative to an xyz coordinate system, the scalar components of Eq. 5.24 are written:

$$a_{x_p} = \frac{Du}{Dt} = u\frac{\partial u}{\partial x} + v\frac{\partial u}{\partial y} + w\frac{\partial u}{\partial z} + \frac{\partial u}{\partial t} \tag{5.25a}$$

$$a_{y_p} = \frac{Dv}{Dt} = u\frac{\partial v}{\partial x} + v\frac{\partial v}{\partial y} + w\frac{\partial v}{\partial z} + \frac{\partial v}{\partial t} \tag{5.25b}$$

$$a_{z_p} = \frac{Dw}{Dt} = u\frac{\partial w}{\partial x} + v\frac{\partial w}{\partial y} + w\frac{\partial w}{\partial z} + \frac{\partial w}{\partial t} \tag{5.25c}$$

Example 5.6
Consider the one-dimensional steady, incompressible flow through the converging channel shown. The velocity field is given by $\vec{V} = V_1[1 + (x/L)]\hat{i}$. Find the x component of acceleration, that is Du/Dt, for a particle moving in the flow field. If we use the method of description of particle mechanics, the position of the particle, located at $x = 0$ at time $t = 0$, will be a function of time, $x_p = f(t)$. Obtain the expression for $f(t)$ and then by taking the second derivative of the function with respect to time, obtain an expression for the x component of the particle acceleration.

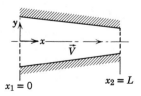

Example Problem 5.6

GIVEN:

Steady, one-dimensional, incompressible flow through the converging channel shown.

$$\vec{V} = V_1 \left(1 + \frac{x}{L}\right) \hat{i}$$

$x_1 = 0$

$x_2 = L$

FIND:

(a) The x component of the acceleration of a particle moving in the flow field.
(b) For the particle located at $x = 0$ at time $t = 0$, obtain an expression for its
   (1) position, x_p, as a function of time.
   (2) x component of acceleration, a_{x_p}, as a function of time.

SOLUTION:

The acceleration of a particle moving in a velocity field is given by

$$\frac{D\vec{V}}{Dt} = u\frac{\partial\vec{V}}{\partial x} + v\frac{\partial\vec{V}}{\partial y} + w\frac{\partial\vec{V}}{\partial z} + \frac{\partial\vec{V}}{\partial t}$$

The x component of the acceleration is given by

$$\frac{Du}{Dt} = u\frac{\partial u}{\partial x} + v\frac{\partial u}{\partial y} + w\frac{\partial u}{\partial z} + \frac{\partial u}{\partial t}$$

For the flow field given,

$$v = w = 0 \qquad u = V_1\left(1 + \frac{x}{L}\right)$$

Therefore, $\dfrac{Du}{Dt} = u\dfrac{\partial u}{\partial x} = V_1\left(1 + \dfrac{x}{L}\right)\dfrac{V_1}{L} = \dfrac{V_1^2}{L}\left(1 + \dfrac{x}{L}\right)$ $\underset{\longleftarrow}{\dfrac{Du}{Dt}}$

$\left\{\begin{array}{l}\text{To determine the acceleration of a particle at any point in the flow field, one} \\ \text{merely substitutes the present location of the particle into the above result.}\end{array}\right\}$

In the second part of this problem we are interested in following a particular particle, namely, the one located at $x = 0$ at time $t = 0$, as it flows through the channel.

The x coordinate that locates this particle will be a function of time, $x_p = f(t)$. Furthermore, $u_p = df/dt$ will be a function of time. The particle will have the velocity corresponding to its location in the velocity field. At time, $t = 0$, the particle is at $x = 0$, and its velocity $u_p = V_1$. At some time later, time $t = t$, the particle will reach the exit, $x = L$; at that time it will have a velocity $u_p = 2V_1$.

To find the expression for $x_p = f(t)$, we write

$$u_p = \frac{dx_p}{dt} = \frac{df}{dt} = V_1\left(1 + \frac{x}{L}\right) = V_1\left(1 + \frac{f}{L}\right)$$

Separating variables,

$$\frac{dt}{(1 + f/L)} = V_1\, dt$$

Since at time $t = 0$, the particle in question was located at $x = 0$, and at time t, this particle is located at $x_p = f$, then

$$\int_0^f \frac{df}{(1 + f/L)} = \int_0^t V_1\, dt$$

$$L \ln\left(1 + \frac{f}{L}\right) = V_1 t$$

$$\ln\left(1 + \frac{f}{L}\right) = \frac{V_1 t}{L}$$

$$1 + \frac{f}{L} = e^{V_1 t/L}$$

and

$$f = L[e^{V_1 t/L} - 1]$$

Then the position of the particle, located at $x = 0$ at time $t = 0$, as a function of time is given by

$$x_p = f(t) = L[e^{V_1 t/L} - 1] \qquad\qquad x_p$$

The x component of acceleration of this particle is given by

$$a_{x_p} = \frac{d^2 x_p}{dt^2} = \frac{d^2 f}{dt^2} = \frac{V_1^2}{L} e^{V_1 t/L} \qquad\qquad a_{x_p}$$

We now have two different ways of expressing the acceleration of the particular particle, that was located at $x = 0$ at time, $t = 0$. Note that although the flow field is steady, when we follow a particular particle, its position and acceleration (and velocity for that matter) are functions of time.

We check to see that both expressions for the acceleration give identical results.

$$a_{x_p} = \frac{V_1^2}{L} e^{V_1 t/L} \qquad\qquad a_{x_p} = \frac{Du}{Dt} = \frac{V_1^2}{L}\left(1 + \frac{x}{L}\right)$$

(a) At time $t = 0$, $x_p = 0$ At time $t = 0$, the particle is at $x = 0$

$$a_{x_p} = \frac{V_1^2}{L} e^0 = \frac{V_1^2}{L} \qquad\text{(a)}\qquad \frac{Du}{Dt} = \frac{V_1^2}{L}(1 + 0) = \frac{V_1^2}{L} \qquad\text{(a)}$$

Check.

(b)   When $x_p = \dfrac{L}{2}$, time $t = t_1$,

$$x_p = \frac{L}{2} = L[e^{V_1 t_1/L} - 1]$$

Therefore, $e^{V_1 t_1/L} = 1.5$, and

$$a_{x_p} = \frac{V_1^2}{L} e^{V_1 t_1/L}$$

$$a_{x_p} = \frac{V_1^2}{L}(1.5) = \frac{1.5V_1^2}{L} \qquad \underleftarrow{\qquad} \text{(b)}$$

(c)   When $x_p = L$, time $t = t_2$,

$$x_p = L = L[e^{V_1 t_2/L} - 1]$$

Therefore, $e^{V_1 t_2/L} = 2$, and

$$a_{x_p} = \frac{V_1^2}{L} e^{V_1 t_2/L}$$

$$a_{x_p} = \frac{V_1^2}{L}(2) = \frac{2V_1^2}{L} \qquad \underleftarrow{\qquad} \text{(c)}$$

At $x = 0.5L$

$$\frac{Du}{Dt} = \frac{V_1^2}{L}(1 + 0.5)$$

$$\frac{Du}{Dt} = \frac{1.5V_1^2}{L} \qquad \underleftarrow{\qquad} \text{(b)}$$

Check.

At $x = L$

$$\frac{Du}{Dt} = \frac{V_1^2}{L}(1 + 1)$$

$$\frac{Du}{Dt} = \frac{2V_1^2}{L} \qquad \underleftarrow{\qquad} \text{(c)}$$

Check.

$\Big\{$This problem illustrates the two different methods of describing the motion of a$\Big\}$
particle.

5-7.3   FORMULATION OF FORCES ACTING ON A FLUID PARTICLE

Having obtained an expression for the acceleration of a fluid element of mass, dm, moving in a velocity field, we can now write Newton's second law as the vector equation

$$d\vec{F} = dm \frac{D\vec{V}}{Dt} = dm \left[u\frac{\partial \vec{V}}{\partial x} + v\frac{\partial \vec{V}}{\partial y} + w\frac{\partial \vec{V}}{\partial z} + \frac{\partial \vec{V}}{\partial t} \right] \qquad (5.26)$$

In terms of scalar component equations we write

$$dF_x = dm \frac{Du}{Dt} = dm \left[u\frac{\partial u}{\partial x} + v\frac{\partial u}{\partial y} + w\frac{\partial u}{\partial z} + \frac{\partial u}{\partial t} \right] \qquad (5.27a)$$

$$dF_y = dm \frac{Dv}{Dt} = dm \left[u\frac{\partial v}{\partial x} + v\frac{\partial v}{\partial y} + w\frac{\partial v}{\partial z} + \frac{\partial v}{\partial t} \right] \qquad (5.27b)$$

$$dF_z = dm \frac{Dw}{Dt} = dm \left[u\frac{\partial w}{\partial x} + v\frac{\partial w}{\partial y} + w\frac{\partial w}{\partial z} + \frac{\partial w}{\partial t} \right] \qquad (5.27c)$$

We need now to obtain a suitable formulation for the force, $d\vec{F}$, or its components dF_x, dF_y, dF_z, acting on the element. Recall that the forces acting on a fluid element may be classified as body forces and surface forces; surface forces include both normal forces and tangential (shear) forces.

We shall consider the x component of the force acting on a differential element of mass, dm, and volume, $d\mathcal{V} = dx\,dy\,dz$. Only those stresses that act in the x direction will give rise to surface forces in the x direction. If the stresses at the center of the differential element are taken to be σ_{xx}, τ_{yx}, and τ_{zx}, then the stresses acting in the x direction on each face of the element (obtained by a Taylor series expansion about the center of the element) are as shown in Fig. 5.9.

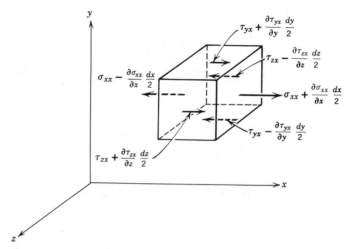

Fig. 5.9 Stresses in the x direction on an element of fluid.

To obtain the net surface force in the x direction, dF_{s_x}, we must sum the forces in the x direction. Thus

$$
dF_{s_x} = \left(\sigma_{xx} + \frac{\partial\sigma_{xx}}{\partial x}\frac{dx}{2}\right)dy\,dz - \left(\sigma_{xx} - \frac{\partial\sigma_{xx}}{\partial x}\frac{dx}{2}\right)dy\,dz
$$

$$
+ \left(\tau_{yx} + \frac{\partial\tau_{yx}}{\partial y}\frac{dy}{2}\right)dx\,dz - \left(\tau_{yx} - \frac{\partial\tau_{yx}}{\partial y}\frac{dy}{2}\right)dx\,dz
$$

$$
+ \left(\tau_{zx} + \frac{\partial\tau_{zx}}{\partial z}\frac{dz}{2}\right)dx\,dy - \left(\tau_{zx} - \frac{\partial\tau_{zx}}{\partial z}\frac{dz}{2}\right)dx\,dy
$$

On simplifying, we obtain

$$dF_{s_x} = \left(\frac{\partial \sigma_{xx}}{\partial x} + \frac{\partial \tau_{yx}}{\partial y} + \frac{\partial \tau_{zx}}{\partial z}\right) dx\,dy\,dz$$

If the body force per unit mass is designated as $\vec{B} = \hat{\imath}B_x + \hat{\jmath}B_y + \hat{k}B_z$, then the component of the body force in the x direction, dF_{B_x}, is given by $dF_{B_x} = B_x\,dm = B_x \rho\,d\Psi$.

Then the net force in the x direction, dF_x, is given by

$$dF_x = dF_{s_x} + dF_{B_x} = \left(\rho B_x + \frac{\partial \sigma_{xx}}{\partial x} + \frac{\partial \tau_{yx}}{\partial y} + \frac{\partial \tau_{zx}}{\partial z}\right) dx\,dy\,dz \quad (5.28a)$$

One can derive similar expressions for the force components in the y and z directions:

$$dF_y = dF_{s_y} + dF_{B_y} = \left(\rho B_y + \frac{\partial \tau_{xy}}{\partial x} + \frac{\partial \sigma_{yy}}{\partial y} + \frac{\partial \tau_{zy}}{\partial z}\right) dx\,dy\,dz \quad (5.28b)$$

$$dF_z = dF_{s_z} + dF_{B_z} = \left(\rho B_z + \frac{\partial \tau_{xz}}{\partial x} + \frac{\partial \tau_{yz}}{\partial y} + \frac{\partial \sigma_{zz}}{\partial z}\right) dx\,dy\,dz \quad (5.28c)$$

5-7.4 DIFFERENTIAL MOMENTUM EQUATION

We have now formulated expressions for the components dF_x, dF_y, and dF_z of the force, $d\vec{F}$, acting on the element of mass, dm. If we substitute these expressions (Eqs. 5.28) for the force components into Eqs. 5.27, we obtain the differential equations of motion.

$$\rho B_x + \frac{\partial \sigma_{xx}}{\partial x} + \frac{\partial \tau_{yx}}{\partial y} + \frac{\partial \tau_{zx}}{\partial z} = \rho\left(\frac{\partial u}{\partial t} + u\frac{\partial u}{\partial x} + v\frac{\partial u}{\partial y} + w\frac{\partial u}{\partial z}\right) \quad (5.29a)$$

$$\rho B_y + \frac{\partial \tau_{xy}}{\partial x} + \frac{\partial \sigma_{yy}}{\partial y} + \frac{\partial \tau_{zy}}{\partial z} = \rho\left(\frac{\partial v}{\partial t} + u\frac{\partial v}{\partial x} + v\frac{\partial v}{\partial y} + w\frac{\partial v}{\partial z}\right) \quad (5.29b)$$

$$\rho B_z + \frac{\partial \tau_{xz}}{\partial x} + \frac{\partial \tau_{yz}}{\partial y} + \frac{\partial \sigma_{zz}}{\partial z} = \rho\left(\frac{\partial w}{\partial t} + u\frac{\partial w}{\partial x} + v\frac{\partial w}{\partial y} + w\frac{\partial w}{\partial z}\right) \quad (5.29c)$$

Equations 5.29 are the differential equations of motion for any fluid satisfying the continuum assumption. Before the equations can be employed in the solution of problems, suitable expressions for the stresses must be obtained.

For Newtonian fluids the stresses may be expressed in terms of velocity gradients and fluid properties as follows[7]:

$$\tau_{xy} = \tau_{yx} = \mu \left(\frac{\partial v}{\partial x} + \frac{\partial u}{\partial y} \right) \tag{5.30a}$$

$$\tau_{yz} = \tau_{zy} = \mu \left(\frac{\partial w}{\partial y} + \frac{\partial v}{\partial z} \right) \tag{5.30b}$$

$$\tau_{zx} = \tau_{xz} = \mu \left(\frac{\partial u}{\partial z} + \frac{\partial w}{\partial x} \right) \tag{5.30c}$$

$$\sigma_{xx} = -p - \frac{2}{3} \mu \nabla \cdot \vec{V} + 2\mu \frac{\partial u}{\partial x} \tag{5.30d}$$

$$\sigma_{yy} = -p - \frac{2}{3} \mu \nabla \cdot \vec{V} + 2\mu \frac{\partial v}{\partial y} \tag{5.30e}$$

$$\sigma_{zz} = -p - \frac{2}{3} \mu \nabla \cdot \vec{V} + 2\mu \frac{\partial w}{\partial z} \tag{5.30f}$$

where p is the local thermodynamic pressure.

If these expressions are introduced into the differential equations of motion (Eqs. 5.29), we obtain

$$\rho \frac{Du}{Dt} = \rho B_x - \frac{\partial p}{\partial x} + \frac{\partial}{\partial x} \left[\mu \left(2 \frac{\partial u}{\partial x} - \frac{2}{3} \nabla \cdot \vec{V} \right) \right] + \frac{\partial}{\partial y} \left[\mu \left(\frac{\partial u}{\partial y} + \frac{\partial v}{\partial x} \right) \right]$$
$$+ \frac{\partial}{\partial z} \left[\mu \left(\frac{\partial w}{\partial x} + \frac{\partial u}{\partial z} \right) \right] \tag{5.31a}$$

$$\rho \frac{Dv}{Dt} = \rho B_y - \frac{\partial p}{\partial y} + \frac{\partial}{\partial x} \left[\mu \left(\frac{\partial u}{\partial y} + \frac{\partial v}{\partial x} \right) \right] + \frac{\partial}{\partial y} \left[\mu \left(2 \frac{\partial v}{\partial y} - \frac{2}{3} \nabla \cdot \vec{V} \right) \right]$$
$$+ \frac{\partial}{\partial z} \left[\mu \left(\frac{\partial v}{\partial z} + \frac{\partial w}{\partial y} \right) \right] \tag{5.31b}$$

$$\rho \frac{Dw}{Dt} = \rho B_z - \frac{\partial p}{\partial z} + \frac{\partial}{\partial x} \left[\mu \left(\frac{\partial w}{\partial x} + \frac{\partial u}{\partial z} \right) \right] + \frac{\partial}{\partial y} \left[\mu \left(\frac{\partial v}{\partial z} + \frac{\partial w}{\partial y} \right) \right]$$
$$+ \frac{\partial}{\partial z} \left[\mu \left(2 \frac{\partial w}{\partial z} - \frac{2}{3} \nabla \cdot \vec{V} \right) \right] \tag{5.31c}$$

[7] The derivation of these results is beyond the scope of this text. A detailed derivation may be found in the following books: J. W. Daily and D. R. F. Harleman, *Fluid Dynamics* (Reading, Mass.: Addison-Wesley, 1966); H. Rouse, *Advanced Mechanics of Fluids* (New York: John Wiley, 1959); H. Schlichting, *Boundary-Layer Theory*, 6th ed. (New York: McGraw-Hill, 1968).

These equations of motion are called the Navier-Stokes equations. The equations are greatly simplified when applied to incompressible flows in which the variations in fluid viscosity can be neglected. Under these conditions the equations reduce to

$$\rho\left(\frac{\partial u}{\partial t} + u\frac{\partial u}{\partial x} + v\frac{\partial u}{\partial y} + w\frac{\partial u}{\partial z}\right) = \rho B_x - \frac{\partial p}{\partial x} + \mu\left(\frac{\partial^2 u}{\partial x^2} + \frac{\partial^2 u}{\partial y^2} + \frac{\partial^2 u}{\partial z^2}\right) \quad (5.32a)$$

$$\rho\left(\frac{\partial v}{\partial t} + u\frac{\partial v}{\partial x} + v\frac{\partial v}{\partial y} + w\frac{\partial v}{\partial z}\right) = \rho B_y - \frac{\partial p}{\partial y} + \mu\left(\frac{\partial^2 v}{\partial x^2} + \frac{\partial^2 v}{\partial y^2} + \frac{\partial^2 v}{\partial z^2}\right) \quad (5.32b)$$

$$\rho\left(\frac{\partial w}{\partial t} + u\frac{\partial w}{\partial x} + v\frac{\partial w}{\partial y} + w\frac{\partial w}{\partial z}\right) = \rho B_z - \frac{\partial p}{\partial z} + \mu\left(\frac{\partial^2 w}{\partial x^2} + \frac{\partial^2 w}{\partial y^2} + \frac{\partial^2 w}{\partial z^2}\right) \quad (5.32c)$$

For the case of frictionless flow ($\mu = 0$) the equations of motion (Eqs. 5.29 or Eqs. 5.32) reduce to Euler's equation,

$$\rho\frac{D\vec{V}}{Dt} = \rho\vec{B} - \nabla p$$

We shall consider the case of frictionless flow in Chapter 6.

summary objectives

After completing study of Chapter 5, you should be able to do the following:

1. Write the differential form of the conservation of mass in (a) vector form, (b) rectangular coordinates, and (c) cylindrical coordinates.

2. Given a velocity field, determine if the field represents a possible incompressible flow.

**3. For a two-dimensional incompressible flow field define the stream function, ψ; given the velocity field, determine the stream function; given the stream function, determine the velocity field.

4. For a fluid particle moving in a flow field, illustrate: translation, rotation, linear deformation, and angular deformation.

**5. Define fluid rotation and irrotational flow. For a two-dimensional, irrotational, incompressible flow field: given the velocity field, determine the velocity potential, ϕ; given the velocity potential, determine the velocity field; show that lines of constant ψ and constant ϕ are orthogonal.

6. For a fluid particle moving in a given velocity field, determine the convective and local accelerations.

** These objectives apply to sections that may be omitted without loss of continuity in the text material.

7. Write the differential form of the momentum equation for viscous flow and state the physical meaning of each term in the equation.

8. Solve those problems at the end of the chapter that relate to the material you have studied.

problems

5.1 A velocity field is given by the expression

$$\vec{V} = Ax\hat{\imath} - Ay\hat{\jmath}$$

where x and y are measured in meters, and A equals 10 m/sec/m. Locate all points in the xy plane where $|\vec{V}|$ is equal to 20 m/sec.

5.2 A velocity field is given by the expression

$$\vec{V} = U\cos\theta\left[1 - \left(\frac{a}{r}\right)^2\right]\hat{\imath}_r - U\sin\theta\left[1 + \left(\frac{a}{r}\right)^2\right]\hat{\imath}_\theta$$

Find all points in the $r\theta$ plane where:

(a) $V_r = 0$
(b) $V_\theta = 0$
(c) $V_r = V_\theta = 0$

5.3 A viscous liquid is sheared between two parallel disks of radius, R, one of which rotates while the other is fixed. The velocity field is purely tangential, and the velocity varies linearly with z from $V_\theta = 0$ at $z = 0$ (the fixed disk) to the velocity of the rotating disk at its surface ($z = h$). Write an expression for the velocity field between the disks.

5.4 Which of the following sets of equations represent possible two-dimensional incompressible flow cases?

(a) $u = x + y; v = x - y$
(b) $u = x + 2y; v = x^2 - y^2$
(c) $u = 4x + y; v = x - y^2$
(d) $u = xt + 2y; v = x^2 - yt^2$
(e) $u = xt^2; v = xyt + y^2$

5.5 Which of the following sets of equations represent possible two-dimensional incompressible flow cases?

(a) $u = 2x^2 + y^2; v = x^3 - x(y^2 - 2y)$
(b) $u = 2xy - x^2 + y; v = 2xy - y^2 + x^2$
(c) $u = xt + 2y; v = xt^2 - yt$
(d) $u = (x + 2y)xt; v = (2x - y)yt$

5.6 Which of the following equations represent possible three-dimensional incompressible flow cases?

(a) $u = x + y + z^2; v = x - y + z; w = 2xy + y^2 + 4$
(b) $u = xyzt; v = -xyzt^2; w = (z^2/2)(xt^2 - yt)$
(c) $u = y^2 + 2xz; v = -2yz + x^2yz; w = \frac{1}{2}x^2z^2 + x^3y^4$

5.7 For a flow in the xy plane, the y component of velocity is given by

$$v = y^2 - 2x + 2y$$

Determine a possible x component for steady, incompressible flow. Is it also valid for unsteady, incompressible flow? Why? How many possible x components are there?

5.8 The three components of velocity in a velocity field are given by

$$u = Ax + By + Cz$$
$$v = Dx + Ey + Fz$$
$$w = Gx + Hy + Jz$$

Determine the relationship among the coefficients A through J, that is necessary if this is to be a possible incompressible flow field.

5.9 Which of the following sets of equations represent possible incompressible flow cases?
(a) $V_r = U \cos \theta$; $V_\theta = -U \sin \theta$
(b) $V_r = -Q/2\pi r$; $V_\theta = k/2\pi r$
(c) $V_r = U \cos \theta [1 - (a/r)^2]$; $V_\theta = -U \sin \theta [1 + (a/r)^2]$

5.10 For an incompressible flow in the $r\theta$ plane, the r component of velocity is given as

$$V_r = -\frac{\Lambda \cos \theta}{r^2}$$

Determine a possible θ component of velocity. How many possible θ components are there?

5.11 The stream function for a certain incompressible flow is given as

$$\psi = Axy$$

Plot several streamlines, including $\psi = 0$. Obtain an expression for the velocity field.

5.12 The stream function for a certain incompressible flow field is given by the expression

$$\psi = -Ur \sin \theta + \frac{Q}{2\pi} \theta$$

Find the point(s) where $|\vec{V}| = 0$, and show that $\psi = 0$ there. Obtain an expression for the velocity field.

5.13 Incompressible flow around a circular cylinder of radius, a, is represented by the stream function

$$\psi = -Ur \sin \theta + \frac{Ua^2 \sin \theta}{r}$$

where U represents the freestream velocity. Show that $V_r = 0$ along the circle, $r = a$. Locate the points along $r = a$, where $|\vec{V}| = U$.

5.14 Consider a flow with velocity components:

$$u = 0, \qquad v = -y^3 - 4z, \qquad w = 3y^2 z$$

(a) Is this one-, two-, or three-dimensional flow?
(b) Demonstrate whether this is an incompressible or compressible flow.
(c) If possible, derive a stream function for this flow.

5.15 Determine the family of ψ functions that will yield the velocity field

$$\vec{V} = (x^2 - y^2)\hat{\imath} - 2xy\hat{\jmath}$$

5.16 An incompressible frictionless flow field is specified by the stream function

$$\psi = -6Ax - 8Ay$$

where $A = 1$ m/sec and x and y are coordinates in meters.
(a) Sketch the streamlines $\psi = 0$ and $\psi = 8$.
(b) Indicate the direction of the resultant velocity vector at the point $(0, 0)$ on the sketch of part (a).
(c) Determine the magnitude of the flowrate between the streamlines passing through the points $(2, 2)$ and $(4, 1)$.

5.17 In a parallel one-dimensional flow in the positive x direction, the velocity varies linearly from zero at $y = 0$ to 100 ft/sec at $y = 4$ ft. Determine an expression for the stream function, ψ. Also determine the y coordinate above which the volume flowrate is half the total between $y = 0$ and $y = 4$ ft.

5.18 A flow is represented by the velocity field

$$\vec{V} = 10x\hat{\imath} - 10y\hat{\jmath} + 30\hat{k}$$

Determine if the field is:
(a) A possible incompressible flow
(b) Irrotational

5.19 A flow is represented by the velocity field

$$\vec{V} = (4x^2 + 3y)\hat{\imath} + (3x - 2y)\hat{\jmath}$$

Determine if the field is:
(a) A possible incompressible flow
(b) Irrotational

5.20 Consider the flow field represented by the stream function

$$\psi = 10xy + 17$$

(a) Is this a possible two-dimensional, incompressible flow?
(b) Is the flow irrotational?

5.21 Consider a laminar flow between closely spaced parallel plates. The velocity profile is given by

$$u = V\frac{y}{h}$$

Show that the rate of rotation of a fluid particle is given by

$$\omega_z = -\frac{V}{2h}$$

5.22 For a fluid in solid body rotation about the origin, the velocity field is given by

$$V_r = 0, \qquad V_\theta = r\omega, \qquad V_z = 0$$

Determine the rotation and vorticity for a particle in this flow field.

5.23 The velocity field near the core of a tornado can be approximated as

$$\vec{V} = -\frac{Q}{2\pi r}\,\hat{\imath}_r + \frac{k}{2\pi r}\,\hat{\imath}_\theta$$

Is this an irrotational flow field?

5.24 A free vortex is a flow field whose streamlines are concentric circles about the origin. Thus

$$V_r = 0 \qquad \text{and} \qquad V_\theta = g(r)$$

for this flow field. Determine $g(r)$ such that the rotation of any fluid particle located at $r > 0$ is zero.

5.25 A flow field is represented by the stream function

$$\psi = x^2 - y^2$$

Show that this flow field is irrotational and determine the velocity potential, ϕ.

5.26 Consider the flow field represented by the potential function, $\phi = -ax + by$.
(a) Is this a possible incompressible flow field?
(b) Is the flow irrotational?
(c) If $a = 10$ m/sec and $b = 10$ m/sec, what is the velocity of a particle at the origin?
(d) What are the coordinates of a particle, initially at the origin, after 10 sec?
(e) Sketch the lines of constant ϕ, labeling $\phi = 0$ and showing direction of increasing ϕ.

5.27 Consider the flow field represented by the potential function,

$$\phi = x^2 - y^2$$

Obtain the stream function that corresponds to this flow field.

5.28 Consider the flow field represented by the potential function,

$$\phi = ax^2 + bxy - ay^2$$

Determine the corresponding stream function and the acceleration of a particle at the point $(x, y) = (1, 0)$.

5.29 Consider the flow field given by

$$\vec{V} = xy^2\hat{\imath} - \tfrac{1}{3}y^3\hat{\jmath} + xy\hat{k}$$

Determine:

(a) The number of dimensions of the flow
(b) If it is a possible incompressible flow
(c) The acceleration of a particle at the point $(x, y, z) = (1, 2, 3)$

5.30 Consider the flow field given by

$$\vec{V} = ax^2 y\hat{\imath} - by\hat{\jmath} + cz^2\hat{k}$$

where $a = 1/m^2 \cdot \sec$, $b = 3/\sec$ and $c = 2/m \cdot \sec$.
Determine:

(a) The number of dimensions of the flow
(b) If it is a possible incompressible flow
(c) The acceleration of a particle at the point $(x, y, z) = (3, 1, 2)$

5.31 A steady, two-dimensional velocity field is given by

$$\vec{V} = Ax\hat{\imath} - Ay\hat{\jmath}$$

where $A = 1 \ \sec^{-1}$. Show that the streamlines for this flow are rectangular hyperbolas, $xy = C$. Plot streamlines that correspond to $C = 0$, 1 and 4 m². Obtain a general expression for the acceleration of a fluid particle in this velocity field. Calculate the acceleration of fluid particles at the points $(x, y) = (\frac{1}{2}, 2)$, $(1, 1)$ and $(2, \frac{1}{2})$, where x and y are measured in meters. Show the acceleration vectors on the streamline plot.

5.32 Consider again the steady, two-dimensional velocity field of Problem 5.31,

$$\vec{V} = Ax\hat{\imath} - Ay\hat{\jmath}$$

where $A = 1 \ \sec^{-1}$. Obtain expressions for the particle coordinates, $x_p = f_1(t)$ and $y_p = f_2(t)$ as functions of time and the initial particle position, (x_0, y_0) at $t = 0$. Determine the time required for a particle to travel from initial position, $(x_0, y_0) = (\frac{1}{2}, 2)$ to positions $(x, y) = (1, 1)$ and $(2, \frac{1}{2})$. Compare the particle accelerations determined by differentiating $f_1(t)$ and $f_2(t)$ with those obtained in Problem 5.31.

5.33 In cylindrical coordinates, the velocity field for a two-dimensional flow is given by

$$\vec{V} = \vec{V}(r, \theta, t)$$

Show that the radial and tangential components of the acceleration of a particle are given by

$$a_r = V_r \frac{\partial V_r}{\partial r} + \frac{V_\theta}{r}\frac{\partial V_r}{\partial \theta} - \frac{V_\theta^2}{r} + \frac{\partial V_r}{\partial t}$$

$$a_\theta = V_r \frac{\partial V_\theta}{\partial r} + \frac{V_\theta}{r}\frac{\partial V_\theta}{\partial \theta} + \frac{V_\theta V_r}{r} + \frac{\partial V_\theta}{\partial t}$$

5.34 A cylindrical tank of radius, $R = 4$ in., is filled with water to a depth of 6 in. The tank is rotated about its vertical axis. During start up, $0 \leq t \leq \tau$, the rate of rotation is given by $\omega = \omega_0 t/\tau$, where $\tau = 2$ sec and the steady state rotational speed, $\omega_0 = 78$ rpm. The no-slip condition requires that fluid particles at the tank wall have zero velocity relative to the wall. Using the results of Problem 5.33, determine, for a particle at the wall:

(a) The acceleration at time, $t = 1$ sec.
(b) The steady state acceleration.

5.35 A velocity field is given by

$$\vec{V} = -\frac{Ay}{\sqrt{x^2 + y^2}}\,\hat{\imath} + \frac{Ax}{\sqrt{x^2 + y^2}}\,\hat{\jmath}$$

(a) Show that this is a possible two-dimensional, incompressible flow.
(b) For the steady state conditions of Problem 5.34, determine the value of A.
(c) Find a_x and a_y for the particle located on the tank wall at $\theta = 60°$.

5.36 The velocity field for steady inviscid flow from left to right over a circular cylinder of radius a is given by

$$\vec{V} = U\cos\theta\left[1 - \left(\frac{a}{r}\right)^2\right]\hat{\imath}_r - U\sin\theta\left[1 + \left(\frac{a}{r}\right)^2\right]\hat{\imath}_\theta$$

Obtain expressions for the acceleration of a fluid particle moving along the stagnation streamline ($\theta = \pi$) and for the acceleration along the cylinder surface ($r = a$). Determine the locations at which these accelerations reach maximum and minimum values.

5.37 Consider the low speed flow of air between parallel disks as shown in Fig. 5.10. Assume that the flow is imcompressible and inviscid, and that the velocity is purely radial and uniform at any section. The flow speed is 15 m/sec at $R = 75$ mm. Simplify the continuity equation to a form applicable to this flow field. Show that a general expression for the velocity field is $\vec{V} = V(R/r)\hat{\imath}_r$ for $r_i \le r \le R$. Calculate the acceleration of a fluid particle at the locations $r = r_i$ and $r = R$.

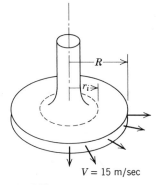

$V = 15$ m/sec

Fig. 5.10

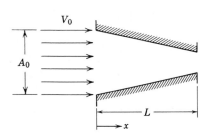

Fig. 5.11

5.38 Consider the incompressible flow of a fluid through a nozzle, as shown in Fig. 5.11. The area of the nozzle is given by $A = A_0(1 - bx)$ and the inlet velocity

problems *263*

varies according to $V_0 = U(1 + at)$ where $A_0 = 1$ ft², $L = 4$ ft, $b = 0.1$ ft⁻¹, $a = 2$ sec⁻¹, and $U = 10$ ft/sec. The flow may be assumed one-dimensional. Find the acceleration of a fluid particle at $x = L/2$ for $t = 0$ and 0.5 sec.

5.39 Consider the one-dimensional, incompressible flow through the circular channel shown in Fig. 5.12. The velocity at section ① is given by $U = U_0 + U_1 \sin \omega t$, where $U_0 = 20$ m/sec, $U_1 = 2$ m/sec, and $\omega = 0.3$ rad/sec. The channel dimensions are $L = 1$ m, $R_1 = 0.2$ m, and $R_2 = 0.1$ m. Determine the particle acceleration at the channel exit. Plot the results as a function of time over a complete cycle.

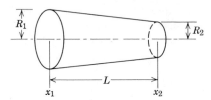

Fig. 5.12

5.40 The circular channel of Problem 5.39 is replaced by a plane channel of uniform width with the same linear dimensions. For the conditions given, determine the particle acceleration at the channel exit and plot the results as a function of time over a complete cycle.

5.41 An "air hockey" puck may be modeled as a circular disk supported by a layer of air that issues from multiple tiny holes in the game table. Assume that a puck floats a distance, $h = 1$ mm, above the table from which air flows at a volumetric flowrate per unit area of table, $q = 0.08$ m³/m² · sec. Obtain an expression for radial flow speed under the puck if the flow is considered to be uniform and incompressible. If the puck diameter is 75 mm, determine the magnitude and location of the maximum radial acceleration experienced by a fluid particle under the puck.

5.42 The temperature, T, in a long tunnel is known to vary approximately as

$$T = T_0 - \alpha e^{-x/L} \sin \frac{2\pi t}{\tau}$$

where T_0, α, L, and τ are constants, and x is measured from the entrance. A particle moves into the tunnel with a constant speed, U. Find the rate of change of temperature experienced by the particle.

Chapter 6

Dynamics of Incompressible Inviscid Flow

All real fluids possess viscosity. However, fluids often behave as though they were inviscid. Therefore, it is useful to investigate the dynamics of an *ideal fluid* that is incompressible and has zero viscosity. The analysis of ideal fluid motions is simpler than that of viscous flows because no shear stresses are present in inviscid flow. Normal stresses are the only stresses that must be considered in the analysis.

6-1 STRESS FIELD IN AN INVISCID FLOW

By applying Newton's second law to the element of mass shown in Fig. 6.1, we can show that the normal stress at a point is the same in all directions, that is, the normal stress at a point is a scalar. Writing Newton's second law in the z direction, we include surface and body forces,

$$dF_z = dF_{S_z} + dF_{B_z} = dma_z$$

Substituting according to the stresses shown in Fig. 6.1,

$$-\sigma_{zz}\, dx\, dy + \sigma_{nn}\, ds\, dx \sin \alpha - \rho g\, \frac{dx\, dy\, dz}{2} = \rho\, \frac{dx\, dy\, dz}{2}\, a_z$$

Since

$$\sin \alpha = \frac{dy}{ds},$$

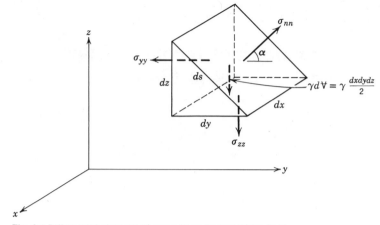

Fig. 6.1 Differential element of mass in an inviscid flow field.

then

$$-\sigma_{zz}\,dx\,dy + \sigma_{nn}\,dx\,dy - \rho g\,\frac{dx\,dy\,dz}{2} = \rho\,\frac{dx\,dy\,dz}{2}\,a_z$$

Dividing by $dx\,dy$, we obtain

$$-\sigma_{zz} + \sigma_{nn} - \rho g\,\frac{dz}{2} = \rho\,\frac{dz}{2}\,a_z$$

Recognizing that dz is vanishingly small, we obtain

$$\sigma_{zz} = \sigma_{nn}$$

Writing Newton's second law in the y direction leads to the relation that $\sigma_{yy} = \sigma_{nn}$. Consequently, we have

$$\sigma_{zz} = \sigma_{yy} = \sigma_{nn}$$

Since the shape of the element (i.e. the surface ds) was chosen arbitrarily, we conclude that for a nonviscous fluid in motion the normal stress at a point is the same in all directions (i.e. a scalar quantity). The normal stress in an inviscid flow is the negative of the thermodynamic pressure, $\sigma_{nn} = -p$ (This result is consistent with Eqs. 5.30 for $\mu = 0$.).

6-2 MOMENTUM EQUATION FOR FRICTIONLESS FLOW: EULER'S EQUATIONS

The equations of motion for frictionless flow, called Euler's equations, are obtained from the general equations of motion (Eqs. 5.29). Since, in a frictionless flow, there can be no shear stresses present and the normal stress is the negative of the thermodynamic pressure, then the equations of motion for a frictionless flow are

$$\rho B_x - \frac{\partial p}{\partial x} = \rho \left(\frac{\partial u}{\partial t} + u\frac{\partial u}{\partial x} + v\frac{\partial u}{\partial y} + w\frac{\partial u}{\partial z} \right) \tag{6.1a}$$

$$\rho B_y - \frac{\partial p}{\partial y} = \rho \left(\frac{\partial v}{\partial t} + u\frac{\partial v}{\partial x} + v\frac{\partial v}{\partial y} + w\frac{\partial v}{\partial z} \right) \tag{6.1b}$$

$$\rho B_z - \frac{\partial p}{\partial z} = \rho \left(\frac{\partial w}{\partial t} + u\frac{\partial w}{\partial x} + v\frac{\partial w}{\partial y} + w\frac{\partial w}{\partial z} \right) \tag{6.1c}$$

We can also write the above equations as a single vector equation

$$\rho\vec{B} - \nabla p = \rho \left(\frac{\partial \vec{V}}{\partial t} + u\frac{\partial \vec{V}}{\partial x} + v\frac{\partial \vec{V}}{\partial y} + w\frac{\partial \vec{V}}{\partial z} \right)$$

or

$$\rho\vec{B} - \nabla p = \rho\left(\frac{D\vec{V}}{Dt}\right) \longrightarrow \vec{a} \leftarrow page\ 249 \tag{6.2}$$

For the case where the only body force is due to gravity, the body force per unit mass, $\vec{B}$, is equal to the gravity vector, $\vec{g}$, and Eq. 6.2 becomes

$$\rho\vec{g} - \nabla p = \rho\frac{D\vec{V}}{Dt} \tag{6.3}$$

If the z coordinate is directed vertically, then, since $\nabla z = \hat{k}$,

$$\rho\vec{g} = -\rho g\hat{k} = -\rho g\,\nabla z$$

and Euler's equation can be written as

$$-\frac{1}{\rho}\nabla p - g\,\nabla z = \frac{D\vec{V}}{Dt} \tag{6.4}$$

The momentum equation for frictionless flow can also be written in cylindrical coordinates. The equations in component form, with gravity the only body force, are

$$g_r - \frac{1}{\rho}\frac{\partial p}{\partial r} = a_r = \frac{\partial V_r}{\partial t} + V_r\frac{\partial V_r}{\partial r} + \frac{V_\theta}{r}\frac{\partial V_r}{\partial \theta} + V_z\frac{\partial V_r}{\partial z} - \frac{V_\theta^2}{r} \tag{6.5a}$$

$$g_\theta - \frac{1}{\rho r}\frac{\partial p}{\partial \theta} = a_\theta = \frac{\partial V_\theta}{\partial t} + V_r\frac{\partial V_\theta}{\partial r} + \frac{V_\theta}{r}\frac{\partial V_\theta}{\partial \theta} + V_z\frac{\partial V_\theta}{\partial z} + \frac{V_r V_\theta}{r} \tag{6.5b}$$

$$g_z - \frac{1}{\rho}\frac{\partial p}{\partial z} = a_z = \frac{\partial V_z}{\partial t} + V_r \frac{\partial V_z}{\partial r} + \frac{V_\theta}{r}\frac{\partial V_z}{\partial \theta} + V_z \frac{\partial V_z}{\partial z} \tag{6.5c}$$

If the z axis is directed vertically upward, then $g_r = g_\theta = 0$ and $g_z = -g$.

Example 6.1

Consider a flow in which the velocity field is given by $\vec{V} = Ax\hat{\imath} + Ay\hat{\jmath} - 2Az\hat{k}$ where $A = 1\ \sec^{-1}$. Show that this is a possible incompressible flow. Assuming that the z axis is vertical, $\rho = 1000\ \text{kg/m}^3$, and all dimensions are in meters, calculate the pressure gradient at the point $(1, 2, 5)$. ·

Example Problem 6.1

GIVEN:

Velocity field, $\vec{V} = Ax\hat{\imath} + Ay\hat{\jmath} - 2Az\hat{k}$, with $A = 1.0\ \sec^{-1}$.
The z axis is vertical and $\rho = 1000\ \text{kg/m}^3$.

FIND:

(a) Show that velocity field represents a possible incompressible flow.
(b) Pressure gradient at $(1, 2, 5)$.

SOLUTION:

Basic equations: $\nabla \cdot \rho\vec{V} + \dfrac{\partial \rho}{\partial t} = 0$

$$\rho\vec{B} - \nabla p = \rho \frac{D\vec{V}}{Dt} \qquad \text{(neglecting viscous forces)}$$

For incompressible flow, the conservation of mass becomes $\nabla \cdot \vec{V} = 0$, or

$$\frac{\partial u}{\partial x} + \frac{\partial v}{\partial y} + \frac{\partial w}{\partial z} = 0$$

With $\vec{V} = Ax\hat{\imath} + Ay\hat{\jmath} - 2Az\hat{k}$, then

$$\frac{\partial(Ax)}{\partial x} + \frac{\partial(Ay)}{\partial y} + \frac{\partial(-2Az)}{\partial z} = A + A - 2A = 0$$

Therefore, the velocity field represents a possible incompressible flow.

If gravity is the only body force and the z axis is directed vertically upward, then

$$\rho\vec{B} = \rho\vec{g} = -\rho g\hat{k}$$

Solving Euler's equation for ∇p,

$$\nabla p = -\rho g \hat{k} - \rho \frac{D\vec{V}}{Dt}$$

$$= -\rho g \hat{k} - \rho \left[\overset{0}{\cancel{\frac{\partial \vec{V}}{\partial t}}} + u \frac{\partial \vec{V}}{\partial x} + v \frac{\partial \vec{V}}{\partial y} + w \frac{\partial \vec{V}}{\partial z} \right]$$

$$= -\rho g \hat{k} - \rho [Ax(A\hat{\imath}) + Ay(A\hat{\jmath}) + (-2Az)(-2A\hat{k})]$$

$$\nabla p = -\rho [A^2 x \hat{\imath} + A^2 y \hat{\jmath} + (4A^2 z + g)\hat{k}]$$

At the point (1, 2, 5),

$$\nabla p = -\frac{1000 \text{ kg}}{\text{m}^3} \left[\left(\frac{1}{\text{sec}}\right)^2 1 \text{ m } \hat{\imath} + \left(\frac{1}{\text{sec}}\right)^2 2 \text{ m } \hat{\jmath} + \left(4\left(\frac{1}{\text{sec}}\right)^2 5 \text{ m} + \frac{9.81 \text{ m}}{\text{sec}^2}\right)\hat{k} \right] \frac{\text{N} \cdot \text{sec}^2}{\text{kg} \cdot \text{m}}$$

$$\nabla p = -1\hat{\imath} - 2\hat{\jmath} - 29.8\hat{k} \text{ kN/m}^2/\text{m} \qquad\qquad \underleftarrow{\hspace{3cm}} \nabla p$$

{This problem illustrates the calculation of the local pressure gradient from} {Euler's equation.}

**6-3 EULER'S EQUATIONS FOR FLUIDS IN RIGID-BODY MOTION

In Chapter 3 we found that if a fluid is accelerated so that there is no relative motion between adjacent layers of the fluid, that is, when the fluid moves without deformation, no shear stresses occur. We were able to determine the pressure variation within the fluid by applying the equations of motion to an appropriate free body. We considered two specific cases. For the case of rectilinear acceleration, we derived the differential equation of motion, Eq. 3.19; in Example Problem 3.10 we applied the equation to a tank of water moving as a solid body. In Example Problem 3.11 we considered the case of a fluid undergoing steady rotation about a vertical axis; we derived the equation of motion for a differential element of fluid undergoing rotation.

Since Euler's equations are the equations of motion for a frictionless flow, that is, a flow in which the shear stresses are zero, they can be employed as the basic governing equations (written in the appropriate coordinate system) to solve frictionless flow problems. It is left to you as an exercise to show that the use of Euler's equations to solve Example Problems 3.10 and 3.11 leads to results identical to those previously obtained,

---

** This section may be omitted without loss of continuity in the text material.

6-4 EULER'S EQUATIONS IN STREAMLINE COORDINATES

In Chapter 2 we pointed out that the notion of a streamline, that is, a line drawn tangent to the velocity vector at every point in the flow field, provides a convenient graphical representation. In steady flow a fluid particle will move along a streamline because, for steady flow, pathlines and streamlines coincide. Thus, in describing the motion of a fluid particle in a steady flow, the distance along a streamline is a logical coordinate to use in writing the equations of motion. "Streamline coordinates" may also be used to describe unsteady flow. Streamlines in unsteady flow give a graphical representation of the instantaneous velocity field.

For simplicity, consider the flow in the yz plane shown in Fig. 6.2. The equations of motion are to be written in terms of the coordinate, s, distance along a streamline and the coordinate, n, distance normal to the streamline. Since the velocity vector must be tangent to the streamline, then the velocity field is given by $\vec{V} = \vec{V}(s, t)$. The pressure at the center of the fluid element is p. If we apply Newton's second law in the streamwise direction, that is, in the s direction, to the fluid element of volume $ds\,dn\,dx$, then neglecting the viscous forces

$$\left(p - \frac{\partial p}{\partial s}\frac{ds}{2}\right)dn\,dx - \left(p + \frac{\partial p}{\partial s}\frac{ds}{2}\right)dn\,dx - \rho g \sin\beta\,dn\,dx\,ds = \rho\,ds\,dn\,dx\,a_s$$

where β is the angle between the tangent to the streamline and the horizontal, and a_s is the acceleration of the fluid particle along the streamline. Simplifying the equation, we obtain

$$-\frac{\partial p}{\partial s} - \rho g \sin\beta = \rho a_s$$

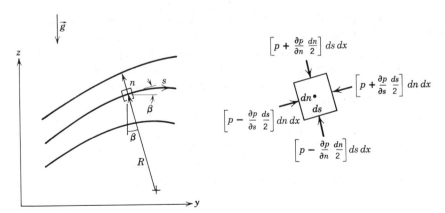

Fig. 6.2 Fluid particle moving along a streamline.

Since $\sin \beta = \partial z / \partial s$, we can write

$$-\frac{1}{\rho}\frac{\partial p}{\partial s} - g\frac{\partial z}{\partial s} = a_s$$

Along any streamline $V_s = V_s(s, t)$, and the total acceleration of a fluid particle in the streamwise direction is given by

$$a_s = \frac{DV_s}{Dt} = \frac{\partial V_s}{\partial t} + V_s\frac{\partial V_s}{\partial s}$$

The velocity is tangent to the streamline and, hence, the subscript, s, on V_s is redundant and can be dropped. Euler's equation in the streamwise direction with the z axis directed vertically is then

$$-\frac{1}{\rho}\frac{\partial p}{\partial s} - g\frac{\partial z}{\partial s} = \frac{\partial V}{\partial t} + V\frac{\partial V}{\partial s} \qquad (6.6a)$$

For steady flow, and neglecting body forces, Euler's equation in the streamwise direction reduces to

$$\frac{1}{\rho}\frac{\partial p}{\partial s} = -V\frac{\partial V}{\partial s} \qquad (6.6b)$$

which indicates that a decrease in velocity is accompanied by an increase in pressure and conversely.[1]

To obtain Euler's equation in a direction normal to the streamlines, we apply Newton's second law in the n direction to the fluid element. Again, neglecting viscous forces, we obtain

$$\left(p - \frac{\partial p}{\partial n}\frac{dn}{2}\right)ds\,dx - \left(p + \frac{\partial p}{\partial n}\frac{dn}{2}\right)ds\,dx - \rho g\cos\beta\,dn\,dx\,ds = \rho a_n\,dn\,dx\,ds$$

where β is the angle between the n direction and the vertical, and a_n is the acceleration of the fluid particle in the n direction. Simplifying the equation, we obtain

$$-\frac{\partial p}{\partial n} - \rho g\cos\beta = \rho a_n$$

Since $\cos\beta = \partial z / \partial n$, we write

$$-\frac{1}{\rho}\frac{\partial p}{\partial n} - g\frac{\partial z}{\partial n} = a_n$$

---

[1] The relationship between variations in pressure and velocity in the streamwise direction for steady incompressible inviscid flow is illustrated in the NCFMF film loop, S-FM038, *Streamwise Pressure Gradient in Inviscid Flow.*

The normal acceleration of the fluid element is toward the center of curvature of the streamline, that is, in the minus n direction; thus in the coordinate system of Fig. 6.2, the familiar centripetal acceleration is written

$$a_n = \frac{-V^2}{R}$$

for steady flow,[2] where R is the radius of curvature of the streamline. Then, Euler's equation normal to the streamline is written for steady flow as

$$\frac{1}{\rho}\frac{\partial p}{\partial n} + g\frac{\partial z}{\partial n} = \frac{V^2}{R} \tag{6.7a}$$

For steady flow in a horizontal plane, Euler's equation normal to a streamline becomes

$$\frac{1}{\rho}\frac{\partial p}{\partial n} = \frac{V^2}{R} \tag{6.7b}$$

which indicates that pressure increases in a direction outward from the center of curvature of the streamlines.[3] In regions where the streamlines are straight, the radius of curvature, R, of the streamlines is infinite and, hence, there is no pressure variation normal to the streamlines.

6-5 BERNOULLI EQUATION—INTEGRATION OF EULER'S EQUATIONS ALONG A STREAMLINE FOR STEADY FLOW

We have written the momentum equations and the continuity equation in differential form. Theoretically, for an incompressible inviscid flow these equations can be solved to give the complete velocity and pressure fields. (If the density is not constant, an additional thermodynamic relation for the density is required.) While, in theory, the equations can be solved, the solution for a particular flow field may be very involved. However, we can integrate Euler's equation readily for steady flow along a streamline. This is most easily accomplished by using the form of Euler's equation along a streamline given by Eq. 6.6a. We shall derive the result in two ways in order to give added physical insight about restrictions on the results.

[2] If the flow were not steady, the streamline pattern could change with time. In that case,

$$a_n = -\frac{V_s^2}{R} + \frac{\partial V_n}{\partial t}$$

[3] The effect of streamline curvature on the pressure gradient normal to a streamline is illustrated in the NCFMF film loop, S-FM037, *Streamline Curvature and Normal Pressure Gradient*.

6-5.1 DERIVATION USING STREAMLINE COORDINATES

Euler's equation for steady flow along a streamline is given by

$$-\frac{1}{\rho}\frac{\partial p}{\partial s} - g\frac{\partial z}{\partial s} = V\frac{\partial V}{\partial s} \tag{6.8}$$

If a fluid particle moves a distance ds along a streamline, then

$$\frac{\partial p}{\partial s}\,ds = dp \qquad \text{(the change in pressure along } s)$$

$$\frac{\partial z}{\partial s}\,ds = dz \qquad \text{(the change in elevation along } s)$$

$$\frac{\partial V}{\partial s}\,ds = dV \qquad \text{(the change in velocity along } s)$$

Then after multiplying Eq. 6.8 by ds, we can write

$$-\frac{dp}{\rho} - g\,dz = V\,dV \qquad \text{(along } s)$$

or

$$\frac{dp}{\rho} + g\,dz + V\,dV = 0 \qquad \text{(along } s)$$

Integration of this equation gives

$$\int\frac{dp}{\rho} + gz + \frac{V^2}{2} = \text{constant} \qquad \text{(along } s) \tag{6.9}$$

Before Eq. 6.9 can be used, we must specify the relation between the pressure, p, and the density, ρ. For the special case of incompressible flow, $\rho = $ constant, and Eq. 6.9 becomes

$$\frac{p}{\rho} + gz + \frac{V^2}{2} = \text{constant} \qquad \text{(along a streamline)} \tag{6.10}$$

Equation 6.10 is called the Bernoulli equation. It is often misused because the restrictions on its derivation are not kept firmly in mind. It is important to remember the assumptions made in deriving the Bernoulli equation. These were:

1. Steady flow.
2. Incompressible flow.
3. Frictionless flow.
4. Flow along a streamline.

The Bernoulli equation is a powerful and useful equation, because it relates pressure changes to velocity and elevation changes along a streamline. However, it gives correct results only when applied to a flow situation where all four of the restrictions are reasonable. Therefore, keep the restrictions firmly in mind when you consider using the Bernoulli equation. (In general, the Bernoulli constant in Eq. 6.10 has different values along different streamlines.)

**6-5.2 DERIVATION USING RECTANGULAR COORDINATES

The vector form of Euler's equation, Eq. 6.4, can also be integrated along a streamline. Since the velocity field, $\vec{V}$, is specified in terms of the rectangular coordinates, x, y, z, it is convenient to employ vector notation.[4] We shall restrict the derivation to steady flow; thus, the end result of our effort should be Eq. 6.9.

For steady flow, Euler's equation in rectangular coordinates becomes

$$-\frac{1}{\rho}\nabla p - g\nabla z = \frac{D\vec{V}}{Dt} = u\frac{\partial \vec{V}}{\partial x} + v\frac{\partial \vec{V}}{\partial y} + w\frac{\partial \vec{V}}{\partial z}$$

Using vector notation

$$u\frac{\partial \vec{V}}{\partial x} + v\frac{\partial \vec{V}}{\partial y} + w\frac{\partial \vec{V}}{\partial z} = (\vec{V}\cdot\nabla)\vec{V} \tag{6.11}$$

(We suggest that you check this equality by expanding the right side of Eq. 6.11 using the familiar dot product operation.) Then, Euler's equation can be expressed as

$$-\frac{1}{\rho}\nabla p - g\nabla z = (\vec{V}\cdot\nabla)\vec{V} \tag{6.12}$$

For steady flow the velocity field is given by $\vec{V} = \vec{V}(x, y, z)$. The streamlines are lines drawn in the flow field tangent to the velocity vector at every point. Recall again that for steady flow, streamlines, pathlines, and streaklines coincide. The motion of a particle along a streamline is governed by Eq. 6.12. In the time increment dt, the particle moves a distance $d\vec{s}$ along the streamline.

If we take the dot product of the terms in Eq. 6.12 with the distance $d\vec{s}$ along the streamline, we obtain a scalar equation relating the pressure, p,

**   This section may be omitted without loss of continuity in the text material.
4   We will have to use the vector identity

$$(\vec{V}\cdot\nabla)\vec{V} = \tfrac{1}{2}\nabla(\vec{V}\cdot\vec{V}) - \vec{V}\times(\nabla\times\vec{V})$$

which may be verified by expanding each side into components (see Appendix D).

velocity, V, and elevation, z, along the streamline. Taking the dot product of $d\vec{s}$ with Eq. 6.12 gives

$$-\frac{1}{\rho}\nabla p \cdot d\vec{s} - g\nabla z \cdot d\vec{s} = (\vec{V} \cdot \nabla)\vec{V} \cdot d\vec{s} \qquad (6.13)$$

where

$$d\vec{s} = dx\,\hat{\imath} + dy\,\hat{\jmath} + dz\,\hat{k} \qquad \text{(along } s)$$

Now we shall evaluate each of the three terms in Eq. 6.13.

$$-\frac{1}{\rho}\nabla p \cdot d\vec{s} = -\frac{1}{\rho}\left[\hat{\imath}\frac{\partial p}{\partial x} + \hat{\jmath}\frac{\partial p}{\partial y} + \hat{k}\frac{\partial p}{\partial z}\right] \cdot [dx\,\hat{\imath} + dy\,\hat{\jmath} + dz\,\hat{k}]$$

$$= -\frac{1}{\rho}\left[\frac{\partial p}{\partial x}dx + \frac{\partial p}{\partial y}dy + \frac{\partial p}{\partial z}dz\right] \qquad \text{(along } s)$$

$$-\frac{1}{\rho}\nabla p \cdot d\vec{s} = -\frac{1}{\rho}dp \qquad \text{(along } s)$$

and

$$-g\nabla z \cdot d\vec{s} = -g\hat{k} \cdot [dx\,\hat{\imath} + dy\,\hat{\jmath} + dz\,\hat{k}]$$

$$-g\nabla z \cdot d\vec{s} = -g\,dz \qquad \text{(along } s)$$

Using a vector identity,[5] we can write the third term as

$$(\vec{V} \cdot \nabla)\vec{V} \cdot d\vec{s} = [\tfrac{1}{2}\nabla(\vec{V} \cdot \vec{V}) - \vec{V} \times (\nabla \times \vec{V})] \cdot d\vec{s}$$

$$= \{\tfrac{1}{2}\nabla(\vec{V} \cdot \vec{V})\} \cdot d\vec{s} - \{\vec{V} \times (\nabla \times \vec{V})\} \cdot d\vec{s}$$

The last term on the right side of this equation is zero, since $\vec{V}$ is parallel to $d\vec{s}$. Consequently,

$$(\vec{V} \cdot \nabla)\vec{V} \cdot d\vec{s} = \tfrac{1}{2}\nabla(\vec{V} \cdot \vec{V}) \cdot d\vec{s} \qquad \text{(along } s)$$

$$= \frac{1}{2}\left[\hat{\imath}\frac{\partial V^2}{\partial x} + \hat{\jmath}\frac{\partial V^2}{\partial y} + \hat{k}\frac{\partial V^2}{\partial z}\right] \cdot [dx\,\hat{\imath} + dy\,\hat{\jmath} + dz\,\hat{k}]$$

$$= \frac{1}{2}\left[\frac{\partial V^2}{\partial x}dx + \frac{\partial V^2}{\partial y}dy + \frac{\partial V^2}{\partial z}dz\right]$$

$$(\vec{V} \cdot \nabla)\vec{V} \cdot d\vec{s} = \tfrac{1}{2}d(V^2) \qquad \text{(along } s)$$

Substituting these three terms into Eq. 6.13 yields

$$\frac{dp}{\rho} + g\,dz + \frac{1}{2}d(V^2) = 0 \qquad \text{(along } s)$$

[5] This identity is proved in Appendix D.

Integrating this equation, we obtain

$$\int \frac{dp}{\rho} + gz + \frac{V^2}{2} = \text{constant} \qquad \text{(along } s\text{)}$$

If the density is constant, we obtain the Bernoulli equation

$$\frac{p}{\rho} + gz + \frac{V^2}{2} = \text{constant} \qquad \text{(along a streamline)}$$

As expected, we see that the last two equations are identical to Eqs. 6.9 and 6.10 derived previously using streamline coordinates. The Bernoulli equation derived using rectangular coordinates is still limited by the restrictions:

1. Steady flow.
2. Incompressible flow.
3. Frictionless flow.
4. Flow along a streamline.

6-5.3 APPLICATIONS

The Bernoulli equation can be applied between any two points on a streamline, provided that the other three restrictions are satisfied. The result is

$$\frac{p_1}{\rho} + \frac{V_1^2}{2} + gz_1 = \frac{p_2}{\rho} + \frac{V_2^2}{2} + gz_2 \qquad (6.14)$$

where subscripts 1 and 2 represent any two points on a streamline. Applications of Eqs. 6.10 and 6.14 to typical flow situations are illustrated in Example Problems 6.2 through 6.4.

In some problems, the flow appears unsteady from one reference frame, but steady from another, which translates in the flow. Since the Bernoulli equation was derived by integrating Newton's second law for a fluid particle, it can be applied in any inertial reference frame (see the discussion of translating frames in Section 4-4.3). The procedure is illustrated in Example Problem 6.5.

Example 6.2

Air flows steadily and at low speed through a horizontal nozzle, discharging to the atmosphere. At the nozzle inlet, the area is 0.1 m². At the nozzle exit, the area is 0.02 m². The flow is essentially incompressible, and frictional effects are negligible. Determine the gage pressure required at the nozzle inlet to produce an outlet speed of 50 m/sec.

Example Problem 6.2

GIVEN:

Flow through a nozzle, as shown. The air flow is steady, incompressible, and frictionless.

FIND:

$p_1 - p_{atm}$.

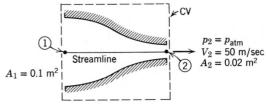

SOLUTION:

Basic equations:

$$\frac{p_1}{\rho} + \frac{V_1^2}{2} + gz_1 = \frac{p_2}{\rho} + \frac{V_2^2}{2} + gz_2$$

$$= 0(1)$$

$$0 = \frac{\partial}{\partial t}\int_{CV} \rho \, d\forall + \int_{CS} \rho \vec{V} \cdot d\vec{A}$$

Assumptions: (1) Steady flow
(2) Incompressible flow
(3) Frictionless flow
(4) Flow along a streamline
(5) $z_1 = z_2$
(6) Uniform flow at sections ① and ②

Apply the Bernoulli equation along a streamline between points ① and ② to evaluate p_1. Then

$$p_1 = p_2 + \frac{\rho}{2}(V_2^2 - V_1^2)$$

Apply the continuity equation to determine V_1, using CV as shown.

$$0 = \{-|\rho V_1 A_1|\} + \{|\rho V_2 A_2|\} \quad \text{or} \quad V_1 A_1 = V_2 A_2$$

so that

$$V_1 = V_2 \frac{A_2}{A_1} = \frac{50 \text{ m}}{\text{sec}} \times \frac{0.02 \text{ m}^2}{0.1 \text{ m}^2} = 10 \text{ m/sec}$$

For air at standard conditions, $\rho = 1.23 \text{ kg/m}^3$.

$$p_1 - p_{atm} = \frac{\rho}{2}(V_2^2 - V_1^2)$$

$$= \frac{1}{2} \times \frac{1.23 \text{ kg}}{\text{m}^3}\left[(50)^2 \frac{\text{m}^2}{\text{sec}^2} - (10)^2 \frac{\text{m}^2}{\text{sec}^2}\right]\frac{\text{N} \cdot \text{sec}^2}{\text{kg} \cdot \text{m}}$$

$$p_1 - p_{atm} = 1.48 \text{ kPa} \qquad\qquad\qquad\qquad p_1 - p_{atm}$$

This problem illustrates a typical application of the Bernoulli equation. Note that if the flow streamlines are straight at the nozzle inlet and exit, the pressure will be uniform at those sections.

Example 6.3

A U-tube acting as a water siphon is shown in the following figure. If the flow is frictionless as a first approximation, and the fluid issues from the bottom of the siphon as a free jet at atmospheric pressure, determine (after listing the necessary assumptions) the velocity of the free jet, in m/sec, and the absolute pressure of the fluid at point A in the flow, in Pa.

A

1 m

8 m

Example Problem 6.3

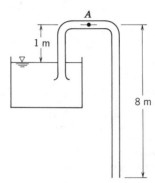

GIVEN:

Water flowing through a siphon as shown.

FIND:

(a) Velocity of water leaving as a free jet.
(b) Pressure at point Ⓐ in the flow.

SOLUTION:

Basic equation:

$$\frac{p}{\rho} + \frac{V^2}{2} + gz = \text{constant}$$

Assumptions: (1) Neglect friction
　　　　　　 (2) Steady flow
　　　　　　 (3) Incompressible flow

Apply the Bernoulli equation between points ① and ②.

$$\frac{p_1}{\rho} + \frac{V_1^2}{2} + gz_1 = \frac{p_2}{\rho} + \frac{V_2^2}{2} + gz_2$$

Since Area $_{\text{reservoir}} \gg$ Area$_{\text{pipe}}$, $V_1 \approx 0$. Also $p_1 = p_2 = p_{\text{atm}}$, so that

$$gz_1 = \frac{V_2^2}{2} + gz_2 \quad \text{and} \quad V_2^2 = 2g(z_1 - z_2)$$

$$V_2 = \sqrt{2g(z_1 - z_2)} = \sqrt{2 \times 9.81 \, \frac{m}{\text{sec}^2} \times 7 \, m} = 11.7 \, m/\text{sec} \qquad V_2$$

To determine the pressure at location Ⓐ, we write the Bernoulli equation between ① and Ⓐ.

$$\frac{p_1}{\rho} + \frac{V_1^2}{2} + gz_1 = \frac{p_A}{\rho} + \frac{V_A^2}{2} + gz_A$$

Again $V_1 \approx 0$ and from conservation of mass $V_A = V_2$. Hence

$$\frac{p_A}{\rho} = \frac{p_1}{\rho} + gz_1 - \frac{V_2^2}{2} - gz_A = \frac{p_1}{\rho} + g(z_1 - z_A) - \frac{V_2^2}{2}$$

$$p_A = p_1 + \rho g(z_1 - z_A) - \rho \frac{V_2^2}{2}$$

$$= \frac{1.01 \times 10^5 \, N}{m^2} + \frac{999 \, kg}{m^3} \times \frac{9.81 \, m}{\text{sec}^2} \times (-1 \, m) \times \frac{N \cdot \text{sec}^2}{kg \cdot m}$$

$$- \frac{1}{2} \times \frac{999 \, kg}{m^3} \times \frac{(11.7)^2 \, m^2}{\text{sec}^2} \times \frac{N \cdot \text{sec}^2}{kg \cdot m}$$

$$p_A = 22.8 \, \text{kPa (absolute)} \qquad\qquad p_A$$

{ This problem illustrates a straightforward application of the Bernoulli equation with elevation changes included. }

Example 6.4

Water flows under a sluice gate on a horizontal bed at the inlet to a flume. Above the gate, the water level is 1.5 ft, and the velocity is negligible. At the vena contracta below the gate, the flow streamlines are straight and the depth is 2 in. Hydrostatic pressure distributions and uniform flow may be assumed at each section; friction is negligible. Determine the flow velocity downstream from the gate, and the discharge in cubic feet per second per foot of width.

Example Problem 6.4

GIVEN:

Flow of water under a sluice gate. Flow is frictionless, uniform at each section, and the pressure distribution is hydrostatic at sections ① and ②.

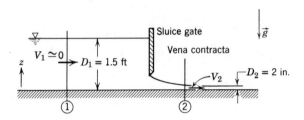

FIND:

(a) V_2.
(b) Q in ft³/sec/ft of width.

SOLUTION:

The flow satisfies all conditions necessary to apply the Bernoulli equation. The question is, what streamline do we use?

Basic equation:
$$\frac{p_1}{\rho} + \frac{V_1^2}{2} + gz_1 = \frac{p_2}{\rho} + \frac{V_2^2}{2} + gz_2$$

Assumptions: (1)   Steady flow
(2)   Incompressible flow
(3)   Frictionless flow
(4)   Flow along a streamline
(5)   Uniform flow at each section
(6)   Hydrostatic pressure distribution

From assumption 6,

$$\frac{dp}{dz} = -\gamma$$

so that

$$p = p_{atm} + \gamma(D - z) \qquad \text{or} \qquad \frac{p}{\rho} = \frac{p_{atm}}{\rho} + g(D - z)$$

Substituting this relation into the Bernoulli equation gives

$$\frac{p_{atm}}{\rho} + g(D_1 - z_1) + \frac{V_1^2}{2} + gz_1 = \frac{p_{atm}}{\rho} + g(D_2 - z_2) + \frac{V_2^2}{2} + gz_2$$

or

$$\frac{V_1^2}{2} + gD_1 = \frac{V_2^2}{2} + gD_2$$

This result implies that $V^2/2 + gD =$ constant, and the constant has the same value along *any* streamline for this flow. Solving for V_2 yields

$$V_2 = \sqrt{2g(D_1 - D_2) + V_1^2}$$

But $V_1^2 \approx 0$, so

$$V_2 = \sqrt{2g(D_1 - D_2)} = \sqrt{2 \times \frac{32.2 \ \frac{\text{ft}}{\text{sec}^2}} \left(1.5 \ \text{ft} - 2 \ \text{in.} \times \frac{\text{ft}}{12 \ \text{in.}}\right)}$$

$$V_2 = 9.23 \ \text{ft/sec} \qquad\qquad\qquad V_2$$

For uniform flow, $Q = VA = VDw$, or $Q/w = $ ⟨discharge cuft/ft·sec⟩

$$\frac{Q}{w} = VD = V_2 D_2 = \frac{9.23 \ \text{ft}}{\text{sec}} \times 2 \ \text{in.} \times \frac{\text{ft}}{12 \ \text{in.}} = 1.58 \ \text{ft}^2/\text{sec}$$

$$\frac{Q}{w} = 1.58 \ \frac{\text{ft}^3/\text{sec}}{\text{ft}} \qquad\qquad\qquad \frac{Q}{w}$$

Example 6.5

A Piper Club flies at 150 km/hr in standard air at an altitude of 1000 m. At a certain point close to the wing, the air speed *relative* to the wing is 60 m/sec. Compute the pressure at this point.

Example Problem 6.5

GIVEN:

Aircraft in flight at 150 km/hr at 1000 m altitude in standard air.

$V_{\text{air}} = 0$

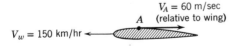

$V_A = 60 \ \text{m/sec}$
A (relative to wing)

Observer

$V_w = 150 \ \text{km/hr}$ ⟵

FIND:

Pressure, p_A, at point A.

SOLUTION:

Flow is unsteady when observed from a fixed frame, that is, by an observer on the ground. However, an observer *on* the wing sees the following steady flow:

$$V_{air} = V_w = 150 \text{ km/hr}$$

Thus the Bernoulli equation can be applied along a streamline in the moving observer's inertial reference frame.

Basic equation:
$$\frac{p_{air}}{\rho} + \frac{V_{air}^2}{2} + gz_{air} = \frac{p_A}{\rho} + \frac{V_A^2}{2} + gz_A$$

Assumptions: (1) Steady flow
(2) Incompressible flow ($V < 100$ m/sec)
(3) Frictionless flow
(4) Flow along a streamline
(5) Neglect Δz

Values for pressure and density may be found from Table A.4. Thus at 1000 m,

$$\frac{p}{p_0} = 0.8870 \quad \text{and} \quad \frac{\rho}{\rho_0} = 0.9075$$

Consequently,

$$p = 0.8870 p_0 = 0.8870 \times \frac{1.01 \times 10^5 \text{ N}}{\text{m}^2} = 8.96 \times 10^4 \text{ N/m}^2$$

and

$$\rho = 0.9075 \rho_0 = 0.9075 \times \frac{1.23 \text{ kg}}{\text{m}^3} = 1.12 \text{ kg/m}^3$$

Solving for p_A,

$$p_A = p_{air} + \frac{\rho}{2}\left(V_{air}^2 - V_A^2\right)$$

$$= \frac{8.96 \times 10^4 \text{ N}}{\text{m}^2}$$

$$+ \frac{1}{2} \times \frac{1.12 \text{ kg}}{\text{m}^3} \left[\left(\frac{150 \text{ km}}{\text{hr}} \times \frac{1000 \text{ m}}{\text{km}} \times \frac{\text{hr}}{3600 \text{ sec}}\right)^2 - \frac{(60)^2 \text{ m}^2}{\text{sec}^2}\right] \frac{\text{N} \cdot \text{sec}^2}{\text{kg} \cdot \text{m}}$$

$$p_A = 88.6 \text{ kPa (absolute)} \qquad\qquad\qquad p_A$$

6-6 STATIC, STAGNATION, AND DYNAMIC PRESSURES

The pressure, p, which we have used in deriving the Bernoulli equation, Eq. 6.10, is the thermodynamic pressure, which is commonly called the static pressure. The static pressure is that pressure which would be measured by an instrument moving with the flow. However, such a measurement is rather difficult to make in a practical situation! How do we measure static pressure experimentally?

In Section 6-5.1 we showed that there was no pressure variation normal to flow streamlines when those streamlines were straight. This fact makes it possible to measure the static pressure in a flowing fluid using a wall pressure "tap," placed in a region where the flow streamlines are straight, as shown in Fig. 6.3a. The pressure tap is a small hole, drilled carefully in the wall, with its axis perpendicular to the surface. If the hole is perpendicular to the duct wall and free from burrs, accurate measurements of static pressure can then be made by connecting the tap to a suitable measuring instrument.

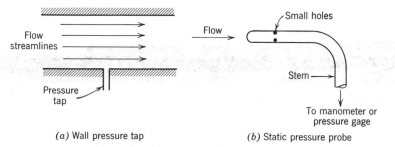

(a) Wall pressure tap   (b) Static pressure probe

Fig. 6.3 Measurement of static pressure.

In a fluid stream far from a wall, or where streamlines are curved, accurate static pressure measurements can be made by careful use of a static pressure probe, shown in Fig. 6.3b. Such probes must be designed so that the measuring holes are placed correctly with respect to the probe tip and stem to avoid erroneous results. In use, the measuring section must be aligned with the local flow direction.

Static pressure probes, such as that shown in Fig. 6.3b, and in a variety of other forms, are available commercially in sizes as small as $\frac{1}{16}$ in. in diameter.

The stagnation pressure is that value obtained when a flowing fluid is decelerated to zero velocity by a frictionless process. In incompressible flow, the Bernoulli equation can be used to relate changes in velocity and pressure along a streamline for such a process. Neglecting elevation differences,

Eq. 6.10 becomes

$$\frac{p}{\rho} + \frac{V^2}{2} = \text{constant}$$

If the static pressure is p at a point in the flow where the velocity is V, then the stagnation pressure, p_0, may be computed from

$$\frac{p_0}{\rho} + \frac{V_0^2}{2} = \frac{p}{\rho} + \frac{V^2}{2}$$

or

$$p_0 = p + \tfrac{1}{2}\rho V^2 \tag{6.15}$$

since the stagnation pressure is obtained where the stagnation velocity, V_0, is zero.

Equation 6.15 is a mathematical statement of the definition of stagnation pressure, valid for incompressible flow. The term $\tfrac{1}{2}\rho V^2$ is generally called the dynamic pressure. Solving for the dynamic pressure gives

$$\tfrac{1}{2}\rho V^2 = p_0 - p$$

and for the velocity

$$V = \sqrt{\frac{2(p_0 - p)}{\rho}} \tag{6.16}$$

Thus if the stagnation pressure and the static pressure could be measured at a point, Eq. 6.16 would give the local flow velocity.

Stagnation pressure is measured in the laboratory using a probe with a hole that faces directly upstream as shown in Fig. 6.4. Such a probe is called a stagnation pressure probe, or pitot tube. Again, the measuring section should be aligned with the local flow direction.

We have seen that static pressure at a point can be measured with a static pressure tap or probe (Fig. 6.3). If we knew the stagnation pressure at the same point, then the flow velocity could be computed from Eq. 6.16. Two possible experimental setups are shown in Fig. 6.5.

In Fig. 6.5a, the static pressure corresponding to point A is read from the wall static pressure tap. The stagnation pressure is measured directly at A

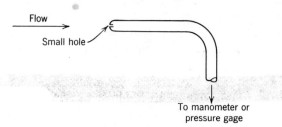

Fig. 6.4 Measurement of stagnation pressure.

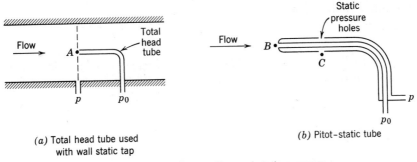

(a) Total head tube used
with wall static tap

(b) Pitot–static tube

Fig. 6.5 Simultaneous measurement of stagnation and static pressure.

by the total head tube, as shown. (The stem of the total head tube is placed downstream from the measurement location to minimize disturbance of the local flow.)

Two probes are often combined, as in the pitot-static tube shown in Fig. 6.5b. The inner tube is used to measure the stagnation pressure at point B, while the static pressure at C is sensed by the small holes in the outer tube. In flow fields where the static pressure variation in the streamwise direction is small, the pitot-static tube may be used to infer the velocity at point B in the flow, by assuming $p_B = p_C$, and using Eq. 6.16. (Note that when $p_B \neq p_C$, this procedure will give erroneous results.)

The definition and calculation of the stagnation pressure for compressible flow will be discussed in Section 9-3.1.

Example 6.6

A pitot probe is inserted in a water flow to measure the flow velocity. The tube is inserted so that it points upstream into the flow and the pressure sensed by the probe is the stagnation pressure. The static pressure is measured at the same location in the flow, using a wall pressure tap. If the fluid in the manometer tube is mercury, determine the flow speed.

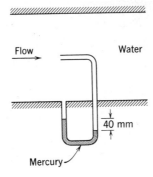

static, stagnation, and dynamic pressures/6-6 *285*

Example Problem 6.6

GIVEN:

A pitot tube inserted in a flow as shown. The flowing fluid is water and the manometer fluid is mercury.

FIND:

The flow speed.

SOLUTION:

Basic equation:

$$\frac{p}{\rho} + \frac{V^2}{2} + gz = \text{constant}$$

Assumptions: (1) Steady flow
(2) Incompressible flow
(3) Flow along a streamline
(4) Frictionless deceleration along stagnation streamline

Writing Bernoulli's equation along the stagnation streamline, taking $\Delta z = 0$, yields

$$\frac{p_0}{\rho} = \frac{p}{\rho} + \frac{V^2}{2}$$

p_0 is the stagnation pressure at the tube opening where the velocity has been reduced, without friction, to zero. Solving for V gives

$$V = \sqrt{\frac{2(p_0 - p)}{\rho}}$$

From the diagram,

$$p_0 - p = \rho_{Hg}gh - \rho_{H_2O}gh = \rho_{H_2O}gh(SG_{Hg} - 1)$$

and

$$V = \sqrt{\frac{2\rho_{H_2O}gh(SG_{Hg} - 1)}{\rho_{H_2O}}} = \sqrt{2gh(SG_{Hg} - 1)}$$

$$= \sqrt{2 \times \frac{9.81 \text{ m}}{\text{sec}^2} \times 40 \text{ mm} \times \frac{\text{m}}{1000 \text{ mm}} (13.6 - 1)}$$

$$V = 3.14 \text{ m/sec} \qquad \qquad \qquad \qquad V$$

{This problem illustrates the use of a pitot tube in determining the velocity at a point.}

6-7 RELATION BETWEEN THE FIRST LAW OF THERMODYNAMICS AND THE BERNOULLI EQUATION

The Bernoulli equation, Eq. 6.10, was obtained by integrating Euler's equation, along a streamline, for steady, incompressible, frictionless flow. Thus Eq. 6.10 was derived from the momentum equation for a fluid particle.

An equation identical in form to Eq. 6.10 (although requiring very different restrictions) may be obtained from the first law of thermodynamics. Our objective in this section is to reduce the energy equation to the form of the Bernoulli equation given by Eq. 6.10. Having arrived at this form, we shall compare the restrictions on the two equations. This procedure will help us to understand more clearly the restrictions on the use of Eq. 6.10.

Consider steady flow in the absence of shear forces. We choose a control volume bounded by streamlines along its periphery. Such a control volume, shown in Fig. 6.6, is often referred to as a stream tube.

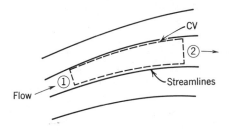

Fig. 6.6 Flow through a stream tube.

Basic equation:

$$\dot{Q} + \overset{=0(1)}{\cancel{\dot{W}_s}} + \overset{=0(2)}{\cancel{\dot{W}_{shear}}} + \overset{=0(3)}{\cancel{\dot{W}_{other}}} = \overset{=0(4)}{\cancel{\frac{\partial}{\partial t} \int_{CV} e\rho \, dV}} + \int_{CS} (e + pv)\rho \vec{V} \cdot d\vec{A} \quad (4.59)$$

$$e = u + \frac{V^2}{2} + gz$$

Restrictions: (1) $\dot{W}_s = 0$
            (2) $\dot{W}_{shear} = 0$
            (3) $\dot{W}_{other} = 0$
            (4) Steady flow
            (5) Uniform flow and properties at each section

Under these restrictions, Eq. 4.59 becomes

$$0 = \left(u_1 + p_1 v_1 + \frac{V_1^2}{2} + gz_1 \right) \{ -|\rho_1 V_1 A_1| \}$$

$$+ \left(u_2 + p_2 v_2 + \frac{V_2^2}{2} + gz_2 \right) \{ |\rho_2 V_2 A_2| \} - \dot{Q}$$

But from continuity under these restrictions,

$$0 = \underbrace{\frac{\partial}{\partial t} \int_{cv} \rho \, d\mathbf{V}}_{= 0(4)} + \int_{cs} \rho \vec{V} \cdot d\vec{A}$$

or

$$0 = \{ -|\rho_1 V_1 A_1| \} + \{ |\rho_2 V_2 A_2| \}$$

That is,

$$\dot{m} = \rho_1 V_1 A_1 = \rho_2 V_2 A_2$$

Also

$$\dot{Q} = \frac{\delta Q}{dt} = \frac{\delta Q}{dm} \frac{dm}{dt} = \frac{\delta Q}{dm} \dot{m}$$

Thus, from the energy equation,

$$0 = \left[\left(p_1 v_1 + \frac{V_1^2}{2} + gz_1 \right) - \left(p_2 v_2 + \frac{V_2^2}{2} + gz_2 \right) \right] \dot{m} + \left(u_2 - u_1 - \frac{\delta Q}{dm} \right) \dot{m}$$

or

$$p_1 v_1 + \frac{V_1^2}{2} + gz_1 = p_2 v_2 + \frac{V_2^2}{2} + gz_2 + \left(u_2 - u_1 - \frac{\delta Q}{dm} \right)$$

If the last term in this equation were neglected, we would have the desired form. Thus we apply the additional restrictions:

(6) $(u_2 - u_1 - \delta Q/dm) = 0$
(7) Incompressible flow, that is, $v_1 = v_2 = 1/\rho =$ constant

Then the energy equation reduces to

$$\frac{p_1}{\rho} + \frac{V_1^2}{2} + gz_1 = \frac{p_2}{\rho} + \frac{V_2^2}{2} + gz_2$$

or

$$\frac{p}{\rho} + \frac{V^2}{2} + gz = \text{constant} \tag{6.17}$$

Equation 6.17 is identical in form to the Bernoulli equation, Eq. 6.10. The Bernoulli equation was derived from momentum considerations (Newton's second law), and is valid for steady, incompressible, frictionless flow along a streamline. Equation 6.17 was obtained by applying the first law of thermodynamics to a stream tube control volume, subject to restrictions 1 through 7 above. Thus the Bernoulli equation (Eq. 6.10) and the identical form of the energy equation (Eq. 6.17) were developed from entirely different models, coming from entirely different basic concepts, and involving different restrictions.

After deriving Eq. 6.17, there might still be some confusion remaining about restriction 6, that is,

$$u_2 - u_1 - \frac{\delta Q}{dm} = 0$$

Obviously, one possibility is that $u_2 = u_1$ and $\delta Q/dm = 0$. Another possibility is that the terms $u_2 - u_1$ and $\delta Q/dm$ are equal. That this is true for incompressible frictionless flow is shown in Example Problem 6.7.

For the special case considered in this section it is true that the first law of thermodynamics reduces to the Bernoulli equation. Since the Bernoulli equation is obtained by integrating Euler's equation (differential form of Newton's second law) for steady, incompressible, frictionless flow along a streamline, for this special case the first law of thermodynamics and Newton's second law do not yield separate information. However, in general, the first law of thermodynamics and Newton's second law are independent equations and must be satisfied separately.

Example 6.7

Consider frictionless, incompressible flow with heat transfer. Show that

$$u_2 - u_1 = \frac{\delta Q}{dm}$$

Example Problem 6.7

GIVEN:

Frictionless, incompressible flow with heat transfer.

SHOW:

$$u_2 - u_1 = \frac{\delta Q}{dm}$$

SOLUTION:

In general, the internal energy, u, can be expressed as $u = u(T, v)$. For incompressible flow, $v = $ constant, and $u = u(T)$. Thus the thermodynamic state of the fluid is determined by the single thermodynamic property, T. The internal energy change for any process, $u_2 - u_1$, depends only on the temperatures at the end states.

From the Gibbs equation, $T\,ds = du + p\,dv$, valid for a pure substance undergoing any process, we obtain

$$T\,ds = du$$

Since the internal energy change, du, is independent of the process, we take a reversible process for which $T\,ds = d(\delta Q/dm) = du$. Therefore,

$$\frac{\delta Q}{dm} = u_2 - u_1$$

or

$$u_2 - u_1 - \frac{\delta Q}{dm} = 0$$

Example 6.8

Water flows steadily from a large open reservoir through the system shown. The discharge at point ③ is to atmospheric pressure. A heater around the pipe adds 100 Btu/lbm to the flow. The flow is assumed to be steady, frictionless, and incompressible. Find the temperature rise of the fluid between points ① and ②.

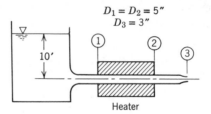

$D_1 = D_2 = 5''$
$D_3 = 3''$

10′

Heater

Example Problem 6.8

GIVEN:

Water flows from a large reservoir through the system shown and discharges to atmospheric pressure. The heater adds 100 Btu/lbm to the flow.

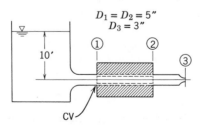

$D_1 = D_2 = 5''$
$D_3 = 3''$

10′

CV

FIND:

The temperature rise of the fluid between points ① and ②.

SOLUTION:

Basic equations:
$$\frac{p}{\rho} + \frac{V^2}{2} + gz = \text{constant}$$

$$0 = \frac{\partial}{\partial t} \overset{= 0(1)}{\cancel{\int_{CV} \rho \, d\forall}} + \int_{CS} \rho \vec{V} \cdot d\vec{A}$$

$$\dot{Q} + \overset{= 0(4)}{\cancel{\dot{W}_s}} + \overset{= 0(4)}{\cancel{\dot{W}_{shear}}} = \overset{= 0(1)}{\cancel{\frac{\partial}{\partial t} \int_{CV} e\rho \, d\forall}} + \int_{CS} \left(u + pv + \frac{V^2}{2} + gz \right) \rho \vec{V} \cdot d\vec{A}$$

Assumptions: (1) Steady flow
(2) Frictionless flow
(3) Incompressible flow
(4) No shaft work, no shear work
(5) Flow along a streamline

Under the assumptions listed, the first law of thermodynamics for the CV shown becomes

$$\dot{Q} = \int_{CS} \left(u + pv + \frac{V^2}{2} + gz \right) \rho \vec{V} \cdot d\vec{A}$$

$$= \int_{A_1} \left(u + pv + \frac{V^2}{2} + gz \right) \rho \vec{V} \cdot d\vec{A} + \int_{A_2} \left(u + pv + \frac{V^2}{2} + gz \right) \rho \vec{V} \cdot d\vec{A}$$

For uniform properties at ① and ②

$$\dot{Q} = -|\rho A_1 V_1| \left(u_1 + p_1 v + \frac{V_1^2}{2} + gz_1 \right) + |\rho V_2 A_2| \left(u_2 + p_2 v + \frac{V_2^2}{2} + gz_2 \right)$$

From conservation of mass $|\rho A_1 V_1| = |\rho A_2 V_2| = \dot{m}$

$$\dot{Q} = \dot{m} \left[u_2 - u_1 + \left(\frac{p_2}{\rho} + \frac{V_2^2}{2} + gz_2 \right) - \left(\frac{p_1}{\rho} + \frac{V_1^2}{2} + gz_1 \right) \right]$$

For frictionless, incompressible, steady flow, along a streamline,

$$\frac{p}{\rho} + \frac{V^2}{2} + gz = \text{constant}$$

Therefore,
$$\dot{Q} = \dot{m}(u_2 - u_1)$$

Dividing by $\dot{m}$ gives

$$\frac{\delta Q}{dm} = u_2 - u_1$$

{This equation could have been written directly from the results of Example Problem 6.7.}

Since for an incompressible fluid $u_2 - u_1 = c(T_2 - T_1)$, then

$$T_2 - T_1 = \frac{u_2 - u_1}{c} = \frac{100 \text{ Btu}}{\text{lbm}} \times \frac{\text{lbm} \cdot \text{R}}{1 \text{ Btu}}$$

$$T_2 - T_1 = 100 \text{ R} \qquad\qquad\qquad T_2 - T_1$$

{This problem illustrates that in general the first law of thermodynamics and the Bernoulli equation are independent equations.}

**6-8 BERNOULLI EQUATION APPLIED TO IRROTATIONAL FLOW

In Section 6-5.1, we integrated Euler's equation along a streamline for steady, incompressible, inviscid flow to obtain the Bernoulli equation

$$\frac{p}{\rho} + \frac{V^2}{2} + gz = \text{constant} \qquad \text{(along a streamline)} \qquad (6.10)$$

Equation 6.10 can be applied between any two points on the same streamline. The value of the constant will vary, in general, from streamline to streamline.

If, in addition to being inviscid, steady and incompressible, the flow field is also irrotational (i.e. the velocity field is such that $2\vec{\omega} = \nabla \times \vec{V} = 0$), we can show that Bernoulli's equation can be applied between any two points in the flow; that is, the value of the constant in Eq. 6.10 is the same for all streamlines. To illustrate this, we start with Euler's equation in vector form,

$$-\frac{1}{\rho}\nabla p - g\nabla z = (\vec{V} \cdot \nabla)\vec{V} \qquad (6.12)$$

Using the vector identity

$$(\vec{V} \cdot \nabla)\vec{V} = \tfrac{1}{2}\nabla(\vec{V} \cdot \vec{V}) - \vec{V} \times (\nabla \times \vec{V})$$

we see that for irrotational flow, since $\nabla \times \vec{V} = 0$, then

$$(\vec{V} \cdot \nabla)\vec{V} = \tfrac{1}{2}\nabla(\vec{V} \cdot \vec{V})$$

** This section may be omitted without loss of continuity in the text material.

and Euler's equation for irrotational flow can be written as

$$-\frac{1}{\rho}\nabla p - g\nabla z = \frac{1}{2}\nabla(\vec{V}\cdot\vec{V}) = \frac{1}{2}\nabla(V^2)$$ (6.18)

In the time increment dt, a fluid particle moves from the vector position $\vec{r}$ to the position $\vec{r} + d\vec{r}$; the displacement, $d\vec{r}$, is an arbitrary infinitesimal displacement in any direction. Taking the dot product of $d\vec{r} = dx\hat{\imath} + dy\hat{\jmath} + dz\hat{k}$ with each of the terms in Eq. 6.18, we have

$$-\frac{1}{\rho}\nabla p \cdot d\vec{r} - g\nabla z \cdot d\vec{r} = \frac{1}{2}\nabla(V^2)\cdot d\vec{r}$$

and hence

$$-\frac{dp}{\rho} - g\,dz = \frac{1}{2}d(V^2)$$

or

$$\frac{dp}{\rho} + g\,dz + \frac{1}{2}d(V^2) = 0$$

Integrating this equation gives

$$\int\frac{dp}{\rho} + gz + \frac{V^2}{2} = \text{constant}$$ (6.19)

For incompressible flow, $\rho = \text{constant}$, and

$$\frac{p}{\rho} + gz + \frac{V^2}{2} = \text{constant}$$ (6.20)

Since $d\vec{r}$ was an arbitrary displacement, then for a steady, incompressible, inviscid flow that is also irrotational, Eq. 6.20 is valid between any two points in the flow field.

**6-9 UNSTEADY BERNOULLI EQUATION— INTEGRATION OF EULER'S EQUATION ALONG A STREAMLINE

It is not necessary to restrict the development of the Bernoulli equation to steady flows. The purpose of this section is to develop the corresponding equation for unsteady flow along a streamline and to illustrate its use.

The momentum equation for frictionless flow was found in Section 6-2 to be

$$-\frac{1}{\rho}\nabla p - g\nabla z = \frac{D\vec{V}}{Dt}$$ (6.4)

** This section may be omitted without loss of continuity in the text material.

Equation 6.4 is a vector equation. It can be converted to a scalar equation by taking the dot product with $d\vec{s}$, where $d\vec{s}$ is an element of distance along a streamline. Thus

$$-\frac{1}{\rho}\nabla p \cdot d\vec{s} - g\,\nabla z \cdot d\vec{s} = \frac{D\vec{V}}{Dt} \cdot d\vec{s} = \frac{DV_s}{Dt}\,ds = V_s\frac{\partial V_s}{\partial s}\,ds + \frac{\partial V_s}{\partial t}\,ds \quad (6.21)$$

The terms become

$$\nabla p \cdot d\vec{s} = dp \quad \text{(the change in pressure along } s)$$

$$\nabla z \cdot d\vec{s} = dz \quad \text{(the change in } z \text{ along } s)$$

$$\frac{\partial V_s}{\partial s}\,ds = dV_s \quad \text{(the change in } V_s \text{ along } s)$$

Substituting into Eq. 6.21, we obtain

$$-\frac{dp}{\rho} - g\,dz = V_s\,dV_s + \frac{\partial V_s}{\partial t}\,ds \tag{6.22}$$

Integrating along a streamline from point 1 to point 2,

$$\int_1^2 \frac{dp}{\rho} + \frac{V_2^2 - V_1^2}{2} + g(z_2 - z_1) + \int_1^2 \frac{\partial V_s}{\partial t}\,ds = 0 \tag{6.23}$$

For incompressible flow, the density is constant. For this special case, Eq. 6.23 becomes

$$\frac{p_1}{\rho} + \frac{V_1^2}{2} + gz_1 = \frac{p_2}{\rho} + \frac{V_2^2}{2} + gz_2 + \int_1^2 \frac{\partial V_s}{\partial t}\,ds \quad \text{(along a streamline)}$$

$$\tag{6.24}$$

To evaluate the integral term in Eq. 6.24, the variation in the quantity $\partial V_s/\partial t$ must be known as a function of s, the distance along the streamline measured from point 1. (For steady flow, $\partial V_s/\partial t = 0$, and Eq. 6.24 reduces to Eq. 6.10.)

Let us review the restrictions on Eq. 6.24. They were:

1. Incompressible flow.
2. Frictionless flow.
3. Flow along a streamline.

Consequently, Eq. 6.24 may be applied to any flow in which these restrictions are compatible with the physical situation.

Application of Eq. 6.24 is illustrated in Example Problem 6.9.

Example 6.9

A long pipe is connected to a large reservoir that is initially filled with water to a depth of 3 m. The pipe is 150 mm in diameter and 6 m long. As a first approximation, friction may be neglected. Determine the flow velocity leaving the pipe as a function of time after a cap is removed from its free end. The reservoir is large enough so that the change in its level may be neglected.

Example Problem 6.9

GIVEN:

Pipe and large reservoir as shown.

FIND:

$V_2(t)$.

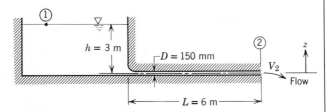

SOLUTION:

Apply the Bernoulli equation to the unsteady flow along a streamline from point ① to point ②.

Basic equation:
$$\frac{\cancel{p_1}}{\rho} + \frac{\cancel{V_1^2}}{2} + gz_1 = \frac{\cancel{p_2}}{\rho} + \frac{V_2^2}{2} + \cancel{gz_2} + \int_1^2 \frac{\partial V_s}{\partial t}\,ds$$

$0(5) \qquad = 0(6)$

Assumptions:
(1) Incompressible flow
(2) Frictionless flow
(3) Flow along a streamline from ① to ②
(4) $p_1 = p_2 = p_{atm}$
(5) $V_1^2 \simeq 0$
(6) $z_2 = 0$
(7) $z_1 = h = $ constant
(8) Neglect velocity in reservoir, except for a small region near the inlet to the tube

Then

$$gz_1 = gh = \frac{V_2^2}{2} + \int_1^2 \frac{\partial V_s}{\partial t}\,ds$$

In view of assumption (8), the integral becomes

$$\int_1^2 \frac{\partial V_s}{\partial t}\,ds \approx \int_0^L \frac{\partial V_s}{\partial t}\,ds$$

In the tube, $V_s = V_2$ everywhere, so that

$$\int_0^L \frac{\partial V_s}{\partial t}\,ds = \int_0^L \frac{dV_2}{dt}\,ds \approx L\frac{dV_2}{dt}$$

Substituting gives

$$gh = \frac{V_2^2}{2} + L\frac{dV_2}{dt}$$

Separating variables,

$$\frac{dV_2}{2gh - V_2^2} = \frac{dt}{2L}$$

Integrating between limits $V = 0$ at $t = 0$ and $V = V_2$ at $t = t$,

$$\int_0^{V_2} \frac{dV}{2gh - V^2} = \left[\frac{1}{\sqrt{2gh}}\tanh^{-1}\left(\frac{V}{\sqrt{2gh}}\right)\right]_0^{V_2} = \frac{t}{2L}$$

Since $\tanh^{-1}(0) = 0$, we obtain

$$\frac{1}{\sqrt{2gh}}\tanh^{-1}\left(\frac{V_2}{\sqrt{2gh}}\right) = \frac{t}{2L}$$

or

$$\frac{V_2}{\sqrt{2gh}} = \tanh\left(\frac{t}{2L}\sqrt{2gh}\right) \qquad\qquad V_2(t)$$

For the given conditions,

$$\sqrt{2gh} = \sqrt{2 \times 9.81\ \frac{m}{sec^2} \times 3\ m} = 7.67\ m/sec$$

and

$$\frac{t}{2L}\sqrt{2gh} = \frac{t}{2} \times \frac{1}{6\ m} \times 7.67\ \frac{m}{sec} = 0.639t$$

The results are then $V_2 = 7.67\tanh(0.639t)$

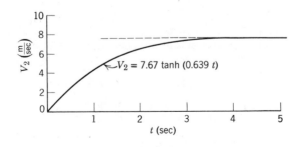

summary objectives

After completing your study of Chapter 6, you should be able to do the following:

1. Write Euler's equations in (a) vector form, (b) rectangular coordinates, (c) cylindrical coordinates, and (d) streamline coordinates.

2. Integrate Euler's equation along a streamline in steady flow to obtain the Bernoulli equation. State the restrictions on the use of the Bernoulli equation.

3. Define static pressure, stagnation pressure, and dynamic pressure.

4. For steady, incompressible, inviscid flow through a stream tube, state the conditions under which the first law of thermodynamics reduces to the Bernoulli equation.

**5. For a steady, inviscid, incompressible flow that is irrotational, show that the Bernoulli equation may be applied between any two points in the flow field.

**6. Write the unsteady Bernoulli equation for flow along a streamline. State the restrictions on the use of the equation.

7. Solve those problems at the end of the chapter that relate to the material you have studied.

problems

6.1   A velocity field is given as

$$\vec{V} = Ay\hat{i} + Ax\hat{j}$$

where $A = 3$ m/sec/m and the coordinates x and y are given in meters. The fluid density is 750 kg/m^3. Calculate the acceleration of a fluid particle at the point $(x, y) = (1, 0)$, and determine the pressure gradient at the same point if $\vec{B} = -g\hat{j}$.

6.2   Consider the flow field with velocity in ft/sec given by

$$\vec{V} = (Axy - Bx^2)\hat{i} + (Axy - By^2)\hat{j}$$

where $A = 2/\text{ft} \cdot \text{sec}$ and $B = 1/\text{ft} \cdot \text{sec}$. The density is 2 slug/ft^3, and the body force per unit mass is $\vec{B} = -g\hat{j}$. Determine the acceleration of a fluid particle and also the pressure gradient at the point $(x, y) = (1, 1)$.

6.3   The x component of velocity in an incompressible flow is given in m/sec by

$$u = Ax; \quad A = 6 \text{ sec}^{-1}$$

At the point $(x, y) = (2, 0)$, the y component of velocity is $v = 0$; $w = 0$ everywhere. Obtain an equation for the y component of velocity, and find the acceleration of a fluid particle at the point $(2, 0)$.

6.4   An incompressible flow field is given by

$$\vec{V} = (Ax + By)\hat{i} - Ay\hat{j}$$

**   These objectives apply to sections that may be omitted without loss of continuity in the text material.

where $A = 1$ sec^{-1}, $B = 2$ sec^{-1} and the coordinates are measured in meters. Find the magnitude and direction of the acceleration of a fluid particle at the point $(x, y) = (1, 2)$.

6.5 A flow of water is described by the following velocity field in ft/sec:

$$\vec{V} = (Ax + Bt)\hat{i} + (-Ay + Bt)\hat{j}$$

where $A = 10$ sec^{-1}, $B = 5$ ft$\cdot$sec^{-2}, x and y are in ft, and t is in sec. Compute the acceleration of a fluid particle at the point $(x, y) = (1, 5)$ at $t = 10$ sec. Evaluate $\partial p/\partial x$ under the same conditions, if $B_x = 0$.

6.6 A velocity field in a fluid with density equal to 1500 kg/m^3, is given in m/sec by

$$\vec{V} = (Ax - By)t\hat{i} + (Ay - Bx)t\hat{j}$$

where $A = 1$ sec^{-2}, $B = 2$ sec^{-2}, x and y are in meters, and t is in sec. Body forces are negligible. Evaluate ∇p at the point $(x, y) = (1, 2)$ at $t = 1$ sec.

6.7 In a frictionless, incompressible flow, the velocity field in m/sec, and the body force are given by

$$\vec{V} = Ax\hat{i} - Ay\hat{j}$$
$$\vec{B} = -g\hat{k}$$

The pressure is p_0 at the point $(x, y, z) = (0, 0, 0)$. Obtain an expression for the pressure field, $p(x, y, z)$.

6.8 For the flow field of Problem 5.23, obtain an expression for the pressure gradient.

6.9 Steady, frictionless, and incompressible flow from right to left over a stationary circular cylinder of radius, a, is given by the field

$$\vec{V} = U\left[\left(\frac{a}{r}\right)^2 - 1\right]\cos\theta\,\hat{i}_r + U\left[\left(\frac{a}{r}\right)^2 + 1\right]\sin\theta\,\hat{i}_\theta$$

Consider flow along the streamline forming the cylinder surface, that is, $r = a$. Express the components of the pressure gradient in terms of angle, θ.

6.10 Example Problem 3.9 read: "As a result of a promotion, you are transferred from your present location. You must transport a fish tank in the back of your station wagon. The tank is $12 \times 24 \times 12$ in. How much water should you leave in the tank to be reasonably sure that it will not spill over during the trip?" Solve this problem using the Euler equations.

6.11 Example Problem 3.10 read: "A cylindrical container, partially filled with liquid, is rotated at constant angular velocity, ω, about its axis. After a short period of time there is no relative motion; that is, the liquid rotates with the cylinder as if the system were a rigid body. Determine the shape of the free surface." Solve this problem using the Euler equations. (The cylinder radius is R.)

6.12 The cylindrical container of Problem 6.11 is initially three-fourths full of liquid. Determine the maximum angular speed at which the container can be spun without spilling liquid over the top. Use the Euler equations.

6.13 Consider a steady fluid motion with rigid body swirl about the z axis. Assume $\vec{g} = -g\hat{k}$. Show that Eqs. 6.5 reduce to

$$\frac{\partial p}{\partial r} = \frac{\rho V_\theta^2}{r}$$

$$\frac{\partial p}{\partial \theta} = 0$$

$$\frac{\partial p}{\partial z} = -\rho g$$

6.14 The flow field in a forced, or rigid body, vortex is given by the expression

$$\vec{V} = \omega r \hat{i}_\theta$$

where $\omega = 10 \, \text{sec}^{-1}$, and r is measured in meters. Assume a frictionless fluid with $\rho = 1000 \, \text{kg/m}^3$. Express the radial pressure gradient, $\partial p/\partial r$, as a function of r, and evaluate the pressure change between $r_1 = 1$ m and $r_2 = 2$ m.

6.15 Air at 20 psia and 100 F flows around a smooth corner at the inlet to a diffuser. The air velocity is 150 ft/sec, and the radius of curvature of the streamlines is 3 in. Determine the magnitude of the centripetal acceleration in "G's" experienced by a fluid particle rounding the corner. Evaluate the pressure gradient, $\partial p/\partial r$.

6.16 The flow field in a free, or irrotational, vortex is given by the expression

$$\vec{V} = \frac{k}{2\pi r} \hat{i}_\theta; \qquad r > 0$$

where r is measured in meters. Assume a frictionless fluid with $\rho = 1000 \, \text{kg/m}^3$, and $k = 20\pi \, \text{m}^2/\text{sec}$. Express the radial pressure gradient, $\partial p/\partial r$, as a function of r, and evaluate the pressure change between $r_1 = 1$ m and $r_2 = 2$ m.

6.17 By expanding the expression $(\vec{V} \cdot \nabla)\vec{V}$ into components, show that

$$\frac{D\vec{V}}{Dt} = (\vec{V} \cdot \nabla)\vec{V} + \frac{\partial \vec{V}}{\partial t}$$

6.18 The flow area of a horizontal air duct is reduced smoothly from 0.75 to 0.25 ft^2. The flowrate is steady at 1.5 lbm/sec of air at 40 psia, 70 F. Frictionless flow may be assumed. Determine the pressure change over the length of duct in which the area is reduced.

6.19 Water flows in a circular pipe. At one section the diameter is 0.3 m, the static absolute pressure is 260 kPa, the velocity is 3 m/sec, and the elevation is 10 m above ground level. The elevation at a section downstream is 0 m, and the pipe diameter is 0.15 m. Find the gage pressure at the downstream section. Frictional effects may be neglected.

6.20 Water flows steadily up the vertical, 0.1 m diameter pipe (Fig. 6.7) and out the nozzle, which is 0.05 m in diameter, discharging to atmospheric pressure. The stream velocity at the nozzle exit must be 20 m/sec. Calculate the gage pressure required at section ① , assuming frictionless flow.

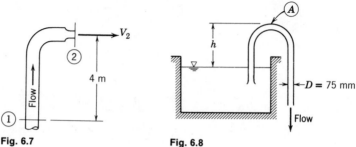

Fig. 6.7   Fig. 6.8

6.21 A siphon is shown in Fig. 6.8. The water may be considered to flow without friction. The water flowrate is 0.03 m^3/sec, its temperature is 20 C, and the pipe diameter is 75 mm. Compute the maximum allowable height, h, so that the pressure at point A is above the vapor pressure of the water.

6.22 Water flows from a very large tank through a 2 in. diameter tube (Fig. 6.9). The dark fluid in the manometer is mercury. Determine the velocity in the pipe and the rate of discharge from the tank.

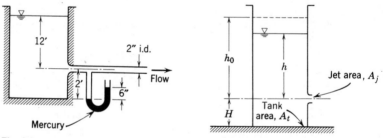

Fig. 6.9   Fig. 6.10

6.23 The tank shown in Fig. 6.10 has a well-rounded orifice with area A_j. At time $t = 0$, the water level is at height h_0. Develop an expression for the water height, h, at any later time, t.

6.24 A smoothly contoured nozzle is connected to the end of a garden hose. At the nozzle inlet, where the velocity is negligible, the water gage pressure is 160 kPa. Pressure at the nozzle exit is atmospheric. Assuming that the water remains in a single stream that has negligible aerodynamic drag, estimate the maximum height above the nozzle outlet that the stream could reach.

6.25 A stream of liquid moving at low speed leaves a nozzle pointed directly downward. The velocity may be considered uniform across the nozzle section, and the effects of friction may be ignored. At the nozzle exit, located at elevation z_0, the jet velocity and area are V_0 and A_0, respectively. Determine the variation of jet area with elevation for $z < z_0$.

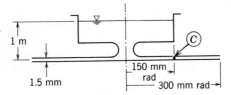

Fig. 6.11

6.26 The flow system of parallel disks shown in Fig. 6.11 contains water. As a first approximation, friction may be neglected. Determine the volumetric flowrate, and the pressure at point Ⓒ.

6.27 Consider the incompressible flow of air between the parallel disks shown in Fig. 6.12. Assume that the flow is purely radial and uniform at any section, and that viscous effects may be neglected. Obtain an expression for the velocity distribution between the disks if $V = 15$ m/sec at $r = R = 75$ mm. Evaluate the magnitude and direction of the net pressure force that acts on the upper plate between r_i and R, if $r_i = R/2$.

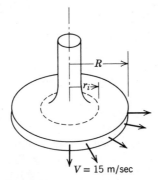

Fig. 6.12

6.28 Obtain an expression for the pressure distribution as a function of radius for the "air hockey" puck of Problem 5.41. Evaluate the net pressure force that acts on the puck.

6.29 Steady, frictionless, and incompressible flow from left to right over a stationary circular cylinder of radius, a, is represented by the velocity field

$$\vec{V} = U\left[1 - \left(\frac{a}{r}\right)^2\right]\cos\theta\,\hat{i}_r - U\left[1 + \left(\frac{a}{r}\right)^2\right]\sin\theta\,\hat{i}_\theta$$

(a) Obtain an expression for the pressure distribution along the streamline forming the cylinder surface, that is, $r = a$.

(b) Determine the locations where the static pressure on the cylinder is equal to the freestream static pressure.

(c) Determine the net pressure force on the cylinder.

6.30 The flow over a Quonset hut (Fig. 6.13) may be approximated by the velocity distribution of Problem 6.29 with $0 \leq \theta \leq \pi$. During a storm the wind velocity reaches 100 km/hr; the outside temperature is 5 C. A barometer inside the hut reads 720 mm of mercury. The hut has a diameter of 6 m and a length of 18 m. Determine the net force tending to lift the hut off its foundation.

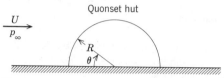

Quonset hut

Fig. 6.13

6.31 Consider the steady, frictionless, incompressible flow of air over the wing of an airplane. The air approaching the wing is at 10 psia, 40 F, and has a velocity of 200 ft/sec relative to the wing. At a certain point in the flow, the gage pressure is −0.40 psi. Calculate the velocity of the air at this point relative to the wing.

6.32 A sea water inlet for reactor cooling is to be located on the outer hull of a nuclear submarine. The maximum submerged speed of the sub is 35 knots. At the location of the inlet, the water speed parallel to the hull is 20 knots when the sub moves at maximum speed. Determine the maximum static pressure that might be expected at the inlet, in N/m².

6.33 A pitot-static tube is used to measure the velocity of moving air at a point in a flow system. To ensure that flow may be assumed incompressible for calculations of engineering accuracy, the velocity is to be maintained at 100 m/sec or less. Properties of standard air may be assumed. Determine the manometer deflection, in millimeters of water, that corresponds to the maximum desirable velocity.

6.34 An open-circuit wind tunnel draws in air from the atmosphere through a well-contoured nozzle. In the test section, where the flow is straight and nearly uniform, a static pressure tap is drilled into the tunnel wall. A manometer connected to the tap shows that the static pressure within the tunnel is 45 mm of water below atmospheric. Assume that the air is incompressible, and at 25 C, 100 kPa (absolute). Calculate the velocity in the wind tunnel test section.

6.35 A jet of air from a nozzle is blown at right angles against a wall in which a pressure tap is located. A manometer connected to the tap reads a pressure of 0.14 in. of mercury above atmospheric pressure. Determine the approximate velocity of the air leaving the nozzle if it is at 40 F and 14.7 psia.

6.36 Water is discharged from the system shown in Fig. 6.14 through the pipes ① and ② .The area of ① is 1.0 in.² and the area of ② is 2.0 in.² There is 1.8 Btu of heat added at the heater per slug of fluid flowing through the heater. Neglect frictional effects and any pressure drop across the heater.
(a) Calculate the volumetric flowrate at section ①.
(b) If area ③ equals area ④ and both are at the same elevation, compute the change in specific internal energy across the heater.
(c) Calculate the volumetric flowrate at section ②.

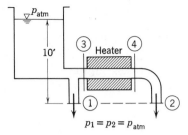

Fig. 6.14

6.37 Consider an incompressible flow of standard air, with velocity field given by

$$\vec{V} = Ax\hat{i} - Ay\hat{j} + B\hat{k}$$

where $A = 10$ sec^{-1}, $B = 30$ m/sec, and the coordinates are in meters. Neglect gravity. Determine the pressure change between the points (0, 0, 0) and (3, 1, 0).

6.38 Consider the flow of water represented by the velocity field

$$\vec{V} = Ay\hat{i} + Ax\hat{j}$$

where $A = 3$ sec^{-1} and the coordinates are in meters. Is it possible to calculate the pressure change between the points (0, 0, 0) and (1, 1, 1)? If possible, do so.

6.39 An incompressible flow field is given as

$$\vec{V} = Ax^2y\hat{i} - Axy^2\hat{j}$$

where $A = 6/\text{m}^2 \cdot \text{sec}$. Evaluate the rotation of the flow. What can be said about the use of the Bernoulli equation for this flow?

6.40 Consider the flow field of water given by

$$\vec{V} = Ax^2y^2\hat{i} - Bxy^3\hat{j}$$

where $A = 3/\text{m}^3 \cdot \text{sec}$ and $B = 2/\text{m}^3 \cdot \text{sec}$.
(a) Determine the stream function for this flow.
(b) Determine the fluid rotation.
(c) Neglecting gravity, is it possible to calculate the pressure difference between the points (0, 0, 0) and (1, 1, 1). If so, do it; if not, why not?

6.41 Determine whether the Bernoulli equation can be applied between different radii for the following vortex flows:

(a) $\vec{V} = \omega r\hat{i}_\theta$

(b) $\vec{V} = \dfrac{k}{2\pi r}\hat{i}_\theta$

6.42 Two circular disks of radius, R, are separated by a distance, b. The upper disk moves toward the lower one at velocity, V. The space between the disks is filled with a frictionless, incompressible fluid, which is squeezed out as the disks come together. Assume that at any radial section, the velocity is uniform across the gap width, b. However, note that b is a function of time. The pressure surrounding the disks is atmospheric. Determine the gage pressure at $r = 0$.

6.43 Apply the unsteady Bernoulli equation to the U-tube manometer of constant area shown in Fig. 6.15. Assume that the manometer is initially deflected, and then released. Obtain a differential equation for l as a function of time.

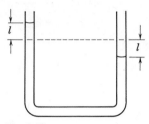

Fig. 6.15

Chapter 7

Dimensional Analysis and Similitude

7-1 INTRODUCTION

Up to now we have concentrated mainly on the analytical aspects of fluid flow. To solve some of the problems, it has been necessary for us to make some assumptions that we recognize are questionable. For example, in applying the control volume equations to the solution of the rocket motion in Example Problem 4.11, we neglected the effect of air resistance (i.e. the drag force) acting on the rocket. In Chapter 6 we made the assumption of frictionless flow in order to obtain approximate solutions to Example Problems 6.2 and 6.3. What do we do when these approximate solutions are not adequate? We could use the complete equations of motion, Eqs. 5.31. Solution of the equations is at best extremely difficult and, in many cases, virtually impossible even by the most sophisticated methods. Consequently, it often is necessary to rely on experimental results.

The history of the development of fluid mechanics has depended heavily on experimental results because so few real flows can be solved exactly by analytical methods alone. Solutions of real problems involve a combination of analysis and experimental information. First, the real physical flow situation is approximated with a mathematical model that is simple enough to yield a solution. Then experimental measurements are made to check the analytical results. Based on the measurements, refinements in the analysis can be made, and so on. The experimental results are an essential link in this iterative design process. Empirical designs developed without analysis or careful review of available experimental data are often high in cost and poor or inadequate in performance.

However, experimental work in the laboratory is both time-consuming and expensive. One obvious goal is to obtain the most information from the fewest experiments. Dimensional analysis is an important tool that often helps us to achieve this goal. The dimensionless parameters that we obtain can also be used to correlate data for succinct presentation using the minimum possible number of plots.

When experimental testing of a full-size prototype is either impossible or prohibitively expensive (as happens so often), the only feasible way of attacking the problem is through model testing in the laboratory. If we are to predict the prototype behavior from measurements on the model, it is obvious that we cannot run just any test on any model. The model flow and the prototype flow must be similar, i.e. related by known scaling laws. We shall investigate the conditions necessary to obtain this similarity of model and prototype flows following the discussion of dimensional analysis.

7-2 NATURE OF DIMENSIONAL ANALYSIS

Most phenomena in fluid mechanics depend in a complex way on geometric and flow parameters. For example, consider the drag force on a stationary smooth sphere immersed in a uniform stream of flowing fluid. What experiments must be conducted to determine the drag force on the sphere? In order to answer this question, we must specify the parameters that are important in determining the drag force. Clearly, we would expect the drag force to depend on the size of the sphere (characterized by the diameter, D), the fluid velocity, V, and the fluid viscosity, μ. In addition, the mass of the fluid as characterized by the density, ρ, might also be important. Representing the drag force by F, we can write the symbolic equation

$$F = f(D, V, \rho, \mu)$$

Although we may have neglected parameters on which the drag force depends, such as surface roughness (or have included parameters on which it does not depend), we have formulated the problem of determining the drag force for a stationary sphere in terms of quantities that are both controllable and measurable in the laboratory.

Let us imagine a series of experiments to determine the form of dependence of F on the variables D, V, ρ, and μ. After building a suitable experimental facility, the work could begin. To obtain a curve of F versus V for fixed values of ρ, μ, and D, we might need tests at 10 values of V. To explore the diameter effect, each test would be repeated for spheres of 10 different diameters. Then the procedure would be repeated 10 times for ρ and μ in turn.

Simple arithmetic shows that 10^4 separate experiments would be needed. If each test takes $\frac{1}{2}$ hour and we work 8 hours per day, the testing will require $2\frac{1}{2}$ years to complete. Needless to say, there also would be some difficulty in presenting the data. By plotting F versus V with D as a parameter for each combination of density and viscosity values, all of the data could be presented on a total of 100 sheets of graph paper. The utility of such results would be limited at best.

Fortunately, we can obtain meaningful results with significantly less effort through the use of dimensional analysis. As we shall see in Section 7-4, all the data for the drag force on a smooth sphere can be plotted as a functional relation between only two nondimensional parameters in the form

$$\frac{F}{\rho V^2 D^2} = f_1 \left(\frac{\rho V D}{\mu} \right)$$

The form of the function still must be determined experimentally. However, rather than conduct 10^4 experiments, we could establish the nature of the function as accurately with only 10 different experiments. The time saved in performing only 10 rather than 10^4 experiments is obvious. Even more important is the greater experimental convenience. No longer must we find fluids with 10 different values of density and viscosity. Nor must we make 10 spheres of different diameters. Instead, only the *ratio* $\rho V D / \mu$ must be varied. This can be accomplished by changing the velocity, for example.

Next, we shall formally state the theorem on which these results are based. Then we shall present a detailed procedure for obtaining appropriate dimensionless parameters for any given physical phenomenon. (Note that this technique is not restricted to use in fluid mechanics.)

7-3   BUCKINGHAM PI THEOREM

Given a physical problem in which the dependent parameter is a function of $n - 1$ independent parameters, we may express the relationship among the variables in functional form as

$$q_1 = f(q_2, q_3, \ldots, q_n)$$

where q_1 is the dependent parameter, and $q_2, q_3, \ldots, q_n$ are the $n - 1$ independent parameters. Mathematically, we can express the functional relationship in the equivalent form

$$g(q_1, q_2, \ldots, q_n) = 0$$

where g is an unspecified function, different from f. For the drag on a sphere we wrote the symbolic equation

$$F = f(D, V, \rho, \mu)$$

We could just as well write

$$g(F, D, V, \rho, \mu) = 0$$

The Buckingham Pi theorem states that: Given a relation among n parameters of the form

$$g(q_1, q_2, \ldots, q_n) = 0$$

then the n parameters may be grouped into $n - m$ independent dimensionless ratios, or Π parameters, expressible in functional form by

$$G(\Pi_1, \Pi_2, \ldots, \Pi_{n-m}) = 0$$

or

$$\Pi_1 = G_1(\Pi_2, \Pi_3, \ldots, \Pi_{n-m})$$

The number m is usually,[1] but not always, equal to the minimum number of independent dimensions required to specify the dimensions of all the parameters $q_1, q_2, \ldots, q_n$.

The theorem does not predict the functional form of G or G_1. The functional relation among the independent, dimensionless Π parameters must be determined experimentally.

A Π parameter is not independent if it can be formed from a product or quotient of the other parameters of the problem. For example, if

$$\Pi_5 = \frac{2\Pi_1}{\Pi_2\Pi_3}$$

or

$$\Pi_6 = \frac{\Pi_1^{3/4}}{\Pi_3^2}$$

then neither Π_5 nor Π_6 is independent of the other dimensionless parameters. Dimensional analysis of a problem is performed in three steps:

1. An appropriate list of parameters is selected.
2. The dimensionless Π parameters are obtained by using the Pi theorem.
3. The functional relation among the Π parameters is determined experimentally.

In the next section, a detailed procedure for applying the Pi theorem is developed.

[1] See Example Problem 7.3.

7-4 DETAILED PROCEDURE FOR USE OF BUCKINGHAM PI THEOREM

7-4.1 SELECTION OF PARAMETERS

Some experience is necessary to select a list that includes all parameters that affect a given flow phenomenon. Students, who do not have this experience, are often troubled by the need to apply engineering judgment in an apparent massive dose. However, it is really difficult to go wrong if a generous selection is made.

If you suspect that a phenomenon depends on a given parameter, include it. If your suspicion is correct, experiments will show that the parameter must be included to get consistent results. If the parameter is an extraneous factor, an extra II parameter may result, but experiments will show that it may be eliminated from consideration.

Therefore, do not be afraid to include *all* the parameters that you feel are important.

7-4.2 PROCEDURE FOR DETERMINING THE II GROUPS

Six steps are given here for determining the II parameters. They are almost foolproof when followed.

Step 1. *List all the parameters involved.* (Let *n* be the number of parameters.) If all the pertinent parameters are not included, some relation may finally be obtained, but it will not give the complete story. If parameters that actually have no effect on the physical phenomenon are included, either the process of dimensional analysis will show that these do not enter the relation sought or one or more dimensionless groups will be obtained that experiments will show to be extraneous.

Step 2. *Select a set of fundamental (primary) dimensions,* for example, MLt or FLt. (Note that for heat transfer problems you may also need *T* for temperature, and in electrical systems, *q* for charge.)

Step 3. *List the dimensions of all parameters in terms of primary dimensions.* (Let *r* be the number of primary dimensions.) Either force or mass may be selected as a primary dimension.

Step 4. *Select from the list of parameters a number of repeating parameters equal to the number of primary dimensions, r, and including all the primary dimensions.* No two repeating parameters should have the same net dimensions differing by only a single exponent; for example, do not include both a length (L) and a moment of inertia of an area (L^4) as repeating parameters. The repeating parameters chosen may appear in all the dimensionless groups obtained; consequently do *not* include the dependent parameter among those selected in this step.

Step 5. *Set up dimensional equations combining the parameters selected in Step 4 with each of the other parameters in turn to form dimensionless groups.* (There will be ($n - m$) equations.) Solve the dimensional equations to obtain the ($n - m$) dimensionless groups.

Step 6. *Check to see that each group obtained is dimensionless.* If mass was initially selected as a primary dimension, it is wise to check the groups using force as a primary dimension, and vice versa.

The functional relationship among the Π parameters must be determined experimentally.

The detailed procedure for determining the dimensionless Π parameters is illustrated in Example Problems 7.1 and 7.2.

Example 7.1

As noted in Section 7-2, the drag force, F, on a smooth sphere depends on the relative velocity, V, the sphere diameter, D, the fluid density, ρ, and the fluid viscosity, μ. Obtain a set of dimensionless groups that can be used to correlate experimental data.

Example Problem 7.1

GIVEN:

$F = f(\rho, V, D, \mu)$ for a smooth sphere.

FIND:

An appropriate set of dimensionless groups.

SOLUTION:

(Circled numbers refer to steps in the procedure for determining dimensionless Π parameters.)

① F V D ρ μ $n = 5$ parameters

②   Select primary dimensions M, L, and t

③   F   V   D   ρ   μ

$$\frac{ML}{t^2} \quad \frac{L}{t} \quad L \quad \frac{M}{L^3} \quad \frac{M}{Lt} \qquad r = 3 \text{ primary dimensions}$$

④   ρ, V, D $m = r = 3$ repeating parameters

⑤   Then $n - m = 2$ dimensionless groups will result. Setting up dimensional equations,

$$\Pi_1 = \rho^a V^b D^c F = \left(\frac{M}{L^3}\right)^a \left(\frac{L}{t}\right)^b (L)^c \left(\frac{ML}{t^2}\right) = M^0 L^0 t^0$$

Equating the exponents of M, L and t

$$
\begin{aligned}
M: &\quad a + 1 = 0 \\
L: &\quad -3a + b + c + 1 = 0 \\
t: &\quad -b - 2 = 0
\end{aligned}
\qquad
\left.
\begin{aligned}
a &= -1 \\
c &= -2 \\
b &= -2
\end{aligned}
\right\}
\quad \text{Therefore, } \Pi_1 = \frac{F}{\rho V^2 D^2}
$$

Similarly,

$$\Pi_2 = \rho^d V^e D^f \mu = \left(\frac{M}{L^3}\right)^d \left(\frac{L}{t}\right)^e (L)^f \left(\frac{M}{Lt}\right) = M^0 L^0 t^0$$

$$
\begin{aligned}
M: &\quad d + 1 = 0 \\
L: &\quad -3d + e + f - 1 = 0 \\
t: &\quad -e - 1 = 0
\end{aligned}
\qquad
\left.
\begin{aligned}
d &= -1 \\
f &= -1 \\
e &= -1
\end{aligned}
\right\}
\quad \text{Therefore, } \Pi_2 = \frac{\mu}{\rho V D}
$$

⑥   Check using F, L, t dimensions

$$\Pi_1 = \frac{F}{\rho V^2 D^2} \quad : \quad F \frac{L^4}{Ft^2} \left(\frac{t}{L}\right)^2 \frac{1}{L^2} = [1]$$

where [ ] means "has dimensions of" and

$$\Pi_2 = \frac{\mu}{\rho V D} \quad : \quad \frac{Ft}{L^2} \frac{L^4}{Ft^2} \frac{t}{L} \frac{1}{L} = [1]$$

The function relationship is $\Pi_1 = f(\Pi_2)$, or

$$\frac{F}{\rho V^2 D^2} = f\left(\frac{\mu}{\rho V D}\right)$$

as noted before. The form of the function, f, must be determined experimentally.

Example 7.2

The pressure drop, Δp, for steady, incompressible viscous flow through a straight horizontal pipe depends on the pipe length, l, the average velocity, $\overline{V}$, the viscosity, μ, the pipe diameter, D, the density, ρ, and the average variation, e, of the inside radius (the average "roughness" height). Determine a set of dimensionless groups that can be used in the correlation of data.

Example Problem 7.2

GIVEN:

$\Delta p = f(\rho, \bar{V}, D, I, \mu, e)$ for flow in a circular pipe.

FIND:

A suitable set of dimensionless groups.

SOLUTION:

(Circled numbers refer to steps in the procedure for determining dimensionless Π parameters.)

① $\quad \Delta p \quad \rho \quad \mu \quad \bar{V} \quad I \quad D \quad e \qquad n = 7$ parameters

② $\quad$ Choose primary dimensions M, L, and t

③ $\quad \Delta p \quad \rho \quad \mu \quad \bar{V} \quad I \quad D \quad e$

$$\frac{M}{Lt^2} \quad \frac{M}{L^3} \quad \frac{M}{Lt} \quad \frac{L}{t} \quad L \quad L \quad L \qquad r = 3 \text{ primary dimensions}$$

④ $\quad \rho, \bar{V}, D \qquad\qquad\qquad\qquad m = r = 3 \text{ repeating parameters}$

⑤ $\quad$ Then $n - m = 4$ dimensionless groups will result. Setting up dimensional equations:

$\Pi_1 = \rho^a \bar{V}^b D^c \Delta p$

$$= \left(\frac{M}{L^3}\right)^a \left(\frac{L}{t}\right)^b (L)^c \left(\frac{M}{Lt^2}\right) = M^0 L^0 t^0$$

$$\left.\begin{array}{ll} M: & 0 = a + 1 \\ L: & 0 = -3a + b + c - 1 \\ t: & 0 = -b - 2 \end{array}\right\} \begin{array}{l} a = -1 \\ b = -2 \\ c = 0 \end{array}$$

Therefore, $\Pi_1 = \rho^{-1} \bar{V}^{-2} D^0 \Delta p = \dfrac{\Delta p}{\rho \bar{V}^2}$

$\Pi_3 = \rho^g \bar{V}^h D^i I$

$$= \left(\frac{M}{L^3}\right)^g \left(\frac{L}{t}\right)^h (L)^i L = M^0 L^0 t^0$$

$$\left.\begin{array}{ll} M: & 0 = g \\ L: & 0 = -3g + h + i + 1 \\ t: & 0 = -h \end{array}\right\} \begin{array}{l} g = 0 \\ h = 0 \\ i = -1 \end{array}$$

Therefore, $\Pi_3 = \dfrac{I}{D}$

$\Pi_2 = \rho^d \bar{V}^e D^f \mu$

$$= \left(\frac{M}{L^3}\right)^d \left(\frac{L}{t}\right)^e (L)^f \frac{M}{Lt} = M^0 L^0 t^0$$

$$\left.\begin{array}{ll} M: & 0 = d + 1 \\ L: & 0 = -3d + e + f - 1 \\ t: & 0 = -e - 1 \end{array}\right\} \begin{array}{l} d = -1 \\ e = -1 \\ f = -1 \end{array}$$

Therefore, $\Pi_2 = \dfrac{\mu}{\rho \bar{V} D}$

$\Pi_4 = \rho^j \bar{V}^k D^l e$

$$= \left(\frac{M}{L^3}\right)^j \left(\frac{L}{t}\right)^k (L)^l L = M^0 L^0 t^0$$

$$\left.\begin{array}{ll} M: & 0 = j \\ L: & 0 = -3j + k + l + 1 \\ t: & 0 = -k \end{array}\right\} \begin{array}{l} j = 0 \\ k = 0 \\ l = -1 \end{array}$$

Therefore, $\Pi_4 = \dfrac{e}{D}$

⑥ Check, using F, L, t dimensions

$$\Pi_1 = \frac{\Delta p}{\rho \bar{V}^2} : \frac{F}{L^2} \frac{L^4}{Ft^2} \frac{t^2}{L^2} = [1]$$

$$\Pi_2 = \frac{\mu}{\rho \bar{V} D} : \frac{Ft}{L^2} \frac{L^4}{Ft^2} \frac{t}{L} \frac{1}{L} = [1]$$

$$\Pi_3 = \frac{l}{D} : \frac{L}{L} = [1]$$

$$\Pi_4 = \frac{e}{D} : \frac{L}{L} = [1]$$

Finally, the functional relationship is

$$\Pi_1 = f(\Pi_2, \Pi_3, \Pi_4)$$

or

$$\frac{\Delta p}{\rho \bar{V}^2} = f\left(\frac{\mu}{\rho \bar{V} D}, \frac{l}{D}, \frac{e}{D}\right)$$

{ Experiments in many laboratories have shown that this relationship correlates }
{ the data well. We shall discuss this result in greater detail in Section 8-8.1. }

7-4.3 COMMENTS ON THE PROCEDURE

The procedure outlined above, where m is taken equal to r (the fewest independent dimensions required to specify the dimensions of all parameters involved) almost always produces the correct number of dimensionless Π parameters. In a few cases, trouble arises because the number of primary dimensions differs when variables are expressed in terms of different systems of dimensions. The value of m can be established with certainty by determining the rank of the dimensional matrix; m is equal to the rank of the dimensional matrix. The procedure is illustrated in Example Problem 7.3.

The $n - m$ dimensionless groups obtained from the Buckingham procedure are independent but not unique. If a different set of repeating parameters is chosen, different groups will result. Only the tradition of common usage has dictated which groups are preferred.

If $n - m = 1$, then a single dimensionless Π parameter is obtained. In this case, the Buckingham Pi theorem indicates that the single Π parameter must be a constant.

Example 7.3

When a small tube is dipped into a pool of liquid, surface tension causes a *meniscus* to form at the free surface, which is elevated or depressed depending on the contact angle at the liquid-solid-gas interface. Experiments indicate that the magnitude of this *capillary effect*, Δh, is a function of the tube diameter, D, liquid specific weight, γ, and surface tension, σ. Determine the number of independent Π parameters that can be formed, and obtain a set.

Example Problem 7.3

GIVEN:

$\Delta h = f(D, \gamma, \sigma)$

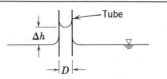
—Tube

Liquid
(Specific weight $= \gamma$
Surface tension $= \sigma$)

FIND:

(a) Number of independent Π parameters.
(b) Evaluate one set.

SOLUTION:

(Circled numbers refer to steps in the procedure for determining dimensionless Π parameters.)

① $\quad \Delta h \quad D \quad \gamma \quad \sigma \qquad n = 4$ parameters
② $\quad$ Choose primary dimensions (use both M, L, t and F, L, t dimensions to illustrate the problem in determining m)

③ (a) M, L, t

$\quad\Delta h \quad D \quad \gamma \quad \sigma$

$\quad L \quad L \quad \dfrac{M}{L^2 t^2} \quad \dfrac{M}{t^2}$

$r = 3$ primary dimensions

(b) F, L, t

$\quad\Delta h \quad D \quad \gamma \quad \sigma$

$\quad L \quad L \quad \dfrac{F}{L^3} \quad \dfrac{F}{L}$

$r = 2$ primary dimensions

Thus we ask, "Is m equal to r?" Let us check the dimensional matrix to find out.

	Δh	D	γ	σ
M	0	0	1	1
L	1	1	-2	0
t	0	0	-2	-2

	Δh	D	γ	σ
F	0	0	1	1
L	1	1	-3	-1

The rank of a matrix is equal to the order of its largest nonzero determinant.

$\begin{vmatrix} 0 & 1 & 1 \\ 1 & -2 & 0 \\ 0 & -2 & -2 \end{vmatrix} = 0 - (1)(-2) + (1)(-2) = 0$

$\begin{vmatrix} 1 & 1 \\ -3 & -1 \end{vmatrix} = -1 + 3 = 2 \neq 0$

$\begin{vmatrix} -2 & 0 \\ -2 & -2 \end{vmatrix} = 4 \neq 0 \qquad \therefore m = 2$
$\qquad\qquad\qquad\qquad\qquad m \neq r$

$\therefore m = 2$
$\quad m = r$

④ $m = 2$. Choose D, γ as repeating parameters.

$m = 2$. Choose D, γ as repeating parameters.

⑤ $n - m = 2$ dimensionless groups will result.

$$\Pi_1 = D^a \gamma^b \, \Delta h$$

$$= (L)^a \left(\frac{M}{L^2 t^2}\right)^b (L) = M^0 L^0 t^0$$

$M:$ $b + 0 = 0$
$L:$ $a - 2b + 1 = 0$ $\left.\right\}$ $\begin{array}{l} b = 0 \\ a = -1 \end{array}$
$t:$ $-2b + 0 = 0$

Therefore, $\Pi_1 = \dfrac{\Delta h}{D}$

$$\Pi_2 = D^c \gamma^d \sigma$$

$$= (L)^c \left(\frac{M}{L^2 t^2}\right)^d \frac{M}{t^2} = M^0 L^0 t^0$$

$M:$ $d + 1 = 0$
$L:$ $c - 2d = 0$ $\left.\right\}$ $\begin{array}{l} d = -1 \\ c = -2 \end{array}$
$t:$ $-2d - 2 = 0$

Therefore, $\Pi_2 = \dfrac{\sigma}{D^2 \gamma}$

$n - m = 2$ dimensionless groups will result.

$$\Pi_1 = D^e \gamma^f \, \Delta h$$

$$= (L)^e \left(\frac{F}{L^3}\right)^f L = F^0 L^0 t^0$$

$F:$ $f = 0$
$L:$ $e - 3f + 1 = 0$ $\left.\right\}$ $e = -1$

Therefore, $\Pi_1 = \dfrac{\Delta h}{D}$

$$\Pi_2 = D^g \gamma^h \sigma$$

$$= (L)^g \left(\frac{F}{L^3}\right)^h \frac{F}{L} = F^0 L^0 t^0$$

$F:$ $h + 1 = 0$
$L:$ $g - 3h - 1 = 0$ $\left.\right\}$ $\begin{array}{l} h = -1 \\ g = -2 \end{array}$

Therefore, $\Pi_2 = \dfrac{\sigma}{D^2 \gamma}$

---

⑥ Check, using F, L, t dimensions

$$\Pi_1 = \frac{\Delta h}{D} : \frac{L}{L} = [1]$$

$$\Pi_2 = \frac{\sigma}{D^2 \gamma} : \frac{F}{L} \frac{1}{L^2} \frac{L^3}{F} = [1]$$

Check, using M, L, t dimensions

$$\Pi_1 = \frac{\Delta h}{D} : \frac{L}{L} = [1]$$

$$\Pi_2 = \frac{\sigma}{D^2 \gamma} : \frac{M}{t^2} \frac{1}{L^2} \frac{L^2 t^2}{M} = [1]$$

Therefore, both systems of dimensions yield the same dimensionless Π parameters. The predicted functional relationship is

$$\Pi_1 = f(\Pi_2)$$

or

$$\frac{\Delta h}{D} = f\left(\frac{\sigma}{D^2 \gamma}\right)$$

7-5 PHYSICAL MEANING OF COMMON DIMENSIONLESS GROUPS

Over the years, several hundred different important dimensionless groups have been identified. Following tradition, each such group has been given the name of a prominent scientist or engineer, usually the one who pioneered its use. In spite of the plethora of names we encounter, several are so important and occur so frequently that we should take time to learn their definitions. Understanding their physical meaning also gives insight into the phenomena we study.

7-5.1 THE REYNOLDS NUMBER

In the 1880s, Osborne Reynolds, the British engineer, studied the transition between laminar and turbulent flow in a tube. He discovered that the parameter (later named after him)

$$Re = \frac{\rho \bar{V} D}{\mu} = \frac{\bar{V} D}{\nu}$$

is a criterion by which the state of a flow may be determined. Later experiments have shown that the Reynolds number is a key parameter for other flow cases as well. Thus, in general,

$$Re = \frac{\rho V L}{\mu} = \frac{V L}{\nu}$$

where L is a characteristic length descriptive of the flow field.

The physical significance of the Reynolds number may be seen more readily by rewriting it in the form

$$Re = \frac{\rho V L}{\mu} = \frac{\rho V L}{\mu} \frac{V}{V} \frac{L}{L} \frac{1}{L/L} = \frac{\rho V^2 L^2}{(\mu V/L)L^2}$$

In its final form, this expression may be interpreted as follows:

$$\rho V^2 L^2 \sim (\text{dynamic pressure}) \times (\text{area}) \sim \text{inertia force}$$

$$\frac{\mu V}{L} L^2 \sim (\text{viscous stress}) \times (\text{area}) \sim \text{viscous force}$$

and

$$Re \sim \frac{\text{inertia forces}}{\text{viscous forces}}$$

Thus the Reynolds number may be considered as a ratio of inertia forces to viscous forces.

7-5.2 THE MACH NUMBER

In the 1870s, the Austrian physicist Ernst Mach introduced the parameter

$$M = \frac{V}{c}$$

where V is the flow speed and c is the local sonic speed. Analysis and experiments have shown that the Mach number is a key parameter that characterizes compressibility effects in a fluid flow.

By rearranging slightly, the Mach number may be written

$$M = \frac{V}{c} = \sqrt{\frac{\rho V^2}{\rho c^2}}$$

which may be interpreted as a ratio of inertia forces to forces due to compressibility. For truly incompressible flow (under some conditions even liquids are quite compressible), $c = \infty$, so that $M = 0$.

7-5.3 THE FROUDE NUMBER

William Froude was a naval architect. Together with his son, Robert Edmund Froude, he discovered that the parameter

$$Fr = \frac{V}{\sqrt{gL}}$$

was significant for flows with free surface effects.

Squaring the Froude number gives

$$Fr^2 = \frac{V^2}{gL} = \frac{\rho V^2 L^2}{\rho g L^3}$$

which may be interpreted as the ratio of inertia forces to gravity forces.

7-5.4 THE EULER NUMBER (PRESSURE COEFFICIENT)

In aerodynamic and other model testing, it is convenient to present pressure data in dimensionless form. The ratio

$$Eu(= C_p) = \frac{\Delta p}{\frac{1}{2}\rho V^2}$$

is formed, where Δp is the local pressure minus the freestream pressure, and ρ and V are properties of the freestream flow. This ratio has been named after Leonhard Euler, the French mathematician who did much of the early analytical work in fluid mechanics.

7-6 FLOW SIMILARITY AND MODEL STUDIES

The Saturn V booster used for manned space missions is a very large vehicle (about 110 meters high!), and enormous forces are applied to it by even a gentle breeze. Failure of the vehicle or its support structure due to wind loading would be catastrophic, so data on expected wind loads were required before the first booster and launch tower were designed and erected. Existing theories for wind loads were not adequate, so that test data were required. The impossibility of testing a full-scale mockup of the booster is obvious. Therefore, model tests were needed.

To be useful, a model test must yield data that can be scaled to obtain the forces, moments, and dynamic loads that would exist on the full-scale prototype. What are the conditions that must be met to insure the similarity of model and prototype flows?

Perhaps the most obvious requirement is that the model and the prototype must be geometrically similar. Geometric similarity requires that the model and the protoype be the same shape, and that all linear dimensions of the model be related to the corresponding dimensions of the prototype by a constant scale factor.

A second requirement is that the model and prototype flows must be kinematically similar. If two flows are kinematically similar, then the velocities at corresponding points in the two flows are in the same direction and are related in magnitude by a constant scale factor. Thus two flows that are kinematically similar also have streamline patterns related by a constant scale factor. Since the boundaries form the bounding streamlines, flows that are kinematically similar must be geometrically similar.

In principle, kinematic similarity would require that a wind tunnel of infinite cross section be used to obtain data for drag on an object, in order to model correctly the performance in an infinite flow field. In practice, this restriction may be relaxed considerably, permitting the use of reasonable size equipment.

Kinematic similarity requires that the regimes of flow be the same for model and prototype. If compressibility or cavitation effects, which may change even the qualitative patterns of flow, are not present in the prototype flow, they must be avoided in the model flow.

When two flows have force distributions such that at corresponding points in the flows, identical types of forces are parallel and are related in magnitude by a constant scale factor at all corresponding points, the flows are dynamically similar.

The requirements for dynamic similarity are most restrictive; two flows must possess both geometric and kinematic similarity in order to be dynamically similar.

To establish the conditions required for complete dynamic similarity, all forces that are important in the flow situation must be considered. Thus the effects of viscous forces, of pressure forces, of surface tension forces, and so on, must be considered. Then test conditions must be established so that all important forces are related by the same scale factor between model and prototype flows. When dynamic similarity exists, data measured in a model flow may be related quantitatively to conditions in the prototype flow. What, then, are the conditions for insuring dynamic similarity between model and prototype flows?

The Buckingham Pi theorem is used to obtain the governing dimensionless groups for a flow phenomenon; to achieve dynamic similarity between geometrically similar flows we must duplicate all but one of these dimensionless groups.

For example, in considering the drag force on a sphere in Example Problem 7.1, we began with

$$F = f(D, V, \rho, \mu)$$

The Buckingham Pi theorem predicted the functional relation

$$\frac{F}{\rho V^2 D^2} = f_1\left(\frac{\rho V D}{\mu}\right)$$

In Section 7-5 we have shown that the dimensionless parameters can be viewed as ratios of forces. Thus in considering a model flow and a prototype flow about a sphere (the flows are geometrically similar), the flows also will be dynamically similar if

$$\left(\frac{\rho V D}{\mu}\right)_{model} = \left(\frac{\rho V D}{\mu}\right)_{prototype}$$

Furthermore, if

$$Re_{model} = Re_{prototype}$$

then

$$\left(\frac{F}{\rho V^2 D^2}\right)_{model} = \left(\frac{F}{\rho V^2 D^2}\right)_{prototype}$$

and the results determined from the model study can be used to predict the drag on the full-scale prototype.

Note that the actual force due to the fluid on the object is not the same in both cases, but its dimensionless value is. Note also that the two tests can be run using different fluids, if desired, as long as the Reynolds numbers are matched. For experimental convenience, test data can be measured in a wind tunnel in air and the results used to predict drag in water, as illustrated in Example Problem 7.4.

Example 7.4
The drag of a sonar transducer is to be predicted, based on wind tunnel test data. The prototype, a 1 ft diameter sphere, is to be towed at 5 knots (nautical miles per hour) in sea water. The model is 6 in. in diameter. Determine the required test speed in air. If the drag of the model at test conditions is 5.58 lbf, estimate the drag of the prototype.

Example Problem 7.4

GIVEN:

Sonar transducer to be tested in a wind tunnel.

FIND:

(a) V_m.
(b) F_p.

$V_p = 5$ knots $-D_p = 1$ ft $\rightarrow F_p$ $\rightarrow V_m$ $-D_m = 6$ in. $\rightarrow F_m = 5.58$ lbf

SOLUTION:

If no cavitation or compressibility effects are present in the prototype or model flows, kinematic similarity may be obtained, provided the wind tunnel cross section is large enough. (Experience shows that $A_{tunnel} > 15 A_{model}$ is satisfactory.) Then

$$\frac{F}{\rho V^2 D^2} = f\left(\frac{\rho V D}{\mu}\right)$$

and the test should be run at

$$Re_{model} = Re_{prototype}$$

to ensure dynamic similarity. For sea water, $\rho = 1.98$ slug/ft^3 and $\nu = 1.4 \times 10^{-5}$ ft^2/sec. At prototype conditions,

$$V_p = \frac{5 \text{ nmi}}{\text{hr}} \times \frac{6080 \text{ ft}}{\text{nmi}} \times \frac{\text{hr}}{3600 \text{ sec}} = 8.44 \text{ ft/sec}$$

$$Re_p = \frac{V_p D_p}{\nu_p} = \frac{8.44 \text{ ft}}{\text{sec}} \times \frac{1 \text{ ft}}{1.4 \times 10^{-5} \text{ ft}^2} = 6.03 \times 10^5$$

The model test conditions must duplicate this Reynolds number. Thus

$$Re_m = \frac{V_m D_m}{\nu_m} = 6.03 \times 10^5$$

For air at STP, $\rho = 0.00238$ slug/ft^3 and $\nu = 1.56 \times 10^{-4}$ ft^2/sec. The wind tunnel must be operated at

$$V_m = Re_m \frac{\nu_m}{D_m} = 6.03 \times 10^5 \times \frac{1.56 \times 10^{-4} \text{ ft}^2}{\text{sec}} \times \frac{1}{0.5 \text{ ft}}$$

$$V_m = 188 \text{ ft/sec} \qquad\qquad \underline{V_m}$$

This value is low enough for compressibility effects to be negligible.

At these test conditions, the model and prototype flows are dynamically similar. Hence

$$\left.\frac{F}{\rho V^2 D^2}\right)_m = \left.\frac{F}{\rho V^2 D^2}\right)_p$$

and

$$F_p = F_m \frac{\rho_p}{\rho_m} \frac{V_p^2}{V_m^2} \frac{D_p^2}{D_m^2} = 5.58 \text{ lbf} \times \frac{1.98}{0.00238} \times \frac{(8.44)^2}{(188)^2} \times \frac{1}{(0.5)^2}$$

$$F_p = 37.4 \text{ lbf} \underset{\longleftarrow}{\hspace{11cm}} F_p$$

If cavitation were expected—that is, if the sonar probe were operated at high speed near the free surface of the sea water—then additional parameters would have to be included.

$$\left\{\begin{array}{l}\text{This problem demonstrates the calculation of prototype values from model test} \\ \text{data.}\end{array}\right\}$$

7-7 SIMILITUDE ESTABLISHED FROM THE DIFFERENTIAL EQUATIONS

A more rigorous and broader approach to determine the conditions under which two different flows are similar is to use the governing differential equations.[2] Similitude or dynamic similarity may be present when two physical phenomena are governed by identical differential equations and boundary conditions. Similitude is obtained when the governing equations and boundary conditions have the same dimensionless form. Dynamic similarity is obtained by duplicating the dimensionless coefficients of the equations between prototype and model.

Establishment of similitude from the differential equations that describe the flow is a rigorous procedure. If one begins from the correct equations and performs each step correctly, one can be sure that all appropriate variables have been included. However, in this text we shall not use this procedure.

In contrast to use of the differential equations, success in the use of the Buckingham Pi theorem is determined by the insight used to select the parameters affecting the problem. If a complete set is chosen, the results will be complete. If one or more important variables is omitted, the results will be without meaning. Additional variables can be included if there is any uncertainty. As more experience is gained with fluid flow phenomena, the selection process becomes easier. Experience also gives more insight into the physical significance of each dimensionless group.

[2] For a detailed discussion of similitude and the use of the governing equations, refer to S. J. Kline, *Similitude and Approximation Theory* (New York: McGraw-Hill, 1965).

summary objectives

After completing study of Chapter 7, you should be able to do the following:

1. Define:

   Reynolds number
   Mach number
   Froude number
   Euler number (pressure coefficient)

   geometric similarity
   kinematic similarity
   dynamic similarity

2. State the Buckingham Pi Theorem.

3. Given a physical problem in which the dependent parameter is a function of specified independent parameters, determine a set of independent dimensionless ratios that characterize the problem.

4. State the conditions under which prototype behavior can be predicted from model tests.

5. Predict results for a prototype from model test data.

6. Solve those problems at the end of the chapter that relate to the material you have studied.

problems

7.1 The Reynolds number for flow in a pipe is defined as $Re = \rho \bar{V} D / \mu$, where $\bar{V}$ is the average velocity and D is the pipe diameter. Using both M, L, t and F, L, t systems of dimensions, show that the Reynolds number is dimensionless.

7.2 On a standard day, sonic speed at sea level is approximately 1120 ft/sec; at 28,000 ft, it is approximately 1000 ft/sec. A jet aircraft is capable of level flight at a Mach number of 1.8 at sea level, and 2.3 at 28,000 ft. Determine the corresponding air speeds in mph.

7.3 For free surface flow in a wide channel, the Froude number is defined as $Fr = V/\sqrt{gD}$. Flow in a laboratory flume occurs at a depth of 100 mm with a speed of 0.5 m/sec or at a depth of 12 mm with a speed of 3 m/sec. Calculate the Froude numbers corresponding to these flow conditions.

7.4 The *Weber number* can be considered the ratio of inertia forces to surface tension forces. It is defined as $We = \rho V^2 L / \sigma$. For wave motion, the characteristic length used is the wavelength, λ. On a water table where the depth is 0.25 in., and the velocity is 0.75 ft/sec, waves are observed with $\lambda = 6$ in. The water temperature is 68 F. Show that the Weber number is dimensionless. Determine its magnitude for the water table flow.

7.5 Cavitation phenomena are known to depend on the *cavitation number*, defined as

$$Ca = \frac{p - p_v}{\frac{1}{2}\rho V^2}$$

where p, ρ, and V are conditions in the liquid stream far ahead of a model, and p_v is the liquid vapor pressure at the test temperature. Consider a model test performed in a water tunnel at 20 C. Determine the flow speed at which the cavitation number equals unity.

7.6 At very low velocities, the drag on an object is independent of fluid density. Thus, the drag, F, on a small sphere is a function only of velocity, fluid viscosity, and sphere diameter, D. Use dimensional analysis to express the drag as a function of these variables. Use μ, V, and D as repeating variables.

7.7 Experiments show that the pressure drop due to flow through a sudden contraction in a circular duct may be expressed as

$$\Delta p = p_1 - p_2 = f(\rho, \mu, \bar{V}, d, D)$$

where the geometric variables are defined in Fig. 7.1. You are asked to organize some experimental data.Obtain the resulting dimensionless parameters, using ρ, $\bar{V}$, and D as repeating variables.

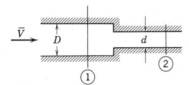

Fig. 7.1

7.8 The power, $\mathcal{P}$, required to drive a fan is believed to depend on the fluid density, the volumetric flowrate, Q, the impeller diameter, D, and angular velocity, ω. Use dimensional analysis to determine the dependence of $\mathcal{P}$ on other variables. Choose ρ, D, and ω as repeating variables.

7.9 The radiator fan in an automobile is a source of considerable noise. The sound power, $\mathcal{P}$ (energy per unit time), radiated by a fan depends on its diameter and angular velocity, and on the fluid density, ρ, and sonic speed, c. Determine the dependence of $\mathcal{P}$ on ω, using dimensional analysis. Choose ρ, ω, and D as repeating variables.

7.10 Small droplets of liquid are formed when a liquid jet breaks up in spray and fuel injection processes. The resulting droplet diameter, d, is thought to depend on the liquid density, viscosity, and surface tension, as well as the jet speed, V, and diameter, D. How many dimensionless ratios are required to characterize this process? Determine these ratios, using ρ, D, and V as repeating variables.

7.11 Assume that the resistance, R, of a flat plate immersed in a fluid depends on the fluid density and viscosity, the velocity, and the width, b, and height, h, of the plate. Find a convenient set of coordinates for organizing data.

7.12 The velocity, V, of a free surface gravity wave in deep water is a function of the wavelength, λ, depth, D, water density, and the acceleration of gravity. Use dimensional analysis to find the functional dependence of V on the other variables. Choose ρ, D, and g as repeating variables. Express V in the simplest form possible.

7.13 The power, $\mathscr{P}$, required to drive a propeller is known to depend on the following variables:

$$V = \text{freestream speed}$$
$$D = \text{propeller diameter}$$
$$\omega = \text{angular speed of propeller}$$
$$\mu = \text{fluid viscosity}$$
$$\rho = \text{fluid density}$$
$$c = \text{speed of sound in fluid}$$

(a) How many dimensionless groups are required to characterize this situation?
(b) Obtain these dimensionless groups (do *not* use μ as a repeating variable).

7.14 Capillary waves are formed on a liquid free surface as a result of surface tension. They have short wavelengths. The speed of a capillary wave depends on the surface tension, wavelength, and liquid density. Use dimensional analysis to express the wave speed as a function of these variables.

7.15 A continuous belt moving vertically through a bath of viscous liquid drags a layer of liquid of thickness, h, along with it. The volumetric rate of liquid loss, Q, is assumed to depend on μ, ρ, g, h, and V, where V is the belt speed. Apply dimensional analysis to predict the form of dependence of Q on other variables. Use ρ, V, and h as repeating variables.

7.16 The power loss, $\mathscr{P}$, in a journal bearing depends on the diameter, D, and clearance, c, of the bearing, in addition to its angular speed, ω. The lubricant viscosity and mean pressure are also important. Determine the functional form of dependence of $\mathscr{P}$ on these parameters.

7.17 The boundary-layer thickness, δ, on a smooth flat plate in an incompressible flow without pressure gradients depends on the freestream velocity, U, the fluid density, ρ, the fluid viscosity, μ, and the distance from the leading edge of the plate, x. Express these variables in dimensionless form.

7.18 A disk rotates near a fixed surface. The radius of the disk is R, and the space between the disk and the surface is filled with a fluid of viscosity, μ. The spacing between the disk and the surface is h, and the disk rotates at angular speed, ω. Find the dependence between torque on the disk, T, and the other variables.

7.19 Approximately 90 percent of the resistance of surface ships is due to gravity wave motion. Gravity waves scale with Froude number, so that in model tests the Froude number is matched between model and prototype. The remaining 10 percent of the resistance is due to viscous forces, which depend on the Reynolds number. Assume, in a ship model test, that Froude numbers are matched. Determine the kinematic viscosity required in a model fluid to match Reynolds numbers for length scales of 1:10, 1:50, and 1:100. Does any liquid exist with the required properties?

7.20 An automobile is to travel through standard air at a speed of 100 km/hr. To determine the pressure distribution, a scale model $\frac{1}{5}$ the length of the full-size car is to be tested in water. Determine the water speed that should be used. What factors must be considered to assure kinematic similarity in the tests?

7.21 The capillary rise, Δh, of a liquid in a circular tube of diameter, D, depends on the surface tension, σ, and the specific weight, γ, of the fluid. The significant variables found from dimensional analysis are

$$\Pi_1 = \Delta h \sqrt{\frac{\gamma}{\sigma}}, \qquad \Pi_2 = \frac{\sigma}{\gamma D^2}$$

The capillary rise for liquid A is 1.0 in. in a tube of 0.010 in. diameter. What will be the rise for liquid B (having the same surface tension but four times the density of A) in a tube of diameter 0.005 in.?

7.22 The drag force, F, on a boat can be expressed through dimensional analysis as follows:

$$\frac{F}{\rho V^2 L^2} = f\left(\frac{\rho V L}{\mu}, \frac{V^2}{Lg}\right)$$

(a) If we wish to perform experiments on a model to determine the drag on the prototype, what relations between the model and the prototype must be satisfied?

(b) If the model is constructed to $\frac{1}{20}$ scale and the prototype is to operate at 10 m/sec, at what speed should the model test be run?

(c) If the prototype is to operate in water at 20 C, what should be the kinematic viscosity of the liquid used in the model test?

7.23 An airship is to operate at 20 m/sec in air at standard conditions. A model is constructed to $\frac{1}{20}$ scale and tested in a wind tunnel at the same air temperature to determine drag.

(a) What criterion should be considered to obtain dynamic similarity?

(b) If the model is tested at 20 m/sec, what pressure should be used in the wind tunnel?

(c) If the model drag force is 250 N, what will be the drag of the prototype?

7.24 The drag of an airfoil at zero angle of attack is a function of the flow density, viscosity, and velocity, in addition to a length parameter. A $\frac{1}{10}$-scale model of an airfoil was tested in a wind tunnel at a Reynolds number of 5.5×10^6, based on chord length. Test conditions in the wind tunnel air stream were 15 C and 10 atm absolute pressure. The prototype airfoil has a chord length of 2 m, and it is to be flown in air at standard conditions. Determine the velocity at which the wind tunnel model was tested, and the corresponding prototype velocity.

7.25 The power input, $\mathcal{P}$, to an impeller depends on the viscosity, μ, density, ρ, and velocity, V, of the fluid and on the diameter, D, and rotational speed, ω, of the impeller. The following nondimensional parameters were found:

$$\Pi_1 = \frac{\mathcal{P}}{\rho \omega^3 D^5}, \qquad \Pi_2 = \frac{\rho V D}{\mu}, \qquad \Pi_3 = \frac{D\omega}{V}$$

Tests of a $\frac{1}{5}$-scale model (operating with the same incompressible fluid at the same temperature as the prototype) required 100 W of power at maximum speed.
(a) Have all the Π parameters been included? If not, how many would you expect?
(b) Determine the maximum power requirement of the prototype.

7.26 The power input to an axial flow pump is thought to depend on the volumetric flowrate, head, speed and diameter of the pump, and density of the fluid, that is

$$\mathscr{P} = f(Q, H, N, D, \rho)$$

where

Q = volumetric flowrate

H = head ($\sim$ energy per unit mass)

N = angular speed

D = diameter

ρ = density

An axial flow pump is required to deliver 25 ft^3/sec of water at a head of 150 ft·lbf/slug. The diameter of the rotor is 1 ft, and it is to be driven at 500 rpm. The prototype is to be modeled on a small test apparatus having a 3 hp, 1000 rpm power supply. For similar performance between the prototype and the model, calculate the head, flowrate, and the diameter of the model.

7.27 The drag force, F, on a submarine moving far below the free surface in an incompressible flow is a function of density, ρ, viscosity, μ, speed, V, and maximum cross-sectional area, A. It has been suggested that dimensional analysis will yield the following results:

$$\frac{F}{\rho V^2 A} = g\left(\frac{\rho V A}{\mu}\right)$$

(a) Is this the correct number of Π parameters? Why or why not?
(b) Are these Π parameters dimensionally correct? If not, correct them.
(c) A submarine model is scaled down so that all lengths are $\frac{1}{10}$ those of the prototype. The model is to be tested in sea water; the model test is to be used in predicting the prototype drag at a speed of 15 knots.
(1) At what speed should the model test be run?
(2) If this speed could be achieved, what would be the relationship between the drag force of the model and that of the prototype?

7.28 In some low speed ranges, vortices are shed from the rear of bluff cylinders placed across the flow. The vortices alternately leave the top and bottom of the cylinder, as shown in Fig. 7.2, causing an alternating force normal to the freestream velocity. The vortex shedding frequency, f, is thought to depend on ρ, V, d, and μ.
(a) Use dimensional analysis to develop a functional relationship for f.
(b) Vortex shedding occurs in standard air on two cylinders with a diameter ratio of 2. Determine the velocity ratio for dynamic similarity, and the ratio of vortex shedding frequencies.

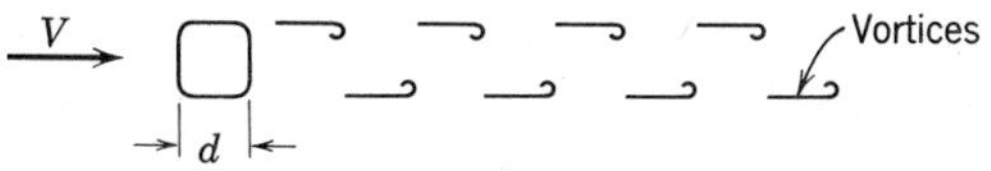

Fig. 7.2

7.29 A model propeller 2 ft in diameter is tested in a wind tunnel. Air approaches the propeller at 150 ft/sec when it rotates at 2000 rpm. The thrust and torque measured under these conditions are 25 lbf and 15 ft·lbf, respectively. A prototype 10 times as large as the model is to be built. At a dynamically similar operating point, the approach air speed is to be 400 ft/sec. Calculate the speed, thrust, and torque of the prototype propeller under these conditions, neglecting the effect of viscosity but including density.

7.30 Independent variables in a turbomachine are the impeller diameter, D, angular speed, ω, and fluid viscosity and density. Dependent properties are volumetric flowrate, Q, head, H ($\sim$ energy per unit mass), and power input, $\mathscr{P}$. Use ρ, μ, and D as repeating variables in a dimensional analysis.

(a) Determine the dimensionless ratios that characterize this problem.

(b) Under what conditions will flows in two different machines be similar?

(c) Determine the speed of operation for machine 2 for the same flow as machine 1 if $D_2/D_1 = 2$ and viscous effects are unimportant. What will be the head ratio under these conditions?

Chapter 8

Incompressible Viscous Flow

PART A.   INTRODUCTION

Viscous flows may be divided into the general categories of internal and external flows. In addition, laminar or turbulent flow regimes are possible, depending on the flow Reynolds number and external conditions. Thus this chapter begins with a brief discussion of internal and external, laminar and turbulent flows.

The rest of the chapter is devoted to the solution of practical flow cases. The objectives of the analysis are to determine the flowrate, pressure loss, and power requirements for flows in pipes and ducts, and to evaluate the lift and drag on bodies moving through fluids. We shall employ the same basic tools we have used before: the basic equations in control volume form that we developed in Chapter 4.

After deriving the differential form of the momentum equation (Eqs. 5.29), we studied the dynamics of inviscid flow in Chapter 6. The resulting differential equations of motion, the Euler equations, were valid for flows in which the shear stresses could be neglected. In this chapter we are interested in flows for which viscous forces are not negligible. Since all fluids possess viscosity, the presence of velocity gradients necessitates the presence of viscous stresses. We shall thus consider flows with both velocity gradients and shear stresses. Such flows were discussed qualitatively in Chapter 2 (Section 2-5.1). It would be helpful for you to reread that section at this point.

8-1 INTERNAL AND EXTERNAL FLOWS

Flows completely bounded by solid surfaces (i.e. duct flows) are called internal flows. Figure 8.1 illustrates laminar flow in the entrance region of a circular pipe. The flow is uniform at the pipe entrance with velocity U_0. Because of the no-slip condition at the wall, we know that the velocity at the wall must be zero along the entire length of the pipe. A boundary layer (Section 2-5.1) develops along the walls of the channel. The solid surface exerts a retarding shear force on the flow; thus the speed of the fluid in the neighborhood of the surface is reduced. At successive sections along the pipe, the effect of the solid surface is felt farther out into the flow.

For incompressible flow, the velocity at the pipe centerline must increase with distance from the inlet in order to satisfy the continuity equation. However, the average velocity at any cross section

$$\bar{V} = \frac{1}{A} \int_{Area} u \, dA$$

must equal U_0, so

$$\bar{V} = U_0 = \text{constant} \tag{8.1}$$

Sufficiently far from the pipe entrance, the boundary layer developing on the pipe wall reaches the pipe centerline. The distance downstream from the entrance to the location at which the boundary layer reaches the centerline is called the entrance length. Beyond the entrance length the velocity profile no longer changes with increasing distance, x, and the flow is fully developed. The actual shape of the fully-developed velocity profile depends on whether the flow is laminar or turbulent. In Fig. 8.1 the profile is shown qualitatively for a laminar flow.

For laminar flow, the entrance length, L, is a function of the Reynolds number,

$$\frac{L}{D} \simeq 0.06 \frac{\rho \bar{V} D}{\mu}$$

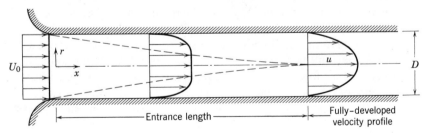

Fig. 8.1 Flow in the entrance region of a pipe.

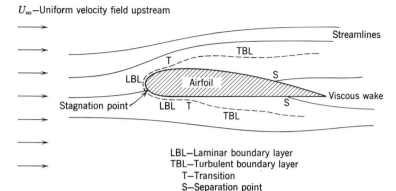

Streamlines

TBL

T

LBL

Airfoil

S

Stagnation point

LBL T

Viscous wake

TBL

S

LBL—Laminar boundary layer
TBL—Turbulent boundary layer
T—Transition
S—Separation point

Fig. 8.2 Details of viscous flow around an airfoil.

where D is the pipe diameter, $\bar{V}$ is the average velocity, ρ is the fluid density, and μ is the fluid viscosity. As pointed out in Chapter 2, laminar flow in a pipe may be expected only for Reynolds numbers less than about 2300. Thus the entrance length for laminar pipe flow may be as long as

$$L \simeq 0.06 \; Re \; D \le (0.06)(2300)D = 138D$$

that is, more than 100 pipe diameters. If the flow is turbulent, enhanced mixing among fluid layers[1] causes more rapid growth of the boundary layer. Experiments show that the mean velocity profile becomes fully developed within 25 to 40 pipe diameters from the entrance. However, the details of the turbulent motion may not be fully developed for 80 or more pipe diameters. Fully-developed internal flows will be treated in Parts B and C of this chapter.

External flows are flows over bodies immersed in an unbounded fluid. The flow over a semi-infinite flat plate (Fig. 2.10) and the flow over a cylinder (Fig. 2.11a) are examples of external flows. These were discussed qualitatively in Chapter 2. A more detailed sketch of viscous flow over an airfoil is shown in Fig. 8.2. Although viscous effects are confined to a thin boundary layer (the boundary-layer thickness in Fig. 8.2 is exaggerated greatly), they are responsible for flow separation and wake formation. The presence of the boundary layer affects the entire pressure distribution around the airfoil.

Forces and moments on objects result from pressure and shear forces, caused by the flow. In Parts D and E of this chapter, we develop approximate

---

[1] This mixing is illustrated extremely well in the introductory portion of the film *Turbulence*, R. W. Stewart, principal.

analyses and correlations of experimental data so that forces and moments can be determined for some simple viscous flow situations.

8-2 LAMINAR AND TURBULENT FLOWS

As discussed previously in Section 2-5.2, the nature of pipe flow (laminar or turbulent) is determined by the value of the Reynolds number,

$$Re = \frac{\rho \bar{V} D}{\mu}$$

One can demonstrate by the classic Reynolds experiment[2] the qualitative difference between the nature of laminar and turbulent flow. In this experiment water flows from a large reservoir through a clear tube. A thin filament of dye injected at the entrance to the tube allows visual observation of the flow. At low flowrates (low Reynolds numbers) the dye injected into the flow remains in a single filament; there is little dispersion of dye because the flow is laminar. A laminar flow is one in which the fluid flows in laminae or layers; there is no macroscopic mixing of adjacent fluid layers.

As the flowrate through the tube is increased, the dye filament becomes unstable and breaks up into a totally random motion; the line of dye is stretched and twisted into myriad entangled threads, and it quickly disperses throughout the entire flow field. This behavior of turbulent flow is due to small, high frequency, velocity fluctuations superimposed on the mean motion of a turbulent flow; mixing of fluid particles from adjacent layers of fluid results in rapid dispersion of the dye.

One can obtain a more quantitative picture of the difference between laminar and turbulent flow by examining the output from a sensitive velocity-measuring device immersed in the flow. If one measures the x component of velocity at a fixed location in a pipe for both laminar and turbulent steady flow, the traces of velocity versus time would be as shown in Fig. 8.3. For steady laminar flow, the velocity at a point remains constant with time. In turbulent flow the velocity trace indicates a random fluctuation of the instantaneous velocity, u, about the time mean velocity, $\bar{u}$. Thus we consider the instantaneous velocity, u, as the sum of the time mean velocity, $\bar{u}$, and the fluctuating component, u', that is,

$$u = \bar{u} + u'$$

Because the flow is steady, the mean velocity, $\bar{u}$, does not vary with time.

[2] This experiment is demonstrated in the film *Turbulence*, R. W. Stewart, principal.

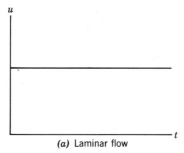

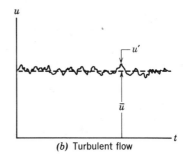

(a) Laminar flow   (b) Turbulent flow

Fig. 8.3 Velocity at a point.

Although many turbulent flows of interest are steady in the mean (i.e. $\bar{u}$ is not a function of time), the presence of the random, high frequency velocity fluctuations makes the analysis of turbulent flows extremely difficult. In a one-dimensional laminar flow, the shear stress is related to the velocity gradient by the simple relation

$$\tau_{yx} = \mu \frac{du}{dy} \tag{2.8}$$

For a turbulent flow in which the mean velocity field is one-dimensional, no such simple relation is valid. Random, three-dimensional velocity fluctuations (u', v' and w') transport momentum across the mean flow streamlines, increasing the effective shear stress. Consequently, in turbulent flow there is no universal relationship between the stress field and the mean velocity field. Thus in turbulent flows we must rely heavily on semi-empirical theories and on experimental data.

PART B. FULLY-DEVELOPED LAMINAR FLOW

There are relatively few viscous flow problems for which we can obtain analytical solutions in closed form, but the method of solution is important. In this section we consider a few classic examples of fully-developed laminar flows. Our interest is in obtaining detailed information about the velocity field. Knowledge of the velocity field permits calculation of shear stress, pressure drop, and flowrate.

Rather than use the complete differential equations of motion (Eqs. 5.32) for the flow of a viscous fluid, we shall derive the governing equations from first principles for each flow field of interest. Since we are interested in the details of the flow field, our aim will be to derive differential equations that describe the flow. In every case we shall begin by applying the familiar control volume formulation of Newton's second law (we assume that you are very familiar with it by now) to a suitably chosen differential control volume.

8-3 FULLY-DEVELOPED LAMINAR FLOW BETWEEN INFINITE PARALLEL PLATES

8-3.1   BOTH PLATES STATIONARY

Fluid in high pressure hydraulic systems often leaks through the annular gap between a piston and cylinder. For very small gaps (typically about 0.005 mm), this flow field may be imagined as flow between infinite parallel plates. To calculate the leakage flowrate, we must first determine the velocity field for such a flow.

Let us consider the fully-developed laminar flow between infinite parallel plates. The plates are separated by a distance, a, as shown in Fig. 8.4. The plates are considered infinite in the z direction, with no variation of any fluid property in this direction. The flow is further assumed to be steady and in-compressible. Before starting our analysis, what do we know about the flow field? For one thing we know that the x component of velocity must be zero at both the upper and lower plates as a result of the no-slip condition at the wall, that is the boundary conditions are:

$$\text{at} \quad y = 0 \quad u = 0$$
$$\text{at} \quad y = a \quad u = 0$$

Since the flow is fully developed, the velocity cannot vary with x and, hence, depends on y only, that is, $u = u(y)$. Furthermore, there is no component of velocity in either the y or z directions (i.e. $v = w = 0$).

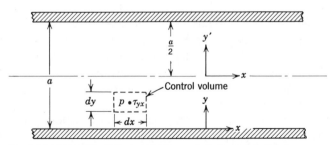

Fig. 8.4 Laminar flow between stationary infinite parallel plates.

For our analysis we select a differential control volume of size $d\forall = dx\,dy\,dz$ and apply the x component of the momentum equation.

Basic equation:

$$=0(3)\quad =0(1)$$

$$F_{S_x} + \overset{\nearrow}{F_{B_x}} = \overset{\nearrow}{\frac{\partial}{\partial t}} \int_{CV} u\rho\,d\forall + \int_{CS} u\rho\vec{V}\cdot d\vec{A} \qquad (4.19a)$$

Assumptions: (1) Steady flow
(2) Fully-developed flow
(3) $F_{B_x} = 0$

For fully-developed flow, the net momentum flux through the control surface is zero. (The momentum flux through the right face of the control surface is equal in magnitude but opposite in sign to the momentum flux through the left face; there is no momentum flux through any of the remaining faces of the control volume.) Since there are no body forces in the x direction, the momentum equation reduces to

$$F_{S_x} = 0 \qquad (8.2)$$

The next step is to sum the forces acting on the control volume in the x direction. We recognize that there are normal forces (pressure forces) acting on the left and right faces; there are tangential forces (shear forces) acting on the top and bottom faces.

If the pressure at the center of the element is p, then the force on the left face is

$$\left(p - \frac{\partial p}{\partial x}\frac{dx}{2}\right) dy\,dz,$$

and the force on the right face is

$$-\left(p + \frac{\partial p}{\partial x}\frac{dx}{2}\right) dy\,dz$$

If the shear stress at the center of the element is τ_{yx}, then the shear force on the bottom face is

$$-\left(\tau_{yx} - \frac{d\tau_{yx}}{dy}\frac{dy}{2}\right) dx\,dz$$

and the shear force on the top face is

$$\left(\tau_{yx} + \frac{d\tau_{yx}}{dy}\frac{dy}{2}\right) dx\,dz$$

Note that in expanding the shear stress, τ_{yx}, in a Taylor series about the center of the element, we have used the total derivative rather than a partial derivative. We did this because we recognized that τ_{yx} is only a function of y, since $u = u(y)$.

Having evaluated the forces acting on each face of the control volume, we can substitute them into Eq. 8.2

$$\left(p - \frac{\partial p}{\partial x}\frac{dx}{2} \right) dy\, dz - \left(p + \frac{\partial p}{\partial x}\frac{dx}{2} \right) dy\, dz - \left(\tau_{yx} - \frac{d\tau_{yx}}{dy}\frac{dy}{2} \right) dx\, dz$$
$$+ \left(\tau_{yx} + \frac{d\tau_{yx}}{dy}\frac{dy}{2} \right) dx\, dz = 0$$

This equation simplifies to

$$-\frac{\partial p}{\partial x} + \frac{d\tau_{yx}}{dy} = 0$$

or

$$\frac{d\tau_{yx}}{dy} = \frac{\partial p}{\partial x} \tag{8.3}$$

Equation 8.3 must be valid for all values of x and y. This requires that

$$\frac{d\tau_{yx}}{dy} = \frac{\partial p}{\partial x} = \text{constant}$$

Integrating this equation, we obtain

$$\tau_{yx} = \left(\frac{\partial p}{\partial x} \right) y + c_1$$

which states that the shear stress varies linearly with y. Since

$$\tau_{yx} = \mu \frac{du}{dy} \tag{2.8}$$

then

$$\mu \frac{du}{dy} = \left(\frac{\partial p}{\partial x} \right) y + c_1$$

and

$$u = \frac{1}{2\mu}\left(\frac{\partial p}{\partial x} \right) y^2 + \frac{c_1}{\mu} y + c_2 \tag{8.4}$$

To evaluate the constants c_1 and c_2 we must use the boundary conditions.

At $y = 0$, $u = 0$. Consequently, $c_2 = 0$. At $y = a$, $u = 0$. Hence

$$0 = \frac{1}{2\mu}\left(\frac{\partial p}{\partial x}\right) a^2 + \frac{c_1}{\mu} a$$

This gives

$$c_1 = -\frac{1}{2}\left(\frac{\partial p}{\partial x}\right) a$$

and hence

$$u = \frac{1}{2\mu}\left(\frac{\partial p}{\partial x}\right) y^2 - \frac{1}{2\mu}\left(\frac{\partial p}{\partial x}\right) ay$$

or

$$u = \frac{a^2}{2\mu}\left(\frac{\partial p}{\partial x}\right)\left[\left(\frac{y}{a}\right)^2 - \left(\frac{y}{a}\right)\right] \tag{8.5}$$

At this point we have the velocity profile. What else can we learn about the flow?

Shear Stress Distribution

The shear stress distribution is given by

$$\tau_{yx} = \left(\frac{\partial p}{\partial x}\right) y + c_1 = \left(\frac{\partial p}{\partial x}\right) y - \frac{1}{2}\left(\frac{\partial p}{\partial x}\right) a$$

$$\tau_{yx} = a\left(\frac{\partial p}{\partial x}\right)\left[\frac{y}{a} - \frac{1}{2}\right] \tag{8.6a}$$

Volumetric Flowrate

The volumetric flowrate is given by

$$Q = \int_A \vec{V} \cdot d\vec{A}$$

For a depth l in the z direction

$$Q = \int_0^a ul\,dy$$

or

$$\frac{Q}{l} = \int_0^a \frac{1}{2\mu}\left(\frac{\partial p}{\partial x}\right)(y^2 - ay)\,dy$$

Thus the volumetric flowrate per depth l is given by

$$\frac{Q}{l} = -\frac{1}{12\mu}\left(\frac{\partial p}{\partial x}\right) a^3 \tag{8.6b}$$

Flowrate as a Function of Pressure Drop

Since $\partial p/\partial x$ is a constant, then

$$\frac{p_2 - p_1}{L} = \frac{-\Delta p}{L} = \frac{\partial p}{\partial x}$$

Substituting into the expression for volumetric flowrate gives

$$\frac{Q}{l} = -\frac{1}{12\mu}\left[\frac{-\Delta p}{L}\right]a^3$$

or

$$\frac{Q}{l} = \frac{a^3 \Delta p}{12\mu L} \tag{8.6c}$$

Average Velocity

The average velocity, $\bar{V}$, is given by

$$\bar{V} = \frac{Q}{A} = -\frac{1}{12\mu}\left(\frac{\partial p}{\partial x}\right)\frac{a^3 l}{la}$$

$$\bar{V} = -\frac{1}{12\mu}\left(\frac{\partial p}{\partial x}\right)a^2 \tag{8.6d}$$

Point of Maximum Velocity

To find the point of maximum velocity, we set du/dy equal to zero and solve for the corresponding value of y. From Eq. 8.5

$$\frac{du}{dy} = \frac{a^2}{2\mu}\left(\frac{\partial p}{\partial x}\right)\left[\frac{2y}{a^2} - \frac{1}{a}\right]$$

Thus

$$\frac{du}{dy} = 0 \quad \text{at} \quad y = \frac{a}{2}$$

At

$$y = \frac{a}{2}, \quad u = u_{max} = -\frac{1}{8\mu}\left(\frac{\partial p}{\partial x}\right)a^2 = \frac{3}{2}\bar{V} \tag{8.6e}$$

Transformation of Coordinates

In deriving the above relations, the origin of coordinates, $y = 0$, was taken at the bottom plate. We could just as easily have taken the origin at the centerline of the channel. If we denote the coordinates with origin at the channel centerline as x, y', the boundary conditions are

$$u = 0 \quad \text{at} \quad y' = \frac{a}{2}$$

$$u = 0 \quad \text{at} \quad y' = -\frac{a}{2}$$

To obtain the velocity profile in terms of x, y', we substitute $y = y' + a/2$ into Eq. 8.5. The result is

$$u = \frac{a^2}{2\mu}\left(\frac{\partial p}{\partial x}\right)\left[\left(\frac{y'}{a}\right)^2 - \frac{1}{4}\right] \tag{8.7}$$

This equation shows that the velocity profile we have determined is parabolic, as shown in Fig. 8.5.

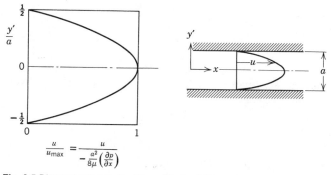

Fig. 8.5 Dimensionless velocity profile for fully-developed laminar flow between infinite parallel plates.

All of the results in this section are valid for laminar flow only, since Newton's law of viscosity was used to relate shear stress to velocity gradient. Experiments show that transition to turbulent flow occurs in this flow case for Reynolds number values (defined as $Re = \rho \bar{V} a/\mu$) greater than approximately 1400. Consequently, the Reynolds number should be checked after using Eqs. 8.6 to assure a valid solution.

Example 8.1

A hydraulic system operates at a gage pressure of 20 MPa and 55 C. The hydraulic fluid is SAE 10W oil (SG = 0.92). A control valve consists of a piston 25 mm in diameter, fitted to a cylinder with a mean radial clearance of 0.005 mm. Determine the leakage flowrate if the gage pressure on the low pressure side of the piston is 1.0 MPa. (The piston is 15 mm long.)

Example Problem 8.1

GIVEN:

Flow of hydraulic oil between piston and cylinder, as shown.
Fluid is SAE 10W oil at 55 C (SG = 0.92).

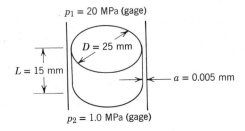

$p_1 = 20$ MPa (gage)

$D = 25$ mm

$L = 15$ mm

$a = 0.005$ mm

$p_2 = 1.0$ MPa (gage)

FIND:

Leakage flowrate, Q.

SOLUTION:

The gap width is very small, so the flow may be modeled as flow between parallel
plates. Equation 8.6c is applied.

Computing equation: $\qquad \dfrac{Q}{l} = \dfrac{a^3 \, \Delta p}{12\mu L}$ $\qquad\qquad$ (8.6c)

Assumptions: (1) Laminar flow
(2) Steady flow
(3) Incompressible flow
(4) Fully-developed flow (note $L/a = 15/0.005 = 3000!$)

The plate width, l, is approximately $l = \pi D$. Thus

$$Q = \frac{\pi D a^3 \, \Delta p}{12\mu L}$$

For SAE 10W oil at 55 C, $\mu = 0.018$ kg/m · sec, from Fig. A.2, Appendix A. Thus

$$Q = \frac{\pi}{12} \times 25 \text{ mm} \times (0.005)^3 \text{ mm}^3 \times \frac{(20-1)\,10^6 \text{ N}}{\text{m}^2} \times \frac{\text{m} \cdot \text{sec}}{0.018\,\text{kg}} \times \frac{1}{15 \text{ mm}} \times \frac{\text{kg} \cdot \text{m}}{\text{N} \cdot \text{sec}^2}$$

$Q = 57.6 \text{ mm}^3/\text{sec}$ $\qquad\qquad\qquad\qquad\qquad\qquad\qquad\qquad\qquad Q$

To assure that flow is laminar, we should also check the flow Reynolds number.

$$\bar{V} = \frac{Q}{A} = \frac{Q}{\pi D a} = \frac{57.6 \text{ mm}^3}{\text{sec}} \times \frac{1}{\pi} \times \frac{1}{25 \text{ mm}} \times \frac{1}{0.005 \text{ mm}} \times \frac{\text{m}}{10^3 \text{ mm}} = 0.147 \text{ m/sec}$$

and

$$Re = \frac{\rho \bar{V} a}{\mu} = \frac{SG \rho_{H_2O} \bar{V} a}{\mu}$$

$$Re = 0.92 \times 999 \frac{kg}{m^3} \times 0.147 \frac{m}{sec} \times 0.005 \, mm \times \frac{m \cdot sec}{0.018 \, kg} \times \frac{m}{10^3 \, mm} = 0.0375$$

Thus flow is surely laminar, since $Re \ll 1400$.

8-3.2 UPPER PLATE MOVING WITH CONSTANT VELOCITY, U

A second laminar flow case of practical importance is flow in a journal bearing. In such a bearing, an inner cylinder, or journal, rotates inside a stationary member. At light loads, the centers of the two members essentially coincide, and the small clearance gap is symmetric. Since the gap is normally small, it is reasonable to "unfold" the bearing and consider the flow field between infinite parallel plates.

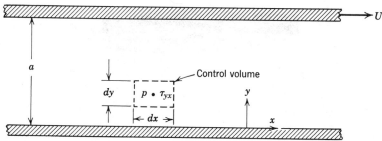

Fig. 8.6 Laminar flow between infinite parallel plates: upper plate moving with constant velocity, U.

Let us now consider a case where the upper plate is moving to the right with a constant velocity, U, as shown in Fig. 8.6. All we have done in going from a stationary upper plate to a moving upper plate is to change one of the boundary conditions. The boundary conditions for the moving plate case are:

$$u = 0 \quad \text{at} \quad y = 0$$
$$u = U \quad \text{at} \quad y = a$$

Since it is only the boundary conditions that have been changed, there is no need to repeat the entire analysis of Section 8-3.1. The analysis leading to Eq. 8.4 is equally valid for the moving plate case. Thus the velocity distribution is given by

$$u = \frac{1}{2\mu}\left(\frac{\partial p}{\partial x}\right) y^2 + \frac{c_1}{\mu} y + c_2 \qquad (8.4)$$

and our only task is to evaluate the constants c_1 and c_2 by employing the appropriate boundary conditions.

At $y = 0$, $u = 0$. Consequently, $c_2 = 0$.

At $y = a$, $u = U$. Consequently,

$$U = \frac{1}{2\mu}\left(\frac{\partial p}{\partial x}\right) a^2 + \frac{c_1}{\mu} a$$

Thus

$$c_1 = \frac{U\mu}{a} - \frac{1}{2}\left(\frac{\partial p}{\partial x}\right) a$$

and

$$u = \frac{1}{2\mu}\left(\frac{\partial p}{\partial x}\right) y^2 + \frac{Uy}{a} - \frac{1}{2\mu}\left(\frac{\partial p}{\partial x}\right) ay$$

$$= \frac{Uy}{a} + \frac{1}{2\mu}\left(\frac{\partial p}{\partial x}\right)(y^2 - ay)$$

$$u = \frac{Uy}{a} + \frac{a^2}{2\mu}\left(\frac{\partial p}{\partial x}\right)\left[\left(\frac{y}{a}\right)^2 - \left(\frac{y}{a}\right)\right] \qquad (8.8)$$

It is reassuring to note that Eq. 8.8 reduces to Eq. 8.5 for a stationary upper plate. From Eq. 8.8, for zero pressure gradient (i.e. for $\partial p/\partial x = 0$) the velocity varies linearly with y. This was the case treated earlier in Chapter 2.

From the velocity distribution of Eq. 8.8 we can obtain additional information about the flow.

Shear Stress Distribution

The shear stress distribution is given by $\tau_{yx} = \mu(du/dy)$,

$$\tau_{yx} = \mu \frac{U}{a} + \frac{a^2}{2}\left(\frac{\partial p}{\partial x}\right)\left[\frac{2y}{a^2} - \frac{1}{a}\right]$$

$$\tau_{yx} = \mu \frac{U}{a} + a\left(\frac{\partial p}{\partial x}\right)\left[\frac{y}{a} - \frac{1}{2}\right] \qquad (8.9a)$$

Volumetric Flowrate

The volumetric flowrate is given by $Q = \int_A \vec{V} \cdot d\vec{A}$. For a depth l in the z direction

$$Q = \int_0^a ul\,dy$$

or

$$\frac{Q}{l} = \int_0^a \left[\frac{Uy}{a} + \frac{1}{2\mu}\left(\frac{\partial p}{\partial x}\right)(y^2 - ay)\right] dy$$

Thus the volumetric flowrate per depth l is given by

$$\frac{Q}{l} = \frac{Ua}{2} - \frac{1}{12\mu}\left(\frac{\partial p}{\partial x}\right)a^3 \tag{8.9b}$$

Average Velocity

The average velocity, $\bar{V}$, is given by

$$\bar{V} = \frac{Q}{A} = l\left[\frac{Ua}{2} - \frac{1}{12\mu}\left(\frac{\partial p}{\partial x}\right)a^3\right]\bigg/ la$$

$$\bar{V} = \frac{U}{2} - \frac{1}{12\mu}\left(\frac{\partial p}{\partial x}\right)a^2 \tag{8.9c}$$

Point of Maximum Velocity

To find the point of maximum velocity, we set du/dy equal to zero and solve for the corresponding value of y. From Eq. 8.8

$$\frac{du}{dy} = \frac{U}{a} + \frac{a^2}{2\mu}\left(\frac{\partial p}{\partial x}\right)\left[\frac{2y}{a^2} - \frac{1}{a}\right]$$

Thus

$$\frac{du}{dy} = 0 \quad \text{at} \quad y = \frac{a}{2} - \frac{U/a}{(1/\mu)(\partial p/\partial x)}$$

There is no simple relation between the maximum velocity, u_{max}, and the mean velocity, $\bar{V}$, for this flow case.

Equation 8.8 suggests that the velocity profile may be treated as a combination of a linear and a parabolic velocity profile; the last term in Eq. 8.8 is identical with Eq. 8.5. The result is a family of velocity profiles, depending on the values of U and $(1/\mu)(\partial p/\partial x)$; a few profiles are sketched in Fig. 8.7. (As shown in Fig. 8.7, some reverse flow, i.e. flow in the negative x direction, can occur when $\partial p/\partial x > 0$.)

Again, all of the results developed in this section are valid for laminar flow only. Experiments show that transition to turbulent flow occurs (for $\partial p/\partial x = 0$) at a Reynolds number value of approximately 1500, where $Re = \rho Ua/\mu$ for this flow case. Not much information is available for the case where the pressure gradient is not zero.

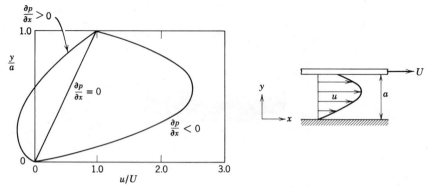

Fig. 8.7 Dimensionless velocity profile for fully-developed laminar flow between infinite parallel plates: upper plate moving with constant velocity, U.

Example 8.2

A crankshaft journal bearing in an automobile engine is lubricated by SAE 30 engine oil at 210 F. The bearing is 3 in. in diameter, has a diametral clearance of 0.0025 in., and rotates at 3600 rpm. It is 1.25 in. wide. The bearing is under no load, so the clearance is symmetric. Determine the torque required to turn the journal, and the power dissipated.

Example Problem 8.2

GIVEN:

Journal bearing, as shown. Note that the gap width, a, is *half* the diametral clearance.

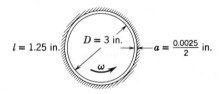

Lubricant is SAE 30 oil at 210 F. Speed: 3600 rpm.

FIND:

(a) Torque, T.
(b) Power dissipated.

SOLUTION:

Torque on the journal is due to viscous shear in the oil film. The gap width is small, so the flow may be modeled as flow between infinite parallel plates:

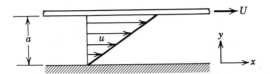

Computing equation:

$$\tau_{yx} = \mu \frac{U}{a} + a \left[\overset{= \,0(6)}{\left[\frac{\partial p}{\partial x} \right]} \right] \left[\frac{y}{a} - \frac{1}{2} \right] \tag{8.9a}$$

Assumptions: (1) Laminar flow
(2) Steady flow
(3) Incompressible flow
(4) Fully-developed flow
(5) Infinite width ($l/a = 1.25/0.00125 = 1000$, so this is a reasonable assumption)
(6) $\partial p/\partial x = 0$ (i.e. flow is symmetric in the actual bearing at no load)

Then

$$\tau_{yx} = \mu \frac{U}{a} = \mu \frac{\omega R}{a} = \mu \frac{\omega D}{2a}$$

For SAE 30 oil at 210 F (99 C), $\mu = 9.6 \times 10^{-3}$ N · sec/m^2 (2.0 × 10^{-4} lbf · sec/ft^2), from Fig. A.2, Appendix A. Thus

$$\tau_{yx} = \frac{2.0 \times 10^{-4}\ \text{lbf} \cdot \text{sec}}{\text{ft}^2} \times \frac{3600\ \text{rev}}{\text{min}} \times \frac{2\pi\ \text{rad}}{\text{rev}} \times \frac{\text{min}}{60\ \text{sec}} \times \frac{3\ \text{in.}}{2} \times \frac{1}{0.00125\ \text{in.}}$$

$$\tau_{yx} = 90.5\ \text{lbf/ft}^2$$

Since $\tau_{yx} > 0$, it acts to the *left* on the upper plate, which is a minus y surface. The total shear force is given by the shear stress times the area. It is applied to the journal surface. Therefore,

$$T = FR = \tau_{yx}\pi DlR = \frac{\pi}{2}\tau_{yx}D^2l$$

$$= \frac{\pi}{2} \times \frac{90.5\ \text{lbf}}{\text{ft}^2} \times \frac{(3)^2\ \text{in.}^2}{} \times \frac{\text{ft}^2}{144\ \text{in.}^2} \times \frac{1.25\ \text{in.}}{}$$

$$T = 11.1\ \text{in.} \cdot \text{lbf} \qquad\qquad T$$

The power dissipated in the bearing is

$$\dot{W} = FU = FR\omega = T\omega$$

$$= \frac{11.1 \text{ in.} \cdot \text{lbf}}{} \times \frac{3600 \text{ rev}}{\text{min}} \times \frac{\text{min}}{60 \text{ sec}} \times \frac{2\pi \text{ rad}}{\text{rev}} \times \frac{\text{ft}}{12 \text{ in.}} \times \frac{\text{hp} \cdot \text{sec}}{550 \text{ ft} \cdot \text{lbf}}$$

$$\dot{W} = 0.634 \text{ hp} \qquad\qquad\qquad \dot{W}$$

To assure laminar flow, check the Reynolds number value.

$$Re = \frac{\rho Ua}{\mu} = \frac{SG \, \rho_{H_2O} Ua}{\mu} = \frac{SG \, \rho_{H_2O} \omega Ra}{\mu}$$

$$= \frac{0.92}{} \times \frac{1.94 \text{ slug}}{\text{ft}^3} \times \frac{(3600)2\pi \text{ rad}}{60 \quad \text{sec}} \times \frac{1.5 \text{ in.}}{} \times \frac{0.00125 \text{ in.}}{} \times \frac{\text{ft}^2}{2.0 \times 10^{-4} \text{ lbf} \cdot \text{sec}}$$

$$\times \frac{\text{ft}^2}{144 \text{ in.}^2} \times \frac{\text{lbf} \cdot \text{sec}^2}{\text{slug} \cdot \text{ft}}$$

$Re = 43.8$

Therefore, the flow is laminar, since $Re \ll 1500$.

8-4 FULLY-DEVELOPED LAMINAR FLOW IN A PIPE

As a final example of fully-developed laminar flow cases, let us consider fully-developed laminar flow in a pipe. Here the flow is axisymmetric and it is therefore most convenient to work in cylindrical coordinates. We shall again employ a differential control volume, but this time since the flow is axisymmetric, the control volume will be a differential annular ring as shown in Fig. 8.8. The annular differential control volume is of length dx and has thickness dr.

For a fully-developed steady flow, the x component of the momentum equation (Eq. 4.19a) reduces to

$$F_{s_x} = 0$$

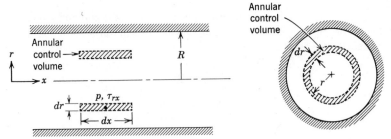

Fig. 8.8 Control volume for analysis of fully-developed laminar flow in a pipe.

The next step is to sum the forces acting on the control volume in the x direction. We know that there are normal forces (pressure forces) acting on the left and right ends of the control volume, and that there are tangential forces (shear forces) acting on the inner and outer cylindrical surfaces.

If the pressure at the center of the annular control volume is p, then the force on the left end is

$$\left(p - \frac{\partial p}{\partial x}\frac{dx}{2}\right) 2\pi r \, dr$$

The force on the right end is

$$-\left(p + \frac{\partial p}{\partial x}\frac{dx}{2}\right) 2\pi r \, dr$$

If the shear stress at the center of the annular control volume is τ_{rx}, then the shear force on the inner cylindrical surface is

$$-\left(\tau_{rx} - \frac{d\tau_{rx}}{dr}\frac{dr}{2}\right) 2\pi \left(r - \frac{dr}{2}\right) dx$$

The shear force on the outer cylindrical surface is

$$\left(\tau_{rx} + \frac{d\tau_{rx}}{dr}\frac{dr}{2}\right) 2\pi \left(r + \frac{dr}{2}\right) dx$$

The summation of the x components of the forces acting on the control volume must be zero. This leads to the condition:

$$-\frac{\partial p}{\partial x} 2\pi r \, dr \, dx + \tau_{rx} \, dr 2\pi \, dx + \frac{d\tau_{rx}}{dr} r \, dr 2\pi \, dx = 0$$

Dividing this equation by $2\pi r \, dr \, dx$ gives

$$\frac{\partial p}{\partial x} = \frac{\tau_{rx}}{r} + \frac{d\tau_{rx}}{dr} = \frac{1}{r}\frac{d(r\tau_{rx})}{dr} \qquad (8.10)$$

Since τ_{rx} is only a function of r (this is the reason for using the total rather than the partial derivative of τ_{rx} in the force components above), we recognize that Eq. 8.10 holds true for all values of r and x only if each side of the equation is equal to a constant. Equation 8.10 can be written as

$$\frac{1}{r}\frac{d(r\tau_{rx})}{dr} = \frac{\partial p}{\partial x} = \text{constant}$$

or

$$\frac{d(r\tau_{rx})}{dr} = r \frac{\partial p}{\partial x}$$

Integrating this equation, we obtain

$$r\tau_{rx} = \frac{r^2}{2}\left(\frac{\partial p}{\partial x}\right) + c_1$$

or

$$\tau_{rx} = \frac{r}{2}\left(\frac{\partial p}{\partial x}\right) + \frac{c_1}{r}$$

Since

$$\tau_{rx} = \mu \frac{du}{dr}$$

then

$$\mu \frac{du}{dr} = \frac{r}{2}\left(\frac{\partial p}{\partial x}\right) + \frac{c_1}{r}$$

and

$$u = \frac{r^2}{4\mu}\left(\frac{\partial p}{\partial x}\right) + \frac{c_1}{\mu}\ln r + c_2 \qquad (8.11)$$

At this point we need to evaluate the constants c_1 and c_2. However, we have only the one boundary condition that $u = 0$ at $r = R$. What do we do? Before throwing in the towel, let us look at the solution for the velocity profile given by Eq. 8.11. Although we do not know the value of the velocity at the pipe centerline, we do know from physical considerations that the velocity must be finite at $r = 0$. The only way that this can be true is for c_1 to be zero. Thus from physical considerations we conclude that $c_1 = 0$, and hence

$$u = \frac{r^2}{4\mu}\left(\frac{\partial p}{\partial x}\right) + c_2$$

The constant c_2 is evaluated by using the available boundary condition at the pipe wall: at $r = R$, $u = 0$. Consequently,

$$0 = \frac{R^2}{4\mu}\left(\frac{\partial p}{\partial x}\right) + c_2$$

This gives

$$c_2 = -\frac{R^2}{4\mu}\left(\frac{\partial p}{\partial x}\right)$$

and hence

$$u = \frac{r^2}{4\mu}\left(\frac{\partial p}{\partial x}\right) - \frac{R^2}{4\mu}\left(\frac{\partial p}{\partial x}\right) = \frac{1}{4\mu}\left(\frac{\partial p}{\partial x}\right)(r^2 - R^2)$$

or

$$u = -\frac{R^2}{4\mu}\left(\frac{\partial p}{\partial x}\right)\left[1 - \left(\frac{r}{R}\right)^2\right] \qquad (8.12)$$

Because we have the velocity profile, there are a number of additional features of the flow that we can obtain.

Shear Stress Distribution

The shear stress is given by

$$\tau_{rx} = \mu \frac{du}{dr} = \frac{r}{2}\left(\frac{\partial p}{\partial x}\right) \tag{8.13a}$$

Volumetric Flowrate

$$Q = \int_A \vec{V} \cdot d\vec{A} = \int_0^R u 2\pi r \, dr$$

$$= \int_0^R \frac{1}{4\mu}\left(\frac{\partial p}{\partial x}\right)(r^2 - R^2) 2\pi r \, dr$$

$$Q = -\frac{\pi R^4}{8\mu}\left(\frac{\partial p}{\partial x}\right) \tag{8.13b}$$

Flowrate as a Function of Pressure Drop

In fully-developed flow, the pressure gradient, $\partial p/\partial x$, is constant. Therefore, $\partial p/\partial x = (p_2 - p_1)/L = -\Delta p/L$. Substituting into Eq. 8.13b for the volumetric flowrate gives

$$Q = -\frac{\pi R^4}{8\mu}\left[\frac{-\Delta p}{L}\right]$$

or

$$Q = \frac{\pi \Delta p R^4}{8\mu L} = \frac{\pi \Delta p D^4}{128\mu L} \tag{8.13c}$$

for laminar flow in a horizontal pipe.

Average Velocity

The average velocity, $\bar{V}$, is given by

$$\bar{V} = \frac{Q}{A} = \frac{Q}{\pi R^2}$$

$$\bar{V} = -\frac{R^2}{8\mu}\left(\frac{\partial p}{\partial x}\right) \tag{8.13d}$$

Point of Maximum Velocity

To find the point of maximum velocity, we set du/dr equal to zero and solve for the corresponding value of r. From Eq. 8.12

$$\frac{du}{dr} = \frac{1}{2\mu}\left(\frac{\partial p}{\partial x}\right)r$$

Thus

$$\frac{du}{dr} = 0 \qquad \text{at} \qquad r = 0$$

At $r = 0$,

$$u = u_{\max} = -\frac{R^2}{4\mu}\left(\frac{\partial p}{\partial x}\right) = 2\bar{V} \tag{8.13e}$$

The parabolic velocity profile, given by Eq. 8.12 for fully-developed laminar pipe flow, was sketched in Fig. 8.1.

Example 8.3

A simple and accurate viscometer can be made from a length of capillary tubing. If the flowrate and pressure drop are measured, and the tube geometry is known, the viscosity can be computed from Eq. 8.13c. A test of a certain liquid in a capillary viscometer gave the following data:

> Flowrate: 880 mm³/sec
> Tube diameter: 0.50 mm
> Tube length: 1 m
> Pressure drop: 1.0 MPa

Determine the viscosity of the liquid.

Example Problem 8.3

GIVEN:

Flow in a capillary viscometer. The flowrate is $Q = 880$ mm³/sec.

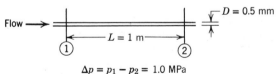

$$\Delta p = p_1 - p_2 = 1.0 \text{ MPa}$$

FIND:

The fluid viscosity.

SOLUTION:

Equation 8.13c may be applied.

Computing equation:
$$Q = \frac{\pi \, \Delta p D^4}{128 \mu L} \qquad \text{(8.13c)}$$

Assumptions: (1) Laminar flow
(2) Steady flow
(3) Incompressible flow
(4) Fully-developed flow

Then

$$\mu = \frac{\pi \, \Delta p D^4}{128 L Q} = \frac{\pi}{128} \times \frac{1.0 \times 10^6 \ \text{N}}{\text{m}^2} \times \frac{(0.50)^4 \ \text{mm}^4}{} \times \frac{\text{sec}}{880 \ \text{mm}^3} \times \frac{1}{10^3 \ \text{mm}}$$

$$\mu = 1.74 \times 10^{-3} \ \text{N} \cdot \text{sec/m}^2 \hspace{4cm} \mu$$

Check the Reynolds number. Assume the fluid density to be similar to that of water, that is, 999 kg/m³.

$$\bar{V} = \frac{Q}{A} = \frac{4Q}{\pi D^2} = \frac{4}{\pi} \times \frac{880 \ \text{mm}^3}{\text{sec}} \times \frac{1}{(0.50)^2 \ \text{mm}^2} \times \frac{\text{m}}{10^3 \ \text{mm}} = 4.48 \ \text{m/sec}$$

Then

$$Re = \frac{\rho \bar{V} D}{\mu} = \frac{999 \ \text{kg}}{\text{m}^3} \times \frac{4.48 \ \text{m}}{\text{sec}} \times \frac{0.50 \ \text{mm}}{} \times \frac{\text{m}^2}{1.74 \times 10^{-3} \ \text{N} \cdot \text{sec}} \times \frac{\text{m}}{10^3 \ \text{mm}} \times \frac{\text{N} \cdot \text{sec}^2}{\text{kg} \cdot \text{m}}$$

$$Re = 1290$$

Consequently, flow is laminar, since $Re < 2300$.

Example 8.4

A viscous, incompressible, Newtonian liquid flows in steady, laminar flow down a vertical wall as shown in the sketch. The thickness, δ, of the liquid

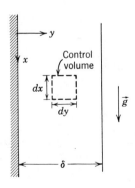

film is constant. Since the free liquid surface is exposed to atmospheric pressure, there is no pressure gradient. Apply the momentum equation to the differential control volume of volume $dx\,dy\,dz$ to derive the velocity distribution through the liquid film.

Example Problem 8.4

GIVEN:

Fully-developed laminar flow of an incompressible, Newtonian liquid down a vertical wall; thickness, δ, of the liquid film is constant. $\partial p/\partial x = 0$.

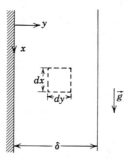

FIND:

Expression for the velocity distribution through the film.

SOLUTION:

The x component of the momentum equation for a control volume is

$$F_{S_x} + F_{B_x} = \frac{\partial}{\partial t} \int_{CV} u\rho\,d\forall + \int_{CS} u\rho\vec{V} \cdot d\vec{A} \qquad (4.19a)$$

For steady flow, $\dfrac{\partial}{\partial t} \displaystyle\int_{CV} u\rho\,d\forall = 0$

For fully-developed flow, $\displaystyle\int_{CS} u\rho\vec{V} \cdot d\vec{A} = 0$

Thus the momentum equation for the present case reduces to

$$F_{S_x} + F_{B_x} = 0$$

The body force, F_{B_x}, is given by $F_{B_x} = \rho g\,d\forall = \rho g\,dx\,dy\,dz$. The surface forces acting on the differential control volume are shear forces on the vertical surfaces. Since $\partial p/\partial x = 0$, there is no net pressure force acting on the element.

If the shear stress at the center of the differential control volume is τ_{yx}, then,

$$\text{shear stress on left face is } \tau_{yx_L} = \left(\tau_{yx} - \frac{d\tau_{yx}}{dy} \frac{dy}{2} \right),$$

and

$$\text{shear stress on right face is } \tau_{yx_R} = \left(\tau_{yx} + \frac{d\tau_{yx}}{dy} \frac{dy}{2} \right)$$

The direction of the shear stress vectors is taken consistent with the sign convention of Section 2-3.2. Thus on the left face, τ_{yx_L} acts acts upward, and on the right face, τ_{yx_R} acts downward.

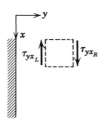

The surface forces are obtained by multiplying each shear stress by the area over which it acts.

Substituting into $F_{S_x} + F_{B_x} = 0$, we obtain

$$-\tau_{yx_L} \, dx \, dz + \tau_{yx_R} \, dx \, dz + \rho g \, dx \, dy \, dz = 0$$

or

$$-\left(\tau_{yx} - \frac{d\tau_{yx}}{dy} \frac{dy}{2} \right) dx \, dz + \left(\tau_{yx} + \frac{d\tau_{yx}}{dy} \frac{dy}{2} \right) dx \, dz + \rho g \, dx \, dy \, dz = 0$$

Simplifying, $\qquad \dfrac{d\tau_{yx}}{dy} + \rho g = 0 \qquad \text{or} \qquad \dfrac{d\tau_{yx}}{dy} = -\rho g$

Since $\qquad\qquad \tau_{yx} = \mu \dfrac{du}{dy}, \qquad \text{then} \qquad \mu \dfrac{d^2 u}{dy^2} = -\rho g$

and

$$\frac{d^2 u}{dy^2} = -\frac{\rho g}{\mu}$$

Integrating with respect to y gives

$$\frac{du}{dy} = -\frac{\rho g}{\mu} y + c_1$$

Integrating again,

$$u = -\frac{\rho g}{\mu} \frac{y^2}{2} + c_1 y + c_2$$

To evaluate the constants c_1 and c_2, we must apply appropriate boundary conditions:

(i) $y = 0$ $u = 0$ (no-slip)

(ii) $y = \delta$ $\dfrac{du}{dy} = 0$ (neglect air resistance)

From boundary condition (i) $c_2 = 0$

From boundary condition (ii) $0 = -\dfrac{\rho g}{\mu}\delta + c_1$ or $c_1 = \dfrac{\rho g}{\mu}\delta$

Hence $\qquad\qquad\qquad\qquad u = -\dfrac{\rho g}{\mu}\dfrac{y^2}{2} + \dfrac{\rho g}{\mu}\delta y$

or

$$u = \frac{\rho g}{\mu}\delta^2 \left[\left(\frac{y}{\delta}\right) - \frac{1}{2}\left(\frac{y}{\delta}\right)^2\right] \qquad u(y)$$

$\left\{\begin{array}{l}\text{The purpose of this problem is to illustrate the application of the momentum}\\ \text{equation to a differential control volume for a fully-developed laminar flow.}\end{array}\right\}$

PART C. FLOW IN PIPES AND DUCTS

Our main purpose in this section is to evaluate the pressure changes that result from incompressible flow in pipes, ducts, and flow systems. The pressure changes in a flow system result from changes in elevation or flow velocity (due to area changes) and from friction. In a frictionless flow, the Bernoulli equation could be used to account for the effects of changes in elevation and flow velocity. Thus the prime concern in the analysis of real flows is to account for the effect of friction. The effect of friction is to decrease the pressure, i.e. to cause a pressure "loss" compared to the ideal, frictionless flow case. To simplify analysis, the "loss" will be divided into *major losses* (due to friction in fully-developed flow in constant area portions of the system) and *minor losses* (due to flow through valves, tees, elbows, and frictional effects in other nonconstant area portions of the system.

In developing the relations for the major losses due to friction in constant area ducts, we shall deal with fully-developed flows, that is, flows in which the velocity profile is unvarying in the direction of flow. The pressure drop that occurs at the entrance of a pipe will be treated as a minor loss.

Since ducts of circular cross section are most common in engineering applications, the basic analysis will be performed for circular geometries. The results can be extended to other geometries by introducing the hydraulic diameter, which is treated in Section 8-9.3. (Compressible flow in ducts will be treated in Chapter 10.)

8-5 VELOCITY PROFILES IN PIPE FLOW

Pipe flow may be either laminar or turbulent, depending on the Reynolds number. For fully-developed laminar pipe flow, the velocity profile is parabolic, as shown in Section 8-4. Thus

$$u = -\frac{R^2}{4\mu}\left(\frac{\partial p}{\partial x}\right)\left[1 - \left(\frac{r}{R}\right)^2\right] \tag{8.12}$$

But $u = u_{max} = U$ at $r = 0$, that is,

$$U = -\frac{R^2}{4\mu}\left(\frac{\partial p}{\partial x}\right)$$

so that

$$\frac{u}{U} = 1 - \left(\frac{r}{R}\right)^2 \tag{8.14}$$

for laminar pipe flow. Furthermore, according to Eq. 8.13e, the average velocity is equal to half the centerline velocity, that is,

$$\frac{\bar{V}}{U} = \frac{1}{2} \tag{8.15}$$

Except for flow of very viscous fluids in small diameter ducts, internal flows in general are turbulent. As noted in the discussion of laminar and turbulent flows (Section 8-2), in turbulent flow there is no universal relationship between the stress field and the mean velocity field. Thus, in turbulent flows, we are forced to rely on experimental data.

The velocity profile for turbulent flow through a smooth pipe can be represented by the empirical "power-law" equation

$$\frac{u}{U} = \left(1 - \frac{r}{R}\right)^{1/n} \tag{8.16}$$

where the exponent, n, varies with the Reynolds number. The value of the exponent, n, is 6 at a Reynolds number of 4.0×10^3. For a Reynolds number of 1.1×10^5, n is equal to 7; the value of n increases to 10 for a Reynolds number of 3.2×10^6. Since the average velocity $\bar{V} = Q/A$ and

$$Q = \int_A \vec{V} \cdot d\vec{A}$$

the ratio of the average velocity to the centerline velocity may be calculated from Eq. 8.16. The result is

$$\frac{\bar{V}}{U} = \frac{2n^2}{(n+1)(2n+1)} \tag{8.17}$$

From Eq. 8.17 we see that as n increases (due to increasing Reynolds number) the ratio of the average velocity to the centerline velocity increases, that is, the velocity profile becomes more blunt, or "fuller." (For $n = 6$, $\bar{V}/U = 0.79$, while for $n = 10$, $\bar{V}/U = 0.87$.)

In Fig. 8.9 the velocity profiles for both laminar flow (Eq. 8.14) and turbulent flow (Eq. 8.16) are plotted for the same value of the Reynolds number ($Re = 4.0 \times 10^3$). Both profiles have the same average velocity and, hence, the same flow rate; they do not have the same centerline velocity. From Fig. 8.9 we also see that the turbulent profile has a much steeper slope at the wall.

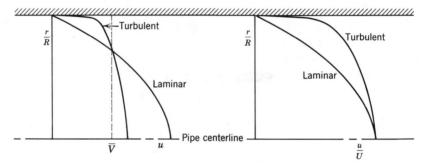

Fig. 8.9 Velocity profiles for laminar and turbulent flow in a smooth circular pipe at Reynolds number, Re = 4 × 10³.

8-6 SHEAR STRESS DISTRIBUTION IN FULLY-DEVELOPED PIPE FLOW

In fully-developed steady flow in a horizontal pipe, be it laminar or turbulent, the pressure drop is balanced only by shear forces at the pipe wall. This can be seen by applying the momentum equation to a cylindrical control volume in the flow, Fig. 8.10. The x component of the momentum equation is

Basic equation:

$$F_{S_x} + \overset{=\,0(1)}{\cancel{F_{B_x}}} = \overset{=\,0(2)}{\cancel{\frac{\partial}{\partial t} \int_{CV} u\rho \, dV}} + \overset{=\,0(3,\,4)}{\cancel{\int_{CS} u\rho \vec{V} \cdot d\vec{A}}} \tag{4.19a}$$

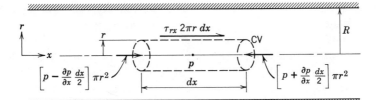

Fig. 8.10 Control volume for analysis of shear stress distribution in fully-developed flow in a circular pipe.

Assumptions: (1) Horizontal pipe, $F_{B_x} = 0$
(2) Steady flow
(3) Incompressible flow
(4) Fully-developed flow

Then

$$F_{S_x} = 0$$

The surface forces acting on the control volume are shown in Fig. 8.10. The pressure at the center of the element has been taken as p; the pressure at each end of the element has been obtained from a Taylor series expansion of p about the center of the element. The shear force acts on the circumferential surface of the element. The direction of the stress has been assumed such that the stress is positive. Thus

$$F_{S_x} = \left(p - \frac{\partial p}{\partial x}\frac{dx}{2}\right)\pi r^2 - \left(p + \frac{\partial p}{\partial x}\frac{dx}{2}\right)\pi r^2 + \tau_{rx}2\pi r\, dx = 0$$

or

$$-\frac{\partial p}{\partial x}\, dx\pi r^2 + \tau_{rx}2\pi r\, dx = 0$$

Therefore,

$$\tau_{rx} = \frac{r}{2}\frac{\partial p}{\partial x} \qquad (8.18)$$

Thus we see that the shear stress on the fluid varies linearly across the pipe, from zero at the centerline to a maximum at the pipe wall. If we denote the wall value of shear stress as τ_w, Eq. 8.18 shows that at the surface of the pipe

Both
Laminar
& turbulent

$$\tau_w = -[\tau_{rx}]_{r=R} = -\frac{R}{2}\frac{\partial p}{\partial x} \qquad (8.19)$$

Equation 8.19 relates the wall shear stress to the axial pressure gradient. The momentum equation was used to derive it, but no assumption was made about a relation between shear stress and the velocity field. Consequently, Eq. 8.19 is applicable to both laminar and turbulent pipe flow.

If we could now relate the shear stress field to the mean velocity field, we could determine analytically the pressure drop over a length of pipe for fully-developed flow. Such a relation between the stress field and the mean velocity field does exist for laminar flow and was used in Section 8-4. The resulting equation, Eq. 8.13c, was first discovered experimentally by Jean Louis Poiseuille, a French physician, and independently by Gotthif H. L. Hagen, a German engineer, in the 1850s.

8-7 ENERGY CONSIDERATIONS IN PIPE FLOW

Thus far in our discussion of viscous flow, we have derived all results by applying the momentum equation for a control volume. We have, of course, also used the control volume formulation of the conservation of mass. Nothing has been said about the conservation of energy—the first law of thermodynamics. Additional insight into the nature of the pressure losses in internal viscous flows can be obtained from the energy equation. Consider, for example, the steady flow through the section of a piping system including a reducing elbow shown in Fig. 8.11. The control volume boundaries are shown as dashed lines. They are normal to the flow at sections ① and ② and coincide with the inside pipe wall elsewhere.

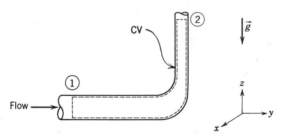

Fig. 8.11 Control volume and coordinates for energy analysis of flow through a 90 degree reducing elbow.

Basic equation:

$$= 0(1) \quad = 0(2) \quad = 0(1) \quad = 0(3)$$

$$\dot{Q} + \dot{W_s} + \dot{W}_{shear} + \dot{W}_{other} = \frac{\partial}{\partial t} \int_{CV} e\rho \, d\mathbf{V} + \int_{CS} (e + pv)\rho \vec{V} \cdot d\vec{A} \quad (4.59)$$

$$e = u + \frac{V^2}{2} + gz$$

358 **8/incompressible viscous flow**

Assumptions: (1) $\dot{W}_s = 0$, $\dot{W}_{\text{other}} = 0$
(2) $\dot{W}_{\text{shear}} = 0$ (although shear stresses are present at the walls of the elbow, the velocities are zero at the walls)
(3) Steady flow
(4) Incompressible flow
(5) Internal energy and pressure uniform across sections ① and②

Then the energy equation reduces to

$$\dot{Q} = \dot{m}(u_2 - u_1) + \dot{m}\left(\frac{p_2}{\rho} - \frac{p_1}{\rho}\right) + \dot{m}g(z_2 - z_1)$$

$$+ \int_{A_2} \frac{V_2^2}{2}\rho V_2\, dA_2 - \int_{A_1} \frac{V_1^2}{2}\rho V_1\, dA_1 \qquad (8.20)$$

Note that we have not assumed the velocity to be uniform at sections ① and ②, since we know for viscous flows the velocity at a section cannot be uniform. However, it is convenient to introduce the average velocity into Eq. 8.20 so that we can eliminate the integral signs. To do this, we define a *kinetic energy flux coefficient*, α, such that

$$\int_A \frac{V^2}{2}\rho V\, dA = \alpha \int_A \frac{\bar{V}^2}{2}\rho V\, dA = \alpha\dot{m}\frac{\bar{V}^2}{2} \qquad (8.21)$$

With this definition of α, Eq. 8.20 can be written

$$\dot{Q} = \dot{m}(u_2 - u_1) + \dot{m}\left(\frac{p_2}{\rho} - \frac{p_1}{\rho}\right) + \dot{m}g(z_2 - z_1) + \dot{m}\left(\frac{\alpha_2\bar{V}_2^2}{2} - \frac{\alpha_1\bar{V}_1^2}{2}\right)$$

Dividing by the mass rate of flow gives

$$\frac{\delta Q}{dm} = u_2 - u_1 + \frac{p_2}{\rho} - \frac{p_1}{\rho} + gz_2 - gz_1 + \frac{\alpha_2\bar{V}_2^2}{2} - \frac{\alpha_1\bar{V}_1^2}{2}$$

Rearranging this equation, we have

$$\left(\frac{p_1}{\rho} + \alpha_1\frac{\bar{V}_1^2}{2} + gz_1\right) = \left(\frac{p_2}{\rho} + \alpha_2\frac{\bar{V}_2^2}{2} + gz_2\right) + (u_2 - u_1) - \frac{\delta Q}{dm} \qquad (8.22)$$

8-7.1 HEAD LOSS

In Eq. 8.22, the term

$$\left(\frac{p}{\rho} + \alpha\frac{\bar{V}^2}{2} + gz\right)$$

represents the mechanical energy per unit mass at a flow cross section. The term $u_2 - u_1 - \delta Q/dm$ is equal to the difference in mechanical energy per unit mass between sections ① and ②. It represents the (irreversible) conversion of mechanical energy at section ① to unwanted thermal energy $(u_2 - u_1)$ and loss of energy via heat transfer $(-\delta Q/dm)$. We identify this group of terms as the total head loss, h_{l_T}. Then

$$\left(\frac{p_1}{\rho} + \alpha_1 \frac{\overline{V}_1^2}{2} + gz_1\right) - \left(\frac{p_2}{\rho} + \alpha_2 \frac{\overline{V}_2^2}{2} + gz_2\right) = h_{l_T} \qquad (8.23)$$

If the flow were frictionless, under the assumptions used to derive Eq. 8.22 the velocity at a section would be uniform ($\alpha_1 = \alpha_2 = 1$) and Bernoulli's equation would predict zero head loss.

In incompressible frictionless flow a change in internal energy can only occur through the effects of heat transfer; there is no conversion of mechanical energy ($p/\rho + V^2/2 + gz$) to internal energy. For viscous flow in a pipe, one effect of friction may be to increase the internal energy of the flow, as shown by Eq. 8.22.

For laminar flow in a pipe (velocity profile given by Eq. 8.12), $\alpha = 2.0$. However, there is no need to use Eq. 8.23 since we can calculate the pressure drop analytically as was done in Section 8-4.

In turbulent pipe flow, the velocity profile is quite flat as shown in Fig. 8.9. We can use Eq. 8.21 together with Eqs. 8.16 and 8.17 to determine the value of α. Substituting the power-law velocity profile of Eq. 8.16 into Eq. 8.21, we obtain (after a little algebra) the expression

$$\alpha = \left[\frac{U}{\overline{V}}\right]^3 \frac{2n^2}{(3+n)(3+2n)} \qquad (8.24)$$

The value of $\overline{V}/U$ is determined from Eq. 8.17. Since the exponent, n, in the power-law profile is a function of the Reynolds number, α also varies with the Reynolds number. For $n = 6$ ($Re = 4.0 \times 10^3$), $\alpha = 1.08$; for $n = 10$ ($Re = 3.2 \times 10^6$), $\alpha = 1.03$. Therefore, α is reasonably close to unity over a large range of Reynolds number. Although the change of kinetic energy is usually small compared to the total head loss, its contribution to head loss may be significant for developing flows. Hence for pipe flow calculations, we use the energy equation in the form

$$\left(\frac{p_1}{\rho} + \alpha_1 \frac{\overline{V}_1^2}{2} + gz_1\right) - \left(\frac{p_2}{\rho} + \alpha_2 \frac{\overline{V}_2^2}{2} + gz_2\right) = h_{l_T} \qquad (8.23)$$

Head loss has dimensions of energy per unit mass $[FL/M]$. In the $FLtT$ system of dimensions, this is equivalent to dimensions of $[L^2/t^2]$.

You might wonder why the energy loss, h_{l_T}, is called a "head" loss. As the empirical science of hydraulics developed in the 19th century, it was common practice to express the energy balance in terms of energy per unit *weight* of flowing liquid (i.e. water) rather than energy per unit *mass* as in Eq. 8.23. To obtain dimensions of energy per unit weight, we divide each term in Eq. 8.23 by the acceleration of gravity, g. Then the net dimensions of h_{l_T} are $[(L^2/t^2)(t^2/L)] = [L]$, or feet of flowing liquid. Since the term head loss is in common use, we shall also use it here. Remember that its physical interpretation is a loss in mechanical energy, expressed per unit mass of flowing fluid.

Equation 8.23 can be used to calculate the pressure difference between any two points in a piping system, provided the head loss, h_{l_T}, is known. We shall turn our attention to calculation of h_{l_T} in the next section.

8-8 CALCULATION OF HEAD LOSS

Recall that total head loss, h_{l_T}, is regarded as the sum of major losses, h_l, due to frictional effects in fully-developed flow in constant area tubes, and minor losses, h_{l_m}, due to entrances, fittings, area changes and so on. Therefore, let us consider the major and minor losses separately.

8-8.1 MAJOR LOSSES: FRICTION FACTOR

The energy balance, expressed by Eq. 8.23, can be used to evaluate the major head loss. For fully-developed flow through a constant area pipe, $h_{l_m} = 0$, and $\alpha_1(\bar{V}_1^2/2) = \alpha_2(\bar{V}_2^2/2)$. Then Eq. 8.23 becomes

$$\frac{p_1 - p_2}{\rho} = g(z_2 - z_1) + h_l \tag{8.25}$$

If the pipe is horizontal, then $z_2 = z_1$ and

$$\frac{p_1 - p_2}{\rho} = \frac{\Delta p}{\rho} = h_l \tag{8.26}$$

Thus the major head loss can be expressed as the pressure loss for fully-developed flow through a horizontal pipe of constant area.

Since head loss represents the energy converted by frictional effects from mechanical to thermal energy, head loss for fully-developed flow in a constant area duct depends only on the details of the flow through the duct. Consequently, the head loss obtained from Eq. 8.26 for fully-developed flow in a horizontal pipe is also valid for flow at the same flowrate in an inclined pipe.

a. Laminar Flow

In laminar flow the pressure drop may be computed analytically for fully-developed flow in a horizontal pipe. Thus from Eq. 8.13c

$$\Delta p = \frac{128\mu L Q}{\pi D^4} = \frac{128\mu L \bar{V}(\pi D^2/4)}{\pi D^4} = 32\frac{L}{D}\frac{\mu\bar{V}}{D}$$

Substituting in Eq. 8.26,

$$h_l = 32\frac{L}{D}\frac{\mu\bar{V}}{\rho D} = \frac{L}{D}\frac{\bar{V}^2}{2}\left(64\frac{\mu}{\rho\bar{V}D}\right) = \left(\frac{64}{Re}\right)\frac{L}{D}\frac{\bar{V}^2}{2} \qquad (8.27)$$

(We shall see the reason for writing h_l in this form shortly.)

b. Turbulent Flow

In turbulent flow we cannot evaluate the pressure drop analytically and, hence, we must resort to experimental results and utilize dimensional analysis as an aid in correlating the experimental data. In fully-developed turbulent flow, the pressure drop, Δp, due to friction in a horizontal constant-area pipe is known to depend on the pipe diameter, D, the pipe length, L, the pipe roughness, e, the average flow velocity, $\bar{V}$, the fluid density, ρ, and the fluid viscosity, μ. In functional form

$$\Delta p = \Delta p(D, L, e, \bar{V}, \rho, \mu)$$

We applied dimensional analysis to this problem in Example Problem 7.2. The results were a correlation of the form

$$\frac{\Delta p}{\rho\bar{V}^2} = f\left(\frac{\mu}{\rho\bar{V}D}, \frac{L}{D}, \frac{e}{D}\right)$$

We recognize that $\mu/\rho\bar{V}D = 1/Re$, so we could just as well write

$$\frac{\Delta p}{\rho\bar{V}^2} = \phi\left(Re, \frac{L}{D}, \frac{e}{D}\right)$$

Substituting from Eq. 8.26, we see that

$$\frac{h_l}{\bar{V}^2} = \phi\left(Re, \frac{L}{D}, \frac{e}{D}\right)$$

Although dimensional analysis predicts the functional relationship, we must resort to experiment to obtain actual values.

Experiments show that the nondimensional head loss is directly proportional to L/D. Hence we can write

$$\frac{h_l}{\bar{V}^2} = \frac{L}{D} \phi_1 \left(Re, \frac{e}{D} \right)$$

Since the function ϕ_1 is still undetermined, it is permissible to introduce a constant into the left side of the above equation. The number $\frac{1}{2}$ is introduced into the denominator such that the head loss is nondimensionalized on the kinetic energy per unit mass of flow. Then

$$\frac{h_l}{\frac{1}{2}\bar{V}^2} = \frac{L}{D} \phi_2 \left(Re, \frac{e}{D} \right)$$

The unknown function $\phi_2(Re, e/D)$ is defined as the friction factor, f, that is,

$$f \equiv \phi_2 \left(Re, \frac{e}{D} \right)$$

and

$$h_l = f \frac{L}{D} \frac{\bar{V}^2}{2} \tag{8.28}$$

The friction factor, f, is determined experimentally. The results, published by L. F. Moody, are shown in Fig. 8.12.

To determine head loss for fully-developed flow with known conditions, the flow Reynolds number is evaluated first. The value of relative roughness, e/D, for the flow is obtained from Fig. 8.13. Then the friction factor, f, is read from the appropriate curve in Fig. 8.12, at the known values of Re and e/D. Finally, the head loss is found using Eq. 8.28.

Several features of Fig. 8.12 require some discussion. The friction factor for laminar flow may be obtained by comparing Eqs. 8.27 and 8.28. Thus

$$h_l = \left(\frac{64}{Re} \right) \frac{L}{D} \frac{\bar{V}^2}{2} = f \frac{L}{D} \frac{\bar{V}^2}{2}$$

Consequently, for laminar flow

$$f_{laminar} = \frac{64}{Re} \tag{8.29}$$

Thus, in laminar flow, the friction factor is a function of Reynolds number only; it is independent of roughness. Although we took no notice of roughness

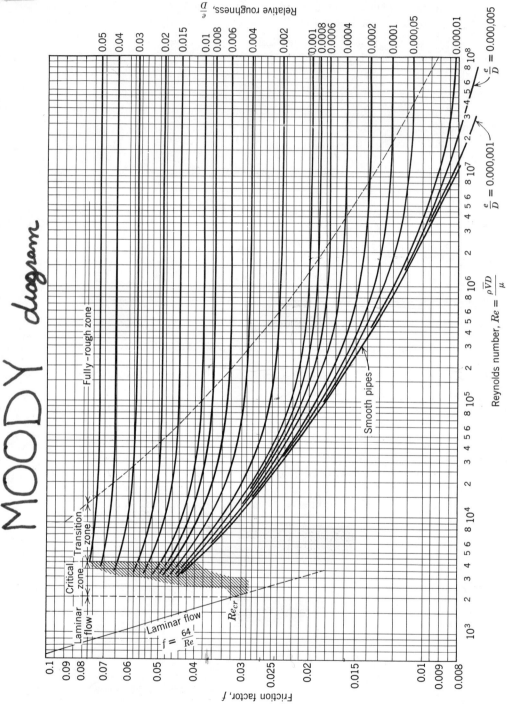

Fig. 8.12 Friction factor for fully-developed flow in circular pipes (data from Ref. 1, used by permission).

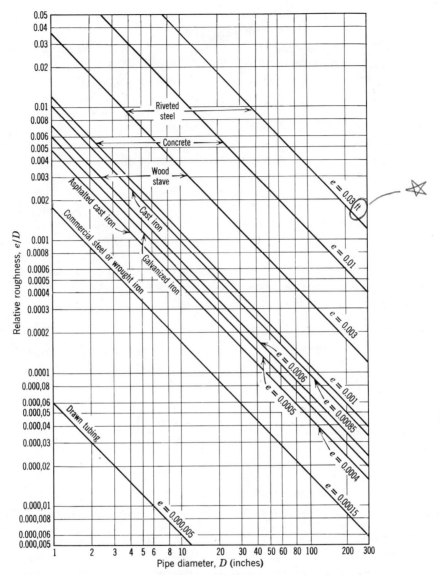

Fig. 8.13 Relative roughness values for pipes of common engineering materials (data from Ref. 1, used by permission).

e is in feet

e = pipe roughness

in deriving Eq. 8.27, experimental results verify that the friction factor is a function only of the Reynolds number in laminar flow.

The Reynolds number in a pipe may be changed most easily by varying the average flow velocity. If the flow in a pipe is originally laminar, increasing the velocity until the critical Reynolds number is reached causes transition to occur; the laminar flow gives way to turbulent flow. The effect of transition on the velocity profile was discussed in Section 8-5. Figure 8.9 shows that the velocity gradient at the tube wall is much larger for turbulent flow than for laminar flow.[3] This change in velocity profile causes the wall shear stress to increase sharply, with the same effect on the friction factor.

As the Reynolds number is increased above the transition value, the velocity profile continues to become fuller, as noted in Section 8-5; the friction factor at first tends to follow the smooth pipe curve, along which friction factor is a function of Reynolds number only. However, as the Reynolds number increases, the velocity profile becomes still fuller. The size of the thin "viscous sublayer" near the tube wall decreases. As roughness elements begin to poke through this layer, the effect of roughness becomes important, and the friction factor becomes a function of the Reynolds number *and* the relative roughness.

At very large Reynolds number, most of the roughness elements on the tube wall protrude through the viscous sublayer; the drag and, hence, the pressure loss, depend only on the size of the roughness elements. This flow regime is termed the "fully-rough" flow regime; friction factor depends on e/D only.

To summarize the preceding discussion, we see that as the Reynolds number is increased, the friction factor will decrease as long as the flow remains laminar. At transition, f increases sharply. In the turbulent flow regime, the friction factor decreases gradually along the smooth pipe curve, and finally levels out at a constant value for extremely large Reynolds number. (The only exception to these trends is that friction factors for pipes with $e/D \gtrsim 0.001$ in turbulent flow fall above the smooth pipe curve.)

Figure 8.13 also needs some explanation. All of the e/D values given are those for new pipes, in relatively good condition. Over long periods of service, corrosion takes place and, particularly in hard water areas, lime deposits and rust scale form on pipe walls. Corrosion can weaken pipes, eventually leading to failure. Deposit formation increases wall roughness appreciably, and also decreases the effective diameter. These factors combine to cause e/D to increase by a factor of 2 or 3 for old pipes. An example is shown in Fig. 8.14.

---

[3] The power law profile (Eq. 8.16) predicts an infinite velocity gradient at the tube wall, which is obviously not correct. However, it fits the central portion of the profile well.

Fig. 8.14 Pipe section removed after 40 years of service as a water line, showing formation of scale.

8-8.2 MINOR LOSSES

The flow in a piping system may be required to pass through a variety of fittings, bends, or abrupt changes in area. Additional head losses are encountered primarily as a result of flow separation. (Energy is eventually dissipated by violent mixing in the separated zones.) These losses will be minor (hence the term minor losses) if the particular piping system includes long lengths of constant area pipe. The minor head loss may be expressed as

$$h_{l_m} = K \frac{\bar{V}^2}{2} \tag{8.30a}$$

where the *loss coefficient*, K, must be determined experimentally for each situation. Minor head loss may also be expressed as

$$h_{l_m} = f\frac{L_e}{D}\frac{\bar{V}^2}{2} \tag{8.30b}$$

where L_e is an *equivalent length* of straight pipe.

Experimental data for minor loss coefficients are plentiful, but they are scattered among a variety of sources. Consequently, data for some commonly encountered situations are summarized below.[4]

a. Inlets and Entrance Length

A poorly designed inlet to a pipe can cause an appreciable pressure drop. If the inlet has sharp corners, flow separation occurs at the corners, and a *vena contracta* is formed.[5] The fluid must accelerate locally to pass through the reduced flow area at the vena contracta. Losses in mechanical energy result from the unconfined mixing as the flow stream decelerates again to fill the pipe. Three basic inlet geometries are shown in Table 8.1.

TABLE 8.1 **Minor Loss Cofficients for Pipe Entrances (data from Ref. 2, except[‡], from Ref. 3)**

Entrance Type	Diagram	Minor Loss Coefficient, K^*
Reentrant		1.0
Square-edged		0.5[‡]
Well rounded[†]		~ 0.04

[*] Based on $h_{l_m} = K(\bar{V}^2/2)$, where $\bar{V}$ is the mean velocity in the pipe.

[†] $r/R \simeq 0.25$.

[4] It is worth noting here that many data found in textbooks have been taken from very old sources, and that sometimes even original sources give different values for the same flow configuration. Some of the data given here reflect the authors' "best guess" values; where uncertainties exist, a range of values is given. In each case, the source of the data we have included is identified.

[5] This behavior is well illustrated in film loop S-FM016, *Flow from a Reservoir to a Duct.*

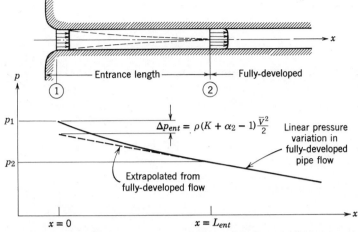

Fig. 8.15 Variation of static pressure in a pipe inlet section, illustrating the definition of Δp_{ent}.

The pressure drop in the entrance length of pipe immediately following an inlet is larger than that in the same length of pipe with fully-developed flow, as shown in Fig. 8.15. In Section 8-1 we noted that fluid in the central region of a pipe is accelerated as the boundary layer thickens on the pipe wall in the entrance region. Because the wall boundary layers are thin, viscous stresses at the pipe wall are also larger in the entrance region than in fully-developed flow. Thus, even with a smoothly contoured inlet, an entrance region has a larger pressure drop than the same length of pipe in fully-developed flow.

The head loss for an inlet and entrance length combination can be evaluated by applying the energy equation. Using the notation shown in Fig. 8.15,

$$\frac{p_1}{\rho} + \alpha_1 \frac{\bar{V}_1^2}{2} + gz_1 = \frac{p_2}{\rho} + \alpha_2 \frac{\bar{V}_2^2}{2} + gz_2 + h_{l_T} \tag{8.23}$$

For a well-contoured inlet, the inlet velocity profile is uniform, so $\alpha_1 = 1$. From continuity, $\bar{V}_1 = \bar{V}_2$. Neglecting the elevation terms and solving for the pressure drop, we obtain

$$p_1 - p_2 = \rho \left[h_{l_T} + (\alpha_2 - 1) \frac{\bar{V}_2^2}{2} \right] \tag{8.31}$$

This equation shows that even for flow without friction or separation ($h_{l_T} = 0$), a pressure drop is required to produce a nonuniform velocity profile with $\alpha_2 > 1$.

Introducing the definition of h_{l_T} into Eq. 8.31, and using $\bar{V}$, yields

$$p_1 - p_2 = \rho\left[h_l + h_{l_m} + (\alpha_2 - 1)\frac{\bar{V}^2}{2}\right] \tag{8.32}$$

The head loss due to friction, h_l, for the same length of pipe with fully-developed flow is

$$h_l = f\frac{L}{D}\frac{\bar{V}^2}{2} \tag{8.28}$$

and the minor loss coefficient, K, is defined by Eq. 8.30a. With these substitutions, Eq. 8.32 becomes

$$p_1 - p_2 = \rho\left[f\frac{L}{D}\frac{\bar{V}^2}{2} + K\frac{\bar{V}^2}{2} + (\alpha_2 - 1)\frac{\bar{V}^2}{2}\right] = \rho\left(f\frac{L}{D} + K + \alpha_2 - 1\right)\frac{\bar{V}^2}{2} \tag{8.33}$$

Values of the minor loss coefficient, K, can be calculated from experimental measurements of pressure drop for inlets using Eq. 8.33. Values of K determined in this way for three pipe entrance geometries are shown in Table 8.1. The data show that entrance losses can be reduced dramatically by rounding the contour of a pipe inlet. (Note from Fig. 8.15 that the entrance length is defined to begin at the start of the constant-diameter pipe section.)

The additional pressure drop due to an entrance, Δp_{ent}, compared to fully-developed flow can be determined from Eq. 8.33. If

$$\Delta p_{f-d} = \rho f\frac{L}{D}\frac{\bar{V}^2}{2}$$

then

$$\Delta p_{\text{ent}} = p_1 - p_2 - \Delta p_{f-d} = \rho(K + \alpha_2 - 1)\frac{\bar{V}^2}{2} \tag{8.34}$$

as shown graphically in Fig. 8.15.

The kinetic energy flux coefficient, α, that appears in Eqs. 8.33 and 8.34 was evaluated in Section 8-7.1 using the power-law approximation to the turbulent velocity profile. Since the power-law exponent, $1/n$, is a weak function of the flow Reynolds number, α also will depend on the Reynolds number. A useful empirical correlation between the power-law exponent and friction factor has been suggested by Nunner (Ref. 4):

$$\frac{1}{n} \simeq \sqrt{f} \tag{8.35}$$

At a typical Reynolds number value of 60,000, Fig. 8.12 shows that f for a smooth pipe is approximately 0.02, giving $1/n = 1/7$ from Eq. 8.35. The corresponding value of α is 1.058. Since α is very nearly equal to unity for most turbulent pipe flows of engineering interest, it is sometimes omitted from calculations of head loss.

b. Enlargements and Contractions

Minor loss coefficients for sudden expansions and contractions in circular ducts are given in Fig. 8.16. Note that both loss coefficients are based on the larger value of $\bar{V}^2/2$. Thus losses for a sudden expansion are based on $\bar{V}_1^2/2$, and those for a contraction are based on $\bar{V}_2^2/2$.

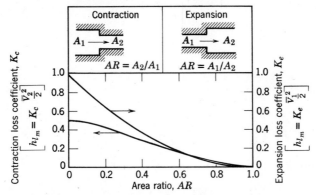

Fig. 8.16 Loss coefficients for flow through sudden area changes (Ref. 5).

Losses due to area change can be reduced somewhat by installing a nozzle or diffuser between the two sections of straight pipe. Data for nozzles are sparse; a few representative values are given in Table 8.2. These data suggest that losses in gradual contractions are independent of area ratio; this is likely to be true only for moderate area ratios.

TABLE 8.2 Loss Coefficients for Gradual Contractions (data from Ref. 3)

Diagram	Included Angle, θ, Degrees	Loss Coefficient, K^*
	30	0.02
	45	0.04
	60	0.07

* Based on $h_{l_m} = K(\bar{V}_2^2/2)$.

Losses in diffusers depend on a number of geometric and flow variables. Diffuser data are most commonly presented in terms of a pressure recovery coefficient, C_p, defined as the ratio of static pressure rise to inlet dynamic pressure, that is,

$$C_p \equiv \frac{p_2 - p_1}{\frac{1}{2}\rho \bar{V}_1^2} \tag{8.36}$$

The definition of C_p may be related to head loss easily. If gravity is neglected, and $\alpha_1 = \alpha_2 = 1.0$, Eq. 8.23 reduces to

$$\left[\frac{p_1}{\rho} + \frac{\bar{V}_1^2}{2} \right] - \left[\frac{p_2}{\rho} + \frac{\bar{V}_2^2}{2} \right] = h_{l_T} = h_{l_m}$$

Thus

$$h_{l_m} = \frac{\bar{V}_1^2}{2} - \frac{\bar{V}_2^2}{2} - \frac{p_2 - p_1}{\rho}$$

$$h_{l_m} = \frac{\bar{V}_1^2}{2}\left[\left(1 - \frac{\bar{V}_2^2}{\bar{V}_1^2} \right) - \frac{p_2 - p_1}{\frac{1}{2}\rho \bar{V}_1^2} \right] = \frac{\bar{V}_1^2}{2}\left[\left(1 - \frac{\bar{V}_2^2}{\bar{V}_1^2} \right) - C_p \right]$$

From continuity, $A_1 \bar{V}_1 = A_2 \bar{V}_2$, so

$$h_{l_m} = \frac{\bar{V}_1^2}{2}\left[1 - \left(\frac{A_1}{A_2} \right)^2 - C_p \right]$$

or

$$h_{l_m} = \frac{\bar{V}_1^2}{2}\left[\left(1 - \frac{1}{(AR)^2} \right) - C_p \right] \tag{8.37}$$

Data for conical diffusers with fully-developed turbulent pipe flow at the diffuser inlet are presented in Fig. 8.17 as a function of geometry. Similar data correlations for plane wall and annular (Ref. 8) and for radial diffusers (Ref. 9) are available in the literature. Diffuser pressure recovery with uniform inlet flow is somewhat better than that for fully-developed inlet flow.

Since the static pressure rises in the direction of flow in a diffuser, the flow may separate from the walls. For some geometries, the outlet flow is distorted; sometimes pulsations occur. The flow regime behavior of plane wall diffusers is illustrated well in the film *Flow Visualization* (S. J. Kline, principal).[6] For wide angle diffusers, vanes or splitters can be used to suppress stall and improve pressure recovery (Ref. 10).

---

[6] Film loop S-FM049, *Flow Regimes in Subsonic Diffusers*, was edited from this film. See also film loop S-FM065, *Wide Angle Diffuser with Suction*.

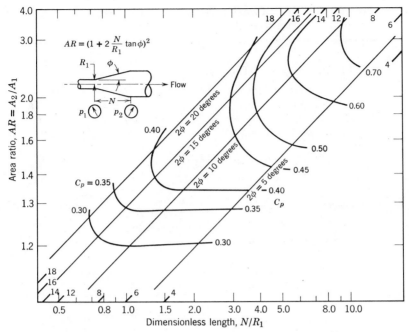

Fig. 8.17 Pressure recovery data for conical diffuser with fully-developed turbulent pipe flow at inlet (data from Ref. 7).

c. Exits

The kinetic energy per unit mass of flow, $\bar{V}^2/2$, is completely dissipated by mixing when flow discharges from a duct into a large reservoir or plenum chamber. The situation corresponds to flow through an abrupt expansion with $AR = 0$, Fig. 8.16. Several examples are sketched and minor loss coefficient values given in Table 8.3. These data suggest that no improvement in exit loss coefficient is possible; addition of a diffuser can reduce the value of $\bar{V}^2/2$ considerably (see Example Problem 8.7).

d. Pipe Bends

The head loss of a bend is larger than that for fully-developed flow through a straight section of equal length. The additional loss is primarily the result of secondary flow,[7] and is expressed most conveniently by an equivalent length of straight pipe. The equivalent length depends on the relative radius

---

[7] Secondary flows are shown in film loop S-FM019, *Secondary Flow in a Bend*.

TABLE 8.3 **Minor Loss Coefficients for Pipe Exits
(data from Ref. 2)**

Exit Type	Diagram	Minor Loss Coefficient, K^*
Projecting pipe		1.0
Square-edged		1.0
Rounded		1.0

* Based on $h_{l_m} = K(\bar{V}^2/2)$.

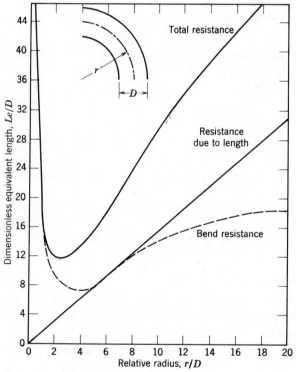

Fig. 8.18 Design chart for resistance of 90 degree bends in circular pipe with fully-developed turbulent flow at inlet (Ref. 2).

of curvature of the bend as shown in Fig. 8.18, for 90 degree bends. Reference 2 also gives an approximate procedure for computing the resistance of bends with other turning angles.

Because they are simple and inexpensive to erect in the field, miter bends are often used, especially in large pipe systems. Design data for miter bends are given in Fig. 8.19.

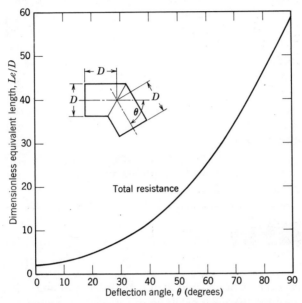

Fig. 8.19 Design chart for resistance of miter bends in circular pipe with fully-developed turbulent flow at inlet (Ref. 2).

$$h_L = f \frac{Le}{D} \frac{\bar{V}^2}{2}$$

e. Valves and Fittings

Losses for flow through valves and fittings are conveniently expressed in terms of an equivalent length of straight pipe. Some representative data are given in Table 8.4.

In practice, insertion losses for fittings and valves vary considerably, depending on the care used in fabricating the pipe system. If burrs from cutting pipe sections are allowed to remain, they cause local flow obstructions, which increase losses appreciably.

Although the losses discussed in this section were termed "minor losses," they can be a large fraction of the overall system loss. Thus a system for which calculations are to be made should be checked carefully to make sure

TABLE 8.4   Representative Dimensionless Equivalent Lengths
(L_e/D) for Valves and Fittings (data from Ref. 2)

Fitting Type	Description	Equivalent Length, L_e/D^*
Globe valve	Fully open	350
Gate valve	Fully open	13
	$\frac{3}{4}$ open	35
	$\frac{1}{2}$ open	160
	$\frac{1}{4}$ open	900
Check valve		50–100
90° std. elbow		30
45° std. elbow		16
90° elbow	Long radius	20
90° street elbow		50
45° street elbow		26
Tee	Flow through run	20
	Flow through branch	60
Return bend	Close pattern	50

* Based on $h_{l_m} = f \dfrac{L_e}{D} \dfrac{\bar{V}^2}{2}$

all losses have been identified and their magnitudes estimated. If calculations
are made carefully, the results will be of satisfactory engineering accuracy,
that is, computed values may be expected to predict actual losses within
plus or minus 10 percent.

8-9   SOLUTION OF PIPE FLOW PROBLEMS

Once the total head loss has been calculated using the methods of Section 8-8,
pipe flow problems can be solved using the energy equation, Eq. 8.23. The
same basic techniques are used, even for complex piping systems, but let us
first consider single-path pipe flow problems.

8-9.1   SINGLE-PATH SYSTEMS

Equation 8.23 is the computing equation for pipe systems. The pressure drop
across a pipe system is a function of flowrate, elevation change, and the
total head loss. The total head loss consists of major losses due to friction

in constant area sections (Eq. 8.28) and minor losses due to fittings, area changes, and so forth (Eqs. 8.30). The pressure drop could be written in the functional form

$$\Delta p = \phi_3(L, Q, D, e, \Delta z, \text{system configuration}, \rho, \mu)$$

The fluid properties may be assumed constant for incompressible pipe flow. The roughness, elevation change, and system configuration depend on the pipe system layout. Once these have been fixed (i.e. for a given system and fluid), the dependence reduces to

$$\Delta p = \phi_4(L, Q, D) \tag{8.38}$$

Equation 8.38 relates four variables. Any one of these may be the unknown quantity in a practical flow situation. Thus four general cases are possible:
(a) L, Q and D known, Δp unknown
(b) $\Delta p, Q$ and D known, L unknown
(c) $\Delta p, L$ and D known, Q unknown
(d) $\Delta p, L$ and Q known, D unknown
   Cases (a) and (b) may be solved directly by applying the continuity and energy equations, and using loss data from Section 8-8. Solutions for cases (c) and (d) make use of the same equations and data but require iteration. Each case is discussed below and illustrated by an example.

a. $L, Q,$ and D Known, Δp Unknown   _FIG 8.12

A friction factor is obtained from the Moody chart using Re and e/D values computed from the given data. The total head loss is computed from Eqs. 8.28 and 8.30. Equation 8.23 is then used to evaluate the pressure drop, Δp. The procedure is illustrated in Example Problems 8.5 through 8.7.

FIG 8.12

b. $\Delta p, Q,$ and D Known, L Unknown

The total head loss is calculated from Eq. 8.23. A friction factor is obtained from the Moody diagram using Re and e/D values computed from the given data. The unknown length is determined by solving Eq. 8.28. The procedure is illustrated in Example Problem 8.8.

c. $\Delta p, L,$ and D Known, Q Unknown

Equation 8.23 is combined with the defining equations for head loss; the result is an expression for $\overline{V}$ (or Q) in terms of the friction factor, f. Most pipe flows of engineering interest have relatively large values of Reynolds number. Thus, even though the Reynolds number (and hence f) cannot be calculated because Q is not known, a good first guess for the friction factor is the value in the fully-rough region of Fig. 8.12. Using the assumed value for f, a first approximation for $\bar{V}$ is calculated. The Reynolds number is computed for this value of $\bar{V}$, and a new value for f and a second approximation for $\bar{V}$ are obtained. Since f is a rather weak function of the Reynolds number, more than two iterations are seldom required for convergence. A flow of this type is evaluated in Example Problem 8.9.

d. Δp, L, and Q Known, D Unknown

When a fluid-handling device is available and the geometry of the piping system is known, the problem is to determine the smallest (and hence least costly) pipe size that can deliver the desired flowrate. Since the pipe diameter is unknown, neither the Reynolds number nor relative roughness can be computed directly, and an iterative solution is required.

TABLE 8.5 **Data for Commercial Wrought Steel Pipe in Standard Wall Thickness (Schedule 40, for cold water service to 125 psig, Ref. 2)**

[handwritten note in margin: Why is diameter always slightly larger than specified?]

Nominal Pipe Size (in.)	Inside Diameter (in.)
$\frac{1}{8}$	0.269
$\frac{1}{4}$	0.364
$\frac{3}{8}$	0.493
$\frac{1}{2}$	0.622
$\frac{3}{4}$	0.824
1	1.049
$1\frac{1}{2}$	1.610
2	2.067
$2\frac{1}{2}$	2.469
3	3.068
$3\frac{1}{2}$	3.548
4	4.026
5	5.047
6	6.065
8	8.071
10	10.020
12	12.090

Calculations begin by assuming a trial value of pipe diameter. The Reynolds number and relative roughness are then calculated using the assumed value of D. A value for the friction factor is obtained from Fig. 8.12. Then the head loss is computed from Eqs. 8.28 and 8.30, and Eq. 8.23 is solved for pressure drop. The resulting trial value is compared to the system requirement.

If the trial value of Δp is too large, calculations are repeated for a larger assumed value of D. If the trial value of Δp is less than the criterion, a smaller assumed value of D should be checked.

In choosing a pipe diameter, it is logical to work with values that are available commercially. Pipe is manufactured in a limited number of standard sizes. Some data for standard pipe sizes are given in Table 8.5. For data on extra heavy (250 psig) or double extra heavy (350 psig) pipes, consult a handbook (e.g. Ref. 2). Note that for standard pipe in nominal sizes larger than 4 in., the actual diameter and nominal diameter are essentially the same. Pipe larger than 12 in. nominal diameter is produced in 14, 16, 18, 20, 22, and 24 in. sizes, and in multiples of 6 in. for still larger sizes.

8.13

Example 8.5

Water at 20 C flows through a 150 mm diameter horizontal pipe ($e/D = 0.0002$) at a volumetric flowrate of 0.1 m³/sec. Determine the pressure drop over a 10 m length of the pipe. The flow is fully developed.

Example Problem 8.5

GIVEN:

Water flowing through a 150 mm diameter horizontal pipe ($e/D = 0.0002$) at a rate of 0.1 m³/sec.

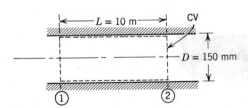

FIND:

Pressure drop over 10 m length.

SOLUTION:

Computing equation:

$$\left(\frac{p_1}{\rho} + \alpha_1 \frac{\bar{V}_1^2}{2} + gz_1\right) - \left(\frac{p_2}{\rho} + \alpha_2 \frac{\bar{V}_2^2}{2} + gz_2\right) = h_{l_T} = h_l + h_{l_m} \qquad (8.23)$$

where

$$h_l = f\frac{L}{D}\frac{\bar{V}^2}{2} \qquad \text{and} \qquad h_{l_m} = K\frac{\bar{V}^2}{2}$$

For incompressible flow, $\rho = c$, and since $A_1 = A_2$, then $\bar{V}_1 = \bar{V}_2$. Since the pipe is horizontal, $z_1 = z_2$. For fully-developed flow in a straight section of pipe, $h_{l_m} = 0$. Thus

$$\frac{p_1}{\rho} - \frac{p_2}{\rho} = h_l = f\frac{L}{D}\frac{\bar{V}^2}{2}$$

or

$$p_1 - p_2 = f\frac{L}{D}\frac{\rho\bar{V}^2}{2}$$

where

$$f = f\left(Re, \frac{e}{D}\right) \qquad \text{and} \qquad Re = \frac{\rho\bar{V}D}{\mu}$$

For water at 20 C, $\rho = 999$ kg/m³ and $\mu = 1.0 \times 10^{-3}$ kg/m · sec, so

$$\bar{V} = \frac{Q}{A} = \frac{Q}{\pi D^2/4} = \frac{4Q}{\pi D^2} = \frac{4}{\pi} \times \frac{0.1 \text{ m}^3}{\text{sec}} \times \frac{1}{(0.15)^2 \text{ m}^2} = 5.66 \text{ m/sec}$$

$$Re = \frac{4\rho Q}{\pi\mu D} = \frac{4}{\pi} \times \frac{999 \text{ kg}}{\text{m}^3} \times \frac{0.1 \text{ m}^3}{\text{sec}} \times \frac{\text{m} \cdot \text{sec}}{1.0 \times 10^{-3} \text{ kg}} \times \frac{1}{0.15 \text{ m}} = 8.48 \times 10^5$$

With $Re = 8.48 \times 10^5$ and $\frac{e}{D} = 0.0002$, from Fig. 8.12, $f = 0.0149$.

Then $p_1 - p_2 = f\frac{L}{D}\rho\frac{\bar{V}^2}{2}$

$$= \frac{0.0149}{} \times \frac{10 \text{ m}}{0.15 \text{ m}} \times \frac{999 \text{ kg}}{\text{m}^3} \times \frac{1}{2} \times \frac{(5.66)^2 \text{ m}^2}{\text{sec}^2} \times \frac{\text{N} \cdot \text{sec}^2}{\text{kg} \cdot \text{m}}$$

$$p_1 - p_2 = 15.9 \text{ kPa}$$

$\longleftarrow$ $\qquad\qquad p_1 - p_2$

$\left\{\begin{array}{l}\text{This problem illustrates the method for calculating the head loss and pressure} \\ \text{drop in a pipe.}\end{array}\right\}$

Example 8.6

What level, h, must be maintained in the reservoir to produce a volumetric flowrate of 0.03 m³/sec of water? The inside diameter of the smooth pipe is 75 mm and the pipe length is 100 m. The loss coefficient, K, for the inlet is $K = 0.5$. The water discharges to the atmosphere.

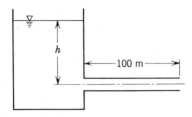

Example Problem 8.6

GIVEN:

Water flow at 0.03 m³/sec through a 75 mm diameter pipe, with $L = 100$ m, attached to a constant-level reservoir. Inlet loss coefficient, $K = 0.5$.

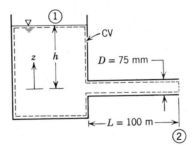

FIND:

Reservoir level, h, to maintain the flow.

SOLUTION:

Computing equation:

$$\left(\frac{p_1}{\rho} + \alpha_1 \frac{\bar{V}_1^2}{2} + gz_1\right) - \left(\frac{p_2}{\rho} + \alpha_2 \frac{\bar{V}_2^2}{2} + gz_2\right) = h_{l_T} = h_l + h_{l_m} \qquad (8.23)$$

where

$$h_l = f\frac{L}{D}\frac{\bar{V}^2}{2} \qquad \text{and} \qquad h_{l_m} = K\frac{\bar{V}^2}{2}$$

For the given problem, $p_1 = p_2 = p_{atm}$, $\bar{V}_1 \simeq 0$, $\bar{V}_2 = \bar{V}$, and $\alpha_2 \simeq 1.0$. If it is assumed that $z_2 = 0$, then $z_1 = h$. Simplifying Eq. 8.23 gives

$$gh - \frac{\bar{V}^2}{2} = f\frac{L}{D}\frac{\bar{V}^2}{2} + K\frac{\bar{V}^2}{2}$$

Then

$$h = \frac{1}{g}\left[f\frac{L}{D}\frac{\bar{V}^2}{2} + K\frac{\bar{V}^2}{2} + \frac{\bar{V}^2}{2}\right] = \frac{\bar{V}^2}{2g}\left[f\frac{L}{D} + K + 1\right]$$

Since $\bar{V} = \frac{Q}{A} = \frac{4Q}{\pi D^2}$, then

$$h = \frac{8Q^2}{\pi^2 D^4 g}\left[f\frac{L}{D} + K + 1\right]$$

Assuming water at 20 C, $\rho = 999$ kg/m^3, $\mu = 1.0 \times 10^{-3}$ kg/m · sec. Thus

$$Re = \frac{\rho \bar{V} D}{\mu} = \frac{4\rho Q}{\pi \mu D}$$

$$Re = \frac{4}{\pi} \times \frac{999 \text{ kg}}{\text{m}^3} \times \frac{0.03 \text{ m}^3}{\text{sec}} \times \frac{\text{m} \cdot \text{sec}}{1.0 \times 10^{-3} \text{ kg}} \times \frac{1}{0.075 \text{ m}} = 5.09 \times 10^5$$

For smooth pipe, from Fig. 8.12, $f = 0.0131$. Then

$$h = \frac{8Q^2}{\pi^2 D^4 g}\left[f\frac{L}{D} + K + 1\right]$$

$$= \frac{8}{\pi^2} \times \frac{(0.03)^2 \text{ m}^6}{\text{sec}^2} \times \frac{1}{(0.075)^4 \text{ m}^4} \times \frac{\text{sec}^2}{9.81 \text{ m}}\left[(0.0131)\frac{100 \text{ m}}{0.075 \text{ m}} + 0.5 + 1\right]$$

$$h = 44.6 \text{ m} \hspace{5cm} h$$

{This problem illustrates the method for calculating the total head loss.}

Example 8.7

Air is supplied to a steel-making process through a 6 in. diameter, smooth circular duct that terminates abruptly into a large plenum chamber. A new young engineer suggests that appreciable power can be saved by replacing the existing duct, which includes two 90° bends (centerline radius = 2 ft), with a combination duct-diffuser. There is room for a diffuser with area ratio, $AR = 1.35$. The proposed new duct would eliminate the elbows, 8 ft of duct, and an additional piece of duct as long as the diffuser. The air speed required in the smooth duct is 150 ft/sec, and the outlet pressure is essentially atmospheric. How much power would be saved by making the changes suggested by the young engineer? The blower efficiency is 80 percent.

Example Problem 8.7

GIVEN:

Flow systems shown. $L_{new} + N + 8 \text{ ft} = L_{old}$.

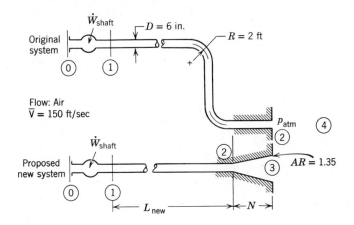

FIND:

Decrease in pumping power for the new system.

SOLUTION:

Determine the head loss for each system, then compute the pumping power using the first law of thermodynamics.

Computing equation:

$$h_{l_T} = h_l + h_{l_m} = f \frac{L}{D} \frac{\bar{V}^2}{2} + h_{l_m}$$

For the original system,

$$h_{l_T} = K_{ent} \frac{\bar{V}_1^2}{2} + f \frac{L_{old}}{D} \frac{\bar{V}_1^2}{2} + 2f \frac{L_{e, bend}}{D} \frac{\bar{V}_1^2}{2} + K_{exit} \frac{\bar{V}_1^2}{2}$$

From Fig. 8.18, with $\dfrac{r}{D} = \dfrac{2}{\left(\frac{1}{2}\right)} = 4$, $\quad \dfrac{L_{e, bend}}{D} = 13.5$

Thus

$$h_{l_T}(\text{old}) = \left[K_{ent} + f \frac{L_{old}}{D} + 2(13.5)f + 1.0 \right] \frac{\bar{V}_1^2}{2} \qquad (1)$$

For the proposed new system,

$$h_{l_T} = K_{ent} \frac{\bar{V}_1^2}{2} + f\frac{L_{new}}{D} \frac{\bar{V}_1^2}{2} + h_{l_m, \text{diffuser}} + K_{exit} \frac{\bar{V}_3^2}{2}$$

The best diffuser with $AR = 1.35$, from Fig. 8.17, has a pressure recovery coefficient, $C_p = 0.40$ at $N/R_1 = 1.5$. Thus $N = (1.5)R_1 = (1.5)(\frac{1}{4}) = 0.375$ ft. From Eq. 8.37,

$$h_{l_m, \text{diffuser}} = \frac{\bar{V}_2^2}{2}\left[1 - \frac{1}{(AR)^2} - C_p\right] = \frac{\bar{V}_2^2}{2}[1 - 0.549 - 0.40] = 0.051\frac{\bar{V}_2^2}{2}$$

The exit loss is

$$h_{l_m, \text{exit}} = K\frac{\bar{V}_2^2}{2} = K\left(\frac{A_2}{A_3}\right)^2\frac{\bar{V}_2^2}{2} = (1)\left(\frac{1}{1.35}\right)^2\frac{\bar{V}_2^2}{2} = 0.549\frac{\bar{V}_2^2}{2}$$

Combining and substituting, with $\bar{V}_1 = \bar{V}_2$,

$$h_{l_T}(\text{new}) = \left[K_{ent} + f\frac{L_{new}}{D} + 0.051 + 0.549\right]\frac{\bar{V}_2^2}{2} = \left[K_{ent} + f\frac{L_{new}}{D} + 0.60\right]\frac{\bar{V}_1^2}{2} \qquad (2)$$

We have included the exit loss in the calculation of total head loss. Therefore, we apply the first law of thermodynamics to a control volume enclosing the entire flow system. The flow leaves the CV at section ④ inside the plenum chamber.

Basic equation:

$$\overset{=0(1)}{} \quad \overset{=0(1)}{} \quad \overset{=0(2)}{}$$

$$\dot{Q} + \dot{W}_s + \cancel{\dot{W}_{shear}} + \cancel{\dot{W}_{other}} = \cancel{\frac{\partial}{\partial t}\int_{CV} e\rho \, d\forall} + \int_{CS}\left(u + \frac{v^2}{2} + gz + \frac{p}{\rho}\right)\rho\vec{V}\cdot d\vec{A}$$

$$(4.59)$$

Assumptions: (1) $\dot{W}_{shear} = \dot{W}_{other} = 0$
(2) Steady flow
(3) Uniform flow at each section; $\alpha \simeq 1.0$
(4) Incompressible flow

Then

$$\dot{W}_s\Big)_{old} = \dot{m}\left[\left(\frac{p_4}{\rho} + \frac{\bar{V}_4^2}{2} + gz_4\right) - \left(\frac{p_0}{\rho} + \frac{\bar{V}_0^2}{2} + gz_0\right)\right] + \dot{m}\left(u_4 - u_0 - \frac{\delta Q}{dm}\right)$$

$$\dot{W}_s\Big)_{old} = \dot{m}\left[\left(\frac{p_4}{\rho} + \frac{\bar{V}_4^2}{2} + gz_4\right) - \left(\frac{p_0}{\rho} + \frac{\bar{V}_0^2}{2} + gz_0\right)\right] + \dot{m}h_{l_T}(\text{old})$$

Similarly for the new system,

$$\dot{W}_{s_{new}} = \dot{m}\left[\left(\frac{p_4}{\rho} + \frac{\bar{V}_4^2}{2} + gz_4\right) - \left(\frac{p_0}{\rho} + \frac{\bar{V}_0^2}{2} + gz_0\right)\right] + \dot{m}h_{l_T}(\text{new})$$

The difference in pumping power is

$$\Delta\dot{W}_s = \dot{W}_{s})_{old} - \dot{W}_{s})_{new} = \dot{m}[h_{l_T}(\text{old}) - h_{l_T}(\text{new})] = \rho\bar{V}_1 A_1 \Delta h_{l_T}$$

Substituting from Eqs. 1 and 2 gives

$$\Delta\dot{W}_s = \rho\bar{V}_1 A_1 \left[f\left(\frac{L_{old} - L_{new}}{D}\right) + 27f + 0.40\right]\frac{\bar{V}_1^2}{2}$$

The friction factor is a function of Reynolds number,

$$Re = \frac{\bar{V}D}{\nu} = 150\,\frac{ft}{sec} \times 0.5\,ft \times \frac{sec}{1.6 \times 10^{-4}\,ft^2} = 469{,}000$$

For smooth pipe, $f = 0.0134$ from Fig. 8.12. Substituting, with $L_{old} - L_{new} = N + 8 = 8.38\,ft$,

$$\Delta\dot{W}_s = \frac{0.00238\,slug}{ft^3} \times 150\,\frac{ft}{sec} \times \frac{\pi\,(0.5)^2\,ft^2}{4}\left[(0.0134)\,\frac{8.38\,ft}{0.5\,ft} + 27(0.0134) + 0.40\right]$$

$$\times \frac{1}{2} \times (150)^2\,\frac{ft^2}{sec^2} \times \frac{lbf \cdot sec^2}{slug \cdot ft} \times \frac{hp \cdot sec}{550\,ft \cdot lbf}$$

$$\Delta\dot{W}_s = 1.41\,hp$$

Since the blower efficiency, $\eta = 0.80$, the change in power would be

$$\Delta\dot{W}_{in} = \frac{\Delta\dot{W}_s}{\eta} = \frac{1.41\,hp}{0.8} = 1.76\,hp \qquad\qquad \Delta\dot{W}_{in}$$

{This problem illustrates use of the empirically determined minor loss coefficients.}

Example 8.8

A compressed air drill requires an air supply of 0.25 kg/sec at a gage pressure of 650 kPa at the drill. The hose from the air compressor to the drill is 40 mm inside diameter. The maximum compressor discharge gage pressure is 690 kPa. Neglect changes in density and any effects due to hose curvature. Air leaves the compressor at 40 C. Calculate the longest hose that may be used.

Example Problem 8.8

GIVEN:

Air flow through a line of length, L, and diameter $D = 40$ mm

$$p_1 = 690 \text{ kPa} \qquad p_2 = 650 \text{ kPa}$$
$$T_1 = 40 \text{ C} \qquad \dot{m} = 0.25 \text{ kg/sec} \qquad \rho \approx \text{constant}$$

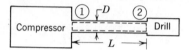

FIND:

Allowable length of hose.

SOLUTION:

Computing equation:

$$\left(\frac{p_1}{\rho} + \alpha_1 \frac{\bar{V}_1^2}{2} + gz_1\right) - \left(\frac{p_2}{\rho} + \alpha_2 \frac{\bar{V}_2^2}{2} + gz_2\right) = h_{l_T} = h_l + h_{l_m} \qquad (8.23)$$

where

$$h_l = f \frac{L}{D} \frac{\bar{V}^2}{2} \qquad h_{l_m} = K \frac{\bar{V}^2}{2}$$

For $\rho = c$, then $\bar{V}_1 = \bar{V}_2$, since $A_1 = A_2$. Since p_1 and p_2 are given, neglect minor losses. Assume that $\alpha_1 = \alpha_2$. Neglect changes in elevation, that is, $z_1 = z_2$. Then Eq. 8.23 can be written

$$\frac{p_1 - p_2}{\rho} = f \frac{L}{D} \frac{\bar{V}^2}{2} \qquad \text{or} \qquad L = \frac{(p_1 - p_2)}{\rho} \frac{2D}{f\bar{V}^2}$$

The density is

$$\rho = \rho_1 = \frac{p_1}{RT_1} = \frac{7.91 \times 10^5 \text{ N}}{\text{m}^2} \times \frac{\text{kg} \cdot \text{K}}{287 \text{ N} \cdot \text{m}} \times \frac{1}{313 \text{ K}} = 8.81 \text{ kg/m}^3$$

From continuity

$$\bar{V} = \frac{\dot{m}}{\rho A} = \frac{4\dot{m}}{\pi\rho D^2} = \frac{4}{\pi} \times \frac{0.25 \text{ kg}}{\text{sec}} \times \frac{\text{m}^3}{8.81 \text{ kg}} \times \frac{1}{(0.04)^2 \text{ m}^2} = 22.6 \text{ m/sec}$$

For air at 40 C, $\mu = 1.8 \times 10^{-5}$ kg/m·sec, so

$$Re = \frac{\rho\bar{V}D}{\mu} = \frac{8.81 \text{ kg}}{\text{m}^3} \times \frac{22.6 \text{ m}}{\text{sec}} \times \frac{0.04 \text{ m}}{} \times \frac{\text{m} \cdot \text{sec}}{1.8 \times 10^{-5} \text{ kg}} = 4.42 \times 10^5$$

Assume smooth pipe, then from Fig. 8.12, $f = 0.0134$. Substituting gives

$$L = \frac{(p_1 - p_2)}{\rho} \frac{2D}{f\bar{V}^2}$$

$$= \frac{0.40 \times 10^5 \text{ N}}{\text{m}^2} \times 2 \times 0.04 \text{ m} \times \frac{\text{m}^3}{8.81 \text{ kg}} \times \frac{1}{0.0134} \times \frac{\sec^2}{(22.6)^2 \text{ m}^2} \times \frac{\text{kg} \cdot \text{m}}{\text{N} \cdot \sec^2}$$

$$L = 53.1 \text{ m} \hspace{5cm} L$$

$\left\{\begin{array}{l}\text{This problem illustrates the method of solving for an unknown pipe length.} \\ \text{Note that the relative change in density for this problem, } \Delta\rho/\rho \simeq \Delta p/p, \text{ is only} \\ \text{about 5 percent. Thus the assumption of incompressible flow is reasonable.}\end{array}\right\}$

Example 8.9

A fire protection system is supplied from a water tower and standpipe 80 ft tall. The longest pipe in the system is 600 ft long, and is made of cast iron about 20 years old. The pipe contains one gate valve; other minor losses may be neglected. The pipe diameter is 4 in. Determine the maximum rate of flow through this pipe, in gallons per minute.

Example Problem 8.9

GIVEN:

Fire protection system, as shown.

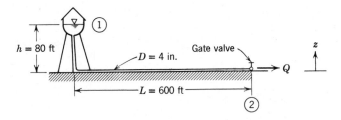

FIND:

Q, gpm.

SOLUTION:

Computing equations:
$$\overbrace{\left(\frac{p_1}{\rho} + \alpha_1 \frac{\bar{V}_1^2}{2} + gz_1\right)}^{\approx 0(2)} - \left(\frac{p_2}{\rho} + \alpha_2 \frac{\bar{V}_2^2}{2} + gz_2\right) = h_{l_T} \quad (8.23)$$

$$h_{l_T} = f \frac{L}{D} \frac{\bar{V}^2}{2} + h_{l_m} = f \frac{(L + L_e)}{D} \frac{\bar{V}^2}{2}$$

Assumptions: (1) $p_1 = p_2 = p_{atm}$

(2) $\bar{V}_1 \simeq 0$, and $\alpha_2 \simeq 1.0$

For a fully-open gate valve, from Table 8.4, $L_e/D = 13$. Then

$$h_{l_T} = f\frac{L}{D}\frac{\bar{V}_2^2}{2} + 13f\frac{\bar{V}_2^2}{2} = g(z_1 - z_2) - \frac{\bar{V}_2^2}{2}$$

or

$$\frac{\bar{V}_2^2}{2}\left[f\left(\frac{L}{D} + 13\right) + 1\right] = g(z_1 - z_2)$$

Solving for $\bar{V}_2$,

$$\bar{V}_2 = \left[\frac{2g(z_1 - z_2)}{f(L/D + 13) + 1}\right]^{1/2}$$

To be conservative, assume the standpipe is the same diameter as the horizontal pipe. Then

$$\frac{L}{D} = \frac{600\ \text{ft} + 80\ \text{ft}}{4\ \text{in.}} \times \frac{12\ \text{in.}}{\text{ft}} = 2040$$

Also

$$z_1 - z_2 = h = 80\ \text{ft}$$

Since $\bar{V}_2$ is not known, we cannot compute Re. But we can assume a value of friction factor in the fully-rough flow region. From Fig. 8.13, $e/D \simeq 0.0025$ for cast iron pipe. Since the pipe is quite old, choose $e/D = 0.005$. Then from Fig. 8.12, guess $f \simeq 0.03$. Then a first approximation to $\bar{V}_2$ is

$$\bar{V}_2 = \left[2 \times 32.2\ \frac{\text{ft}}{\text{sec}^2} \times 80\ \text{ft} \times \frac{1}{0.03(2040 + 13) + 1}\right]^{1/2} = 9.07\ \text{ft/sec}$$

Now check the value assumed for f.

$$Re = \frac{\rho\bar{V}D}{\mu} = \frac{\bar{V}D}{\nu} = 9.07\ \frac{\text{ft}}{\text{sec}} \times \frac{\text{ft}}{3} \times \frac{\text{sec}}{1.2 \times 10^{-5}\ \text{ft}^2} = 2.52 \times 10^5$$

For $e/D = 0.005$, $f = 0.031$ from Fig. 8.12. Using this value,

$$\bar{V}_2 = \left[2 \times 32.2\ \frac{\text{ft}}{\text{sec}^2} \times 80\ \text{ft} \times \frac{1}{0.031(2040 + 13) + 1}\right]^{1/2} = 8.93\ \text{ft/sec}$$

Thus convergence is satisfactory. The volumetric flowrate is

$$Q = \bar{V}_2 A = \bar{V}_2\frac{\pi D^2}{4} = \frac{8.93\ \text{ft}}{\text{sec}} \times \frac{\pi}{4}\left(\frac{1}{3}\right)^2\ \text{ft}^2 \times \frac{7.48\ \text{gal}}{\text{ft}^3} \times \frac{60\ \text{sec}}{\text{min}}$$

$$Q = 350\ \text{gpm} \hspace{4cm} Q$$

{This problem illustrates the procedure for solving pipe flow problems in which the flowrate is unknown. Note that the velocity and, hence, the flowrate, is essentially proportional to $1/\sqrt{f}$. Doubling the value of e/D to account for aging reduced the flowrate by about 10 percent.}

Example 8.10

Spray heads in an agricultural spraying system are to be supplied with water through 500 ft of drawn aluminum tubing from an engine-driven pump. In its most efficient operating range, the pump output is 1500 gpm at a discharge pressure not exceeding 65 psig. For satisfactory operation, the sprinklers must operate at 30 psig or higher pressure. Minor losses and elevation changes may be neglected. Determine the smallest standard pipe size that can be used.

Example Problem 8.10

GIVEN:

Water supply system, as shown.

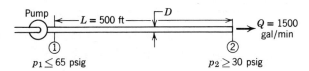

FIND:

Smallest standard D.

SOLUTION:

Δp, L, and Q are known. D is unknown, so iteration will be required to determine the minimum standard diameter that satisfies the pressure drop constraint at the given flowrate. The maximum allowable pressure drop is

$$\Delta p_{max} = p_{1\,max} - p_{2\,min} = (65 - 30) \text{ psi} = 35 \text{ psi}$$

Computing equations:

$$\left(\frac{p_1}{\rho} + \alpha_1 \frac{\bar{V}_1^2}{2} + gz_1\right) - \left(\frac{p_2}{\rho} + \alpha_2 \frac{\bar{V}_2^2}{2} + gz_2\right) = h_{l_T} \qquad (8.23)$$

$$= 0\,(3)$$

$$h_{l_T} = h_l + h_{l_m} = f \frac{L}{D} \frac{\bar{V}^2}{2}$$

Assumptions: (1) Steady flow
(2) Incompressible flow
(3) $h_{l_T} = h_l$, that is, $h_{l_m} = 0$
(4) $z_1 = z_2$
(5) $\bar{V}_1 = \bar{V}_2$; $\alpha_1 \simeq \alpha_2$

Then

$$\Delta p = p_1 - p_2 = f \frac{L}{D} \frac{\rho \bar{V}^2}{2}$$

Since trial values of D are to be assumed, it is convenient to substitute $\bar{V} = Q/A = 4Q/\pi D^2$, so that

$$\Delta p = f \frac{L}{D} \frac{\rho}{2} \left(\frac{4Q}{\pi D^2} \right)^2 = \frac{8 f L \rho Q^2}{\pi^2 D^5} \tag{1}$$

The Reynolds number is needed to find f. In terms of Q,

$$Re = \frac{\rho \bar{V} D}{\mu} = \frac{\bar{V} D}{\nu} = \frac{4Q}{\pi D^2} \frac{D}{\nu} = \frac{4Q}{\pi \nu D}$$

Finally, Q must be converted to cubic feet per second.

$$Q = \frac{1500 \text{ gal}}{\text{min}} \times \frac{\text{min}}{60 \text{ sec}} \times \frac{\text{ft}^3}{7.48 \text{ gal}} = 3.34 \text{ ft}^3/\text{sec}$$

For an initial guess, take nominal 4 in. (4.026 in. i.d.) pipe:

$$Re = \frac{4Q}{\pi \nu D} = \frac{4}{\pi} \times \frac{3.34 \text{ ft}^3}{\text{sec}} \times \frac{\text{sec}}{1.2 \times 10^{-5} \text{ ft}^2} \times \frac{1}{4.026 \text{ in.}} \times \frac{12 \text{ in.}}{\text{ft}} = 1.06 \times 10^6$$

From Fig. 8.12, for drawn tubing, $f = 0.0117$, and

$$\Delta p = \frac{8 f L \rho Q^2}{\pi^2 D^5} = \frac{8}{\pi^2} \times 0.0117 \times 500 \text{ ft} \times \frac{1.94 \text{ slug}}{\text{ft}^3} \times (3.34)^2 \frac{\text{ft}^6}{\text{sec}^2}$$

$$\times \frac{1}{(4.026)^5 \text{ in.}^5} \times \frac{1728 \text{ in.}^3}{\text{ft}^3} \times \frac{\text{lbf} \cdot \text{sec}^2}{\text{slug} \cdot \text{ft}}$$

$$\Delta p = 168 \text{ lbf/in.}^2 > \Delta p_{\text{max}}$$

Since this value is too large, try $D = 6$ in. (actually 6.065 in. i.d.):

$$Re = \frac{4}{\pi} \times \frac{3.34 \text{ ft}^3}{\text{sec}} \times \frac{\text{sec}}{1.2 \times 10^{-5} \text{ ft}^2} \times \frac{1}{6.065 \text{ in.}} \times \frac{12 \text{ in.}}{\text{ft}} = 7.01 \times 10^5$$

From Fig. 8.12, for drawn tubing, $f = 0.0124$, and

$$\Delta p = \frac{8}{\pi^2} \times 0.0124 \times 500 \text{ ft} \times \frac{1.94 \text{ slug}}{\text{ft}^3} \times (3.34)^2 \frac{\text{ft}^6}{\text{sec}^2}$$

$$\times \frac{1}{(6.065)^5 \text{ in.}^5} \times \frac{(12)^3 \text{ in.}^3}{\text{ft}^3} \times \frac{\text{lbf} \cdot \text{sec}^2}{\text{slug} \cdot \text{ft}}$$

$$\Delta p = 22.9 \text{ lbf/in.}^2 < \Delta p_{\text{max}}$$

Since this value is less than the allowable pressure drop, we should check a 5 in. (nominal) pipe. With an actual i.d. of 5.047 in.,

$$Re = \frac{4}{\pi} \times \frac{3.34 \text{ ft}^3}{\text{sec}} \times \frac{\text{sec}}{1.2 \times 10^{-5} \text{ ft}^2} \times \frac{1}{5.047 \text{ in.}} \times \frac{12 \text{ in.}}{\text{ft}} = 8.43 \times 10^5$$

From Fig. 8.12, $f = 0.012$, and

$$\Delta p = \frac{8}{\pi^2} \times 0.012 \times 500 \text{ ft} \times \frac{1.94 \text{ slug}}{\text{ft}^3} \times \frac{(3.34)^2 \text{ ft}^6}{\text{sec}^2}$$

$$\times \frac{1}{(5.047)^5 \text{ in.}^5} \times \frac{(12)^3 \text{ in.}^3}{\text{ft}^3} \times \frac{\text{lbf} \cdot \text{sec}^2}{\text{slug} \cdot \text{ft}}$$

$$\Delta p = 55.5 \text{ lbf/in.}^2 > \Delta p_{max}$$

Thus the criterion for pressure drop is satisfied for a minimum nominal diameter of 6 in. pipe.   D

> This problem illustrates the procedure for solving pipe flow problems when the diameter is unknown. Note from Eq. 1 that the pressure drop in turbulent pipe flow is proportional to f/D^5. The variation of f is small, so Δp at constant flowrate is approximately proportional to $1/D^5$.

**8-9.2 MULTIPLE-PATH SYSTEMS

In many practical situations, such as water supply or fire protection systems, complex pipe networks must be analyzed. The basic techniques developed in Section 8-9.1 also may be used to analyze flow in multiple-path pipe systems. The procedure is analogous to that used in solving direct current electric circuits. A representative pipe system is shown in Fig. 8.20.

The pipe system shown in Fig. 8.20 has two nodes, labeled A and B, and three branches. The total flowrate into the system must be distributed among the branches. Consequently, the flowrate through each branch is unknown.

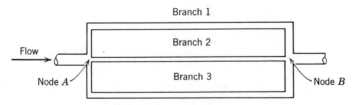

Fig. 8.20 Simple multiple-path pipe flow system.

** This section may be omitted without loss of continuity in the text material.

However, the pressure drop for each branch is the same, that is, equal to $p_A - p_B$. This information is sufficient to permit an iterative solution for the flowrate in each branch, as shown in Example Problem 8.11.

The fluid flowrate and pressure drop are, respectively, analogous to the current and voltage in an electric circuit. However, the simple linear relation between voltage and current given by Ohm's law does not apply to the fluid flow system. Instead, the flow pressure drop is approximately proportional to the square of the flowrate. This nonlinearity makes iterative solutions necessary, and the resulting calculations can be quite lengthy and tedious. A number of schemes have been developed for use with a digital computer. For an example, see Reference 11.

Example 8.11

The irrigation system of Example 8.10 is to be extended, holding the total flowrate constant at 1500 gpm, by adding three branches, each made from 3 in. pipe, as shown in the sketch below. The sprinkler at the end of each branch has a nozzle of 1.5 in. minimum diameter. Minor losses at elbows should be included, but elevation terms may be neglected. Determine the flowrate in each branch, the pressure required at Ⓐ, and the pressure applied to each sprinkler nozzle.

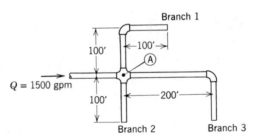

Example Problem 8.11

GIVEN:

Pipe network shown schematically below.

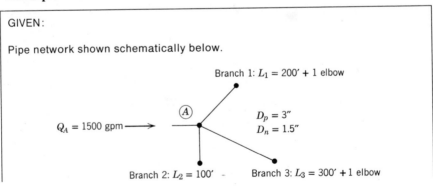

FIND:

(a) Q_1, Q_2, Q_3.
(b) p_A.
(c) p_1, p_2, p_3 at nozzle inlets.

SOLUTION:

Apply Eq. 8.23 to each branch. Choose subscript n for nozzle exit, subscript p for pipe.

Computing equations:

$$\frac{p_A}{\rho} + \alpha \overbrace{\frac{\bar{V}_A^2}{2}}^{\approx\,0(1)} + g \cancel{z_A} = \cancel{\frac{p_n}{\rho}} + \alpha_n \frac{\bar{V}_n^2}{2} + g \cancel{z_n} + h_{l_T};\ h_{l_T} = f\frac{L}{D}\frac{\bar{V}_p^2}{2} + f\frac{L_e}{D}\frac{\bar{V}_p^2}{2}$$

Assumptions: (1) $\bar{V}_A^2 \ll \bar{V}_n^2;\ \alpha_n \simeq 1.0$
(2) $z_A \simeq z_n$
(3) $p_n = p_{atmosphere}$
(4) Neglect loss due to flow split at Ⓐ.

Then for each branch of the flow system, using gage pressure at Ⓐ,

$$\frac{p_A}{\rho} = \frac{\bar{V}_n^2}{2} + f\frac{L}{D}\frac{\bar{V}_p^2}{2} + f\frac{L_e}{D}\frac{\bar{V}_p^2}{2}$$

But for incompressible flow, $\bar{V}_p A_p = \bar{V}_n A_n$, so $\bar{V}_n^2 = \bar{V}_p^2 (A_p/A_n)^2 = \bar{V}_p^2 (D_p/D_n)^4$, and

$$\frac{p_A}{\rho} = \frac{\bar{V}_p^2}{2}\left[\left(\frac{D_p}{D_n}\right)^4 + f\left(\frac{L}{D} + \frac{L_e}{D}\right)\right]$$

or

$$\bar{V}_p = \sqrt{\frac{2p_A}{\rho}}\left[\frac{1}{\left(\frac{D_p}{D_n}\right)^4 + f\left(\frac{L}{D} + \frac{L_e}{D}\right)}\right]^{1/2}$$

The corresponding flowrate for each branch is

$$Q_p = \bar{V}_p A_p = A_p \sqrt{\frac{2p_A}{\rho}}\left[\frac{1}{\left(\frac{D_p}{D_n}\right)^4 + f\left(\frac{L}{D} + \frac{L_e}{D}\right)}\right]^{1/2} \tag{1}$$

To find f, we must determine the pipe Reynolds number. If flow is equally split as a first approximation, then $Q_p \simeq 500$ gpm per branch, and

$$Re = \frac{\bar{V}_p D_p}{\nu} = \frac{4Q_p}{\pi \nu D_p}$$

$$= \frac{4}{\pi} \times \frac{500\ \text{gal}}{\text{min}} \times \frac{\text{sec}}{1.2 \times 10^{-5}\ \text{ft}^2} \times \frac{1}{3\ \text{in.}} \times \frac{\text{min}}{60\ \text{sec}} \times \frac{\text{ft}^3}{7.48\ \text{gal}} \times \frac{12\ \text{in.}}{\text{ft}}$$

$$Re = 4.73 \times 10^5$$

From Fig. 8.12, for smooth pipe, $f \simeq 0.0133$. For an elbow, $L_e/D = 30$, from Table 8.4. Using these values, we can obtain a first approximation for Q_p for each branch. Substituting into Eq. 1,

$$Q_{p_1} \simeq A_p \sqrt{\frac{2p_A}{\rho}} \left[\frac{1}{\left(\frac{3.0}{1.5}\right)^4 + 0.0133 \left(200\text{ ft} \times \frac{1}{3\text{ in.}} \times \frac{12\text{ in.}}{\text{ft}} + 30 \right)} \right]^{1/2}$$

$$Q_{p_1} \simeq 0.192 A_p \sqrt{\frac{2p_A}{\rho}}$$

$$Q_{p_2} \simeq A_p \sqrt{\frac{2p_A}{\rho}} \left[\frac{1}{\left(\frac{3.0}{1.5}\right)^4 + 0.0133 \left(100\text{ ft} \times \frac{1}{3\text{ in.}} \times \frac{12\text{ in.}}{\text{ft}} \right)} \right]^{1/2}$$

$$Q_{p_2} \simeq 0.217 A_p \sqrt{\frac{2p_A}{\rho}}$$

and

$$Q_{p_3} \simeq A_p \sqrt{\frac{2p_A}{\rho}} \left[\frac{1}{\left(\frac{3.0}{1.5}\right)^4 + 0.0133 \left(300\text{ ft} \times \frac{1}{3\text{ in.}} \times \frac{12\text{ in.}}{\text{ft}} + 30 \right)} \right]^{1/2}$$

$$Q_{p_3} \simeq 0.176 A_p \sqrt{\frac{2p_A}{\rho}}$$

From continuity,

$$Q_A = Q_{p_1} + Q_{p_2} + Q_{p_3} \simeq (0.192 + 0.217 + 0.176) A_p \sqrt{\frac{2p_A}{\rho}}$$

or

$$Q_A \simeq 0.585 A_p \sqrt{\frac{2p_A}{\rho}}$$

Thus

$$\frac{Q_{p_1}}{Q_A} \simeq \frac{0.192 A_p \sqrt{\frac{2p_A}{\rho}}}{0.585 A_p \sqrt{\frac{2p_A}{\rho}}} = 0.328$$

or

$$Q_{p_1} \simeq (0.328)1500\text{ gpm} = 492\text{ gpm}$$

Similarly,

$$\frac{Q_{p_2}}{Q_A} \simeq \frac{0.217}{0.585} = 0.371; \quad Q_{p_2} \simeq 557\text{ gpm}$$

$$\frac{Q_{p_3}}{Q_A} \simeq \frac{0.176}{0.585} = 0.301; \quad Q_{p_3} \simeq 452\text{ gpm}$$

Better approximations for the Reynolds number and friction factor for each branch may now be calculated.

$$Re_1 = \frac{Q_{p_1}}{500 \text{ gpm}} \times \frac{4.73 \times 10^5}{500} \simeq \frac{492}{500} \times \frac{4.73 \times 10^5}{500} = 4.65 \times 10^5$$

From Fig. 8.12, $f_1 \simeq 0.0133$. Similarly,

$$Re_2 \simeq \frac{557}{500} \times \frac{4.73 \times 10^5}{500} = 5.27 \times 10^5 \, ; f_2 \simeq 0.0130$$

and

$$Re_3 \simeq \frac{452}{500} \times \frac{4.73 \times 10^5}{500} = 4.28 \times 10^5 \, ; f_3 \simeq 0.0136$$

Substituting these values into Eq. 1, we obtain

$$Q_{p_1} \simeq 0.192 A_p \sqrt{\frac{2p_A}{\rho}}$$

$$Q_{p_2} \simeq 0.217 A_p \sqrt{\frac{2p_A}{\rho}}$$

and

$$Q_{p_3} \simeq 0.175 A_p \sqrt{\frac{2p_A}{\rho}}$$

From continuity,

$$Q_A = Q_{p_1} + Q_{p_2} + Q_{p_3} \simeq (0.192 + 0.217 + 0.175) A_p \sqrt{\frac{2p_A}{\rho}}$$

or

$$Q_A \simeq 0.584 A_p \sqrt{\frac{2p_A}{\rho}}$$

Solving for the individual flowrates,

$$Q_{p_1} \simeq \frac{0.192}{0.584} \times 1500 \text{ gpm} = 493 \text{ gpm}$$

$$Q_{p_2} \simeq \frac{0.217}{0.584} \times 1500 \text{ gpm} = 557 \text{ gpm}$$

and

$$Q_{p_3} \simeq \frac{0.175}{0.584} \times 1500 \text{ gpm} = 449 \text{ gpm}$$

Q_1, Q_2, Q_3

Solving Eq. 1 for p_A, gives

$$p_A = \frac{\rho}{2} \left(\frac{Q_p}{A_p}\right)^2 \left[\left(\frac{D_p}{D_n}\right)^4 + f\left(\frac{L}{D} + \frac{L_e}{D}\right)\right]$$

Substituting values for Branch ①,

$$p_A = \frac{1}{2} \times \frac{1.94 \text{ slug}}{\text{ft}^3} \left(\frac{493 \text{ gal}}{\text{min}} \times \frac{4}{\pi (3)^2 \text{ in.}^2} \times \frac{1}{7.48 \text{ gal}} \times \frac{\text{ft}^3}{} \times \frac{\text{min}}{60 \text{ sec}} \times \frac{144 \text{ in.}^2}{\text{ft}^2} \right)^2$$

$$\times \left[\left(\frac{3.0}{1.5} \right)^4 + 0.0133 \left(\frac{200 \text{ ft}}{} \times \frac{1}{3 \text{ in.}} \times \frac{12 \text{ in.}}{\text{ft}} + 30 \right) \right] \frac{\text{lbf} \cdot \text{sec}^2}{\text{slug} \cdot \text{ft}} \times \frac{\text{ft}^2}{144 \text{ in.}^2}$$

$$p_A = 91.2 \text{ lbf/in.}^2$$

The values for Branches ② and ③ by similar calculations are 91.4 and 90.5 psig, respectively. (The slight differences are due to rounding the flowrate values.) Thus

$$p_A \approx 91.0 \text{ psig} \qquad\qquad\qquad p_A$$

The pressure at the inlet to each sprinkler nozzle may be calculated from the energy equation. The equation between Ⓐ and the nozzle inlet, section ①, is

$$\frac{p_A}{\rho} + \frac{\bar{V}_A^2}{2} + gz_A = \frac{p_i}{\rho} + \frac{\bar{V}_p^2}{2} + gz_i + h_{l_T}; \quad h_{l_T} = f\frac{L}{D}\frac{\bar{V}_p^2}{2} + f\frac{L_e}{D}\frac{\bar{V}_p^2}{2}$$

From continuity $\bar{V}_A = Q_A/A_A$ and $\bar{V}_p = Q_p/A_p$, so

$$p_i = p_A + \frac{\rho}{2} \left[\left(\frac{Q_A}{A_A} \right)^2 - \left(f\frac{L}{D} + f\frac{L_e}{D} + 1 \right) \left(\frac{Q_p}{A_p} \right)^2 \right]$$

or

$$p_i = p_A + \frac{\rho}{2} \left(\frac{Q_A}{A_A} \right)^2 \left\{ 1 - \left[f\left(\frac{L}{D} + \frac{L_e}{D} \right) + 1 \right] \left(\frac{Q_p}{Q_A} \right)^2 \left(\frac{A_A}{A_p} \right)^2 \right\}$$

Substituting values for Branch ①,

$$p_{i_1} = \frac{91.0 \text{ lbf}}{\text{in.}^2} + \frac{1}{2} \times \frac{1.94 \text{ slug}}{\text{ft}^3} \left(\frac{1500 \text{ gal}}{\text{min}} \times \frac{4}{\pi (6)^2 \text{ in.}^2} \times \frac{1}{7.48 \text{ gal}} \times \frac{\text{ft}^3}{} \times \frac{\text{min}}{60 \text{ sec}} \times \frac{144 \text{ in.}^2}{\text{ft}^2} \right)^2$$

$$\times \left\{ 1 - \left[0.0133 \left(\frac{200 \text{ ft}}{} \times \frac{1}{3 \text{ in.}} \times \frac{12 \text{ in.}}{\text{ft}} + 30 \right) + 1 \right] \left(\frac{493}{1500} \right)^2 \left(\frac{6}{3} \right)^4 \right\} \frac{\text{lbf} \cdot \text{sec}^2}{\text{slug} \cdot \text{ft}}$$

$$\times \frac{\text{ft}^2}{144 \text{ in.}^2}$$

$$p_{i_1} = 52.3 \text{ lbf/in.}^2$$

Similar calculations for Branches ② and ③ give

$$p_{i_2} = 66.3 \text{ lbf/in.}^2 \quad \text{and} \quad p_{i_3} = 43.3 \text{ lbf/in.}^2 \qquad p_i$$

$\left\{ \begin{array}{l} \text{This problem illustrates the general method used to solve multiple-path pipe} \\ \text{flow problems.} \end{array} \right\}$

**8-9.3 NONCIRCULAR DUCTS

The empirical correlations for pipe flow may also be used for computations involving noncircular ducts, provided their cross sections are not too exaggerated. Thus ducts of square or rectangular cross section may be treated if the ratio of height to width is less than about 3 or 4.

The correlations for turbulent pipe flow developed in Section 8-8 are extended for use with noncircular geometries by introducing the hydraulic diameter,

$$D_h \equiv \frac{4A}{P_w} \qquad (8.39)$$

in place of the diameter, D. In Eq. 8.39, A is the cross-sectional area, and P_w is the *wetted perimeter*, that is, the length of wall in contact with the flowing fluid at any cross section. The factor 4 is introduced so that the hydraulic diameter will equal the duct diameter for a circular geometry. For a circular duct, $A = \pi D^2/4$ and $P_w = \pi D$, so that

$$D_h = \frac{4A}{P_w} = \frac{4 \left(\dfrac{\pi}{4} \right) D^2}{\pi D} = D$$

For a rectangular duct of width, b, and height, h, then $A = bh$ and $P_w = 2(b + h)$, so

$$D_h = \frac{4bh}{2(b + h)}$$

If the *aspect ratio, ar*, is defined as $ar = h/b$, then

$$D_h = \frac{2h}{1 + ar}$$

for rectangular ducts. For a square duct, $ar = 1$, and $D_h = h$.

As noted, the hydraulic diameter concept can be applied in the approximate range $\frac{1}{3} < ar < 3$. Under these conditions, the correlations of Section 8-8 give acceptably accurate results for rectangular ducts. Since such ducts are easy and cheap to fabricate from sheet metal, they are commonly used in air conditioning, heating, and ventilating applications. Extensive data on losses for air flow are available (e.g. see Refs. 6 and 12).

Losses due to secondary flows increase rapidly for more extreme geometries, so the correlations are not applicable to wide, flat ducts, or to ducts

** This section may be omitted without loss of continuity in the text material.

of triangular or other irregular shapes. One must resort to use of experimental data for specific situations when precise design information is required.

PART D.   BOUNDARY LAYERS

8-10   THE BOUNDARY-LAYER CONCEPT

The concept of a boundary layer was first introduced by Ludwig Prandtl, a German aerodynamicist, in 1904.[8]

Prior to Prandtl's historic breakthrough, the science of fluid mechanics had been developing in two rather different directions. Theoretical hydrodynamics evolved from Euler's equations (published by Leonhard Euler in 1755) of motion for a nonviscous fluid. Since the results of hydrodynamics contradicted many experimental observations, practicing engineers developed their own empirical art of hydraulics. This was based on experimental data and differed significantly from the purely mathematical approach of theoretical hydrodynamics.

Although the complete equations describing the motion of a viscous fluid (the Navier–Stokes equations, Eqs. 5.31, developed by Navier, 1827, and independently by Stokes, 1845) were known prior to Prandtl, the mathematical difficulties in solving these equations (except for a few simple cases) prohibited a theoretical treatment of viscous flows. Prandtl showed that many viscous flows can be analyzed by dividing the flow into two regions, one close to solid boundaries, the other covering the rest of the flow. Only in the thin region adjacent to a solid boundary (the boundary layer) is the effect of viscosity important. In the region outside of the boundary layer, the effect of viscosity is negligible and the fluid may be treated as inviscid.

The boundary-layer concept provided the link that had been missing between theory and practice. Furthermore, the boundary-layer concept permitted the solution of viscous flow problems that would have been impossible through application of the Navier–Stokes equations to the complete flow field.[9] Thus the introduction of the boundary-layer concept marked the beginning of the modern era of fluid mechanics.

The development of a boundary layer on a solid surface was discussed in Section 2-5.1. The development of a laminar boundary layer on a flat plate was illustrated in Fig. 2.10. In the boundary layer both viscous and inertia forces are important. Consequently, it is not surprising that the Reynolds

---

[8] Prandtl presented the boundary-layer concept in a paper, "Fluid Motion with Very Small Friction (in German)," before the Mathematical Congress in Heidelberg.
[9] Today, computer solutions of the Navier–Stokes equations are common.

number (which represents the ratio of inertia to viscous forces) is significant in characterizing boundary-layer flows. The characteristic length used in the Reynolds number is taken either as the length in the flow direction over which the boundary layer has developed, or as some measure of the boundary-layer thickness.

As for flow in a duct, flow in a boundary layer may be laminar or turbulent. However, there is no unique value of the Reynolds number at which transition from laminar to turbulent flow occurs in a boundary layer. Among the factors that affect boundary-layer transition are: pressure gradient, surface roughness, heat transfer, body forces, and free stream disturbances. Detailed consideration of these effects is beyond the scope of this text.

In many real flow situations, a boundary layer develops over a long, essentially flat surface. Examples include flow over ship and submarine hulls, aircraft wings, and atmospheric motions over flat terrain. Since the basic features of all these flows are illustrated in the simpler case of flow over a flat plate, let us consider this first.

For incompressible flow over a smooth flat plate (zero pressure gradient) in the absence of heat transfer, transition from laminar to turbulent flow in the boundary layer can be delayed to a Reynolds number, $Re_x = \rho U x / \mu$, in the range $3 \times 10^6 < Re_x < 4 \times 10^6$ if external disturbances are minimized. For air at standard conditions, with a free stream velocity, $U = 100$ m/sec, this corresponds to a length, x, along the plate of $0.43 < x$ (m) < 0.58. A qualitative picture of the boundary-layer growth over a flat plate is shown in Fig. 8.21. The boundary layer is laminar for a short distance downstream from the leading edge; transition occurs over a region of the plate rather than at a single line across the plate. The transition region extends downstream to the location where the boundary-layer flow becomes completely turbulent. In the qualitative picture of Fig. 8.21, we have shown the turbulent boundary layer growing at a faster rate than the laminar layer. In later

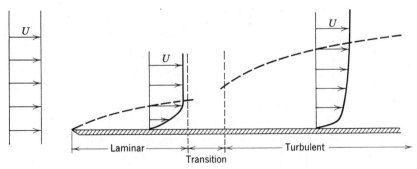

Fig. 8.21 Boundary layer on a flat plate (vertical thickness exaggerated greatly).

sections of this chapter we shall show that this is indeed true. The velocity profiles in the laminar and turbulent boundary layers are qualitatively similar to the respective profiles found in pipe flow (Fig. 8.9).

8-11 DISPLACEMENT THICKNESS

The boundary layer is the region adjacent to a solid surface in which viscous forces are important. Since the velocity profile merges smoothly and asymptotically into the free stream, the boundary-layer thickness, δ, is difficult to measure. The thickness, δ, is usually defined as the distance from the surface to the point where the velocity is within 1 percent of the freestream velocity.

The effect of the boundary layer is to displace the streamlines in the flow outside the boundary layer away from the wall. A boundary-layer displacement thickness, δ^*, can be defined as the distance by which the solid surface would have to be displaced to maintain the same mass flowrate in a hypothetical frictionless flow. The displacement thickness concept is illustrated in Fig. 8.22. The actual, viscous, velocity profile (a) has been replaced by the hypothetical, frictionless profile (b) displaced by a distance, δ^*, from the wall. The displacement thickness, δ^*, is such that the mass flowrate for the hypothetical profile is equal to that for the actual profile. For a plate of width, b:

(i)   The mass flowrate for the actual flow (with viscous boundary layer) is given by

$$\dot{m}_a = \int_0^\infty \rho u b \, dy$$

(ii)   the mass flowrate for the hypothetical frictionless flow is given by

$$\dot{m}_b = \int_{\delta^*}^\infty \rho U b \, dy$$

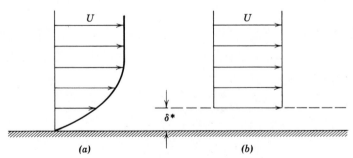

Fig. 8.22 Definition of displacement thickness, δ^*.

Since the displacement thickness, δ^*, is defined such that $\dot{m}_a = \dot{m}_b$, then

$$\int_0^\infty \rho u b \, dy = \int_{\delta^*}^\infty \rho U b \, dy$$

For incompressible flow we can write

$$\int_0^\infty u \, dy = \int_{\delta^*}^\infty U \, dy$$

Then

$$\int_0^\infty u \, dy = \int_0^\infty U \, dy - \int_0^{\delta^*} U \, dy$$

and

$$\int_0^{\delta^*} U \, dy = \int_0^\infty (U - u) \, dy$$

$$U\delta^* = \int_0^\infty (U - u) \, dy$$

Finally,

$$\delta^* = \int_0^\infty \left(1 - \frac{u}{U}\right) dy \qquad (8.40)$$

or

$$\delta^* \simeq \int_0^\delta \left(1 - \frac{u}{U}\right) dy$$

since $u \simeq U$ and, hence, the integrand is essentially zero for $y \geq \delta$. Application of the displacement thickness concept is illustrated in Example Problem 8.12.

Example 8.12
Air at standard conditions is flowing between the parallel flat plates shown below. The velocity is uniform at the entrance and has a magnitude, $U_0 = 25$ m/sec. A boundary-layer trip on each wall at the channel inlet assures that a turbulent boundary layer develops from the leading edge. The boundary-layer velocity profile and thickness may be approximated by the expressions

$$\frac{u}{U} = \left(\frac{y}{\delta}\right)^{1/7} \qquad \frac{\delta}{x} = 0.370 \left(\frac{\nu}{U_0 x}\right)^{1/5}$$

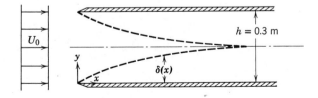

where U is the core velocity. Note that U is a function of x. The width of the plates, w, is much larger than the distance between the plates, h, and therefore end effects may be neglected. Determine the pressure drop between the inlet and a point 5 m downstream of the inlet.

Example Problem 8.12

GIVEN:

Air flow between parallel plates.

Boundary-layer thickness on each wall is given by $\dfrac{\delta}{x} = 0.370 \left(\dfrac{v}{U_0 x}\right)^{1/5}$

The velocity profile in the boundary layer is given by

$$\frac{u}{U} = \left(\frac{y}{\delta}\right)^{1/7}$$

$U_0 = 25$ m/sec.   Uniform plate width, w.

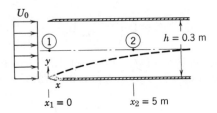

FIND:

Pressure drop, $p_1 - p_2$.

SOLUTION:

Let us first check to see if the boundary layer has reached the channel centerline at section ②. At $x = 5$ m,

$$\delta_2 = 0.370 x_2 \left(\frac{v}{U_0 x_2}\right)^{1/5}$$

$$\delta_2 = 0.370 \times 5 \text{ m} \left(\frac{1.45 \times 10^{-5} \text{ m}^2}{\text{sec}} \times \frac{\text{sec}}{25 \text{ m}} \times \frac{1}{5 \text{ m}}\right)^{1/5} = 0.0759 \text{ m}$$

Since $\delta_2 < h/2$, the flow is not fully developed and we can apply the Bernoulli equation along the center streamline between sections ① and ②.

Basic equation:
$$\frac{p_1}{\rho} + \frac{V_1^2}{2} + g\cancel{z}_1 = \frac{p_2}{\rho} + \frac{V_2^2}{2} + g\cancel{z}_2$$

Assumptions: (1) Steady flow
(2) Incompressible flow
(3) Flow along a streamline
(4) No friction outside boundary layer
(5) $z_1 = z_2$

Then

$$p_1 - p_2 = \frac{\rho}{2}(V_2^2 - V_1^2) = \frac{\rho V_1^2}{2}\left[\left(\frac{V_2}{V_1}\right)^2 - 1\right] = \frac{\rho U_0^2}{2}\left[\left\{\frac{U(x_2)}{U_0}\right\}^2 - 1\right]$$

We recognize that in order to satisfy continuity, the centerline velocity must increase in the direction of flow.

In order to determine U_2, we employ the notion of the displacement thickness. The continuity equation can be written

$$\rho U_1 A_1 = \rho U_2 A_2$$

For a plate width, w, then $A_1 = wh$, and $A_2 = w(h - 2\delta_2^*)$. Thus

$$\rho U_0 hw = \rho U_2(h - 2\delta_2^*)w$$

so that

$$\frac{U_2}{U_0} = \frac{h}{h - 2\delta_2^*}$$

The task now is to calculate δ_2^*.

$$\delta_2^* = \int_0^\delta \left(1 - \frac{u}{U}\right)dy$$

Setting $\eta = y/\delta$, then $dy = \delta\, d\eta$, and

$$\delta_2^* = \delta_2 \int_0^1 \left(1 - \frac{u}{U}\right)d\eta = \delta_2 \int_0^1 (1 - \eta^{1/7})\,d\eta = \delta_2\left[\eta - \frac{7}{8}\eta^{8/7}\right]_0^1 = \frac{\delta_2}{8}$$

$$\delta_2^* = \frac{0.0759\ m}{8} = 0.00949\ m$$

$$\frac{U_2}{U_0} = \frac{h}{h - 2\delta_2^*} = \frac{0.3\ m}{0.3\ m - 2(0.00949\ m)} = 1.0675$$

$$p_1 - p_2 = \frac{\rho U_0^2}{2}\left[\left(\frac{U_2}{U_0}\right)^2 - 1\right] = \frac{1}{2} \times \frac{1.23\ kg}{m^3} \times \frac{(25)^2\ m^2}{sec^2}[(1.0675)^2 - 1]\frac{N \cdot sec^2}{kg \cdot m}$$

$$p_1 - p_2 = 53.6\ Pa \qquad\qquad\qquad p_1 - p_2$$

{This example illustrates the use of the displacement thickness concept.}

8-12   MOMENTUM INTEGRAL EQUATION

Since all real fluids are viscous, a boundary layer will develop on all solid surfaces when there is flow over a body. Thus adjacent to any solid surface there is a region in which viscous forces are important. The natural question that arises is "How large is this viscous region?" In this section we shall develop an analysis that will enable us to determine a good approximation for the boundary-layer thickness as a function of distance along a body. Rather than solve the complete differential equations of motion for the boundary layer, we shall again apply the integral equations to a differential control volume. Our aim is to develop an equation that will enable us to predict (at least approximately) the manner in which the boundary layer grows as a function of distance along a body. We shall derive a relation that may be applied to both laminar and turbulent flow.

8-12.1   APPLICATION OF THE BASIC EQUATIONS

Consider the incompressible, steady flow over a solid surface. The boundary-layer thickness, δ, grows in some manner with increasing distance, x. For our analysis we choose a differential control volume of length, dx, width, dz, and height, $\delta(x)$, as shown in Fig. 8.23.

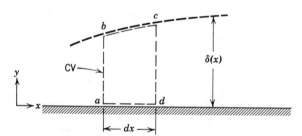

Fig. 8.23 Differential control volume in a boundary layer.

We wish to determine the boundary-layer thickness, δ, as a function of x. In looking at the differential control volume, $abcd$, we see that there will be mass flow across surfaces ab and cd. What about surface bc? Will there be a mass flow across this surface? In our earlier discussion of boundary layers (Chapter 2), we found that the edge of the boundary layer is not a streamline and, hence, there will be mass flow across the surface bc. Since the control surface ad is adjacent to a solid boundary, there will not be a flow across ad. Before considering the forces acting on the control volume and the momentum fluxes through the control surface, let us apply the continuity equation to determine the mass flux through each portion of the control surface.

a. Continuity Equation

Basic equation:

$$0 = \underset{\substack{\nearrow \\ = 0(1)}}{\frac{\partial}{\partial t}} \int_{CV} \rho \, dV + \int_{CS} \rho \vec{V} \cdot d\vec{A} \qquad (4.13)$$

Assumptions: (1) Steady flow
(2) Two-dimensional flow

Then

$$0 = \int_{CS} \rho \vec{V} \cdot d\vec{A} = \dot{m}_{ab} + \dot{m}_{bc} + \dot{m}_{cd}$$

or

$$\dot{m}_{bc} = -\dot{m}_{ab} - \dot{m}_{cd}$$

Now let us evaluate these terms:

Surface   *Mass Flux*

ab   Surface *ab* is located at $x = x$. Since the flow is two-dimensional (no variation with z) the mass flux through *ab* is

$$\dot{m}_{ab} = -\left\{ \int_0^\delta \rho u \, dy \right\} dz$$

cd   Surface *cd* is located at $x + dx$. Expanding $\dot{m}$ in a Taylor series about the location $x = x$, we obtain

$$\dot{m}_{x+dx} = \dot{m}_x + \frac{\partial \dot{m}}{\partial x}\bigg]_x dx$$

and hence

$$\dot{m}_{cd} = \left\{ \int_0^\delta \rho u \, dy + \frac{\partial}{\partial x}\left[\int_0^\delta \rho u \, dy \right] dx \right\} dz$$

bc   Thus, for surface *bc*, we obtain

$$\dot{m}_{bc} = -\left\{ \frac{\partial}{\partial x}\left[\int_0^\delta \rho u \, dy \right] dx \right\} dz$$

Now let us consider the momentum fluxes and forces associated with control volume *abcd*. These are related by the momentum equation.

b. Momentum Equation

Apply the x component of the momentum equation to the control volume $abcd$:

Basic equation:

$$\overset{= 0(3)}{} \quad \overset{= 0(1)}{}$$

$$F_{S_x} + F_{B_x} = \frac{\partial}{\partial t} \int_{CV} u\rho \, dV + \int_{CS} u\rho \vec{V} \cdot d\vec{A} \qquad (4.19a)$$

Assumption: (3) $F_{B_x} = 0$

Then

$$F_{S_x} = mf_{ab} + mf_{bc} + mf_{cd}$$

where mf is the momentum flux.

In applying this equation to our differential control volume, $abcd$, we must obtain expressions for the momentum flux through the control surface and also the surface forces acting on the control volume. Let us consider the momentum flux first and again consider each segment of the control surface.

Surface	Momentum Flux (mf)
ab	Surface ab is located at $x = x$. Since the flow is two-dimensional, the x momentum flux through ab is

$$mf_{ab} = -\left\{ \int_0^\delta u\rho u \, dy \right\} dz$$

cd	Surface cd is located at $x + dx$. Expanding the x momentum flux (mf) in a Taylor series about the location $x = x$, we obtain

$$mf_{x+dx} = mf_x + \frac{\partial mf}{\partial x}\bigg]_x dx$$

or

$$mf_{cd} = \left\{ \int_0^\delta u\rho u \, dy + \frac{\partial}{\partial x}\left[\int_0^\delta u\rho u \, dy \right] dx \right\} dz$$

bc	Since the mass crossing surface bc has velocity U, the momentum flux across bc is given by

$$mf_{bc} = U\dot{m}_{bc}$$

$$mf_{bc} = -U \left\{ \frac{\partial}{\partial x}\left[\int_0^\delta \rho u \, dy \right] dx \right\} dz$$

From the above we can evaluate the momentum flux through the control surface,

$$\int_{CS} u\rho \vec{V} \cdot d\vec{A} = -\left\{\int_0^\delta u\rho u\, dy\right\}dz + \left\{\int_0^\delta u\rho u\, dy\right\}dz$$

$$+ \left\{\frac{\partial}{\partial x}\left[\int_0^\delta u\rho u\, dy\right]dx\right\}dz - U\left\{\frac{\partial}{\partial x}\left[\int_0^\delta \rho u\, dy\right]dx\right\}dz$$

Collecting terms,

$$\int_{CS} u\rho \vec{V} \cdot d\vec{A} = \left\{\frac{\partial}{\partial x}\left[\int_0^\delta u\rho u\, dy\right]dx - U\frac{\partial}{\partial x}\left[\int_0^\delta \rho u\, dy\right]dx\right\}dz$$

Now that we have a suitable expression for the x momentum flux through the control surface, let us turn our attention to the surface forces acting on the control volume in the x direction. For convenience the differential control volume has been redrawn in Fig. 8.24. In analyzing the x component forces acting on the control volume we recognize that normal forces act on three surfaces of the control surface. In addition, there is clearly a shear force acting on surface ad. Since the velocity gradient goes to zero at the edge of the boundary layer, there is no shear force acting along the surface bc.

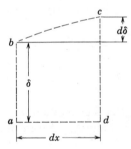

Fig. 8.24 Differential control volume.

Surface   *Force*

ab If the pressure at $x = x$ is p, then the force acting on surface ab is given by

$$F_{ab} = p\delta\, dz$$

(The boundary layer is very thin; its thickness has been greatly exaggerated in all the sketches we have made. Because it is thin, pressure variations in the y direction may be neglected.)

cd

Expanding the pressure, p, in a Taylor series, the pressure at $x + dx$ is given by

$$p_{x+dx} = p + \frac{\partial p}{\partial x}\bigg]_x dx$$

The force on surface *cd* is then given by

$$F_{cd} = -\left(p + \frac{\partial p}{\partial x}\bigg]_x dx\right)(\delta + d\delta)\, dz$$

bc

The average pressure acting over the surface *bc* is

$$p + \frac{1}{2}\frac{\partial p}{\partial x}\bigg]_x dx$$

Then the x component of the normal force acting over *bc* is given by

$$F_{bc} = \left(p + \frac{1}{2}\frac{\partial p}{\partial x}\bigg]_x dx\right) d\delta\, dz$$

ad

The shear force acting on *ad* is given by

$$F_{ad} = -\tau_w\, dx\, dz$$

Summing the x component of each force acting on the control volume, we obtain

$$F_{S_x} = \left\{ -\frac{\partial p}{\partial x}\delta\, dx - \overset{\simeq\,0}{\cancel{\frac{1}{2}\frac{\partial p}{\partial x} dx\, d\delta}} - \tau_w\, dx \right\} dz$$

where we note that $dx\, d\delta \ll \delta\, dx$, and so neglect the second term.

Substituting the expressions for $\int_{CS} u\rho \vec{V}\cdot d\vec{A}$ and F_{S_x} into the momentum equation, we obtain

$$\left\{ -\frac{\partial p}{\partial x}\delta\, dx - \tau_w\, dx \right\} dz = \left\{ \frac{\partial}{\partial x}\left[\int_0^\delta u\rho u\, dy\right] dx - U\frac{\partial}{\partial x}\left[\int_0^\delta \rho u\, dy\right] dx \right\} dz$$

Dividing this equation by $dx\, dz$ gives

$$-\delta\frac{\partial p}{\partial x} - \tau_w = \frac{\partial}{\partial x}\int_0^\delta u\rho u\, dy - U\frac{\partial}{\partial x}\int_0^\delta \rho u\, dy \qquad (8.41)$$

Equation 8.41 is a "momentum integral" equation that gives a relation between the x components of the forces acting in a boundary layer and the momentum flux. In order to use this equation to estimate the boundary-layer thickness as a function of x, we must:

1. Determine a first approximation to the pressure gradient, $\partial p/\partial x$. This is determined from inviscid flow theory (the pressure gradient that would exist in the absence of a boundary layer). The pressure in the boundary layer is related to the free stream velocity, U, using the Bernoulli equation.
2. Assume some reasonable velocity profile shape inside the boundary layer.
3. Relate the wall shear stress to the velocity field.

8-12.2 SPECIAL CASE—FLOW OVER A FLAT PLATE

To illustrate the method of using Eq. 8.41, we consider the special case of flow over a flat plate, for which $U = $ constant. From Bernoulli's equation we see that for this case, $p = $ constant, and thus $\partial p/\partial x = 0$.

The momentum integral equation then reduces to

$$\tau_w = U \frac{\partial}{\partial x} \int_0^\delta \rho u \, dy - \frac{\partial}{\partial x} \int_0^\delta u \rho u \, dy$$

Since $U = $ constant, we can write

$$\tau_w = \frac{\partial}{\partial x} \int_0^\delta U \rho u \, dy - \frac{\partial}{\partial x} \int_0^\delta u \rho u \, dy$$

The right side can be combined into a single term

$$\tau_w = \frac{\partial}{\partial x} \int_0^\delta \rho u (U - u) \, dy$$

or

$$\tau_w = U^2 \frac{\partial}{\partial x} \int_0^\delta \rho \frac{u}{U} \left(1 - \frac{u}{U} \right) dy$$

For incompressible flow,

$$\tau_w = \rho U^2 \frac{\partial}{\partial x} \int_0^\delta \frac{u}{U} \left(1 - \frac{u}{U} \right) dy \qquad (8.42)$$

The velocity distribution, u/U, in the boundary layer is normally specified as a function of y/δ. (Note that u/U is dimensionless and δ is a function of x only.) Consequently, it is convenient to change the variable of integration from y to y/δ. Defining

$$\eta = \frac{y}{\delta}$$

then

$$dy = \delta \, d\eta$$

and the momentum integral equation for zero pressure gradient is written

$$\tau_w = \rho U^2 \frac{d\delta}{dx} \int_0^1 \frac{u}{U}\left(1 - \frac{u}{U}\right) d\eta \qquad (8.43)$$

Equation 8.43 (or Eq. 8.42) was obtained by applying the basic equations (i.e. continuity and momentum) to a differential control volume. Reviewing the assumptions we made in the derivation, we see that these equations are subject to the restrictions:

1. Steady flow.
2. Incompressible flow.
3. Two-dimensional flow.
4. No body forces.
5. Flat plate flow (i.e. $\partial p / \partial x = 0$).

However, we have not made any specific assumption relating the wall shear stress, τ_w, to the velocity field. Thus, Eqs. 8.42 and 8.43 are valid for either a laminar or turbulent boundary-layer flow on a flat plate.

8-13 USE OF THE MOMENTUM INTEGRAL EQUATION FOR ZERO PRESSURE GRADIENT FLOW

The momentum integral equation for zero pressure gradient flow is given by

$$\tau_w = \rho U^2 \frac{\partial}{\partial x} \int_0^\delta \frac{u}{U}\left(1 - \frac{u}{U}\right) dy \qquad (8.42)$$

or

$$\tau_w = \rho U^2 \frac{d\delta}{dx} \int_0^1 \frac{u}{U}\left(1 - \frac{u}{U}\right) d\eta, \quad \text{where } \eta = \frac{y}{\delta} \qquad (8.43)$$

We wish to solve this equation for the boundary-layer thickness as a function of x. In order to do this, we must:

1. Assume a velocity distribution in the boundary layer, that is, a functional relationship

$$\frac{u}{U} = f\left(\frac{y}{\delta}\right)$$

(a) The assumed velocity distribution should satisfy certain physical boundary conditions:

$$\text{at} \quad y = 0, \quad u = 0$$

$$\text{at} \quad y = \delta, \quad u = U$$

$$\text{at} \quad y = \delta, \quad \frac{\partial u}{\partial y} = 0$$

(b) Note that once the velocity distribution has been assumed, then the numerical value of the integral in Eq. 8.43 is simply

$$\int_0^1 \frac{u}{U}\left(1 - \frac{u}{U}\right) d\eta = a \text{ constant} = \beta$$

and the momentum integral equation becomes

$$\tau_w = \rho U^2 \frac{d\delta}{dx} \beta$$

2. Obtain an expression for τ_w in terms of δ. This will then permit us to solve for $\delta(x)$.

8-13.1 LAMINAR FLOW

For laminar flow over a flat plate, a reasonable assumption for the velocity profile is a polynomial in y:

$$u = a + by + cy^2$$

The physical boundary conditions are:

$$\text{at} \quad y = 0, \qquad u = 0$$
$$\text{at} \quad y = \delta, \qquad u = U$$

$$\text{at} \quad y = \delta, \qquad \frac{\partial u}{\partial y} = 0$$

Evaluating the constants, a, b, and c gives

$$\frac{u}{U} = 2\left(\frac{y}{\delta}\right) - \left(\frac{y}{\delta}\right)^2 = 2\eta - \eta^2 \qquad (8.44)$$

The wall shear stress for laminar flow is given by

$$\tau_w = \mu \left.\frac{\partial u}{\partial y}\right)_{y=0}$$

Substituting the assumed velocity profile, Eq. 8.44, into this expression for τ_w gives

$$\tau_w = \mu \left.\frac{\partial u}{\partial y}\right]_{y=0} = \mu \left.\frac{U\partial(u/U)}{\delta\partial(y/\delta)}\right]_{y/\delta=0} = \left.\frac{\mu U}{\delta}\frac{d(u/U)}{d\eta}\right]_{\eta=0}$$

or

$$\tau_w = \left.\frac{\mu U}{\delta}\frac{d}{d\eta}(2\eta - \eta^2)\right]_{\eta=0} = \frac{2\mu U}{\delta}$$

We are now in a position to apply the momentum integral equation

$$\tau_w = \rho U^2 \frac{d\delta}{dx} \int_0^1 \frac{u}{U}\left(1 - \frac{u}{U}\right) d\eta \qquad (8.43)$$

Substituting for τ_w and u/U, we obtain

$$\frac{2\mu U}{\delta} = \rho U^2 \frac{d\delta}{dx} \int_0^1 (2\eta - \eta^2)(1 - 2\eta + \eta^2)\, d\eta$$

or

$$\frac{2\mu U}{\delta\rho U^2} = \frac{d\delta}{dx} \int_0^1 (2\eta - 5\eta^2 + 4\eta^3 - \eta^4)\, d\eta$$

Integrating and substituting limits yields

$$\frac{2\mu}{\delta\rho U} = \frac{2}{15}\frac{d\delta}{dx}$$

or

$$\delta\, d\delta = \frac{15\mu}{\rho U}\, dx$$

which is a differential equation for δ. Integrating again

$$\frac{\delta^2}{2} = \frac{15\mu}{\rho U}x + c$$

If it is assumed that $\delta = 0$ at $x = 0$, then $c = 0$ and thus

$$\delta = \sqrt{\frac{30\mu}{\rho U}}\, x$$

or

$$\frac{\delta}{x} = \sqrt{\frac{30\mu}{\rho U x}} = \frac{5.48}{\sqrt{Re_x}} \qquad (8.45)$$

Equation 8.45 shows that the ratio of laminar boundary-layer thickness to distance along a flat plate varies inversely with the square root of length Reynolds number. It has the same form as an exact solution derived from the complete differential equations of motion by H. Blasius in 1908.[10] Remarkably, Eq. 8.45 is only in error (the constant is too large) by about 10 percent compared to the exact solution (Ref. 13).

[10] H. Blasius, "Boundary layers in fluids with small friction (in German)", Z. Math. u. Phys., **56**, 1, 1908. An English translation is available as NACA Technical Memorandum No. 1256.

Once we know the boundary-layer thickness, all details of the flow may be determined. The wall shear stress, or "skin friction," coefficient is defined as

$$C_f \equiv \frac{\tau_w}{\frac{1}{2}\rho U^2}$$

Substituting from the velocity profile and Eq. 8.45 gives

$$C_f = \frac{\tau_w}{\frac{1}{2}\rho U^2} = \frac{2\mu(U/\delta)}{\frac{1}{2}\rho U^2} = \frac{4\mu}{\rho U \delta} = 4\frac{\mu}{\rho U x}\frac{x}{\delta} = 4\frac{1}{Re_x}\frac{\sqrt{Re_x}}{5.48}$$

Finally,

$$C_f = \frac{0.730}{\sqrt{Re_x}} \qquad (8.46)$$

With knowledge of τ_w, the viscous drag on the surface can be evaluated by integration over the area of the flat plate.

From Eq. 8.45 we can also compute the thickness of the laminar boundary layer at transition (Section 8-10). At $Re_x = 3 \times 10^6$, $x = 0.43$ m for air at standard conditions, with $U = 100$ m/sec. Thus

$$\frac{\delta}{x} = \frac{5.48}{\sqrt{Re_x}} = \frac{5.48}{\sqrt{3 \times 10^6}} = 0.00316; \qquad \delta = 1.36 \text{ mm}$$

At $Re_x = 4 \times 10^6$, $x = 0.58$ m and

$$\frac{\delta}{x} = \frac{5.48}{\sqrt{Re_x}} = \frac{5.48}{\sqrt{4 \times 10^6}} = 0.00274; \qquad \delta = 1.59 \text{ mm}$$

In both of these cases, we have seen that the boundary-layer thickness was less than 1 percent of x. These calculations confirm that viscous effects are confined to a very narrow layer near the surface of a body.

8-13.2 TURBULENT FLOW

For turbulent flow over a flat plate, a reasonable assumption for the velocity profile is

$$\frac{u}{U} = \left(\frac{y}{\delta}\right)^{1/7} = \eta^{1/7}$$

However, this profile does not hold in the immediate vicinity of the wall, since at the wall it predicts $du/dy = \infty$. Consequently, we cannot use this profile in the definition of τ_w to obtain an expression for τ_w in terms of δ as we did for laminar flow. For turbulent flow we will use the experimentally determined result (Ref. 13):

$$\tau_w = 0.0225 \rho U^2 \left(\frac{\nu}{U\delta}\right)^{1/4} \tag{8.47}$$

We are now in a position to apply the momentum integral equation

$$\tau_w = \rho U^2 \frac{d\delta}{dx} \int_0^1 \frac{u}{U}\left(1 - \frac{u}{U}\right) d\eta \tag{8.43}$$

Substituting for τ_w and u/U and integrating, we obtain

$$0.0225 \left(\frac{\nu}{U\delta}\right)^{1/4} = \frac{d\delta}{dx} \int_0^1 \eta^{1/7}(1 - \eta^{1/7}) \, d\eta = \frac{7}{72}\frac{d\delta}{dx}$$

Thus we obtain a differential equation for δ:

$$\delta^{1/4} \, d\delta = 0.231 \left(\frac{\nu}{U}\right)^{1/4} dx$$

Integrating,

$$\frac{4}{5} \delta^{5/4} = 0.231 \left(\frac{\nu}{U}\right)^{1/4} x + c$$

If it is assumed that $\delta \simeq 0$ at $x = 0$ (this is equivalent to assuming turbulent flow from the leading edge), then $c = 0$ and

$$\delta = 0.370 \left(\frac{\nu}{U}\right)^{1/5} x^{4/5}$$

or

$$\frac{\delta}{x} = 0.370 \left(\frac{\nu}{Ux}\right)^{1/5} = \frac{0.370}{Re_x^{1/5}} \tag{8.48}$$

Using Eq. 8.47, we obtain the skin friction coefficient in terms of δ:

$$C_f = \frac{\tau_w}{\frac{1}{2}\rho U^2} = 0.0450 \left(\frac{\nu}{U\delta}\right)^{1/4}$$

Substituting for δ, we obtain

$$C_f = \frac{\tau_w}{\frac{1}{2}\rho U^2} = \frac{0.0577}{Re_x^{1/5}} \tag{8.49}$$

Experiments show that Eq. 8.49 predicts turbulent skin friction on a flat plate within about 3 percent for $5 \times 10^5 < Re_x < 10^7$. This agreement is remarkable in view of the approximate nature of our analysis.

In concluding this section, let us point out that use of the momentum integral equation is an approximate technique for predicting boundary-layer development. The agreement we have obtained with experimental results shows that it is an effective method, which gives us considerable insight into the general behavior of boundary layers.

Example 8.13

Consider two-dimensional laminar flow along a flat plate. The velocity profile in the boundary layer is assumed to be sinusoidal, that is,

$$\frac{u}{U} = \sin\left(\frac{\pi}{2}\frac{y}{\delta}\right)$$

Find expressions for:

(a) The rate of growth of δ as a function of x.
(b) The displacement thickness, δ^*, as a function of x.
(c) The total friction force on a plate of length, L, and width, b.

Example Problem 8.13

GIVEN:

Two-dimensional, laminar flow along a flat plate. The boundary-layer velocity profile is

$$\frac{u}{U} = \sin\left(\frac{\pi}{2}\frac{y}{\delta}\right) \qquad \text{for } 0 \le y \le \delta$$

and

$$\frac{u}{U} = 1 \qquad \text{for } y > \delta$$

FIND:

(a) $\delta(x)$.
(b) $\delta^*(x)$.
(c) Total friction force on a plate of length, L, and width, b.

SOLUTION:

For flat plate flow, $\partial p/\partial x = 0$ and

$$\tau_w = \rho U^2 \frac{d\delta}{dx} \int_0^1 \frac{u}{U}\left(1 - \frac{u}{U}\right)d\eta \qquad (8.43)$$

Substituting $\dfrac{u}{U} = \sin\dfrac{\pi}{2}\eta$, then

$$\tau_w = \rho U^2 \frac{d\delta}{dx} \int_0^1 \sin\frac{\pi}{2}\eta \left(1 - \sin\frac{\pi}{2}\eta\right) d\eta$$

$$= \rho U^2 \frac{d\delta}{dx} \int_0^1 \left(\sin\frac{\pi}{2}\eta - \sin^2\frac{\pi}{2}\eta\right) d\eta$$

$$= \rho U^2 \frac{d\delta}{dx}\frac{2}{\pi}\left[-\cos\frac{\pi}{2}\eta - \frac{1}{2}\frac{\pi}{2}\eta + \frac{1}{4}\sin\pi\eta\right]_0^1$$

$$= \rho U^2 \frac{d\delta}{dx}\frac{2}{\pi}\left[0 + 1 - \frac{\pi}{4} + 0 + 0 - 0\right]$$

$$\tau_w = 0.137\rho U^2 \frac{d\delta}{dx} = \beta\rho U^2 \frac{d\delta}{dx}; \ \beta = 0.137$$

Now

$$\tau_w = \mu\left.\frac{\partial u}{\partial y}\right)_{y=0} = \mu\frac{U}{\delta}\left.\frac{\partial(u/U)}{\partial(y/\delta)}\right]_{y=0} = \mu\frac{U}{\delta}\frac{\pi}{2}\left.\cos\frac{\pi}{2}\eta\right]_{\eta=0} = \frac{\pi\mu U}{2\delta}$$

Therefore,

$$\tau_w = \frac{\pi\mu U}{2\delta} = 0.137\rho U^2 \frac{d\delta}{dx}$$

Separating variables gives

$$0.0872\frac{\rho U}{\mu}\delta\,d\delta = dx$$

Integrating,

$$0.0872\frac{\rho U}{\mu}\frac{\delta^2}{2} = x + c$$

But $c = 0$, since $\delta = 0$ at $x = 0$, so

$$\delta = \sqrt{\frac{2}{0.0872}}\sqrt{\frac{x\mu}{\rho U}}$$

or

$$\frac{\delta}{x} = 4.79\sqrt{\frac{\mu}{\rho U x}} = \frac{4.79}{\sqrt{Re_x}} \qquad\qquad \delta(x)$$

The displacement thickness, δ^*, is given by

$$\delta^* = \delta\int_0^1 \left(1 - \frac{u}{U}\right) d\eta$$

$$= \delta\int_0^1 \left(1 - \sin\frac{\pi}{2}\eta\right) d\eta = \delta\left[\eta + \frac{2}{\pi}\cos\frac{\pi}{2}\eta\right]_0^1$$

$$\delta^* = \delta\left[1 - 0 + 0 - \frac{2}{\pi}\right] = \delta\left[1 - \frac{2}{\pi}\right]$$

Since, from part (a),

$$\frac{\delta}{x} = \frac{4.79}{\sqrt{Re_x}}$$

then

$$\frac{\delta^*}{x} = \left(1 - \frac{2}{\pi}\right)\frac{4.79}{\sqrt{Re_x}} = \frac{1.74}{\sqrt{Re_x}} \qquad\qquad \delta^*(x)$$

The total friction force on one side of the plate is given by

$$F = \int_{A_p} \tau_w \, dA$$

Since $dA = b \, dx$ and $0 \le x \le L$, then

$$F = \int_0^L \tau_w b \, dx$$

Now

$$\tau_w = \rho U^2 \frac{d\delta}{dx} \int_0^1 \frac{u}{U}\left(1 - \frac{u}{U}\right) d\eta = \rho U^2 \frac{d\delta}{dx} \beta$$

and

$$F = \int_0^L \tau_w b \, dx = \int_0^L \rho U^2 \beta \frac{d\delta}{dx} b \, dx = \rho U^2 b \beta \int_0^L \frac{d\delta}{dx} \, dx = \rho U^2 \beta b \delta_L$$

From part (a), $\beta = 0.137$ and $\delta_L = \dfrac{4.79L}{\sqrt{Re_L}}$, so

$$F = \frac{0.656\rho U^2 bL}{\sqrt{Re_L}} \qquad\qquad F$$

{ This problem illustrates the application of the momentum integral equation to a flat plate, laminar, boundary-layer flow. }

8-14 PRESSURE GRADIENTS IN BOUNDARY-LAYER FLOW

We have restricted our discussion of boundary-layer flows to flow over a flat plate, that is, to flow in which the pressure gradient is zero. The momentum integral equation for this case was given as

$$\tau_w = \rho U^2 \frac{\partial}{\partial x} \int_0^\delta \frac{u}{U}\left(1 - \frac{u}{U}\right) dy \qquad (8.42)$$

Recall that in deriving this equation, no assumption was made regarding the flow in the boundary layer; the equation is valid for both laminar and turbulent boundary layers.

Let us again consider the flow over a flat plate (Fig. 8.25). Equation 8.42 indicates that the wall shear stress is balanced by a decrease in fluid momentum. Thus the velocity profiles change as we move along the plate. The boundary-layer thickness continues to increase and the fluid close to the wall is continually being slowed down (i.e. losing momentum).

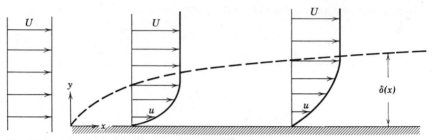

Fig. 8.25 Laminar boundary layer on a flat plate.

One question of interest is, "Will the fluid close to the wall ever be brought to rest?" That is, for the case of $\partial p/\partial x = 0$, is it possible that $\partial u/\partial y)_{y=0} = 0$?[11]

If we consider the wall shear stress distributions obtained for flat plates, we found that for laminar flow

$$\frac{\tau_w(x)}{\rho U^2} = \frac{\text{constant}}{\sqrt{Re_x}}$$

and for turbulent flow

$$\frac{\tau_w(x)}{\rho U^2} = \frac{\text{constant}}{(Re_x)^{1/5}}$$

Recalling that $\tau_w = \mu(\partial u/\partial y)_{y=0}$, we can then say that for any finite length plate, $\partial u/\partial y)_{y=0}$ will never be zero. The point on a solid boundary at which $\partial u/\partial y = 0$ is defined as the point of *separation*. Consequently, we can conclude that for $\partial p/\partial x = 0$, the flow will not separate, that is, the fluid layer in the neighborhood of a solid surface cannot be brought to zero velocity.

---

[11] Note that if $\dfrac{\partial u}{\partial y}\Big)_{y=0} = 0$, then the fluid layer near the wall will have a zero velocity, since

$$u\Big)_{0+dy} = u_0 + \frac{\partial u}{\partial y}\Big)_{y=0} dy$$

and $u_0 = 0$ from the no-slip condition.

8-14.1 EFFECT OF PRESSURE GRADIENT ON FLOW: SEPARATION

The pressure gradient is said to be adverse if the pressure increases in the direction of flow, i.e. if $\partial p/\partial x > 0$. When $\partial p/\partial x < 0$, that is, when the pressure decreases in the direction of flow, the pressure gradient is said to be favorable.

Consider the flow through a channel of variable cross section as shown in Fig. 8.26. To simplify our discussion, we consider the flow along the straight wall.

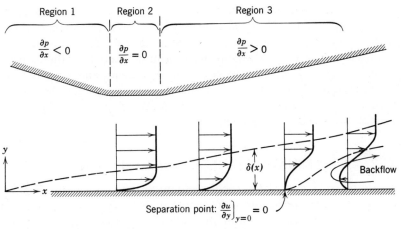

Fig. 8.26 Boundary-layer flow with pressure gradient.

If we consider the forces acting on a fluid particle close to the solid boundary, we see that there is a net retarding shear force on the particle no matter what the sign of the pressure gradient. For $\partial p/\partial x = 0$, the result is a decrease in momentum, but as we have already shown it is not sufficient to bring the particle to rest. Since $\partial p/\partial x < 0$ in Region 1, the pressure behind the particle (aiding its motion) is greater than that opposing its motion; hence the particle is "sliding down a pressure hill," without danger of being slowed to zero velocity. However, in attempting to flow through Region 3, the particle encounters an adverse pressure gradient, $\partial p/\partial x > 0$, and the particle must "climb a pressure hill." The fluid particle could be brought to rest, thus causing the neighboring fluid to be deflected away from the boundary; when this occurs, the flow is said to separate from the surface. The point on the boundary where $\partial u/\partial y)_{y=0} = 0$ is called the point of separation. Just downstream from the point of separation, the flow direction in the separated region is actually opposite to the main flow direction. The low energy fluid in the separated region is forced back upstream by the increased pressure downstream.

Thus we see that an adverse pressure gradient, $\partial p / \partial x > 0$, is a necessary condition for separation. Does this mean that if $\partial p / \partial x > 0$, we will have separation? No, it does not. We have not shown that $\partial p / \partial x > 0$ will always lead to separation, but rather we have reasoned that separation cannot occur unless $\partial p / \partial x > 0$. This conclusion can be shown rigorously using the complete differential equations of motion for boundary-layer flow (Reference 13, p. 122).

The turbulent boundary layer has a much fuller, or more blunt velocity profile than the laminar boundary layer, Fig. 8.21. Therefore, the turbulent velocity profile at the same freestream speed contains much more momentum. Separation occurs when the momentum of fluid layers near the surface is reduced to zero by the combined action of pressure and viscous forces. Because the turbulent layer has more momentum, it is better able to resist separation in an adverse pressure gradient. We shall discuss some consequences of this behavior in Section 8-15.3.

8-14.2 DETERMINATION OF PRESSURE GRADIENT

Since the pressure gradient has such a pronounced effect on the flow behavior, it is important that we be able to determine analytically the magnitude of the pressure gradient for a given flow situation. In fact, to calculate the growth and to predict the behavior of a boundary layer, we must have an expression for the pressure gradient.

For both external and internal flows we can obtain a first approximation for the pressure gradient from ideal flow theory, that is, from the pressure variation in the flow of a frictionless (inviscid) fluid under the same conditions. As pointed out in Chapter 5, for frictionless irrotational flow (i.e. potential flow), the stream function, ψ, and the velocity potential, ϕ, satisfy Laplace's equation. These together with the Euler equations provide the basis for determining the pressure distribution. However, a detailed discussion of potential flow is beyond the scope of this text.[12]

---

[12] An introduction to potential flow is presented in many fluid mechanics texts, for example: A. G. Hansen, *Fluid Mechanics* (New York: Wiley, 1967); W. H. Li and S. H. Lam, *Principles of Fluid Mechanics* (Reading, Mass.: Addison-Wesley, 1964); R. H. Sabersky, A. J. Acosta, and E. G. Hauptmann, *Fluid Flow*, 2nd ed. (New York: Macmillan, 1971).

Anyone interested in a detailed study of potential flow theory may find the following books of interest: V. L. Streeter, *Fluid Dynamics* (New York: McGraw-Hill, 1948); H. R. Vallentine, *Applied Hydrodynamics* (London: Butterworths, 1959); L. M. Milne-Thomson, *Theoretical Hydrodynamics*, 4th ed. (New York: Macmillan, 1960); J. M. Robertson, *Hydrodynamics in Theory and Application* (Englewood Cliffs, N.J.: Prentice-Hall, 1965).

Since we have had some experience with internal flows, let us consider the incompressible flow through the plane-wall diffuser shown in Fig. 8.27. We are interested in determining the pressure distribution along the diffuser, and as a first approximation we will use the pressure distribution obtained for the flow of an inviscid fluid.

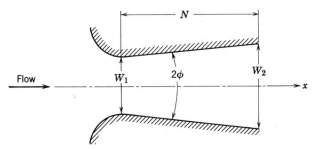

Fig. 8.27 Plane-wall diffuser.

Let us apply the x component of the Euler equations to this flow. Basic equation:

$$\overset{=\,0(2)}{\cancel{\rho B_x}} - \frac{\partial p}{\partial x} = \rho\left[\overset{=\,0(3)}{\cancel{\frac{\partial u}{\partial t}}} + u\frac{\partial u}{\partial x} + v\overset{=\,0(4)}{\cancel{\frac{\partial u}{\partial y}}} + w\overset{=\,0(4)}{\cancel{\frac{\partial u}{\partial z}}}\right] \qquad (6.1a)$$

Assumptions: (1) Inviscid flow
(2) $B_x = 0$
(3) Steady flow
(4) One-dimensional flow: $u = V(x)$, $\partial/\partial y = \partial/\partial z = 0$
(5) Incompressible flow

Then

$$\frac{\partial p}{\partial x} = -\rho V\frac{\partial V}{\partial x}$$

or

$$\frac{dp}{dx} = -\rho V\frac{dV}{dx} \qquad (8.50)$$

since all properties are functions of x only. Under these assumptions, the continuity equation becomes

$$0 = \overset{=\,0(3)}{\cancel{\frac{\partial}{\partial t}\int_{cv}\rho\,d\Psi}} + \int_{cs}\rho\vec{V}\cdot d\vec{A} \qquad (4.13)$$

or

$$\rho VA = \dot{m} = \text{constant}$$

Then

$$VA = \frac{\dot{m}}{\rho} = \text{constant} \qquad (8.51)$$

Differentiating with respect to x,

$$V\frac{dA}{dx} + A\frac{dV}{dx} = 0$$

or

$$V\frac{dV}{dx} = -\frac{V^2}{A}\frac{dA}{dx} \qquad (8.52)$$

Substituting Eq. 8.52 into Eq. 8.50, we obtain

$$\frac{dp}{dx} = \rho\frac{V^2}{A}\frac{dA}{dx}$$

Replacing V from Eq. 8.51 gives

$$\frac{dp}{dx} = \frac{\dot{m}^2}{\rho A^3}\frac{dA}{dx} \qquad (8.53)$$

For a two-dimensional, plane-wall diffuser of uniform depth, b, the cross-sectional area varies linearly with x. From the geometry of Fig. 8.27,

$$A = Wb = (W_1 + 2x\tan\phi)b$$

Hence

$$\frac{dA}{dx} = 2b\tan\phi$$

which is constant for any given diffuser geometry. Then from Eq. 8.53, the pressure gradient is porportional to $1/A^3$ as shown in Fig. 8.28. Thus the pressure rises steeply near the diffuser inlet but less steeply near the outlet section.

As discussed in Section 8-8.2, it is convenient to express the pressure rise through a diffuser in terms of a dimensionless pressure recovery coefficient, C_p, where

$$C_p = \frac{p_2 - p_1}{\frac{1}{2}\rho\overline{V}_1^2} \qquad (8.36)$$

Applying the Bernoulli equation to the diffuser flow of Fig. 8.27,

$$\frac{p_1}{\rho} + \frac{V_1^2}{2} = \frac{p_2}{\rho} + \frac{V_2^2}{2}$$

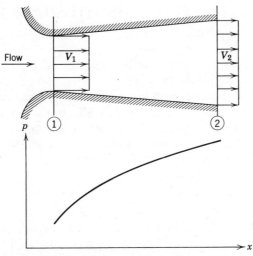

Fig. 8.28 Pressure distribution for steady, one-dimensional inviscid incompressible flow in a plane-wall diffuser.

or

$$p_2 - p_1 = \frac{\rho V_1^2}{2} - \frac{\rho V_2^2}{2} = \frac{\rho V_1^2}{2}\left[1 - \left(\frac{V_2}{V_1}\right)^2\right]$$

Thus for ideal flow,

$$C_{p_i} = 1 - \left(\frac{V_2}{V_1}\right)^2$$

The velocity ratio can be eliminated, using Eq. 8.51, to obtain

$$C_{p_i} = 1 - \left(\frac{A_1}{A_2}\right)^2 = 1 - \frac{1}{AR^2} \tag{8.54}$$

where AR is the area ratio of the diffuser. As Eq. 8.54 shows, the ideal pressure recovery coefficient, designated C_{p_i}, for steady, incompressible, inviscid flow through a diffuser is solely a function of the diffuser geometry. However, in practice, frictional effects are not negligible. The actual pressure recovery coefficient, C_p, is always less than the ideal value, C_{p_i}. The adverse pressure gradient may cause flow separation and severely distorted outflow.[13] Experimental data, such as those shown in Fig. 8.17, must be used for design.

[13] Flow behavior in plane-wall diffusers is a function of diffuser geometry as shown in the film, *Flow Visualization*, S. J. Kline, principal.

PART E. FLUID FLOW ABOUT IMMERSED BODIES

Whenever there is relative motion between a solid body and the fluid in which it is immersed, the body experiences a net force, $\vec{F}$, due to the action of the fluid. In general, the infinitesimal force, $d\vec{F}$, acting on an element of surface area will be neither normal nor parallel to the element. This can be seen clearly when one considers the nature of the surface forces that contribute to the force, $\vec{F}$. If the body is moving through a viscous fluid, then both shear and pressure forces act on the body, that is,

$$\vec{F} = \int_{\text{body surface}} d\vec{F} = \int_{\text{body surface}} d\vec{F}_{\text{shear}} + \int_{\text{body surface}} d\vec{F}_{\text{pressure}}$$

The resultant force, $\vec{F}$, can be resolved into components parallel and perpendicular to the direction of motion. The component of force parallel to the direction of motion is called the drag force, F_D, and the force component perpendicular to the direction of motion is called the lift force, F_L.

Recognizing that

$$d\vec{F}_{\text{shear}} = \vec{\tau}_w \, dA$$

and

$$d\vec{F}_{\text{pressure}} = -p \, d\vec{A}$$

one might be inclined to think that drag and lift could be evaluated analytically. This proves not to be so; there are very few cases in which the lift and drag can be determined without recourse to experimental results. As we have seen, the presence of an adverse pressure gradient often leads to separation; flow separation prohibits the analytical determination of the force acting on a body. Therefore, for most shapes of interest, we must resort to the use of experimentally measured coefficients for lift and drag computations.

8-15 DRAG

The drag force is the component of force on a body acting parallel to the direction of motion. In discussing the need for experimental results in fluid mechanics (Chapter 7), we considered the problem of determining the drag force, F_D, on a smooth sphere of diameter, d, moving through a viscous, incompressible fluid with speed, V; the fluid density and viscosity were ρ and μ, respectively. The drag force, F_D, was written in the functional form

$$F_D = f_1(d, V, \mu, \rho)$$

Application of the Buckingham Pi theorem resulted in two dimensionless Π parameters that can be written in functional form as

$$\frac{F_D}{\rho V^2 d^2} = f_2\left(\frac{\rho V d}{\mu}\right)$$

Note that d^2 is proportional to the cross-sectional area ($A = \pi d^2/4$) and, hence, we could write

$$\frac{F_D}{\rho V^2 A} = f_3\left(\frac{\rho V d}{\mu}\right) = f_3(Re) \tag{8.55}$$

Although Eq. 8.55 was obtained for a sphere, the form of the equation is valid for incompressible flow over any body; the characteristic length employed in the Reynolds number depends on the particular body shape.

The drag coefficient, C_D, is defined as

$$C_D \equiv \frac{F_D}{\frac{1}{2}\rho V^2 A} \tag{8.56}$$

Force Due to drag

The number $\frac{1}{2}$ has been inserted (as was done in the defining equation for the friction factor) so as to form the familiar dynamic pressure. Then, Eq. 8.55 can be written as

$$C_D = f(Re) \tag{8.57}$$

Note that we have not considered compressibility or free surface effects in this discussion of the drag force. Had these been included, we would have obtained the functional form

$$C_D = f(Re, Fr, M)$$

At this point we shall consider the drag force on several bodies for which Eq. 8.57 is valid. The total drag force is the sum of the friction drag and the pressure drag. However, the drag coefficient is a function only of the Reynolds number.

8-15.1 FLOW OVER A FLAT PLATE PARALLEL TO THE FLOW: FRICTION DRAG

This flow situation has been considered in detail in Section 8-13. Since the pressure gradient is zero, the total drag is equal to the friction drag. Thus

$$\text{drag} = \int_{\text{plate surface}} \tau_w \, dA$$

and

$$C_D = \frac{F_D}{\frac{1}{2}\rho V^2 A} = \frac{\int_{PS} \tau_w \, dA}{\frac{1}{2}\rho V^2 A} \tag{8.58}$$

The drag coefficient for a flat plate parallel to the flow depends on the shear stress distribution along the plate. If the boundary layer were entirely laminar or entirely turbulent, the drag coefficient could be calculated using suitable expressions for τ_w developed in Section 8-13. For a boundary layer that is initially laminar and undergoes transition at some location on the plate, an empirical correlation for C_D must be used (Reference 13, p. 600).

The variation in drag coefficient for a flat plate parallel to the flow is shown qualitatively in Fig. 8.29; a range of transition curves is shown. The Reynolds number at which transition occurs depends on a combination of factors, such as surface roughness and freestream disturbances, as pointed out in Section 8-10. Transition tends to occur earlier (i.e. at lower Reynolds number) as surface roughness or freestream turbulence is increased. Figure 8.29 shows that drag is less, for a given length of plate, when laminar flow is maintained over the longest possible distance.

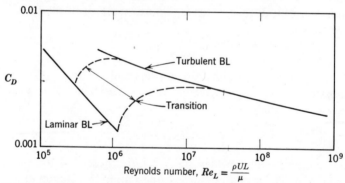

Fig. 8.29 Variation of drag coefficient with Reynolds number for a smooth flat plate parallel to the flow.

8-15.2 FLOW OVER A FLAT PLATE NORMAL TO THE FLOW: PRESSURE DRAG

In flow over a flat plate normal to the flow (Fig. 8.30), we see that the wall shear stress does not contribute to the drag force. The drag is given by

$$F_D = \int_{\text{surface}} p \, dA$$

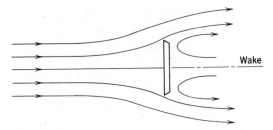

Fig. 8.30 Flow over a flat plate normal to the flow.

For this geometry the flow separates from the edges of the plate; there is backflow in the low energy wake of the plate. While the pressure over the rear surface of the plate is essentially constant, its magnitude cannot be determined analytically. Consequently, we must resort to experiments to determine the drag coefficient.

The drag coefficient for a finite plate normal to the flow depends on the ratio of plate width to height and on the Reynolds number. For values of Re (based on height) greater than about 1000, the drag coefficient is essentially independent of the Reynolds number. The variation of C_D with the ratio of plate width to height (b/h) is shown in Fig. 8.31. (The ratio b/h is defined as the *aspect ratio* of the plate.) For $b/h = 1.0$, the drag coefficient is a minimum at $C_D = 1.18$; this value is just slightly higher than that for a circular disk ($C_D = 1.17$) at large Reynolds number.

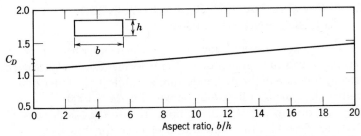

Fig. 8.31 Variation of drag coefficient with aspect ratio for a flat plate of finite width normal to the flow with $Re_h > 1000$ (Ref. 14).

The drag coefficient for all objects with sharp edges is essentially independent of the Reynolds number because the separation points are fixed by the geometry of the object. Drag coefficients for a few selected objects are given in Table 8.6.

TABLE 8.6   Drag Coefficient Data for Selected Objects ($Re \gtrsim 1000$)*

Object	Diagram		$C_D(Re \gtrsim 10^3)$
Square cylinder		$b/h = \infty$	2.05
		$b/h = 1$	1.05
Disk			1.17
Ring			1.20†
Hemisphere (open end facing flow)			1.42
Hemisphere (open end facing downstream)			0.38
C-section (open)			2.30
			1.20

* Data from Ref. 14.
† Based on ring area.

8-15.3   FLOW OVER A SPHERE AND CYLINDER: FRICTION AND PRESSURE DRAG

We have looked at two special flow cases in which either friction or pressure drag was the sole form of drag present. In the former case the drag coefficient was a strong function of the Reynolds number while in the latter case C_D was essentially independent of Reynolds number for $Re \gtrsim 1000$.

In the case of flow over a sphere, both friction drag and pressure drag contribute to the total drag. The drag coefficient for flow over a sphere is shown in Fig. 8.32 as a function of Reynolds number.

At very low Reynolds number,[14] $Re \leq 1$, there will be no flow separation from a sphere; the wake is laminar and the drag is predominantly friction drag. Stokes has shown analytically for very low Reynolds number flows, where inertia forces may be neglected, that drag force on a sphere of diameter,

[14] See the film, *The Fluid Dynamics of Drag*, A. H. Shapiro, principal, or Ref. 15, for a good discussion of drag on spheres and other shapes. Another excellent film is *Low Reynolds Number Flows*, Sir G. I. Taylor, principal. See also Reference 16.

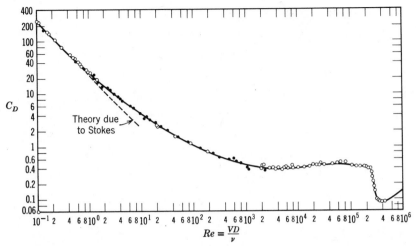

Fig. 8.32 Drag coefficient of a sphere as a function of Reynolds number (Ref. 13).

d, moving at speed, V, through a fluid of viscosity, μ, is given by

$$F_D = 3\pi\mu V d$$

The drag coefficient, C_D, defined by Eq. 8.56 is then

$$C_D = \frac{24}{Re}$$

As shown in Fig. 8.32 this expression agrees with experimental values at low Reynolds number, but begins to deviate significantly from the experimental data for $Re > 1.0$.

As the Reynolds number is increased up to about 1000, the drag coefficient drops continuously. As a result of flow separation the drag is a combination of friction and pressure drag. The relative contribution of friction drag decreases with increasing Reynolds number; at $Re \simeq 1000$, the friction drag is approximately 5 percent of the total drag.

In the range of Reynolds number, $10^3 < Re < 2 \times 10^5$, the drag coefficient curve is relatively flat. The drag coefficient undergoes a rather sharp drop at a Reynolds number of approximately 2×10^5. Experiments show that for $Re < 2 \times 10^5$ the boundary layer on the forward portion of the sphere is laminar. Separation of the boundary layer occurs just upstream of the sphere midsection; a relatively wide turbulent wake is present downstream from the sphere. In the separated region behind the sphere the pressure is essentially constant and lower than the pressure over the forward portion of the sphere (Fig. 8.33). It is this pressure difference that is the main contributor to the drag.

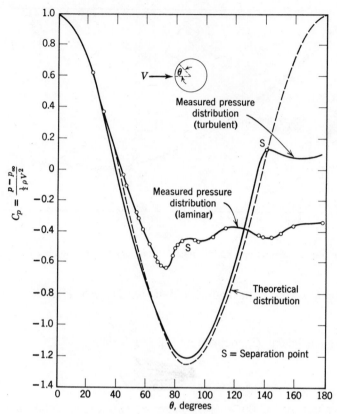

Fig. 8.33 Pressure distribution around a sphere for laminar and turbulent boundary-layer flow, compared to inviscid flow (Ref. 16).

For Reynolds numbers larger than about 2×10^5, transition occurs in the boundary layer on the forward portion of the sphere. The point of separation is located downstream of the center of the sphere and the size of the wake is decreased. The net pressure force on the sphere is reduced (Fig. 8.33), and the drag coefficient decreases abruptly.

A turbulent boundary layer, since it has more momentum than a laminar boundary layer, can better resist an adverse pressure gradient, as discussed in Section 8-14.1. Consequently, turbulent boundary layer flow is desirable on a blunt body, because it delays separation and thus reduces the pressure drag.

Transition in the boundary layer is affected by roughness of the sphere surface and turbulence in the flow stream. Therefore the reduction in drag associated with a turbulent boundary layer does not occur at a unique value of Reynolds number. Experiments with smooth spheres in a flow with low turbulence level show that transition may be delayed to a critical Reynolds number, Re_D, of about 4×10^5. For rough surfaces and/or highly turbulent freestream flow, transition can occur at a critical Reynolds number as low as 1×10^5.

The drag coefficient with turbulent boundary-layer flow is about 5 times less than that for laminar flow near the critical Reynolds number. The corresponding reduction in drag force can appreciably affect the range of a sphere (e.g. a golf ball). The "dimples" on a golf ball are designed to "trip" the boundary layer and, thus, to guarantee turbulent boundary-layer flow and minimum drag. To illustrate this effect more graphically, we obtained samples of golf balls without dimples a few years ago. One of our students volunteered to hit some drives with the smooth balls. In 50 tries with each type of ball, the average distance with the standard balls was 215 yards; the average with the smooth balls was only 125 yards!

Adding roughness elements to a sphere can also suppress local oscillations in location of the transition between laminar and turbulent flow in the boundary layer. These oscillations can lead to variations in drag and to random fluctuations in lift (see Section 8-16). In baseball the "knuckle ball" pitch is intended to behave erratically, to keep the batter confused. By throwing the ball with almost no spin, the pitcher relies on the seams to cause transition in an unpredictable fashion, as the ball moves on its way to the batter. This causes the desired variation in the flight path of the ball.

The drag coefficient for flow over a circular cylinder is shown in Fig. 8.34. The variation of C_D with Reynolds number shows the same characteristics as

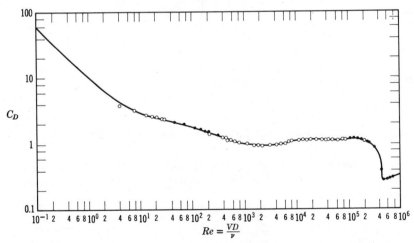

Fig. 8.34 Drag coefficient for circular cylinders as a function of Reynolds number (Ref. 13).

observed in the flow over a sphere, but the values of C_D are about twice as large.

Example 8.14

A cylindrical chimney 1 m in diameter and 25 m tall is exposed to a uniform 50 km/hr wind at standard atmospheric conditions. End effects and gusts may be neglected. Estimate the bending moment at the base of the chimney due to wind forces.

Example Problem 8.14

GIVEN:

Cylindrical chimney, $D = 1$ m, $L = 25$ m in uniform flow with

$$V = 50 \text{ km/hr} \qquad p = 101 \text{ kPa} \qquad T = 15 \text{ C}$$

Neglect end effects.

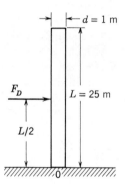

FIND:

Bending moment at bottom of chimney.

SOLUTION:

The drag coefficient is given by $C_D = \dfrac{F_D}{\frac{1}{2}\rho V^2 A}$, and thus $F_D = \frac{1}{2}\rho V^2 A C_D$.

Since the force per unit length is uniform over the entire length, the resultant force, F_D, will act at the midpoint of the chimney. Hence the moment about the chimney base is

$$M_0 = F_D \frac{L}{2} = \frac{L}{4}\rho V^2 A C_D$$

$$V = \frac{50 \text{ km}}{\text{hr}} \times \frac{10^3 \text{ m}}{\text{km}} \times \frac{\text{hr}}{3600 \text{ sec}} = 13.9 \text{ m/sec}$$

For air at standard conditions, $\rho = 1.23$ kg/m^3, and $\mu = 1.78 \times 10^{-5}$ kg/m·sec. Thus

$$Re = \frac{\rho VD}{\mu} = \frac{1.23 \text{ kg}}{m^3} \times \frac{13.9 \text{ m}}{\text{sec}} \times 1 \text{ m} \times \frac{m \cdot \text{sec}}{1.78 \times 10^{-5} \text{ kg}} = 9.61 \times 10^5$$

From Fig. 8.34, $C_D = 0.35$. Thus

$$M_0 = \frac{L}{4}\rho V^2 A C_D = \frac{L}{4}\rho V^2 L\, D C_D = \frac{L^2}{4}\rho V^2\, D C_D$$

$$= \frac{1}{4} \times \frac{(25)^2 \text{ m}^2}{} \times \frac{1.23 \text{ kg}}{m^3} \times \frac{(13.9)^2 \text{ m}^2}{\text{sec}^2} \times 1 \text{ m} \times 0.35 \times \frac{N \cdot \text{sec}^2}{\text{kg} \cdot \text{m}}$$

$$M_0 = 1.30 \times 10^4 \text{ N} \cdot \text{m} \qquad\qquad\qquad M_0$$

Example 8.15

A class AA fuel dragster weighing 1,600 lbf attains a speed of 240 mph in the quarter mile. Immediately after passing through the timing lights, the driver opens the drag chute. The chute area is 25 ft^2, and it has a constant drag coefficient of 1.2. Air and rolling resistance of the car may be neglected. The local air density is 0.0024 slug/ft^3. Find the time required for the machine to decelerate to 100 mph.

Example Problem 8.15

GIVEN:

Dragster weighing 1,600 lbf, moving with speed, $V = 240$ mph, is slowed by the drag force on a chute.

   chute area, $A = 25$ ft^2 drag coefficient, $C_D = 1.2$ $\rho_{air} = 0.0024$ slug/ft^3

Air and rolling resistance of the car may be neglected.

FIND:

Time required for the machine to decelerate to 100 mph.

SOLUTION:

Taking the car as a system and writing Newton's second law in the direction of motion gives

$$-F_D = ma = m\frac{dV}{dt}$$

$V_0 = 240$ mph

$V_f = 100$ mph

Since

$$C_D = \frac{F_D}{\frac{1}{2}\rho V^2 A}$$

then

$$F_D = \tfrac{1}{2}C_D\rho V^2 A$$

Substituting into Newton's second law,

$$-\frac{1}{2}C_D\rho V^2 A = m\,\frac{dV}{dt}$$

Separating variables and integrating,

$$-\frac{1}{2}C_D\rho\frac{A}{m}\int_0^t dt = \int_{V_0}^{V_f}\frac{dV}{V^2}$$

$$-\frac{1}{2}C_D\rho\frac{A}{m}\,t = -\frac{1}{V}\Big]_{V_0}^{V_f} = -\frac{1}{V_f}+\frac{1}{V_0} = -\frac{(V_0 - V_f)}{V_f V_0}$$

Finally,

$$t = \frac{(V_0 - V_f)}{V_f V_0}\frac{2m}{C_D\rho A} = \frac{(V_0 - V_f)}{V_f V_0}\frac{2W}{C_D\rho Ag}$$

$$= \frac{(240 - 100)\text{ mph}}{} \times 2 \times 1600\text{ lbf} \times \frac{1}{100\text{ mph}} \times \frac{\text{hr}}{240\text{ mi}} \times \frac{1}{1.2} \times \frac{\text{ft}^3}{0.0024\text{ slug}}$$

$$\times \frac{1}{25\text{ ft}^2} \times \frac{\sec^2}{32.2\text{ ft}} \times \frac{\text{slug}\cdot\text{ft}}{\text{lbf}\cdot\sec^2} \times \frac{\text{mi}}{5280\text{ ft}} \times \frac{3600\text{ sec}}{\text{hr}}$$

$$t = 5.49\text{ sec} \hspace{8cm} t$$

8-15.4 STREAMLINING

The extent of the separated flow region behind many of the objects discussed in the previous section can be reduced by streamlining, or fairing, the body shape. The objective of streamlining is to reduce the adverse pressure gradient that occurs behind the point of maximum thickness on the body, as pointed out in Section 2-5.1. This delays boundary-layer separation and thus reduces the pressure drag. However, addition of a faired tail section increases the surface area of the body; this causes skin friction drag to increase. The optimum streamlined shape is thus the shape that gives minimum total drag. These effects are discussed at length in the film series, *The Fluid Dynamics of Drag.*[15]

Advanced theories are as yet unable to predict the location of separation analytically in most cases. Thus in general it is not possible to determine

[15] A. H. Shapiro, principal.

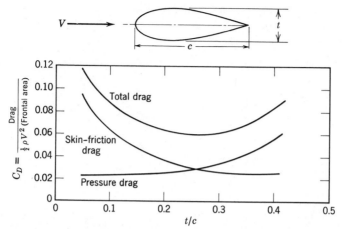

Fig. 8.35 Drag coefficient on a streamlined strut as a function of thickness ratio, showing contribution of skin friction and pressure to total drag (Ref. 17).

optimum bodies by purely analytical means, and we must rely on experimental data or on wind tunnel tests of proposed designs. Let us look at two general classes of results before going on to discuss lift.

The pressure gradient around a "teardrop" shape (i.e. a "streamlined" cylinder) is less severe than that around a cylinder of circular section. The trade-off between pressure and friction drag for this case is illustrated by the results presented in Fig. 8.35, for tests at $Re_c = 4 \times 10^5$. From the figure, the minimum drag coefficient is $C_D \simeq 0.06$, which occurs when $t/c \simeq 0.25$. This value is approximately 20 percent of the minimum drag coefficient for a circular cylinder of the same thickness! Consequently, a streamlined strut about 5 times the thickness of a cylindrical strut could be used with no penalty in aerodynamic drag.

Pressure distribution and drag data[16] for two symmetric airfoils of infinite span and 15 percent thickness at zero angle of attack are presented in Fig. 8.36. Transition on the conventional (NACA 0015) airfoil takes place where the pressure gradient becomes adverse, at $x/c = 0.13$, that is, near the point of maximum thickness. Thus most of the airfoil surface is covered with a turbulent boundary layer; the drag coefficient is $C_D \simeq 0.0061$. The point of maximum thickness has been moved aft on the airfoil (NACA $66_2 - 015$) designed

[16] Note that drag coefficients for airfoils are based on the planform area, that is,

$$C_D = \frac{F_D}{\frac{1}{2}\rho V^2 A_p}$$

where A_p is the maximum projected wing area.

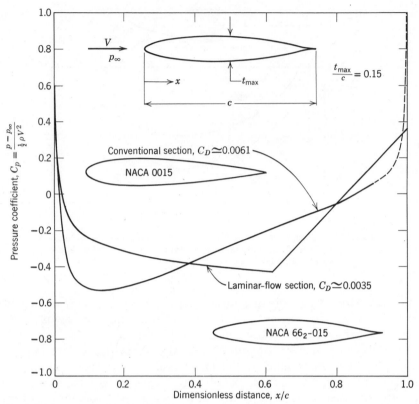

Fig. 8.36 Theoretical pressure distributions at zero angle of attack for two airfoil sections of 15 percent thickness ratio (data from Ref. 18).

for laminar flow. The boundary layer is maintained in the laminar state by the favorable pressure gradient to $x/c = 0.63$. Thus the bulk of the flow is laminar; $C_D \simeq 0.0035$ for this section.

Tests in special wind tunnels have shown that laminar flow can be maintained up to length Reynolds numbers as high as $Re_x = 30 \times 10^6$ by appropriate profile shaping. Because they have favorable drag characteristics, laminar-flow airfoils are used in the design of most modern subsonic aircraft.

8-16 LIFT

As defined previously, lift is the component of the resultant aerodynamic force perpendicular to the direction of fluid motion. One of the most commonly observed examples of dynamic lift is the flow over an airfoil.[17] The lift coefficient, C_L, is defined as

$$C_L \equiv \frac{F_L}{\frac{1}{2}\rho V^2 A_p} \tag{8.59}$$

The lift and drag coefficients for an airfoil are functions of both the Reynolds number and the angle of attack; the angle of attack, α, is the angle between the airfoil chord and the freestream velocity vector. For an airfoil the maximum cross-sectional area at right angles to the flow changes with angle of attack. Consequently, in defining the lift and drag coefficients of an airfoil, the planform area, A_p (the maximum projected area of the wing), is used.

Lift and drag coefficient data for typical conventional and laminar-flow profiles are plotted in Fig. 8.37, for a Reynolds number of 9×10^6, based on chord length. Both sections are cambered to give design lift coefficients of about $C_L = 0.3$, so the lift coefficients at zero angle of attack are not zero. As the angle of attack is increased, the lift coefficients increase smoothly until a maximum value is reached. Further increases in angle of attack produce a sudden decrease in C_L. The airfoil is said to have *stalled* when C_L drops in this fashion.

Airfoil stall results when flow separation occurs over a major portion of the upper surface of the airfoil. As the angle of attack is increased, the stagnation point moves back along the lower surface of the airfoil, as shown schematically in Fig. 8.38. The flow on the upper surface then must accelerate sharply to round the nose of the airfoil.[18] The minimum pressure becomes lower, and it moves forward on the upper surface. A severe adverse pressure gradient appears following the point of minimum pressure; finally it causes the flow to separate completely from the upper surface: the airfoil stalls.

Movement of the minimum pressure point and accentuation of the adverse pressure gradient are responsible for the sudden increase in C_D at an angle of attack of about $1.5°$ for the laminar-flow section, which is apparent in Fig. 8.37. The sudden rise in C_D is due to early transition from laminar to turbulent boundary-layer flow on the upper surface. Aircraft with laminar-flow sections are designed to cruise in the low drag region.

Because laminar-flow sections have very sharp leading edges, all of the effects we have described are exaggerated, and they stall at lower angles of attack than conventional sections, as shown in Fig. 8.37. The maximum possible lift coefficient, $C_{L_{max}}$, is also less for laminar-flow sections.

---

[17] Flow over an airfoil is shown in the film loop S-FM045, *Velocities near an Airfoil.*
[18] Flow patterns and pressure distributions for airfoil sections are shown in the film loops, S-FM117, *Subsonic Flow Patterns and Pressure Distributions for an Airfoil* and S-FM118, *Laminar-Flow versus Conventional Airfoils.*

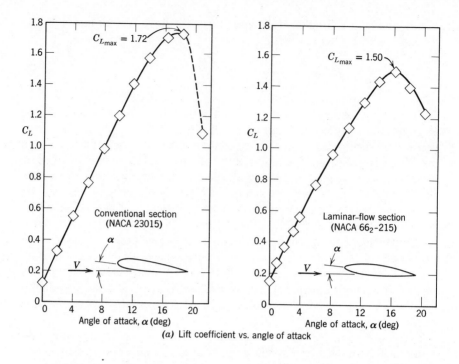

(a) Lift coefficient vs. angle of attack

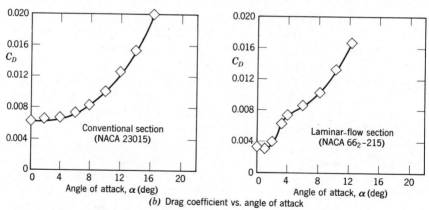

(b) Drag coefficient vs. angle of attack

Fig. 8.37 Lift and drag coefficients versus angle of attack for two airfoil sections of 15 percent thickness ratio (data from Ref. 18).

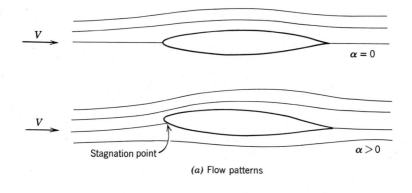

(a) Flow patterns

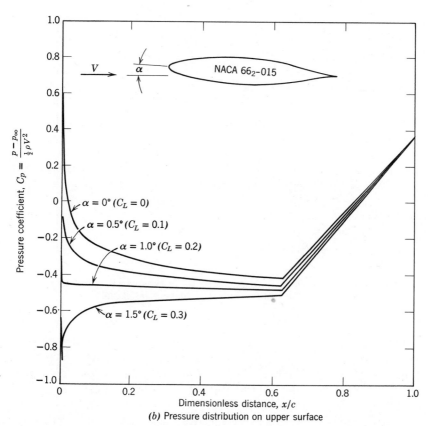

(b) Pressure distribution on upper surface

Fig. 8.38 Effect of angle of attack on flow pattern and theoretical pressure distribution for a laminar-flow airfoil of 15 percent thickness ratio (data from Ref. 18).

Polar plots (i.e. plots of C_L versus C_D) are often used to present airfoil data. Such a plot is given in Fig. 8.39 for the two sections we have been discussing. The value of the lift/drag ratio, C_L/C_D, at the design lift coefficient, $C_L \simeq 0.3$, is shown for both sections. For an aircraft of given mass, the power required for level flight at fixed speed is inversely proportional to the lift/drag ratio. Thus the advantages of laminar-flow designs may be seen clearly.

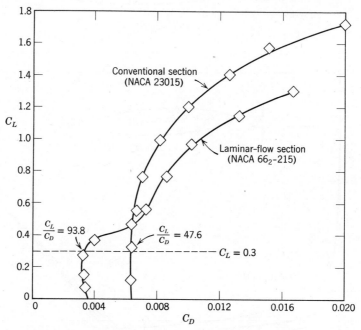

Fig. 8.39 Lift-drag polars for two airfoil sections of 15 percent thickness ratio with design lift coefficients near 0.3 (data from Ref. 18).

Note that all of the airfoil data presented have been for two-dimensional sections, that is, slices from airfoils of infinite span. End effects, which occur on airfoils of finite span, reduce the lift coefficient and cause the drag coefficient to increase. Thus the lift/drag ratio (L/D) values that can be achieved in practice are less than those obtained from two-dimensional test results.

The parameter used to define the effective span of an airfoil is the aspect ratio, defined as

$$ar \equiv \frac{A_p}{c^2}$$

where A_p is the planform area and c is the chord. Thus for a rectangular planform of span S,

$$ar = \frac{Sc}{c^2} = \frac{S}{c}$$

The maximum lift/drag ratio for a modern low-drag section may be as high as 400 for infinite aspect ratio. A sailplane (glider) with $ar = 40$ might have $L/D = 40$, and a typical light plane ($ar \sim 12$) might have $L/D \sim 20$ or so. Two examples of rather poor shapes are lifting bodies used for reentry from the upper atmosphere, and water skis, which are *hydro*foils of low aspect ratio. For both of these shapes, L/D values are less than unity.

Mother nature is well aware of the effects of aspect ratio on aerodynamic performance. Soaring birds, such as the albatross or California condor, have thin wings of long span. Birds that must maneuver quickly to catch their prey, such as owls, have wings of relatively short span, but large area, which gives low wing loading and thus high maneuverability.

For finite airfoils, lift reduction results from changes in the flow pattern arising from end effects. Increased drag results from *downwash* velocities induced by *trailing vortices*, shown schematically in Fig. 8.40. The downwash velocities tend to reduce lift and increase drag by reducing the effective angle of attack. (They are suppressed when an airplane, especially a low-wing craft, is close to the ground.) The trailing vortices are a result of leakage flows around the wingtips,[19] from the high pressure area on the bottom surface to the low pressure region above. They may be very strong and persistent,

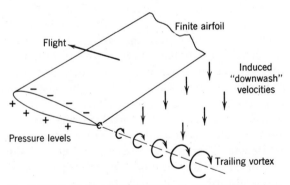

Fig. 8.40 Schematic representation of the trailing vortex system of a finite airfoil.

[19] Formation of trailing vortices is shown clearly in the film loops, S-FM024, *Wing Tip Vortex* and S-FM071, *Flow near Tip of Lifting Wing*.

present a hazard to light planes for distances of 5 to 10 *miles* behind the larger plane. Velocities exceeding 200 mph have been measured in trailing vortices from large aircraft.[20]

As we have seen, aircraft can be fitted with low-drag airfoils to give excellent performance at cruise conditions. However, since the maximum lift coefficient is low for thin airfoils, some additional effort must be expended to obtain acceptably low landing speeds. In steady state flight conditions, the lift must equal the aircraft weight. Thus, from Eq. 8.59,

$$W = F_L = C_L \frac{1}{2} \rho V^2 A$$

The minimum flight speed is obtained when $C_L = C_{L_{max}}$. Solving for V_{min},

$$V_{min} = \sqrt{\frac{2W}{\rho C_{L_{max}} A}} \tag{8.60}$$

According to Eq. 8.60, the minimum landing speed can be reduced by increasing either $C_{L_{max}}$ or the wing area. Two basic techniques are available for controlling these variables: variable geometry wing sections (e.g. through the use of flaps) or boundary-layer control techniques.

Flaps are movable portions of a wing surface that may be extended during landing and takeoff to increase the effective wing area. The effects on lift and drag of two typical flap configurations are shown in Fig. 8.41, as applied to an NACA 23012 airfoil section. The maximum lift coefficient for this section is increased from 1.52 in the "clean" condition to 3.48 with double-slotted flaps. From Eq. 8.60 the corresponding reduction in landing speed would be 34 percent.

Figure 8.41 shows that drag is increased substantially by high-lift devices. From Fig. 8.41*b*, the drag at $C_{L_{max}}$ ($C_D \simeq 0.28$) with double slotted flaps is about five times larger than the drag at $C_{L_{max}}$ ($C_D \simeq 0.055$) for the clean airfoil.

Although the details of boundary-layer control techniques[21] are beyond the scope of this text, the basic purpose of all of them is to delay separation or reduce drag, by adding momentum to the boundary layer through blowing, or by removing low-momentum boundary-layer fluid by suction. Many examples of practical boundary-layer control systems may be seen on commercial transport aircraft at your local airport. One of the more sophisticated systems, on the Boeing 727 transport, is shown in Fig. 8.42. On this aircraft, leading edge devices are used in conjunction with triple-slotted trailing edge flaps, to achieve a value of $C_{L_{max}}$ in excess of 3.6.

---

[20] Sforza, P. M., "Aircraft Vortices: Benign or Baleful?" *Space/Aeronautics*, **53**, 4, April 1970, pp. 42–49. See also the University of Iowa film, *Form Drag, Lift and Propulsion*.

[21] See the excellent film, *Boundary-Layer Control*, D. C. Hazen, principal, for a reviewing of these techniques. The film loops, S-FM119, *Reduction of Airfoil Friction Drag by Suction* and S-FM120, *Some Methods for Increasing Lift Coefficient*, are edited from this film.

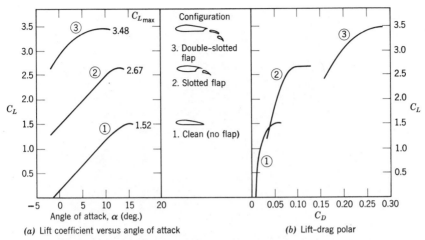

(a) Lift coefficient versus angle of attack (b) Lift-drag polar

Fig. 8.41 Effect of flaps on aerodynamic characteristics of NACA 23012 airfoil section (data from Ref. 18).

Fig. 8.42 Application of high-lift boundary-layer control devices to jet transport aircraft for reduction of landing speed (photographs courtesy of Boeing Airplane Company). (a) The wing of the Boeing 727 is one of the most mechanized of those in commercial service today. During the approach to landing, huge triple-slotted, trailing edge flaps roll out from under the wing and deflect downward to increase lift. A section of the leading edge near the outboard end of the wing slides forward to open a slot that keeps the air flow close to the wing's upper surface. At the leading edge of the wing near the root, a Kruger flap drops down from under the wing, increasing the effective radius of the leading edge to prevent flow separation. After touchdown, spoilers (not shown) pop up in front of each flap to kill the lift and ensure that the plane remains on the ground despite the lift-augmenting devices.

(b)

Fig. 8.42 (b) The view from the passenger's seat in a modern airliner—this one, a Boeing 707—includes a remarkably good view of the increasingly complex machinery needed to make air behave as it passes over the wing. Both the inboard and outboard trailing edge flaps (shown here fully deflected) are equipped with double slots near the pivot. The slots duct air from the under surface to the upper surface of the flap, creating a streamwise jet that causes the air flow to cling more closely to the upper surface. The two rows of short blades in front of the flaps are vortex generators whose purpose is to churn up the slow-moving layers of air next to the surface. This encourages the flow to follow the wing surface more closely. The raised panels just above the flaps are lift-destroying devices called spoilers. Normally, they are used to control lift for landing and maneuvering. They are raised here to help the plane descend from high altitude.

Another method of boundary-layer control, which we shall discuss briefly, is the use of moving surfaces to reduce skin friction effects on the boundary layer. This method is hard to apply to practical devices, because of geometric and weight complications, but it is very important in recreation. Most golfers, tennis players, Ping-Pong enthusiasts, and baseball pitchers can attest to this!

It has long been known that a spinning projectile in flight is affected by a force perpendicular to the direction of motion and to the spin axis. This effect, known as the Magnus effect, is responsible for the systematic drift of artillery shells. Tennis and Ping-Pong players use spin to control the trajectory and bounce of a shot. In golf, a drive can leave the tee at 275 ft/sec or more, with a backspin of 9000 rpm! This spin provides significant aerodynamic lift that substantially increases the carry of a drive. It is also largely responsible for hooking and slicing, when shots are not hit squarely. The baseball pitcher uses spin to throw a curve ball.

Flow about a spinning sphere is shown in Fig. 8.43a. The spin alters the pressure distribution and also affects the location of boundary-layer separation. Separation is delayed on the upper surface of the sphere in Fig. 8.43a, and it occurs earlier on the lower surface. The pressure is reduced on the upper surface and increased on the lower surface; the wake is deflected downward as shown. The pressure forces cause a lift in the direction shown; spin in the opposite direction would produce a negative lift, that is, a downward force. The force is directed perpendicular to both V and the spin axis.

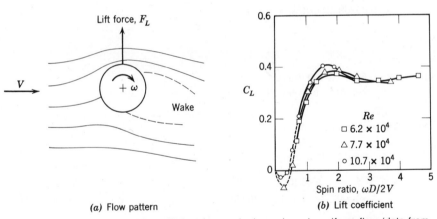

(a) Flow pattern (b) Lift coefficient

Fig. 8.43 Flow pattern and lift coefficient for a spinning sphere in uniform flow (data from Ref. 17).

Example 8.16

An airplane is designed according to the following specifications.

$$\text{weight} = 3000 \text{ lbf}$$
$$\text{wing area} = 300 \text{ ft}^2$$
$$\text{takeoff speed} = 100 \text{ ft/sec}$$

Model tests show that the lift and drag coefficients vary with the angle of attack of the wing according to the approximate relations:

$$C_L = 0.35(1 + 0.2\alpha)$$
$$C_D = 0.008(1 + \alpha)$$

for small α, where α is the angle of attack measured in degrees. The atmospheric density is 0.00238 slug/ft^3. Find the angle of attack that ensures takeoff at the desired speed, and the power required for takeoff.

Example Problem 8.16

GIVEN:

Aircraft with specifications:

weight = 3000 lbf	$C_L = 0.35(1 + 0.2\alpha)$
wing area = 300 ft^2	$C_D = 0.008(1 + \alpha)$
takeoff speed = 100 ft/sec	where α is angle of attack in degrees
ρ_{air} = 0.00238 slug/ft^3	

FIND:

(a) α for takeoff at 100 ft/sec.
(b) Power required for takeoff.

SOLUTION:

To ensure takeoff, lift force must be equal to weight.

$$C_L = \frac{F_L}{\frac{1}{2}\rho V^2 A} = \frac{3000 \text{ lbf}}{\frac{1}{2}} \times \frac{\text{ft}^3}{0.00238 \text{ slug}} \times \frac{\text{sec}^2}{(100)^2 \text{ ft}^2} \times \frac{1}{300 \text{ ft}^2} \times \frac{\text{slug} \cdot \text{ft}}{\text{lbf} \cdot \text{sec}^2}$$

$$C_L = 0.84$$

Since $C_L = 0.35(1 + 0.2\alpha)$, then

$$\alpha = \left[\frac{C_L}{0.35} - 1\right]\frac{1}{0.2} = \left[\frac{0.84}{0.35} - 1\right]\frac{1}{0.2} = 7° \xleftarrow{\hspace{3cm}} \alpha$$

Power required for takeoff = $F_D V$

$$C_D = \frac{F_D}{\frac{1}{2}\rho V^2 A} \qquad F_D = \frac{1}{2}C_D\rho V^2 A$$

For $\alpha = 7°$,

$$C_D = 0.008(1 + 7) = 0.064$$

Power $= F_D V = \frac{1}{2} C_D \rho V^2 A V$

$$= \frac{1}{2}(0.064) \times \frac{0.00238 \text{ slug}}{\text{ft}^3} \times \frac{(100)^2 \text{ ft}^2}{\text{sec}^2} \times \frac{300 \text{ ft}^2}{1} \times \frac{100 \text{ ft}}{\text{sec}} \times \frac{\text{lbf} \cdot \text{sec}^2}{\text{slug} \cdot \text{ft}} \times \frac{\text{hp} \cdot \text{sec}}{550 \text{ ft} \cdot \text{lbf}}$$

Power $= 41.5$ hp $\longleftarrow$ Power

PART F. FLOW MEASUREMENT

The choice of a flow metering device is influenced by the accuracy required, cost, complication, ease of reading or data reduction, and service life. Ordinarily, the simplest and cheapest device for the desired accuracy should be chosen. Accordingly, a few simple methods of flow measurement are discussed first.

8-17 SIMPLE METHODS

For steady liquid flows, simple tanks can be used to determine flowrate by measuring the volume or mass of liquid collected during a known time interval. If the time interval is long enough to be measured accurately, extremely precise values of flowrate may be determined in this way.

For gas flows, compressibility must be considered in making volumetric measurements. The densities of gases are also generally too small to permit accurate determination of mass flowrate directly. However, a volumetric sample can often be collected by displacing a "bell," or inverted jar over water or other liquid (the pressure is held constant by means of counter-weights). If volumetric or mass determinations are set up carefully, no calibration is required; this is a great advantage of these methods.

Another simple method, but one that does require calibration, makes use of the radial pressure gradient caused by streamline curvature. Design of a simple "elbow flowmeter" is illustrated in Example Problem 8.17.

Example 8.17
The flowrate of air at standard conditions in a flat duct is to be determined by installing pressure taps across a bend. The duct is 0.3 m deep and 0.1 m wide. The inner radius of the bend is 0.25 m. The velocity profile is assumed uniform. If the measured pressure difference between the taps is 40 mm of water, compute the approximate flowrate.

Example Problem 8.17

GIVEN:

Flow through elbow, as shown.

$$p_2 - p_1 = \rho_{H_2O}\, g\, \Delta h$$

where Δh = 40 mm H_2O
Flow uniform. Air at STP.

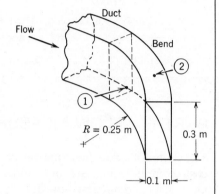

FIND:

Q.

SOLUTION:

Apply Euler's n component equation across the flow streamlines.

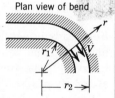

Plan view of bend

Basic equation:
$$\frac{\partial p}{\partial r} = \frac{\rho V^2}{r}$$

Assumptions: (1) Frictionless flow
(2) Incompressible flow
(3) Uniform flow at measurement section

For this flow, $p = p(r)$, so
$$\frac{\partial p}{\partial r} = \frac{dp}{dr} = \frac{\rho V^2}{r}$$

or
$$dp = \rho V^2 \frac{dr}{r}$$

Integrating
$$p_2 - p_1 = \rho V^2 \ln r \Big]_{r_1}^{r_2} = \rho V^2 \ln\frac{r_2}{r_1}$$

Then
$$V = \left[\frac{p_2 - p_1}{\rho \ln (r_2/r_1)}\right]^{1/2}$$

But $\Delta p = p_2 - p_1 = \rho_{H_2O} g \, \Delta h$, so

$$V = \left[\frac{\rho_{H_2O} g \, \Delta h}{\rho \ln (r_2/r_1)} \right]^{1/2}$$

$$= \left[\frac{999 \text{ kg}}{\text{m}^3} \times \frac{9.81 \text{ m}}{\text{sec}^2} \times 0.04 \text{ m} \times \frac{\text{m}^3}{1.23 \text{ kg}} \times \frac{1}{\ln (0.35 \text{ m}/0.25 \text{ m})} \right]^{1/2}$$

$$V = 30.8 \text{ m/sec}$$

For uniform flow

$$Q = VA = \frac{30.8 \text{ m}}{\text{sec}} \times 0.1 \text{ m} \times 0.3 \text{ m} = 0.924 \text{ m}^3/\text{sec} \qquad Q$$

$\left\{ \begin{array}{l} \text{In actual applications, the velocity profile in a channel bend will not be uniform.} \\ \text{Thus elbow flowmeters require calibration.} \end{array} \right\}$

8-18 FLOWMETERS FOR INTERNAL FLOWS

Most nonmechanical flowmeters for internal flow (except the laminar flow element, Section 8-18.4) are based on acceleration of a fluid stream through some form of nozzle as shown schematically in Fig. 8.44. Flow separation at the sharp edge of the nozzle throat causes a recirculation zone to form as shown by the dashed lines downstream from the nozzle. The mainstream flow continues to accelerate from the nozzle throat to form a *vena contracta* at section ② and then decelerates again to fill the duct. At the vena contracta the flow area passes through a minimum, the flow streamlines are essentially straight, and the pressure is uniform across the channel section.

The ideal flowrate may be related to the pressure drop by applying the continuity and Bernoulli equations. Then empirical correction factors may be applied to obtain the actual flowrate.

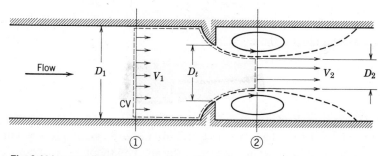

Fig. 8.44 Internal flow through a generalized nozzle, showing control volume used for analysis.

Basic equations:

$$0 = \overbrace{\frac{\partial}{\partial t}\int_{CV} \rho \, d\Psi}^{=0(1)} + \int_{CS} \rho \vec{V} \cdot d\vec{A} \tag{4.13}$$

$$\frac{p_1}{\rho} + \frac{V_1^2}{2} + g\cancel{z_1} = \frac{p_2}{\rho} + \frac{V_2^2}{2} + g\cancel{z_2} \tag{6.10}$$

Assumptions:
(1) Steady flow
(2) Incompressible flow
(3) Flow along a streamline
(4) No friction
(5) Uniform velocity at sections ① and ②
(6) No streamline curvature at section ① or ②, so pressure is uniform across those sections
(7) $z_1 = z_2$

Then from the Bernoulli equation,

$$p_1 - p_2 = \frac{\rho}{2}(V_2^2 - V_1^2) = \frac{\rho V_2^2}{2}\left[1 - \left(\frac{V_1}{V_2}\right)^2\right]$$

and from continuity

$$0 = \{-|\rho V_1 A_1|\} + \{|\rho V_2 A_2|\}$$

or

$$V_1 A_1 = V_2 A_2 \qquad \text{or} \qquad \left(\frac{V_1}{V_2}\right)^2 = \left(\frac{A_2}{A_1}\right)^2$$

Substituting gives

$$p_1 - p_2 = \frac{\rho V_2^2}{2}\left[1 - \left(\frac{A_2}{A_1}\right)^2\right]$$

Solving for the ideal velocity, V_2,

$$V_2 = \sqrt{\frac{2(p_1 - p_2)}{\rho[1 - (A_2/A_1)^2]}} \tag{8.61}$$

As shown in Fig. 8.44, the flow area at the vena contracta, or section of minimum flow area, where the flow streamlines are essentially straight, is considerably smaller than at the nozzle throat. The theoretical velocity can be written in terms of the meter throat area, A_t, by defining the contraction coefficient, C_c, as

$$C_c \equiv \frac{A_{\text{vena contracta}}}{A_{\text{throat}}} \tag{8.62}$$

Then the ideal velocity V_2 is given by

$$V_2 = \sqrt{\frac{2(p_1 - p_2)}{\rho[1 - C_c^2(A_t/A_1)^2]}} \tag{8.63}$$

As a result of friction the actual velocity at the vena contracta will be less than the ideal velocity. Frictional effects are taken into account by defining a velocity coefficient, C_v, as

$$C_v \equiv \frac{V_{2\ \text{actual}}}{V_{2\ \text{ideal}}} \tag{8.64}$$

Then the actual mass flowrate is given by

$$\dot{m}_{\text{actual}} = \rho A_2 V_{2\ \text{actual}} = \rho C_c A_t C_v V_{2\ \text{ideal}}$$

or

$$\dot{m}_{\text{actual}} = \rho A_t C_c C_v \sqrt{\frac{2(p_1 - p_2)}{\rho[1 - C_c^2(A_t/A_1)^2]}} = \frac{C_c C_v A_t \sqrt{2\rho(p_1 - p_2)}}{\sqrt{[1 - C_c^2(A_t/A_1)^2]}} \tag{8.65}$$

The flow coefficient, K, is defined as

$$K = \frac{C_c C_v}{\sqrt{[1 - C_c^2(A_t/A_1)^2]}} \tag{8.66}$$

such that

$$\dot{m}_{\text{actual}} = K A_t \sqrt{2\rho(p_1 - p_2)} \tag{8.67}$$

For nozzles and venturi meters, the section of minimum flow area is located at the throat. There is no vena contracta and $C_c = 1$. For these cases

$$K = \frac{C_v}{\sqrt{1 - (A_t/A_1)^2}} = \frac{C_v}{\sqrt{1 - \beta^4}} \tag{8.68}$$

where $\beta = D_t/D_1$, and

$$\dot{m}_{\text{actual}} = K A_t \sqrt{2\rho(p_1 - p_2)} = \frac{C_v}{\sqrt{1 - \beta^4}} A_t \sqrt{2\rho(p_1 - p_2)} \tag{8.69}$$

The factor $1/\sqrt{1 - \beta^4}$, called the "velocity of approach" factor, is a function of geometry alone.

As we have noted, selection of a flowmeter depends on factors such as cost, accuracy, need for calibration, and ease of installation and maintenance.

TABLE 8.7 Characteristics of Orifice, Flow Nozzle, and Venturi Flowmeters

Flowmeter Type	Diagram	Head Loss	Cost
Orifice	D_1 D_t Flow	High	Low
Flow nozzle	D_1 D_2 Flow	Intermediate	Intermediate
Venturi	D_1 D_2 Flow	Low	High

Some of these factors are compared for orifice plate, flow nozzle, and venturi meters in Table 8.7.

The flow coefficients reported in the literature have been measured with a fully-developed turbulent velocity distribution at the meter inlet (Section ①). If a flowmeter is to be installed downstream from a valve, elbow or other disturbance, a straight section of pipe must be placed in front of the meter. Approximately 10 diameters of straight pipe are required for venturi meters, and up to 40 diameters for orifice plate or flow nozzle meters. When a meter has been properly installed, the flowrate may be computed from Eq. 8.67, after choosing an appropriate value for the empirical flow coefficient, K, defined in Eq. 8.66. Some design data for incompressible flow are given in the next few sections. The same basic methods can be extended to compressible flows, but these will not be treated here. For complete details, see Reference 19.

8-18.1 THE ORIFICE PLATE

The orifice plate (Fig. 8.45) is a thin plate that may be clamped between pipe flanges. Since its geometry is simple, it is low in cost and easy to install or replace. The sharp edge of the orifice will not foul with scale or suspended matter. However, suspended matter can build up at the inlet side of a concentric orifice in a horizontal pipe; an eccentric orifice may be placed flush with the bottom of the pipe to avoid this difficulty. The primary disadvantages

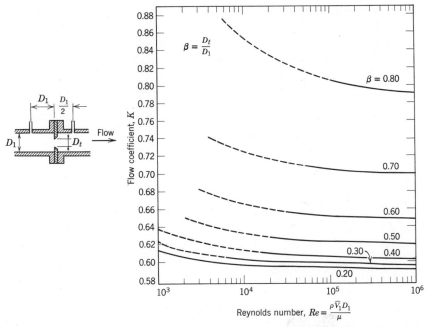

Fig. 8.45 Flow coefficients for concentric orifices with pressure taps as shown (data from Ref. 19).

of the orifice are its limited capacity, and the high head loss due to the uncontrolled expansion downstream from the metering element.

Pressure taps may be placed at the vena contracta, or $\frac{1}{2}$ diameter downstream from the orifice plate, or in the flanges (see Ref. 19 for details). Since the location of the pressure taps influences the empirically determined flow coefficient, one must select handbook values of K consistent with the location of pressure taps.

Some values of flow coefficient for concentric orifices with taps 1 pipe diameter upstream and $\frac{1}{2}$ diameter downstream are given in Fig. 8.45 as a function of Reynolds number and diameter ratio, β. Recommended design practice is to use orifice diameters such that $0.25 < \beta < 0.90$.

8-18.2 THE FLOW NOZZLE

Flow nozzles may be used as metering elements in either plenums or ducts, as shown in Fig. 8.46. As shown in Fig. 8.46, the nozzle section is approximately a quarter ellipse. Design details and locations for pressure taps may be found in Reference 19.

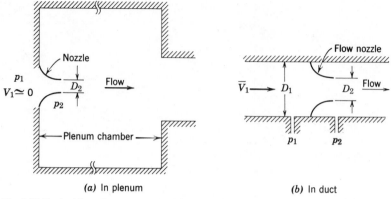

(a) In plenum   (b) In duct

Fig. 8.46 Typical installations of nozzle flowmeters.

a. Plenum Installations

For plenum installation, the approach velocity, V_1, is essentially zero, so $D_1 \to \infty$, and $\beta = 0$. For a nozzle, C_c is unity, and the flowrate may be computed from Eq. 8.69.

For plenum installation, nozzles may be fabricated from spun aluminum, molded fiberglass, or other inexpensive materials. Thus they are simple and cheap to make and install. Since the plenum pressure is equal to p_2, the location of the downstream pressure tap is not critical. Meters suitable for a wide range of flowrates may be made by installing several nozzles in a plenum. At low flowrates, most of them may be plugged with rubber balls or other handy objects. For larger flowrates, more nozzles may be used.

For plenum nozzles, typical velocity coefficient values are in the range, $0.95 < C_v < 0.99$; the larger values apply at high Reynolds numbers. Thus the mass rate of flow can be computed within approximately plus or minus 2 percent using Eq. 8.69 with $K = C_v = 0.97$.

b. Pipe Installation

Equation 8.69, repeated below, must be used with an experimental value for K to compute the mass flowrate through a flow nozzle in a pipe.

$$\dot{m}_{\text{actual}} = KA_t\sqrt{2\rho(p_1 - p_2)} \tag{8.69}$$

The flow coefficient is a function of both Reynolds number and diameter ratio as shown in Fig. 8.47.

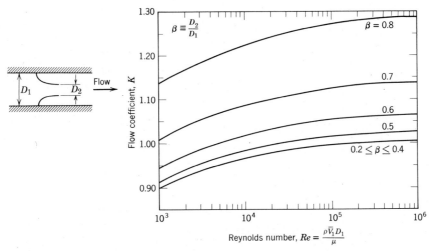

Fig. 8.47 Flow coefficients for ASME long radius flow nozzles (data from Ref. 19).

Figure 8.47 shows that K is essentially constant for large Reynolds number $(Re_{D_1} > 2 \times 10^5)$. Thus at high flowrates, the flowrate may be computed directly. At lower flowrates, where K is a weak function of the Reynolds number, iteration may be required.

For some configurations, K values larger than unity are possible. The reason is that the flow coefficient includes the velocity of approach factor, $1/\sqrt{1 - \beta^4}$, which is always greater than one.

Flow nozzles are intermediate between orifice plates and venturi meters in cost and ease of installation. Their head loss is less than that of an orifice of the same diameter ratio, because no vena contracta is present.

8-18.3 THE VENTURI

The venturi meter was sketched in Table 8.7. It is generally made from a casting that is machined to close tolerances to assure performance duplicating that of the standard design. As a result, venturi meters are bulky, heavy, and expensive. However, the conical difiuser section downstream from the throat gives excellent pressure recovery; the overall head loss is therefore low. The venturi meter is also self-cleaning due to its smooth internal contour.

Experiments show that the velocity coefficient for venturi meters is independent of diameter ratio, over the range $0.25 < \beta < 0.75$. The variation of C_v with Reynolds number is shown in Fig. 8.48. Note that C_v is constant for Reynolds numbers larger than about 2×10^5.

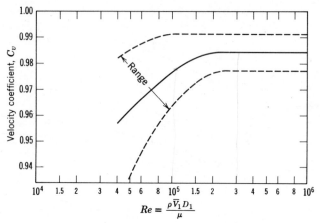

Fig. 8.48 Velocity coefficients for venturi meters, $0.25 < \beta < 0.75$ (data from Ref. 19).

8-18.4 THE LAMINAR FLOW ELEMENT

The orifice, flow nozzle, and venturi all produce a pressure drop proportional to the square of the flowrate, according to Eq. 8.67. In practice, a meter size must be chosen to accommodate the largest flowrate expected. Because the pressure drop versus flowrate relationship is nonlinear, the range of flowrate that can be measured accurately is limited.

The laminar flow element[22] is designed to produce a pressure drop directly proportional to flowrate. The laminar flow element (LFE) contains a section subdivided into many passages, each small enough in diameter to assure fully-developed laminar flow. As shown in Section 8-4, the pressure drop in laminar duct flow is directly proportional to the flowrate. However, the relationship between the pressure drop and the flowrate also depends upon the fluid viscosity, which is a strong function of temperature. Therefore the fluid temperature must be known to obtain accurate metering with a LFE.

The LFE costs approximately as much as a venturi, but it is much lighter and smaller. Thus the LFE is becoming widely used in applications where compactness and extended range are important.

Example 8.18

An air flowrate of 1 m³/sec at standard conditions is expected in a 0.25 m diameter duct. An orifice meter is to be installed to measure the rate of flow. The manometer available to make the measurement has a maximum range

---

[22] Patented and manufactured by Meriam Instrument Co., 10920 Madison Ave., Cleveland, Ohio 44102.

of 150 mm of water. What diameter orifice plate should be used with taps placed 1 diameter upstream and $\frac{1}{2}$ diameter downstream from the plate? Approximately what head loss would be predicted if flow following the vena contracta is treated as a sudden expansion, and $C_c \simeq 0.60$?

Example Problem 8.18

GIVEN:

Flow through duct and orifice as shown.

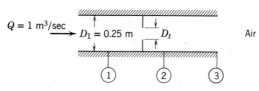

$$(p_1 - p_2)_{max} = 150 \text{ mm H}_2\text{O}$$

FIND:

(a) D_t.
(b) Head loss between sections ① and ③.

SOLUTION:

The orifice plate may be designed using Eq. 8.67 and data from Fig. 8.45.

Computing equation: $\dot{m}_{actual} = KA_t\sqrt{2\rho(p_1 - p_2)}$ (8.67)

Since $A_t/A_1 = (D_t/D_1)^2 = \beta^2$,

$$\dot{m}_{actual} = K\beta^2 A_1\sqrt{2\rho(p_1 - p_2)}$$

or

$$K\beta^2 = \frac{\dot{m}_{actual}}{A_1\sqrt{2\rho(p_1 - p_2)}} = \frac{\rho Q}{A_1\sqrt{2\rho(p_1 - p_2)}} = \frac{Q}{A_1}\sqrt{\frac{\rho}{2(p_1 - p_2)}}$$

$$= \frac{Q}{A_1}\sqrt{\frac{\rho}{2g\rho_{H_2O}\,\Delta h}}$$

$$= \frac{1 \text{ m}^3}{\text{sec}} \times \frac{4}{\pi\,(0.25)^2\text{ m}^2}\left[\frac{1}{2} \times \frac{1.23 \text{ kg}}{\text{m}^3} \times \frac{\text{sec}^2}{9.81 \text{ m}} \times \frac{\text{m}^3}{999 \text{ kg}} \times \frac{1}{0.15 \text{ m}}\right]^{1/2}$$

$$K\beta^2 = 0.417 \quad \text{or} \quad K = \frac{0.417}{\beta^2}$$ (1)

Since K is a function of both β and Re_{D_1}, we must iterate to find β. The duct Reynolds number is

$$Re_{D_1} = \frac{\rho \bar{V}_1 D_1}{\mu} = \frac{\rho (Q/A_1) D_1}{\mu} = \frac{4Q}{\pi \nu D_1}$$

$$Re_{D_1} = \frac{4}{\pi} \times \frac{1 \text{ m}^3}{\text{sec}} \times \frac{\text{sec}}{1.45 \times 10^{-5} \text{ m}^2} \times \frac{1}{0.25 \text{ m}} = 3.51 \times 10^5$$

Guess $\beta = 0.8$. From Fig. 8.45, K should be 0.77. From Eq. 1,

$$K = \frac{0.417}{(0.8)^2} = 0.652$$

Thus our guess for β is too large. Guess $\beta = 0.75$. From Fig. 8.45, K should be 0.72. From Eq. 1,

$$K = \frac{0.417}{(0.75)^2} = 0.741$$

Thus our guess for β is too small. Guess $\beta = 0.76$. From Fig. 8.45, K should be 0.725. From Eq. 1,

$$K = \frac{0.417}{(0.76)^2} = 0.722$$

This is satisfactory agreement. $\beta \simeq 0.76$ and

$$D_t = \beta D_1 = 0.76(0.25 \text{ m}) = 0.19 \text{ m} \qquad\qquad D_t$$

To evaluate the head loss, apply Eq. 8.23 between sections ① and ③.

Computing equation:

$$\left[\frac{p_1}{\rho} + \alpha_1 \frac{\bar{V}_1^2}{2} + gz_1\right] - \left[\frac{p_3}{\rho} + \alpha_3 \frac{\bar{V}_3^2}{2} + gz_3\right] = h_{l_T} \qquad (8.23)$$

Assumptions: (1) $\alpha_1 \bar{V}_1^2 = \alpha_3 \bar{V}_3^2$
(2) Neglect Δz

Then

$$h_{l_T} = \frac{p_1 - p_3}{\rho} = \frac{p_1 - p_2 - (p_3 - p_2)}{\rho} \qquad (2)$$

The pressure at ③ may be found by applying the x component of the momentum equation to a control volume between sections ② and ③.

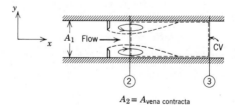

$A_2 = A_{\text{vena contracta}}$

$$= 0(3) = 0(4)$$

Basic equation: $\quad F_{S_x} + F_{B_x} = \dfrac{\partial}{\partial t}\displaystyle\int_{CV} u\rho\, d\forall + \displaystyle\int_{CS} u\rho \vec{V} \cdot d\vec{A}$ (4.19a)

Assumptions: (3) $F_{B_x} = 0$
(4) Steady flow
(5) Uniform flow at sections ② and ③
(6) Pressure uniform across duct at sections ② and ③
(7) No friction

Then

$$(p_2 - p_3)A_1 = u_2\{-|\rho \bar{V}_2 A_2|\} + u_3\{|\rho \bar{V}_3 A_3|\} = (u_3 - u_2)\rho Q = (\bar{V}_3 - \bar{V}_2)\rho Q$$

or

$$p_3 - p_2 = (\bar{V}_2 - \bar{V}_3)\frac{\rho Q}{A_1}$$

Now $\bar{V}_3 = Q/A_1$, and

$$\bar{V}_2 = \frac{Q}{A_2} = \frac{Q}{C_c A_t} = \frac{Q}{C_c \beta^2 A_1}$$

Thus

$$p_3 - p_2 = \frac{\rho Q^2}{A_1^2}\left[\frac{1}{C_c \beta^2} - 1\right]$$

$$p_3 - p_2 = \frac{1.23\ \text{kg}}{\text{m}^3} \times \frac{(1)^2\ \text{m}^6}{\text{sec}^2} \times \frac{4^2}{\pi^2}\frac{1}{(0.25)^4\ \text{m}^4}\left[\frac{1}{0.6(0.76)^2} - 1\right]\frac{\text{N} \cdot \text{sec}^2}{\text{kg} \cdot \text{m}}$$

$$p_3 - p_2 = 962\ \text{N/m}^2$$

The diameter ratio, β, was selected to give the maximum manometer deflection. Thus

$$p_1 - p_2 = \rho_{H_2O}\, g\, \Delta h = \frac{999\ \text{kg}}{\text{m}^3} \times \frac{9.81\ \text{m}}{\text{sec}^2} \times 0.15\ \text{m} \times \frac{\text{N} \cdot \text{sec}^2}{\text{kg} \cdot \text{m}} = 1470\ \text{N/m}^2$$

Substituting into Eq. 2 gives

$$h_{l_T} = \frac{p_1 - p_3}{\rho} = \frac{p_1 - p_2 - (p_3 - p_2)}{\rho}$$

$$h_{l_T} = \frac{(1470 - 962)\ \text{N}}{\text{m}^2} \times \frac{\text{m}^3}{1.23\ \text{kg}} = 413\ \text{N} \cdot \text{m/kg} \qquad\qquad h_{l_T}$$

{ This problem illustrates flowmeter calculations and shows use of the momentum equation to compute the pressure rise in a sudden expansion. }

8-19 MECHANICAL FLOWMETERS

In specialized applications, particularly for remote use or recording, mechanical flowmeters may be specified. Common examples include household water and natural gas meters, which are generally calibrated to read directly in units of the product used, or gasoline metering pumps at the local filling station, which measure total flow and automatically compute the dollar value. A large variety of such meters is available commercially. Manufacturers' literature should be consulted for design and installation details.

Float-type meters may be used to give a direct indication of flowrate for both liquids and gases. An example is shown in Fig. 8.49. In operation, the ball or float is carried upward in the tapered clear tube by the flowing fluid until the drag force and float weight are in equilibrium. Such meters are available with factory calibration for a number of common fluids and ranges of flowrate.

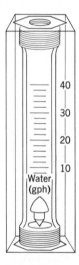

Fig. 8.49 Float-type mechanical flowmeter (courtesy Dwyer Instrument Co., Michigan City, Indiana).

The final type of meter that we shall mention is the turbine flowmeter. In this form of flowmeter, a free-running turbine or vaned impeller is mounted in a cylindrical section of tube (Fig. 8.50). The rate of rotation of the impeller is closely proportional to the volumetric flowrate over a wide range. This linearity is a principal advantage of the turbine flowmeter.

Rotational speed of the turbine element can be sensed using a magnetic pickup external to the meter. This sensing method therefore requires no penetrations or seals in the duct. Thus turbine flowmeters can be used safely to measure flowrates in corrosive or toxic fluids. The electrical signal can be displayed or recorded easily, or integrated to provide total flow.

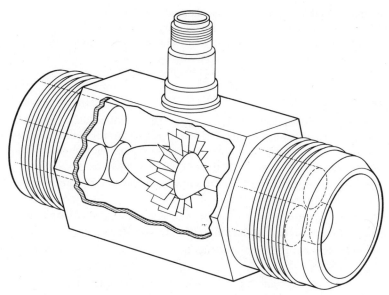

Fig. 8.50 Turbine-type mechanical flowmeter (courtesy Potter Aeronautical Corp., Union, New Jersey).

8-20 TRAVERSING METHODS

In some practical situations, for example, in air handling or refrigeration equipment, it is either impractical or impossible to install a fixed flowmeter. In many such cases it is possible to obtain flowrate data using a traversing technique.

To make a flowrate measurement by traverse, the duct cross section is subdivided into segments of equal area. The fluid velocity is measured at the center of each such segment using a pitot or total head tube or a suitable anemometer. Then the volumetric flowrate is approximated for each segment by the product of velocity and area. The flowrate through the entire duct is the sum of these segmental flowrates. (Reference 6 contains details of accepted procedures for flowrate measurements by the traverse method.)

summary objectives

After completing study of Chapter 8, you should be able to do the following:

1. Define:
   internal flow
   external flow
   major, minor head loss
   **hydraulic diameter

** These objectives apply to sections that may be omitted without loss of continuity in the text material.

entrance length

boundary-layer thickness

fully-developed flow

displacement thickness

laminar flow

pressure gradient (favorable, adverse)

turbulent flow

separation

kinetic energy flux coefficient

skin friction coefficient

momentum flux coefficient

drag

head loss

lift

2. For fully-developed laminar flows, apply the control volume formulation of Newton's second law to a suitably chosen differential control volume to determine: the velocity distribution, the shear stress distribution, the volumetric flowrate, the average velocity, and the location of the maximum velocity.

3. For fully-developed turbulent flow in a pipe, with velocity distribution represented by a power-law profile determine $\bar{V}/U$.

4. For fully-developed flow in a pipe determine the wall shear stress and the shear stress variation in the flow in terms of the pressure gradient.

5. Write the first law of thermodynamics in a form suitable for the solution of pipe flow problems.

6. Use the Moody diagram to determine the friction factor for fully-developed pipe flow.

7. Solve single-path pipe flow system problems for each of the four cases discussed in Section 8-9.1.

**8. Utilize the basic techniques of Section 8-9.1 to analyze multiple-path pipe flow systems.

9. Beginning with the momentum integral equation for zero pressure gradient flow, develop expressions for $\delta(x)$, $\delta^*(x)$, $\tau_w(x)$, $C_f(x)$; determine the total friction force on a flat plate placed parallel to the flow.

10. Utilize the displacement thickness concept to approximate the pressure drop in the entrance region of a channel.

11. Determine a first approximation for the pressure gradient in a channel.

12. Determine the drag and lift forces for bodies in external flow.

13. Determine the mass flowrate from the pressure drop measured, using an orifice plate, flow nozzle, or venturi meter.

14. Solve those problems at the end of the chapter that relate to the material you have studied.

problems

8.1 Approximately how far from the entrance of a 10 mm diameter pipe would flow be fully developed if the Reynolds number is 1500, based on average velocity?

8.2 Standard air enters a 0.3 m diameter duct. The volumetric flowrate is 2 m³/min. Determine whether the flow is laminar or turbulent. Estimate the entrance length required for fully-developed flow to be established.

8.3 The velocity profile for flow between stationary parallel plates is given by

$$u = ay(h - y)$$

where a is a constant, h is the total gap width between plates, and y is the distance measured upward from the lower plate. Determine the ratio $\bar{V}/u_{max}$.

8.4 An incompressible fluid flows between two infinite stationary parallel plates. The velocity profile is given by

$$u = u_{max}(Ay^2 + By + C)$$

where A, B, and C are constants, and y is measured from the center of the gap. The total gap width is h units. Use appropriate boundary conditions to express the constants in terms of h. Develop an expression for volumetric flowrate per unit depth.

8.5 A viscous oil flows steadily between parallel plates. The flow is laminar and fully developed. The velocity profile is given by

$$u = -\frac{h^2}{8\mu}\frac{\partial p}{\partial x}\left[1 - \left(\frac{2y}{h}\right)^2\right]$$

where the total gap width between the plates, $h = 3$ mm, and y is measured from the centerline of the gap. The oil viscosity is 0.5 N·sec/m², and the pressure gradient is -1200 N/m²/m. Find the magnitude and direction of the shear stress on the upper plate, and the volumetric flowrate through the channel, per meter of width.

8.6 A fluid flows steadily between two parallel plates. The distance between the plates is h. The velocity profile for fully-developed laminar flow is given by

$$u = -\frac{h^2}{8\mu}\frac{\partial p}{\partial x}\left[1 - \left(\frac{2y}{h}\right)^2\right]$$

(a) Derive an equation for the shear stress as a function of y. Plot this function.
(b) For $\mu = 2.4 \times 10^{-5}$ lbf·sec/ft², $\partial p/\partial x = -4.0$ lbf/ft²/ft, and $h = 0.05$ in., calculate the maximum shear stress in lbf/ft².

8.7 A block of mass M = 5 kg slides down a plane inclined at 30 degrees from the horizontal. The block rides on an oil film 0.025 mm thick. The oil viscosity is 0.28 N·sec/m², and the terminal velocity of the block is 0.3 m/sec. Find the area of the block surface in contact with the oil film.

8.8 A sealed journal bearing is formed from concentric cylinders. The inner and outer radii are 25 and 26 mm, respectively; the journal length is 100 mm, and it turns at 2800 rpm. The gap is filled with oil in laminar motion. The velocity profile is linear across the gap. The torque needed to turn the journal is 0.2 N·m.

Calculate the viscosity of the oil. Will the torque increase or decrease with time? Why?

8.9 Water at 60 C flows between two large flat plates. The lower plate moves to the left at a speed of 0.3 m/sec. The plate spacing is 3 mm, and the flow is laminar. Determine the pressure gradient required to produce zero net flow at a cross section.

8.10 Water at 60 F flows between parallel plates with gap width $b = 0.01$ ft. The upper plate moves with velocity $U = 1$ ft/sec in the positive x direction. The pressure gradient is $\partial p / \partial x = -0.06$ lbf/ft^2/ft. Locate the point of maximum velocity and determine its magnitude (let $y = 0$ at the bottom plate). Sketch the velocity and shear stress distributions. Determine the volume of flow that passes a given cross section ($x = $ constant) in 10 sec.

8.11 A film of water (at 15 C) in steady, laminar motion runs down a long slope, inclined 30° below the horizontal. The thickness of the film is 0.8 mm. Assume that flow is fully developed, and at zero pressure gradient. Determine the surface shear stress, and the volumetric flowrate per unit width.

8.12 A continuous belt (Fig. 8.51) passing upward through a chemical bath at velocity, U_0, picks up a liquid film of thickness, h, density, ρ, and viscosity, μ. Gravity tends to make the liquid drain down, but the movement of the belt keeps the fluid from running off completely. Assume that the flow is fully-developed laminar flow with zero pressure gradient, and that the atmosphere produces no shear at the outer surface of the film. State clearly the boundary conditions to be satisfied by the velocity at $y = 0$ and $y = h$. Obtain an expression for the velocity profile.

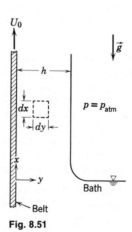

Fig. 8.51

8.13 The velocity distribution for fully-developed laminar flow in a pipe is given by

$$u = -\frac{R^2}{4\mu} \frac{\partial p}{\partial z} \left[1 - \left(\frac{r}{R} \right)^2 \right]$$

Determine the radial distance from the pipe axis at which the velocity equals the average velocity.

8.14 Consider first water and then SAE 10 W lubricating oil flowing at 40 C in a 6 mm diameter tube. Determine the maximum flowrate (and the corresponding pressure gradient, $\partial p / \partial x$) for each fluid at which laminar flow would be expected.

8.15 A tube 17.6 in. long, with an inside diameter of 0.030 in., is used as a capillary viscometer. Calibration tests are made using water at 60 F, and a flowrate of 1 cm^3/sec is measured for an applied pressure drop of 10 psi. (Assume that the pressure drop in the entrance length is twice that for the same length of fully-developed flow.) Determine the percentage error in viscosity that would result if Eq. 8.13c were used directly to compute it, without considering the entrance length.

8.16 Consider fully-developed laminar flow in a circular pipe. Use a cylindrical control volume as shown in Fig. 8.52. Indicate the forces acting on the control volume. Using the momentum equation, develop an expression for the velocity distribution.

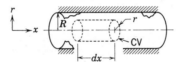

Fig. 8.52

8.17 Consider fully-developed laminar flow in the annulus between two concentric pipes (Fig. 8.53). The inner pipe is stationary, and the outer pipe moves in the x direction with speed, V_0. Assume the axial pressure gradient to be zero ($\partial p / \partial x = 0$). Obtain a general expression for the shear stress, τ, as a function of the radius, r, in terms of a constant, C_1. Obtain a general expression for the velocity profile, $V(r)$, in terms of two constants, C_1 and C_2. Evaluate the constants, C_1 and C_2.

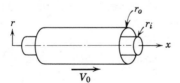

Fig. 8.53

8.18 Consider the velocity profile for fully-developed laminar pipe flow, Eq. 8.14, and the empirical "power-law" profile for turbulent pipe flow, Eq. 8.16. Assume that $n = 7$ for the turbulent profile. Determine the value of r/R for each profile at which u is equal to the average velocity, $\bar{V}$.

8.19 The velocity profile for turbulent flow through smooth pipes is often represented by the empirical equation

$$\frac{u}{U} = \left[1 - \frac{r}{R}\right]^{1/n}$$

Show that the ratio of average to centerline velocities is given by

$$\frac{\bar{V}}{U} = \frac{2n^2}{(n+1)(2n+1)}$$

Evaluate the ratio for $n = 7$ (corresponding to a Reynolds number of 110,000) and for $n = 8.8$ ($Re \simeq 1.1 \times 10^6$). Using the values for n given in Section 8-5, plot $\bar{V}/U$ as a function of Reynolds number.

8.20 Kerosine at 70 F flows in a smooth tube with an inside diameter of 1 in. The flow Reynolds number is 4000. For laminar flow, the pressure gradient is found to be $\partial p/\partial x = -0.2$ lbf/ft^2/ft, while for turbulent flow, $\partial p/\partial x = -0.5$ lbf/ft^2/ft. Plot the variation of shear stress as a function of radius for both flow conditions.

8.21 The kinetic energy flux coefficient, α, is defined by Eq. 8.21. Using the "power-law" turbulent velocity profile, show that

$$\alpha = \left[\frac{U}{\bar{V}}\right]^3 \frac{2n^2}{(3+n)(3+2n)}$$

Evaluate α for $n = 7$.

8.22 A momentum flux coefficient, β, may be defined as

$$\int_A u\rho u \, dA = \beta \int_A \bar{V}\rho u \, dA = \beta \dot{m} \bar{V}$$

Evaluate β for a laminar velocity profile, Eq. 8.14, and for a "power-law" turbulent velocity profile, Eq. 8.16 (choose $n = 7$).

8.23 Water flows in a constant area pipeline; the pipe diameter is 50 mm and the average velocity of the flow is 1.5 m/sec. At the pipe inlet the gage pressure is 590 kPa. The outlet of the pipe is at an elevation 25 m higher than the inlet; the outlet pressure is atmospheric. Determine the head loss between the inlet and outlet of the pipe.

8.24 The pipe of Problem 8.23 is placed on a horizontal surface. The flowrate and outlet pressure are to remain the same. Compute the inlet pressure for this new condition.

8.25 Resistance to fluid flow can be defined by analogy to Ohm's law for electric current. Thus resistance to flow is given by the ratio of pressure drop (driving potential) to volumetric flowrate (current). Show that resistance to laminar flow is given by

$$\text{resistance} = \frac{128\mu L}{\pi D^4}$$

which is independent of the flowrate. Find the maximum pressure drop for which this relation is valid for a tube 50 mm long and 0.25 mm inside diameter for both kerosine and castor oil at 40 C.

8.26 A 15 m length of new wrought iron pipe of 25 mm inside diameter is to be used in a horizontal position to convey water at 15 C. The average velocity of the water in the pipe is 5 m/sec. Determine the volumetric flowrate and friction factor for the flow.

8.27 Water flows through a galvanized iron pipe at a flowrate of 0.2 m³/sec. The inside diameter of the pipe is 150 mm, and the water temperature is 20 C. Compute the friction factor for the flow.

8.28 An empirical correlation for friction factor in turbulent flow in smooth pipes was developed by H. Blasius in 1911. He found that

$$f = \frac{0.316}{Re^{1/4}}, \qquad Re \leq 100{,}000$$

correlated data well. Show that in turbulent flow, the predicted pressure drop is proportional to $(\bar{V})^{7/4}$ when the Blasius correlation is used. How does pressure drop depend on the tube diameter at a given flowrate?

8.29 A correlating equation for flow in the fully-rough flow regime is

$$f = \frac{1}{4[0.57 - \log_{10}(e/D)]^2}$$

as originally obtained by von Kármán. Compare values obtained from this equation with those from the fully-rough region of the Moody chart for values of e/D equal to 0.01, 0.001, and 0.0001.

8.30 A smooth, 3 in. diameter pipe carries water horizontally at 150 F at a mass flowrate of 0.006 slug/sec. The pressure drop is observed to be 0.065 lbf/ft² per 100 ft of pipe. From the Moody chart, the friction factor could be chosen as 0.021 or 0.042. Which is correct?

8.31 Air at standard conditions flows through a sudden expansion in a circular duct. The upstream and downstream duct diameters are 3 and 9 in., respectively. The pressure downstream is 0.25 in. of water *higher* than that upstream. Determine the average velocity of the air approaching the expansion, and the volumetric flowrate.

8.32 Water flows through a 50 mm diameter tube that suddenly contracts to a 25 mm diameter. The pressure drop across the contraction is 3.4 kPa. Determine the volumetric flowrate.

8.33 Air flows out of a clean room test chamber through a 150 mm diameter duct. The original duct had a square-edged entrance, but this has been replaced with a well-rounded one. The pressure in the chamber is 2.5 mm of water above ambient. Losses due to friction are negligible compared to the entrance and exit losses. Determine the increase in volumetric flowrate that results from the change in entrance contour.

8.34 Space has been found for a conical diffuser 0.45 m long in the clean room ventilation system described in Problem 8.33. The best diffuser of this size is to be used. Assume that data from Fig. 8.17 may be used. Determine the appropriate diffuser angle and area ratio for this installation and predict the volumetric flowrate that will be delivered after it is installed.

8.35 Water at 20 C flows through a 0.1 m (internal diameter) concrete drainage pipe at a rate of 15 kg/sec. Determine the pressure drop per 100 m of horizontal pipe.

8.36 Air at 15 C flows through a straight, 0.3 m diameter, smooth duct 50 m long. The flowrate is 0.6 m³/sec, and the pressure is the same at both ends of the duct. Determine the change in elevation between the inlet and outlet.

8.37 Water at 78 F flows in a pipe whose inside diameter is 1.2 in. The flowrate is 0.04 ft³/sec. Determine the slope that the pipe must have to maintain constant pressure along its length. If the temperature remains constant, determine the heat transfer per 100 ft of pipe.

8.38 Water flows from a large reservoir as shown in Fig. 8.54. The pipe is of cast iron, with an inside diameter of 0.2 m. The flowrate is 0.14 m³/sec, and the discharge is to atmospheric pressure. The mean temperature for the flow is 10 C; the entire system is insulated. Determine the gage pressure, p_1, required to produce the flow. Calculate the temperature rise between the liquid surface and the exit.

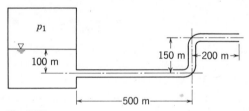

Fig. 8.54

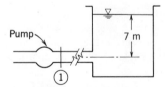

Fig. 8.55

8.39 Water from a pump flows through a 0.25 m diameter pipe for a distance of 5 km from the pump discharge to a reservoir open to the atmosphere (Fig. 8.55). The level of the water in the reserovir is 7 m above the pump discharge, and the average velocity of the fluid in the pipe is 3 m/sec. Calculate the pressure at the pump discharge.

8.40 Water is to flow by gravity from one reservoir to a lower one through a straight, inclined pipe. The flowrate required is 0.007 m³/sec, the pipe diameter is 50 mm and the total length is 250 m. Each reservoir is open to the atmosphere. Neglecting minor losses, calculate the difference in level required to maintain this flowrate.

8.41 Shown in Fig. 8.56 is a water flow system with a variable elevation reservoir, B. Determine the water level in reservoir B so that no water flows into or out of the reservoir. The velocity in the 12 in. diameter pipe is 10 ft/sec. Neglect all minor losses.

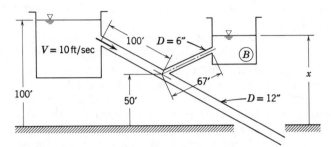

Fig. 8.56

8.42 Two reservoirs are connected by three clean cast-iron pipes in series, $L_1 = 600$ m, $D_1 = 0.3$ m, $L_2 = 900$ m, $D_2 = 0.4$ m, $L_3 = 1500$ m, and $D_3 = 0.45$ m. When the discharge is 0.11 m³/sec of water at 15 C, determine the difference in elevation between the reservoirs.

8.43 Lightweight crude oil ($SG = 0.855$ with viscosity similar to SAE 30 oil) is pumped horizontally through a 1 mile length of 12 in. diameter pipe. The average roughness size is 0.01 in. The flowrate is 4500 gpm. Calculate the horsepower required to drive the pump if it is 75 percent efficient.

8.44 Kerosine at 60 C flows through a pipe system in a refinery (Fig. 8.57) at the rate of 2.3 m³/min. The pipe is commercial steel, with an inside diameter of 0.15 m. The gage pressure in the reactor vessel is 90 kPa. Determine the total length, L, of the straight pipe in the system.

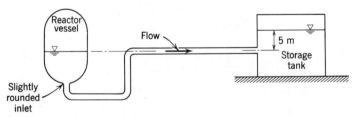

Fig. 8.57

8.45 Gasoline flows in a long, underground pipeline at a constant temperature of 15 C. Two pumping stations at the same elevation are located 13 km apart. The pressure drop between the stations is 1.4 MPa. The pipeline is made from

0.6 m pipe. Although made from commercial steel, age and corrosion have raised the pipe roughness to approximately that for galvanized iron. The specific gravity of the gasoline is 0.68. Compute the volumetric rate of flow of gasoline through the pipe.

8.46   A fire nozzle is supplied through 300 ft of 1.5 in. diameter, smooth, rubber-lined hose. Water from a hydrant is supplied to a booster pump on board the pumper truck at 50 psig. At design conditions, the pressure at the nozzle inlet is 100 psig, and the pressure drop along the hose is 33 psi per 100 ft of length. Determine:
(a)   The design flowrate.
(b)   The nozzle exit velocity, assuming no losses in the nozzle.
(c)   The power required to drive the booster pump if its efficiency is 70 percent.

8.47   Heavy crude oil ($SG = 0.925$ and $v = 1.1 \times 10^{-3}$ ft²/sec) is pumped through a pipeline laid on flat ground. The line is made from steel pipe with 24 in. inside diameter and has a wall thickness of $\frac{1}{2}$ in. The allowable tensile stress in the pipe wall is limited to 40,000 psi by corrosion considerations. At the same time it is important to keep the oil under pressure to ensure that gases remain in solution. The minimum recommended pressure is 75 psia. The pipeline carries a flow of 400,000 barrels (in the petroleum industry, a "barrel" is 42 gal) per day. Determine the maximum spacing between pumping stations. Compute the power added to the oil at each pumping station.

8.48   A siphon is shown in Fig. 8.58. Assume that the only head loss occurs at the tube inlet, which may be considered a reentrant entrance. The tube inside diameter is 75 mm. The fluid is water at 15 C. Determine the volumetric flowrate through the siphon. Compute the pressure at point A.

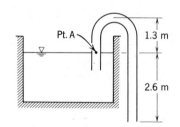

Fig. 8.58

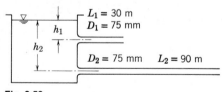

Fig. 8.59

8.49   The tank in Fig. 8.59 has two lengths of 75 mm galvanized pipe attached at the levels shown. The flow may be assumed to be in the fully-rough region, and the entrances are well-rounded. Determine the ratio of h_2 to h_1 that will cause the same flowrate in each pipe line. Compute the minimum value of h_1 that will produce flow in the fully-rough region.

8.50 Two reservoirs containing water are connected by a constant area, galvanized iron pipe that has one right-angle bend. The surface pressure at the upper reservoir is atmospheric whereas the gage pressure at the lower reservoir surface is 70 kPa. The pipe diameter is 75 mm. Assume that the only significant losses occur in the pipe and bend (Fig. 8.60). Determine the magnitude and direction of the volumetric flowrate.

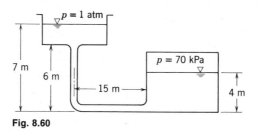

Fig. 8.60

8.51 You are watering your lawn with an *old* hose. Because of the buildup of lime deposits over the years, the 0.75 in. i.d. hose now has an average roughness height of 0.022 in. One 50 ft length of the hose, attached to your spigot, delivers 20 gal of water (60 F) per min. Compute the pressure at the spigot, in psi. Assuming the pressure at the spigot remains constant, estimate the delivery if two 50 ft lengths of the hose are connected together.

8.52 A circular tank 1.5 m in diameter is filled to a depth of 5 m with water at 15 C. A smooth tube 7 m long is attached to a well-rounded entrance at the tank bottom. The 50 mm diameter tube discharges to atmosphere 5 m below the tank bottom. Estimate the time required for the tank level to drop 1.5 m. (You may wish to use the computer to solve this problem. If so, the Blasius correlation for friction factor for turbulent flow,

$$f = \frac{0.316}{Re^{1/4}}$$

might come in handy.)

8.53 A spray system for a sewage treatment plant is shown in Fig. 8.61. Water at 60 F is pumped through a spray arm. The effective flow area of each nozzle is

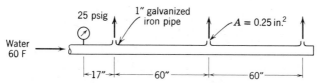

Fig. 8.61

0.25 in.^{2} The pipe inside diameter is 1 in., and the material is galvanized iron. Determine the flowrate of water through the spray arm.

8.54   A hydraulic press is powered by a remote high pressure pump. The gage pressure at the pump outlet is 20 MPa, whereas the pressure required for the press is 19 MPa (gage), at a flowrate of 0.032 m^3/min. The press and pump are to be connected by 50 m of smooth, drawn steel tubing. The fluid is SAE 10 W oil at 40 C. Determine the minimum tubing diameter that may be used.

8.55   A pump is located 15 ft to one side and 12 ft above a reservoir (Fig. 8.62). The pump is designed for a flowrate of 100 gpm. For satisfactory operation, the suction head at the pump inlet must not be lower than -20 ft of water. Determine the smallest standard commercial steel pipe that will give the required performance.

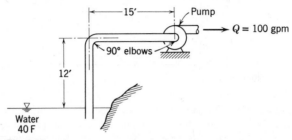

Fig. 8.62

8.56   A new industrial plant requires a water flow of 5.7 m^3/min. The gage pressure in the water main, located in the street 50 m from the plant, is 800 kPa. The supply line will require installation of 4 elbows in a total length of 65 m. The gage pressure required in the plant is 500 kPa. What size galvanized iron line should be installed?

8.57   A swimming pool has a partial flow filtration system. Water at 75 F is pumped from the pool through the system shown in Fig. 8.63. The pump delivers 30 gpm at 60 psig. The pipe is galvanized iron ($e/D \simeq 0.009$). The pressure loss through the filter is given approximately by

$$\Delta p = 0.6Q^2$$

where Δp is in psi and Q is in gallons per minute. Determine the flow through each branch of the system.

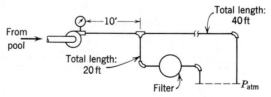

Fig. 8.63

8.58   The results of Example Problem 8.11 indicated that the pipes used were too small to produce the same pressure at each nozzle inlet. Rework the problem using the same geometry, but with $3\frac{1}{2}$ in. smooth pipes.

8.59   In a certain air-conditioning installation, a flowrate of 35 m³/min of air at standard conditions is required. A smooth sheet metal duct 0.3 m square is to be used. Determine the pressure drop for a 30 m horizontal duct run.

8.60   Consider flow of standard air at 35 m³/min. Compare the pressure drop per unit length of a round duct with that for rectangular ducts of aspect ratio 1, 2, and 3. Assume all ducts are smooth, with a cross-sectional area of 0.1 m².

8.61   Determine the minimum size smooth rectangular duct with an aspect ratio of 2 that will pass 80 m³/min of standard air with a head loss of 30 mm of water per 30 m of duct.

8.62   The head versus capacity curve for a certain fan may be approximated by the equation

$$h = 30 - 10^{-7}(Q^2)$$

where h is the output static head in inches of water and Q is the air flowrate in ft³/min. The fan outlet dimensions are 8×16 in. Determine the air flowrate delivered by the fan into a 200 ft straight length of 8×16 in. rectangular duct.

8.63   A submarine moves at 20 knots through sea water at 7 C. Assume that near the bow, the boundary layer behaves as though on a flat plate. Determine the range of distance from the bow that transition from laminar to turbulent flow in the boundary layer might be expected.

8.64   An airplane cruises at 300 knots at an altitude of 10 km on a standard day. Assume that the boundary layers on the wing surfaces behave as on a flat plate. Determine the expected extent of laminar flow in the wing boundary layers.

8.65   Aircraft and missiles flying at high altitude may have regions of laminar flow that become turbulent at lower altitudes at the same speed. Explain a possible mechanism for this effect. Support your answer with calculations based on standard atmosphere data.

8.66   The most general sinusoidal velocity profile for laminar boundary-layer flow on a flat plate is

$$u = A \sin(By) + C$$

State three boundary conditions applicable to the laminar boundary-layer velocity profile. Evaluate the constants A, B, and C.

8.67   Velocity profiles in laminar boundary layers are often approximated by the equations

Linear: $\dfrac{u}{U} = \dfrac{y}{\delta}$

Parabolic: $\dfrac{u}{U} = 2\left(\dfrac{y}{\delta}\right) - \left(\dfrac{y}{\delta}\right)^2$

Cubic: $\dfrac{u}{U} = \dfrac{3}{2}\left(\dfrac{y}{\delta}\right) - \dfrac{1}{2}\left(\dfrac{y}{\delta}\right)^3$

Sinusoidal: $\dfrac{u}{U} = \sin\dfrac{\pi}{2}\left(\dfrac{y}{\delta}\right)$

Compare the shapes of these velocity profiles by plotting y/δ (on the ordinate) versus u/U (on the abscissa).

8.68 The velocity profile in a turbulent boundary layer is often approximated by the "power-law" equation

$$\frac{u}{U} = \left(\frac{y}{\delta}\right)^{1/7}$$

Compare the shape of this profile with the parabolic laminar boundary layer profile (Problem 8.67) by plotting y/δ (on the ordinate) versus u/U (on the abscissa) for both profiles.

8.69 Transition from laminar to turbulent boundary-layer flow actually occurs over a finite length of surface, during which the velocity profile and wall shear stress adjust from laminar to turbulent forms. A useful approximation during transition is that the momentum flux within the boundary layer remains constant. Assuming constant momentum flux, show that

$$\delta_{\text{turbulenl}} = \tfrac{72}{105}\delta_{\text{laminar}}$$

for transition from a parabolic laminar velocity profile to a "$\frac{1}{7}$ power" turbulent velocity profile.

8.70 Evaluate the ratio δ^*/δ for each of the laminar boundary-layer velocity profiles given in Problem 8.67.

8.71 Evaluate the ratio δ^*/δ for the "power-law" form used to represent the turbulent velocity profile, that is,

$$\frac{u}{U} = \left(\frac{y}{\delta}\right)^{1/7}$$

Compare with the value for the cubic, laminar, boundary-layer velocity profile given in Problem 8.67.

8.72 Consider a laminar boundary layer on a flat plate with a velocity profile given by

$$\frac{u}{U} = \frac{3}{2}\eta - \frac{\eta^3}{2}; \qquad \eta = \frac{y}{\delta}$$

For this profile

$$\frac{\delta}{x} = \frac{4.64}{\sqrt{Re_x}}$$

Determine an expression for δ^*/x.

8.73   Air flows in the entrance region of a square duct, as shown in Fig. 8.64. The velocity at the inlet is uniform, $U_0 = 30$ m/sec, and the duct is 80 mm square. At a section 0.3 m downstream from the entrance, the displacement thickness, δ^*, on each wall measures 1.0 mm. Determine the pressure change between sections ① and ②.

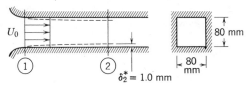

Fig. 8.64

8.74   Air at standard conditions is flowing over a thin flat plate which is 1 m long and 0.3 m wide. The flow is uniform at the leading edge of the plate. The velocity profile in the boundary layer is assumed to be linear, and the freestream velocity is $U = 30$ m/sec. Treat the flow as two-dimensional, that is, assume flow conditions are independent of z (Fig. 8.65). Using the control volume $abcd$, shown by the dashed lines, compute the mass flowrate across surface ab. Determine the magnitude and direction of the x component of the force required to hold the plate stationary.

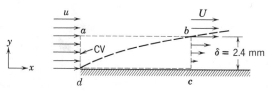

Fig. 8.65

8.75   Rework Problem 8.74 with the same information, except assume the parabolic velocity profile,

$$\frac{u}{U} = 2\left(\frac{y}{\delta}\right) - \left(\frac{y}{\delta}\right)^2$$

at section bc, and $\delta = 3.8$ mm.

8.76   Water flows over a flat plate at a freestream speed of 0.5 ft/sec. There is no pressure gradient and the laminar boundary layer is 0.25 in. thick. Assume a sinusoidal velocity profile.

$$\frac{u}{U} = \sin\frac{\pi}{2}\left(\frac{y}{\delta}\right)$$

Derive an equation for the shear stress at any location within the boundary layer. For the flow conditions given, compute the local wall shear stress and skin friction coefficient.

8.77   When a cubic velocity profile for the laminar boundary layer,

$$\frac{u}{U} = \frac{3}{2}\eta - \frac{1}{2}\eta^3; \qquad \eta = \frac{y}{\delta}$$

is used in the momentum integral equation, the variation of δ with x is found to be

$$\frac{\delta}{x} = \frac{4.64}{\sqrt{Re_x}}$$

For a laminar boundary layer with cubic profile, obtain an expression for the local skin friction coefficient,

$$C_f = \frac{\tau_w}{\frac{1}{2}\rho U^2}$$

in terms of distance and flow properties.

8.78   Air at standard conditions flows over a flat plate. The freestream speed is 15 m/sec. Find δ and τ_w, at $x = 1$ m from the leading edge for (a) completely laminar flow (assume a parabolic velocity profile), and (b) completely turbulent flow (assume a "$\frac{1}{7}$ power" velocity profile).

8.79   The velocity profile in a laminar boundary-layer flow at zero pressure gradient is to be approximated by the linear expression,

$$\frac{u}{U} = \eta; \qquad \eta = \frac{y}{\delta}$$

Use the momentum integral equation with this profile to obtain an expression for the ratio δ/x, and the skin friction coefficient, C_f.

8.80   The velocity profile in a laminar boundary-layer flow at zero pressure gradient is to be approximated by the cubic expression

$$\frac{u}{U} = \frac{3}{2}\eta - \frac{1}{2}\eta^3; \qquad \eta = \frac{y}{\delta}$$

Use the momentum integral equation with this profile to obtain an expression for the ratio δ/x and the skin friction coefficient, C_f.

8.81   The velocity profile in a turbulent boundary-layer flow at zero pressure gradient is to be approximated by the "$\frac{1}{6}$ power" profile expression,

$$\frac{u}{U} = \eta^{1/6}; \qquad \eta = \frac{y}{\delta}$$

Use the momentum integral equation with this profile to obtain an expression for the ratio δ/x and the skin friction coefficient, C_f. Compare with the results obtained in Section 8-13.2 for the "$\frac{1}{7}$ power" profile.

8.82   Repeat Problem 8.81, using the "$\frac{1}{8}$ power" profile expression.

8.83 Figure 8.66 shows two hypothetical boundary-layer velocity profiles. Calculate the momentum flux of each profile. If the two profiles were subjected to the same pressure gradient conditions, which would be most likely to separate first? Why?

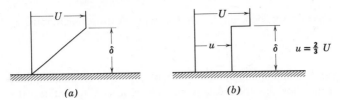

(a)   (b) $u = \frac{2}{3} U$

Fig. 8.66

8.84 A plane-wall diffuser is sketched in Fig. 8.67. We wish to compare the flow of an ideal fluid ($\mu = 0$) and a real fluid in such a passage.

Consider first the case where $\phi = 0$, that is, a straight channel. What can be said of the pressure gradient for the real and ideal fluids? Which fluid gives the higher value of p_2?

Now consider a case where ϕ is not equal to zero, but is small enough to avoid separation. Again, what can be said of the pressure gradient for real and ideal fluids? Which case results in the highest exit pressure?

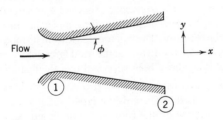

Fig. 8.67

8.85 A uniform flow of standard air at 60 m/sec enters a plane-wall diffuser with negligible boundary-layer thickness. The inlet width is 75 mm. The diffuser walls diverge slightly to accommodate the boundary-layer growth, so that the pressure gradient is negligible. Flat plate boundary-layer behavior may be assumed. Explain why the Bernoulli equation is applicable to this flow. Estimate the diffuser width 1.2 m downstream from the entrance.

8.86 Water at 15 C flows over a flat plate at a speed of 1 m/sec. The plate is 0.4 m long and 1 m wide. The boundary layer on each surface of the plate is laminar. Assume the velocity profile is approximated by a linear expression, for which

$$\frac{\delta}{x} = \frac{3.46}{\sqrt{Re_x}}$$

Determine the drag force on the plate.

8.87 Equation 8.49 gives the local shear stress, τ_w, in terms of Re_x for turbulent boundary-layer flow at zero pressure gradient. The total skin friction coefficient is defined as

$$\bar{C}_f = \frac{\text{drag}}{\frac{1}{2}\rho U^2 bL}$$

where b is the width and L is the length of a flat plate. Show, using Eq. 8.49, that

$$\bar{C}_f = \frac{0.072}{(UL/\nu)^{1/5}}$$

for turbulent boundary-layer flow with the "$\frac{1}{7}$ power" velocity profile. (Experiments show that the constant should be "adjusted" to a value of 0.074.)

8.88 A flat-bottomed barge, which is 25 m long, 10 m wide, and submerged to a depth of 1.5 m, is to be pushed up a river at the rate of 8 km/hr. Estimate the power required to overcome skin friction if the water temperature is 15 C.

8.89 A sheet of plastic material $\frac{3}{8}$ in. thick, with specific gravity, SG = 1.5, is dropped into a large tank containing water. The sheet is 2 ft high and 3 ft wide. It falls vertically. Estimate the terminal velocity of the sheet, assuming that the only drag is due to skin friction, and that the boundary layers are turbulent from the leading edge.

8.90 As a part of the 1976 bicentennial celebration an enterprising group hung a giant American flag (59 m high and 112 m wide) from the suspension cables of the Verrazano Narrows Bridge. They were apparently reluctant to make holes in the flag to alleviate the wind force and hence they effectively had a flat plate normal to the flow. The flag tore loose from its mountings when the wind speed reached 16 km/hr. Estimate the wind force acting on the flag at this wind speed. Should they have been surprised that the flag blew down?

8.91 A rotary mixer is constructed from two circular disks as shown in Fig. 8.68. The mixer is rotated at 60 rpm in a large vessel containing a brine solution (SG = 1.1). The drag on the rods, and motion induced in the liquid may be neglected. Estimate the minimum torque and power required to drive the mixer.

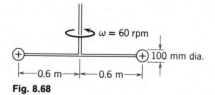

Fig. 8.68

8.92 The vertical component of the landing speed of a parachute is to be less than 6 m/sec. The parachute may be treated as an open hemisphere. The total mass of chute and jumper is 120 kg. Determine the minimum diameter of the open parachute.

8.93 An F-4 aircraft is slowed after landing by dual parachutes deployed from the rear. Each parachute is 20 ft in diameter, and may be assumed to have the same drag coefficient as an open hemisphere facing upstream. The F-4 weighs 32,000 lbf, and lands at 160 knots. Estimate the time required to decelerate the aircraft to 100 knots, assuming that the brakes are not used and the drag of the aircraft is negligible.

8.94   A vehicle is built to try for the land speed record at the Bonneville Salt Flats, elevation 4400 ft. The engine delivers 500 hp to the rear wheels, and careful streamlining has resulted in a drag coefficient of 0.3, based on a 15 ft^2 frontal area. Compute the theoretical maximum ground speed of the car (a) in still air, and (b) with a 20 mph head wind.

8.95   A typical full-size American sedan (1977 Ford LTD) has a frontal area of 23.4 ft^2, and a drag coefficient of 0.5. Plot a curve of horsepower required to overcome drag versus road speed in standard air. If rolling resistance is 1.5 percent of curb weight (4500 lbf), determine the speed at which the aerodynamic force exceeds frictional resistance. How much power is required to cruise at 55 mph, and at 70 mph?

8.96   At low Reynolds number ($Re < 1$), drag of a sphere can be predicted analytically from the complete equations of motion. The result is called Stokes' flow, and the drag is given by $F_D = 3\pi\mu DV$, where μ is the fluid viscosity, and D and V are the sphere diameter and relative velocity, respectively. Show that for Stokes flow, $C_D = 24/Re$.

A small sphere ($D = 6$ mm) is observed to fall through castor oil at a terminal speed of 60 mm/sec. The temperature is 20 C. Compute the drag coefficient for the sphere. Determine the density of the sphere. If dropped in water, would the sphere fall slower or faster? Why?

8.97   A spherical hydrogen-filled balloon 0.6 m in diameter exerts an upward force on a restraining string of 1.3 N when held stationary in standard air with no wind. With a wind speed of 3 m/sec, the string holding the balloon makes an angle of 60° with the horizontal. Calculate the drag coefficient of the balloon under these conditions, neglecting the weight of the string.

8.98   An antique airplane carries 15 m of external guy wires stretched normal to the direction of motion. The diameter of the wires is 6 mm. Basing calculations on two-dimensional flow around the wires, what power saving may be effected by removing the wires if the speed is 150 km/hr in standard air at sea level?

8.99   A water tower consists of a 12 m diameter sphere on top of a vertical tower 30 m tall. Estimate the bending moment exerted on the base of the tower due to the aerodynamic force imposed by a 100 km/hr wind on a standard day. Interference at the joint between the sphere and tower may be neglected.

8.100   The CB antenna on a car is about 8 mm in diameter and 2 m long. Estimate the torque that tends to snap it off if the car is driven at 125 km/hr on a standard day.

8.101   A spherical balloon contains helium and ascends through standard air. The mass of the balloon and its payload is 150 kg. Determine the required diameter if it is to ascend at 3 m/sec.

8.102 The guy wires on the antique plane in Problem 8.98 are to be streamlined rather than removed. Data from Fig. 8.35 may be used to select an optimum faired shape for the guys. Evaluate the maximum power saving that results from fairing the wires.

8.103 A light plane has a 10 m effective wingspan and a 1.8 m chord. It was originally designed to use a conventional (NACA 23015) airfoil section. With this airfoil its cruising speed on a standard day near sea level is 225 km/hr. A conversion to a laminar-flow (NACA 66_2–215) section airfoil is proposed. Determine the cruising speed that could be achieved with the new airfoil section for the same power.

8.104 An aircraft is flying in level flight at a speed of 250 km/hr through air at standard conditions. The lift coefficient at this speed is 0.4 and the drag coefficient is 0.0065. The mass of the aircraft is 850 kg. Calculate the effective lift area for the craft.

8.105 Consider a kite of mass 0.2 kg as a flat plate with an area of 1 m². It is flown in standard air moving horizontally at 10 m/sec. The kite makes an angle of 5° with the horizontal. Assume the lift coefficient is given by the equation, $C_L = 2\pi \sin \alpha$, where α is the angle of attack. If the string makes an angle of 60° with the horizontal, determine the tension in the string.

8.106 Turbojet-powered aircraft operate most efficiently at high altitudes. Because the density is low at high altitude, higher lift coefficients are needed to support the aircraft weight. The stalling speed at altitude is also high.

Consider an aircraft fitted with NACA 66_2–215 airfoil section wings. At sea level and 400 knots air speed, the required lift coefficient is 0.2. Boundary-layer control devices cannot be used at air speeds above 240 knots. Determine the stalling speed of this aircraft when flown at 10 km altitude on a standard day.

8.107 The foils of a hydrofoil type watercraft have an effective area of 0.7 m². Their coefficients of lift and drag are 1.6 and 0.5, respectively. The total mass of the craft in running trim is 1800 kg. Determine the minimum speed at which the craft is supported by the hydrofoils. At this speed, find the power required to overcome water resistance. If the craft is fitted with a 110 kW engine, estimate its top speed.

8.108 An airplane with an effective lift area of 25 m² is fitted with airfoils of NACA 23012 section (Fig. 8.41). The maximum flap setting that can be used on takeoff corresponds to configuration ② in Fig. 8.41. Determine the maximum gross mass possible for the airplane it its takeoff speed is 150 km/hr (neglect added lift due to ground effect).

8.109 An airplane with mass of 4500 kg is flown at constant elevation and speed on a circular path at 250 km/hr. The flight circle has a radius of 1000 m. The plane has an effective lifting area of 22 m², and is fitted with NACA 23015 section airfoils. Determine the drag on the aircraft, and the power required.

8.110 A tennis ball weighs 2 oz and is 2.5 in. in diameter. A typical forehand shot is hit at a speed of 60 ft/sec, with a top spin of 5500 rpm. Determine the aerodynamic

8.111 The mass flowrate in a water flow system determined by collecting the discharge over a timed interval is 0.3 kg/sec. The scales used can be read to the nearest 0.05 kg, and the stopwatch is accurate to 0.2 sec. Estimate the precision with which the flowrate can be calculated for time intervals of (a) 10 sec, and (b) 1 min.

8.112 An "ergometer" is used to measure the rate at which air is consumed by a research subject running on a treadmill. The atmospheric temperature and pressure are 23 C and 752 mm of mercury, respectively. An inverted bell, which is used to measure airflow, is 0.45 m in diameter. During a 30 sec test run, the bell rises 43 mm. Determine the rate at which the subject consumes oxygen.

8.113 Water at 150 F flows through a 3 in. diameter orifice installed in a 6 in. inside diameter pipe. The pressure taps are located 1 pipe diameter upstream and $\frac{1}{2}$ pipe diameter downstream from the orifice plate. The flowrate is 300 gpm. Determine the pressure difference between the taps.

8.114 Kerosine at 40 C flows through a 0.3 m line in a refinery. The flowrate is not expected to exceed 1200 kg/sec. A manometer with a range of 1 m of water is available for use with an orifice meter. Specify a recommended orifice diameter for use with this system. What minimum rate of flow could be measured within 10 percent accuracy if the manometer least count is 1 mm of water?

8.115 Airflow in a test of an internal combustion engine is to be measured using a flow nozzle installed in a plenum. The engine displacement is 1.6 liters and its maximum operating speed is 6000 rpm. The maximum pressure drop across the nozzle should not exceed 0.25 m of water, to avoid loading the engine. The manometer can be read to ± 0.5 mm of water. Determine the flow nozzle diameter that should be specified. Find the minimum rate of airflow that can be metered to ± 2 percent using this setup.

8.116 Consider a flow nozzle installation in a pipe (Fig. 8.69). Apply the basic equations to the control volume indicated, to show that the head loss across the meter can be expressed in dimensionless form as the head loss coefficient,

$$C_l = \frac{p_1 - p_3}{p_1 - p_2} = \frac{1 - A_2/A_1}{1 + A_2/A_1}$$

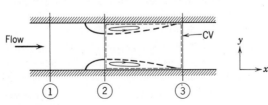

Fig. 8.69

8.117 A venturi meter with a 75 mm diameter throat is placed in a 150 mm line carrying water at 25 C. The pressure drop between the upstream tap and the venturi throat is 300 mm of mercury. Compute the rate of flow.

8.118 Air flows through the venturi meter described in Problem 8.117. Assume that the upstream pressure is 400 kPa, and that the temperature is everywhere constant at 20 C. Determine the maximum possible mass flowrate of air for which the assumption of incompressible flow is a valid engineering approximation. Compute the corresponding pressure reading on a mercury manometer.

references

1. Moody, L. F., "Friction Factors for Pipe Flow," *Transactions of the ASME, 66*, 8, November 1944, pp. 671–684.

2. *Flow of Fluids through Valves, Fittings, and Pipe.* Crane Industrial Products Group, 4100 S. Kedzie Avenue, Chicago, Illinois 60632, Technical Paper No. 410, 1957.

3. Hamilton, J. B., "The Suppression of Intake Losses by Various Degrees of Rounding," University of Washington, Seattle, Washington, Experiment Station Bulletin 51, 1929.

4. Hinze, J. O., *Turbulence*, 2nd ed. New York: McGraw-Hill, 1975.

5. Streeter, V. L., ed., *Handbook of Fluid Dynamics.* New York: McGraw-Hill, 1961.

6. *ASHRAE Guide and Data Book—Equipment Volume.* New York: American Society of Heating, Refrigeration and Air Conditioning Engineers, 1974.

7. Cockrell, D. J. and Bradley, C. I., "The Response of Diffusers to Flow Conditions at Their Inlet," Paper No. 5, *Symposium on Internal Flows*, University of Salford, Salford, England, April 1971, pp. A32–A41.

8. Sovran, G. and Klomp, E. D., "Experimentally Determined Optimum Geometries for Rectlinear Diffusers with Rectangular, Conical or Annular Cross-Section," in *Fluid Mechanics of Internal Flow*, G. Sovran, ed. Amsterdam: Elsevier, 1967, pp. 270–319.

9. Feiereisen, W. J., "An Experimental Investigation of Incompressible Flow without Swirl in R-Radial Diffusers," Ph. D. Thesis, School of Mechanical Engineering, Purdue University, West Lafayette, Indiana, June 1971.

10. Reneau, L. R., Johnston, J. P., and Kline, S. J., "Performance and Design of Straight, Two-Dimensional Diffusers," *Transactions of the ASME, Journal of Basic Engineering, 89D*, 1, March 1967, pp. 141–150.

11. Lam, C. F. and Wolla, M. L., "Computer Analysis of Water Distribution Systems: Part I, Formulation of Equations," *Journal of the Hydraulics Division, Proceedings of the American Society of Civil Engineers, 98,* HY 2, February 1972, pp. 335–344.

12. *Aerospace Applied Thermodynamics Manual.* New York (now Warrendale, Pennsylvania): Society of Automotive Engineers, 1969.

13. Schlichting, H., *Boundary-Layer Theory*, 6th ed. New York: McGraw-Hill, 1968.

14. Hoerner, S. F., *The Fluid Dynamics of Drag*, 2nd ed. Midland Park, New Jersey: Published by the author, 1965.

15. Shapiro, A. H., *Shape and Flow, The Fluid Dynamics of Drag*. New York: Anchor, 1961 (paperback).

16. Fage, A., "Experiments on a Sphere at Critical Reynolds Numbers," Great Britain, Aeronautical Research Council, *Reports and Memoranda No. 1766*, 1937.

17. Goldstein, S., ed., *Modern Developments in Fluid Dynamics*, Vols. I and II. Oxford: Clarendon Press, 1938. (Also reprinted in paperback by Dover, New York, 1967.)

18. Abbott, I. H. and von Doenhoff, A. E., *Theory of Wing Sections, Including a Summary of Airfoil Data*. New York: Dover, 1959 (paperback).

19. *Flowmeters, Their Theory and Application*. New York: American Society of Mechanical Engineers, 1959.

Chapter 9

Introduction to Compressible Flow

In Chapter 4, we developed control volume formulations of the basic equations. In incompressible flows the two variables of principal interest were pressure and velocity. The continuity and momentum equations provided the two independent relations needed to solve for these variables. In Chapter 8, the energy equation was used to identify the losses in mechanical energy due to friction in duct flows.

"Compressible" flow implies appreciable variations in density throughout a flow field. Compressibility becomes important at high flow velocities. Large changes in velocity involve large pressure changes; for gas flows, these pressure changes are accompanied by significant variations in both density and temperature. Since two additional variables are encountered in treating compressible flow, two additional equations are needed. Both the energy equation and an equation of state must be applied to solve problems in compressible flow.

In our study of compressible fluid flow, we shall deal primarily with the steady, one-dimensional flow of an ideal gas. Although many real flows of interest are more complex, these restrictions will allow us to concentrate on the effects of basic flow processes.

A review of the thermodynamics necessary for the study of compressible flows, including the equation of state and "$T\,ds$" equations, is presented in the next section.

9-1 REVIEW OF THERMODYNAMICS

The pressure, density, and temperature of a substance may be related by an *equation of state*. Although many substances are quite complex in behavior, experience shows that most gases of engineering interest, at moderate pressure and temperature, are well represented by the *ideal gas* equation of state,

$$p = \rho R T \qquad (9.1)$$

where R is a constant for each gas.[1] Actually,

$$R = \frac{R_u}{M_m}$$

where R_u is the universal gas constant, $R_u = 8314$ N·m/kgmole·K (1544 ft·lbf/lbmole·R) and M_m is the molecular mass of the gas. Although no real substance behaves exactly as an ideal gas,[2] Eq. 9.1 is in error by less than 1 percent for air at room temperature for pressures as high as 30 atm. For air at 1 atm, the equation is less than 1 percent in error for temperatures as low as 140 K.

The ideal gas has other features that are simple and useful. In general, the internal energy of a substance may be expressed as $u = u(v, T)$. Then

$$du = \left(\frac{\partial u}{\partial T}\right)_v dT + \left(\frac{\partial u}{\partial v}\right)_T dv$$

where $v = 1/\rho$, the specific volume. The specific heat at constant volume is defined as $c_v \equiv (\partial u/\partial T)_v$, so that

$$du = c_v \, dT + \left(\frac{\partial u}{\partial v}\right)_T dv$$

You may recall from your earlier course in thermodynamics that for any substance that follows the equation of state, $p = \rho RT$, $(\partial u/\partial v)_T = 0$ and hence $u = u(T)$. Consequently,

$$du = c_v \, dT \qquad (9.2)$$

for an ideal gas, which means that internal energy and temperature changes may be related if c_v is known. Furthermore, since $u = u(T)$, then $c_v = c_v(T)$.

The enthalpy of a substance is defined as $h \equiv u + p/\rho$. For an ideal gas, $p = \rho RT$, and hence $h = u + RT$. Since, for an ideal gas $u = u(T)$, then h also must be a function of temperature alone.

---

[1] For air, $R = 287$ N·m/kg·K (53.3 ft·lbf/lbm·R).
[2] See, for example, M. J. Zucrow and J. D. Hoffman, *Gas Dynamics*, Vol. 1 (New York: John Wiley, 1976), Chapter 1.

To obtain a relation between h and T, we express h in its most general form as

$$h = h(p, T)$$

Then

$$dh = \left(\frac{\partial h}{\partial T}\right)_p dT + \left(\frac{\partial h}{\partial p}\right)_T dp$$

and

$$dh = c_p dT + \left(\frac{\partial h}{\partial p}\right)_T dp$$

from the definition, $c_p \equiv (\partial h/\partial T)_p$. We have shown that for an ideal gas h is a function of T only. Consequently, $(\partial h/\partial p)_T = 0$ and

$$dh = c_p dT \tag{9.3}$$

Again, since h is a function of T alone, Eq. 9.3 requires that c_p be a function of T only for an ideal gas.

The specific heats for an ideal gas have been shown to be functions of temperature only. We should also note that their difference is a constant. From

$$h = u + RT$$

we can write

$$dh = du + R\,dT$$

Combining this with Eq. 9.3, and using Eq. 9.2, we can write

$$dh = c_p dT = du + R\,dT = c_v dT + R\,dT$$

Then

$$c_p - c_v = R \tag{9.4}$$

The ratio of specific heats is defined as

$$k \equiv \frac{c_p}{c_v} \tag{9.5}$$

By using the definition of k, Eq. 9.4 can be solved for either c_p or c_v in terms of k and R. Thus

$$c_p = \frac{kR}{k-1} \tag{9.6a}$$

and

$$c_v = \frac{R}{k-1} \tag{9.6b}$$

For an ideal gas, the specific heats are functions of temperature. Within reasonable temperature ranges, the specific heats of an ideal gas may be treated as constants for calculations of engineering accuracy. Under these conditions

$$u_2 - u_1 = \int_{u_1}^{u_2} du = \int_{T_1}^{T_2} c_v \, dT = c_v(T_2 - T_1) \qquad (9.7a)$$

$$h_2 - h_1 = \int_{h_1}^{h_2} dh = \int_{T_1}^{T_2} c_p \, dT = c_p(T_2 - T_1) \qquad (9.7b)$$

These equations obviously may be used to advantage in simplifying analyses.

Values of M_m, c_p, c_v, R, and k for common gases are given in Appendix A, Table A.6.

The property entropy is extremely useful in the analysis of compressible flows. State diagrams, particularly the temperature-entropy (Ts) diagram, are valuable aids in the physical interpretation of analytical results. Since we shall make extensive use of the Ts diagram in solving compressible flow problems, it is worthwhile to review briefly some of the useful relationships involving the property entropy.[3]

Entropy is defined by the equation

$$\Delta S \equiv \int_{\text{rev}} \frac{\delta Q}{T} \qquad \text{or} \qquad dS = \left(\frac{\delta Q}{T}\right)_{\text{rev}} \qquad (9.8)$$

The inequality of Clausius, deduced from the second law, states that

$$\oint \frac{\delta Q}{T} \leq 0$$

As a consequence of the second law these results can be extended to

$$dS \geq \frac{\delta Q}{T} \qquad \text{or} \qquad T \, dS \geq \delta Q \qquad (9.9a)$$

For *reversible* processes, the equality holds, that is

$$T \, ds = \frac{\delta Q}{dm} \qquad \text{(reversible process)} \qquad (9.9b)$$

The inequality holds for *irreversible* processes,

$$T \, ds > \frac{\delta Q}{dm} \qquad \text{(irreversible process)} \qquad (9.9c)$$

For an *adiabatic* process, $\delta Q/dm \equiv 0$. Thus

$$ds = 0 \qquad \text{(reversible adiabatic process)}$$

and

$$ds > 0 \qquad \text{(irreversible adiabatic process)}$$

---

[3] For a detailed discussion see, for example, G. J. Van Wylen and R. E. Sonntag, *Fundamentals of Classical Thermodynamics (SI Version)*, 2nd ed. (New York: John Wiley, 1976), Chapters 6 and 7.

A useful relationship among properties (p, v, T, s, u) can be obtained from a joint consideration of the first and second laws. The result is the familiar $T\,ds$ equation

$$T\,ds = du + p\,dv \qquad (9.10a)$$

This is a relationship among properties, valid for all processes between equilibrium states. Although it is based on the first and second laws, in itself it is a statement of neither.

An alternate form of Eq. 9.10a can be obtained by substituting

$$du = d(h - pv) = dh - p\,dv - v\,dp$$

to obtain

$$T\,ds = dh - v\,dp \qquad (9.10b)$$

For an ideal gas, the entropy change can be readily evaluated from the $T\,ds$ equations

$$ds = \frac{du}{T} + \frac{p}{T}\,dv = c_v\frac{dT}{T} + R\frac{dv}{v}$$

$$ds = \frac{dh}{T} - \frac{v}{T}\,dp = c_p\frac{dT}{T} - R\frac{dp}{p}$$

For constant specific heats, the equations may be integrated to yield

$$s_2 - s_1 = c_v\ln\frac{T_2}{T_1} + R\ln\frac{v_2}{v_1}$$

$$s_2 - s_1 = c_p\ln\frac{T_2}{T_1} - R\ln\frac{p_2}{p_1}$$

For the special case of an isentropic process, $ds = 0$, and the $T\,ds$ equations reduce to

$$0 = du + p\,dv$$
$$0 = dh - v\,dp$$

For an ideal gas, we have

$$0 = c_v\,dT + p\,dv$$
$$0 = c_p\,dT - v\,dp$$

Solving for dT gives

$$dT = \frac{v\,dp}{c_p} = -\frac{p\,dv}{c_v}$$

or

$$\frac{dp}{p} + \frac{c_p}{c_v}\frac{dv}{v} = \frac{dp}{p} + k\frac{dv}{v} = 0$$

Integrating (for $k = $ constant),

$$\ln p + k \ln v = \ln c$$

or

$$\ln p + \ln v^k = \ln c$$

Taking antilogarithms, this equation reduces to

$$pv^k = \text{constant} \tag{9.11a}$$

or

$$\frac{p}{\rho^k} = \text{constant} \tag{9.11b}$$

Equations 9.11 are property relations for an ideal gas undergoing an isentropic process.

Qualitative information that is useful in drawing state diagrams also can be obtained from the $T\,ds$ equations. To complete our review of thermodynamic fundamentals, the slopes of lines of constant pressure and of constant volume on the Ts diagram are evaluated in Example Problem 9.2.

Example 9.1

Air flows through a long duct of constant area at a rate of 0.15 kg/sec. A short section of the duct is cooled by liquid nitrogen that surrounds the duct. The rate of heat loss in this section is 15.0 kJ/sec from the air. The absolute pressure, temperature, and velocity at the entrance to the cooled section are 188 kPa, 440 K, and 210 m/sec, respectively. At the outlet, the absolute pressure and temperature are 213 kPa and 351 K. Compute the duct cross-sectional area, and the changes in enthalpy, internal energy, and entropy for this flow.

Example Problem 9.1

GIVEN:

Air flows steadily through a short section of constant-area duct that is cooled by liquid nitrogen.

$T_1 = 440$ K $\qquad\qquad$ $T_2 = 351$ K

$p_1 = 188$ kPa (absolute) $\qquad$ $p_2 = 213$ kPa (absolute)

$V_1 = 210$ m/sec

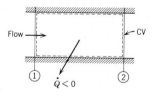

FIND:

(a) The duct area, (b) Δh, (c) Δu, and (d) Δs

SOLUTION:

The duct area may be found from the continuity equation.

$$= 0(1)$$

Basic equation:
$$0 = \frac{\partial}{\partial t} \int_{CV} \rho \, d\forall + \int_{CS} \rho \vec{V} \cdot d\vec{A} \qquad (4.13)$$

Assumptions: (1) Steady flow
(2) Uniform flow at each section
(3) Ideal gas

Then
$$0 = \{-|\rho_1 V_1 A_1|\} + \{|\rho_2 V_2 A_2|\}$$

or
$$\dot{m} = \rho_1 V_1 A = \rho_2 V_2 A$$

since $A = A_1 = A_2 = $ constant. Using the ideal gas relation, $p = \rho R T$, we find

$$\rho_1 = \frac{p_1}{RT_1} = \frac{1.88 \times 10^5 \text{ N}}{\text{m}^2} \times \frac{\text{kg} \cdot \text{K}}{287 \text{ N} \cdot \text{m}} \times \frac{1}{440 \text{ K}} = 1.49 \text{ kg/m}^3$$

From continuity

$$A = \frac{\dot{m}}{\rho_1 V_1} = \frac{0.15 \text{ kg}}{\text{sec}} \times \frac{\text{m}^3}{1.49 \text{ kg}} \times \frac{\text{sec}}{210 \text{ m}} = 4.79 \times 10^{-4} \text{ m}^2 \qquad \underleftarrow{A}$$

For an ideal gas, $dh = c_p \, dT$, so

$$\Delta h = h_2 - h_1 = \int_{T_1}^{T_2} c_p \, dT = c_p (T_2 - T_1)$$

$$\Delta h = \frac{1.00 \text{ kJ}}{\text{kg} \cdot \text{K}} \times (351 - 440) \text{ K} = -89.0 \text{ kJ/kg} \qquad \underleftarrow{\Delta h}$$

Also $du = c_v \, dT$, so

$$\Delta u = u_2 - u_1 = \int_{T_1}^{T_2} c_v \, dT = c_v (T_2 - T_1)$$

$$\Delta u = \frac{0.716 \text{ kJ}}{\text{kg} \cdot \text{K}} \times (351 - 440) \text{ K} = -63.7 \text{ kJ/kg} \qquad \underleftarrow{\Delta u}$$

The entropy change may be obtained from the $T \, ds$ equation

$$T \, ds = dh - v \, dp$$

$$ds = \frac{dh}{T} - \frac{v \, dp}{T} = c_p \frac{dT}{T} - R \frac{dp}{p}$$

or

$$\Delta s = s_2 - s_1 = \int_{T_1}^{T_2} c_p \frac{dT}{T} - \int_{p_1}^{p_2} R \frac{dp}{p} = c_p \ln \frac{T_2}{T_1} - R \ln \frac{p_2}{p_1}$$

$$= \frac{1.00}{kg \cdot K} \frac{kJ}{kg \cdot K} \ln \left(\frac{351}{440} \right) - \frac{0.287}{kg \cdot K} \frac{kJ}{kg \cdot K} \ln \left(\frac{2.13 \times 10^5}{1.88 \times 10^5} \right)$$

$$\Delta s = -0.262 \text{ kJ/kg} \cdot K \qquad\qquad\qquad\qquad \Delta s$$

$\left\{ \begin{array}{l} \text{The purpose of this problem was to review the calculation of thermodynamic} \\ \text{properties for an ideal gas.} \end{array} \right\}$

Example 9.2

For an ideal gas, determine the slope of (a) a constant volume line, and (b) a constant pressure line, in the Ts plane.

Example Problem 9.2

GIVEN:

An ideal gas.

FIND:

(a) Slope of constant volume line in Ts plane.
(b) Slope of constant pressure line in Ts plane.

SOLUTION:

The $T\,ds$ equations may be applied.

$$T\,ds = du + p\,dv \qquad\qquad (9.10a)$$
$$T\,ds = dh - v\,dp \qquad\qquad (9.10b)$$

Substituting for an ideal gas, $du = c_v\,dT$ and $dh = c_p\,dT$, we obtain

$$T\,ds = c_v\,dT + p\,dv$$
$$T\,ds = c_p\,dT - v\,dp$$

For a constant volume process, $dv = 0$. From the first equation

$$\left. \frac{dT}{ds} \right)_{\substack{\text{constant} \\ \text{volume}}} = \left. \frac{\partial T}{\partial s} \right)_v = \frac{T}{c_v} \qquad\qquad \text{constant volume slope}$$

For a constant pressure process, $dp = 0$. From the second equation

$$\left.\frac{dT}{ds}\right)_{\substack{\text{constant}\\\text{pressure}}} = \left.\frac{\partial T}{\partial s}\right)_p = \frac{T}{c_p} \qquad \text{constant pressure slope}$$

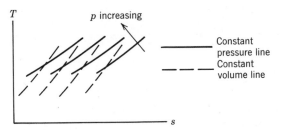

T

p increasing

———— Constant pressure line

- - - - Constant volume line

s

⎧ Note that the slope of each line is proportional at any point to the absolute ⎫
⎨ temperature. Furthermore, at any point a constant volume line has a slope ⎬
⎩ $c_p/c_v = k$ times larger than a line of constant pressure. ⎭

9-2 PROPAGATION OF SOUND WAVES

9-2.1 SPEED OF SOUND

The terms *supersonic* and *subsonic* are familiar terms; they refer to velocities that are, respectively, greater than and less than the speed of sound. The speed of sound (or of a pressure wave of infinitesimal strength) is an important characteristic parameter in a compressible flow. We have previously (Chapters 2 and 7) introduced the Mach number, $M = V/c$, the ratio of the local flow speed to the local speed of sound, as an important nondimensional parameter characterizing compressible flows. Before proceeding with our study of compressible flows, it is logical to obtain a general relation for calculating the sonic speed.

Consider the propagation of a sound wave of infinitesimal strength into an undisturbed medium as shown in Fig. 9.1a. We are interested in relating the speed of propagation of the wave, c, to fluid property changes across the wave. If the pressure and density in the undisturbed medium ahead of the wave are denoted by p and ρ, the passage of the wave will cause them to undergo infinitesimal changes and to become $p + dp$ and $\rho + d\rho$. Since the wave is propagating into a stationary fluid, the velocity ahead of the wave, V_x, is zero. The magnitude of the velocity behind the wave, $V_x + dV_x$, will then simply be dV_x; in Fig. 9.1a, the direction of the motion behind the wave has been assumed to the left.

Unfortunately, the flow of Fig. 9.1a appears unsteady to a stationary observer, observing the motion of the wave from a fixed point on the ground.

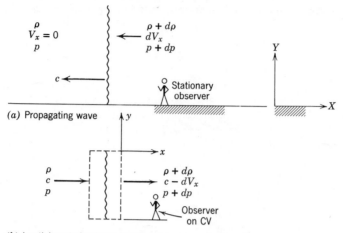

(a) Propagating wave

(b) Inertial control volume moving with wave, velocity c

Fig. 9.1 Propagating sound wave showing control volume chosen for analysis.

However, we can put a moving control volume around a segment of the wave, as shown in Fig. 9.1b; the flow will then appear steady to an observer located *on* this inertial control volume. The velocity approaching the control volume will be c and the velocity leaving, $c - dV_x$.

The basic equations may be applied to the differential control volume shown in Fig. 9.1b (we use V_x for the x component of velocity to avoid confusion with internal energy, u):

a. Continuity Equation

Basic equation:

$$0 = \overset{\overset{=0(1)}{\cancel{\partial}}}{\partial t} \int_{CV} \rho \, d\mathbb{V} + \int_{CS} \rho \vec{V} \cdot d\vec{A} \tag{4.13}$$

Assumptions: (1) Steady flow
(2) Uniform flow at each section

Then

$$0 = \{-|\rho c A|\} + \{|(\rho + d\rho)(c - dV_x)A|\} \tag{9.12a}$$

or

$$0 = -\rho \cancel{c} A + \rho \cancel{c} A - \rho \, dV_x A + d\rho c A - \overset{\simeq 0}{\cancel{d\rho \, dV_x A}}$$

494   9/introduction to compressible flow

Neglecting the product of differentials, $d\rho\, dV_x$, compared to either $d\rho$ or dV_x, we obtain

$$0 = -\rho A\, dV_x + cA\, d\rho$$

or

$$dV_x = \frac{c}{\rho}\, d\rho \qquad (9.12b)$$

b. Momentum Equation

Basic equation:

$$F_{S_x} + \overset{=0(3)}{\cancel{F_{B_x}}} - \overset{=0(4)}{\cancel{\int_{CV} a_{rf_x}\rho\, dV}} = \overset{=0(1)}{\cancel{\frac{\partial}{\partial t} \int_{CV} V_x \rho\, dV}} + \int_{CS} V_x \rho \vec{V}_{xyz} \cdot d\vec{A} \quad (4.35a)$$

Assumptions: (3) $F_{B_x} = 0$
(4) $a_{rf_x} = 0$

The surface forces acting on the infinitesimal control volume are

$$F_{S_x} = dR_x + pA - (p + dp)A$$

where dR_x represents all forces applied to the horizontal portions of the control surface shown in Fig. 9.1b. However, since we consider only a portion of a moving sound wave, $dR_x = 0$, because there is no relative motion along the wave. Thus the surface force simplifies to

$$F_{S_x} = -A\, dp$$

Substituting into the basic equation gives

$$-A\, dp = c\{-|\rho cA|\} + (c - dV_x)\{|(\rho + d\rho)(c - dV_x)A|\}$$

Using the continuity equation in the form of Eq. 9.12a, reduces this to

$$-A\, dp = c\{-|\rho cA|\} + (c - dV_x)\{|\rho cA|\}$$
$$= (-c + c - dV_x)\{|\rho cA|\}$$
$$-A\, dp = -\rho cA\, dV_x$$

or

$$dV_x = \frac{1}{\rho c}\, dp \qquad (9.12c)$$

Combining Eqs. 9.12b and 9.12c, we obtain

$$dV_x = \frac{c}{\rho}\, d\rho = \frac{1}{\rho c}\, dp$$

from which

$$dp = c^2 \, d\rho$$

or

$$c^2 = \frac{dp}{d\rho}$$

To evaluate the derivative of a thermodynamic property, we must specify the property to be held constant during differentiation. For the present case, the limit at vanishing strength of a sound wave will be

$$\lim_{\text{strength} \to 0} \frac{dp}{d\rho} = \left.\frac{\partial p}{\partial \rho}\right)_{s = \text{contant}}$$

Perhaps a more physical justification for the assumption of isentropic propagation is that an infinitesimal pressure change is reversible. Hence, since there is too little time for heat transfer, the process is reversible and adiabatic. Thus the speed of propagation of a sound wave is given by

$$c = \sqrt{\left.\frac{\partial p}{\partial \rho}\right)_s}$$

For an ideal gas, the pressure and density in isentropic flow are related by

$$\frac{p}{\rho^k} = \text{constant} \qquad\qquad (9.11b)$$

as shown previously. Taking logarithms,

$$\ln p - k \ln \rho = \ln c$$

and differentiating

$$\frac{dp}{p} - k \frac{d\rho}{\rho} = 0$$

Therefore,

$$\left.\frac{\partial p}{\partial \rho}\right)_s = k \frac{p}{\rho}$$

But $p/\rho = RT$, so finally

$$c = \sqrt{kRT} \qquad\qquad (9.13)$$

for an ideal gas.

The important feature of sound propagation in an ideal gas, as shown by Eq. 9.13, is that the speed of sound is a function of temperature only. The variation in atmospheric temperature, T, with altitude on a standard day has been discussed in Chapter 3. The corresponding variation in c is computed in Example Problem 9.3 and sketched as a function of altitude.

Example 9.3

Compute the speed of sound at sea level in standard air. Evaluate the speed of sound and plot for altitudes to 40,000 ft.

Example Problem 9.3

GIVEN:

Air under standard atmospheric conditions.

FIND:

(a) The speed of sound at sea level.
(b) Plot the speed of sound for altitudes to 40,000 ft.

SOLUTION:

Assume an ideal gas.

Computing equation:
$$c = \sqrt{kRT}$$

From Table 3.1, the temperature at sea level on a standard day is 59 F. Thus

$$c = \left(1.4 \times \frac{53.3 \text{ ft} \cdot \text{lbf}}{\text{lbm} \cdot \text{R}} \times (460 + 59)\text{R} \times \frac{32.2 \text{ lbm}}{\text{slug}} \times \frac{\text{slug} \cdot \text{ft}}{\text{lbf} \cdot \text{sec}^2}\right)^{1/2}$$

$$c = 1120 \text{ ft/sec} \qquad\qquad\qquad\qquad c$$

Temperatures at various altitudes may be found from Fig. 3.3. The resulting sound speeds are plotted below:

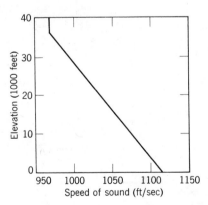

From the plot we see that the speed of sound in air on a standard day varies from 1120 ft/sec at sea level to 968 ft/sec at altitudes between 36,000 and 40,000 ft.

9-2.2 TYPES OF FLOW—THE CONE

Flows for which $M < 1$ are termed *subsonic*, while those for which $M > 1$ are *supersonic*. The transition region between subsonic and supersonic flow, i.e. flows in which $0.9 \gtrsim M \gtrsim 1.1$, is often referred to as *transonic*. Although most flows within our experience are subsonic, there are important practical

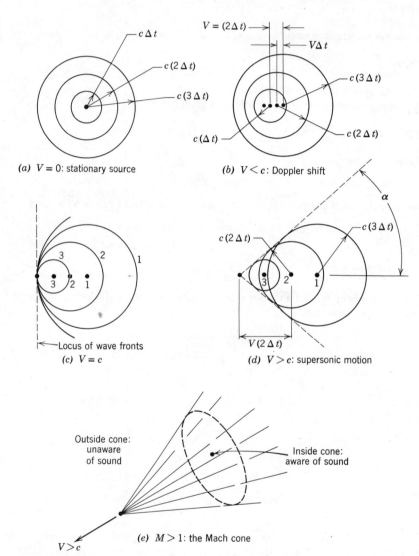

(a) $V = 0$: stationary source

(b) $V < c$: Doppler shift

(c) $V = c$

(d) $V > c$: supersonic motion

(e) $M > 1$: the Mach cone

Fig. 9.2 Propagation of sound waves from a moving source: the Mach cone.

cases where $M \geq 1$ occurs in a flow field. Perhaps the most obvious are supersonic aircraft, transonic flows in aircraft compressors and fans, and stream turbines. Yet another flow regime, *hypersonic* flow ($M \gtrsim 5$), is of interest in missile and reentry vehicle design. Some important qualitative differences between subsonic and supersonic flows can be deduced from the properties of a simple moving sound source.

Consider a point source capable of emitting an instantaneous infinitesimal disturbance. These disturbances propagate in all directions with speed, c. At any time, t, the location of the wave front from the disturbance emitted at time, t_0, will be represented by a sphere with radius, $c(t - t_0)$, whose center is coincident with the location of the disturbance at time, t_0.

We are interested in determining the nature of the disturbance propagation for different speeds of the moving source. Four cases are shown in Fig. 9.2:

1. $V = 0$. The sound pattern propagates uniformly in all directions. At the instant Δt after emission, any given sound pulse is located at radius $c \Delta t$ from the source. At the instant $2 \Delta t$, of course, the radius is $c(2 \Delta t)$. Any given wave front is spherical in shape; all wave fronts are concentric spheres.

2. $0 < V < c$. The concentricity of the wave pattern is lost. Any individual wave front is spherical, but each successive sound is emitted from a different position, $V \Delta t$ distant from the previous position.

If you imagine the circles shown to be the amplitude peaks of a sinusoidal tone, the same qualitative picture holds for a moving source of continuous sound. If this source moves with the constant speed, V, we can imagine the whole pattern shown in Fig. 9.2b carried along with the emitter. Thus a stationary observer would hear more peaks per unit time as the source approaches him than after it passes. This is known as the "Doppler effect." (Have you ever heard a fast-moving train whistle through a crossing?)

3. $V = c$. The locus of the leading surfaces of all waves will be a plane at the source, perpendicular to the path of motion. No sound wave can travel in front of the source. Consequently, an observer in front of the source will not hear it approaching.

4. $V > c$. In this case, the locus of the leading surfaces of the sound waves will be a cone. Again, no sound will be heard in front of this cone.

The cone angle can be related to the Mach number at which the source moves. From the geometry of Fig. 9.2d,

$$\sin \alpha = \frac{c}{V} = \frac{1}{M}$$

or

$$\alpha = \sin^{-1}\left(\frac{1}{M}\right) \tag{9.14}$$

The cone depicted in Fig. 9.2e is termed the Mach cone,[4] and angle α is referred to as the Mach angle. The regions inside and outside the cone are sometimes called the *zone of action* and the *zone of silence*, respectively.

9-3 REFERENCE STATE: LOCAL ISENTROPIC STAGNATION PROPERTIES

If we wish to describe the state of a fluid at any point in a flow field, we must specify two independent intensive thermodynamic properties (usually pressure and temperature), plus the fluid velocity at the point in question.[5]

In our discussion of compressible flow, we shall find it very convenient to employ the stagnation state as a reference state. The stagnation state is the state characterized by a condition of zero velocity; the stagnation properties at any point in a flow field are those properties that would exist at that point in the flow if the velocity at the point were reduced to zero. Consider a point in a flow field at which the temperature is T, the pressure, p, and the velocity, V. Thus the stagnation state at that point in the flow field would be characterized by the stagnation pressure, p_0, stagnation temperature, T_0, and zero velocity.

Before we can proceed to calculate the stagnation properties, we must specify the process by which the fluid is imagined to undergo this deceleration to zero velocity.

Recall in the discussion of incompressible flow in Chapter 6, integration of Euler's equation for frictionless, incompressible flow along a streamline led to the Bernoulli equation

$$\frac{p}{\rho} + \frac{V^2}{2} + gz = \text{constant} \tag{6.10}$$

Then for incompressible flow, a frictionless deceleration process to zero velocity led to a stagnation pressure, p_0, given by

$$p_0 = p + \tfrac{1}{2}\rho V^2 \tag{6.15}$$

For compressible flow we again employ a frictionless deceleration process; in addition we specify that the deceleration process be adiabatic. In short we specify an isentropic deceleration process. This then leads us to the definition of local isentropic stagnation properties:

---

[4] See the Shell film, *Transonic Flight*, for a well-presented discussion of the Mach cone.

[5] You will recall from your thermodynamics course that the state of a pure substance, in the absence of motion, gravity, and surface, magnetic, or electrical effects, is defined by two independent intensive thermodynamic properties.

Local isentropic stagnation properties are those properties that would be obtained at any point in a flow field if the fluid at that point were decelerated from local conditions to zero velocity following a frictionless, adiabatic, that is, *isentropic*, process.

Isentropic stagnation properties are reference properties that may be determined at any point in a fluid flow. The variations in these properties from point to point in a flow field give information about the flow process between points. This will become apparent in our discussion of one-dimensional flow cases. However, before proceeding let us first develop equations for computing the local isentropic stagnation properties for the flow of an ideal gas.

9-3.1 LOCAL ISENTROPIC STAGNATION PROPERTIES FOR THE FLOW OF AN IDEAL GAS

The stagnation properties at a point are the properties that would be obtained if the fluid at that point were decelerated isentropically to zero velocity. In order to calculate the stagnation properties, we must look at the deceleration process. At the beginning of the process the conditions are those corresponding to the actual flow at the point (velocity, V, pressure, p, and temperature, T, etc.); at the end of the process the velocity is zero and the conditions are the corresponding stagnation properties (stagnation pressure, p_0, and stagnation temperature, T_0, etc.).

We need to develop an expression describing the relationship among the fluid properties during the process. Since both the initial and final properties are specified, it makes sense to develop the relationship among properties in differential form. Then this can be integrated to obtain expressions for the stagnation conditions in terms of the initial conditions (the conditions corresponding to the actual flow at the point).

The deceleration process is shown schematically in Fig. 9.3. We are interested in determining the stagnation properties for the flow at point ①.

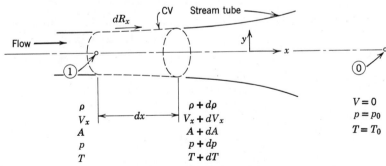

Fig. 9.3 Compressible flow in an infinitesimal streamtube.

To determine a relationship among the fluid properties during the deceleration process, we shall apply the continuity and momentum equations to the stationary differential stream tube control volume shown.

a. Continuity Equation

Basic equation:

$$0 = \overset{=0(1)}{\cancel{\frac{\partial}{\partial t} \int_{CV} \rho \, d\mathrm{V}}} + \int_{CS} \rho \vec{V} \cdot d\vec{A} \qquad (4.13)$$

Assumptions: (1) Steady flow

(2) Uniform flow at each section

Then

$$0 = \{-|\rho V_x A|\} + \{|(\rho + d\rho)(V_x + dV_x)(A + dA)|\}$$

or

$$\rho V_x A = (\rho + d\rho)(V_x + dV_x)(A + dA) \qquad (9.15a)$$

b. Momentum Equation

Basic equation:

$$F_{S_x} + \overset{=0(3)}{\cancel{F_{B_x}}} - \overset{=0(4)}{\cancel{\int_{CV} a_{rf_x} \rho \, d\mathrm{V}}} = \overset{=0(1)}{\cancel{\frac{\partial}{\partial t} \int_{CV} V_x \rho \, d\mathrm{V}}} + \int_{CS} V_x \rho \vec{V}_{xyz} \cdot d\vec{A} \qquad (4.35a)$$

Assumptions: (3) $F_{B_x} = 0$

(4) $a_{rf_x} = 0$

(5) Frictionless flow

The surface forces acting on the infinitesimal control volume are

$$F_{S_x} = dR_x + pA - (p + dp)(A + dA)$$

The force dR_x is applied along the streamtube boundary as shown in Fig. 9.3. There the average pressure is $p + dp/2$, and the area component in the x direction is dA. There is no friction. Thus

$$F_{S_x} = \left(p + \frac{dp}{2}\right)dA + pA - (p + dp)(A + dA)$$

or

$$F_{S_x} = \overset{\simeq 0}{\cancel{p \, dA}} + \overset{\simeq 0}{\cancel{\frac{dp \, dA}{2}}} + \cancel{pA} - \cancel{pA} - dpA - \cancel{p \, dA} - \overset{}{\cancel{dp \, dA}}$$

Substituting this result into the momentum equation,

$$-dpA = V_x\{-|\rho V_x A|\} + (V_x + dV_x)\{|(\rho + d\rho)(V_x + dV_x)(A + dA)|\}$$

which may be simplified using Eq. 9.15a, to obtain

$$-dpA = (-V_x + V_x + dV_x)(\rho V_x A)$$

Finally,

$$dp = -\rho V_x dV_x = -\rho d\left(\frac{V_x^2}{2}\right)$$

or

$$\frac{dp}{\rho} + d\left(\frac{V_x^2}{2}\right) = 0 \tag{9.15b}$$

Equation 9.15b is a relation among properties during the deceleration process. In developing this relation, we have specified a frictionless deceleration process. Before we can integrate this relation between the initial and final (stagnation) states, we must specify the relation that exists between the pressure, p, and the density, ρ, along the process path.

Since the deceleration process is isentropic (i.e. both frictionless and adiabatic), then during the process p and ρ for an ideal gas are related by the expression

$$\frac{p}{\rho^k} = \text{constant} \tag{9.11b}$$

Our task now is to integrate Eq. 9.15b subject to this relation. Along the stagnation streamline there is only a single component of velocity; V_x is the magnitude of the velocity, and hence we can drop the subscript and write

$$\frac{dp}{\rho} + d\left(\frac{V^2}{2}\right) = 0 \tag{9.15c}$$

From $p/\rho^k = \text{constant} = C$, we can write

$$p = C\rho^k \quad \text{and} \quad \rho = p^{1/k}C^{-1/k}$$

Then, from Eq. 9.15c

$$-d\left(\frac{V^2}{2}\right) = \frac{dp}{\rho} = p^{-1/k}C^{1/k}\,dp$$

We can integrate this equation between the initial state and the corresponding stagnation state

$$-\int_V^0 d\left(\frac{V^2}{2}\right) = C^{1/k}\int_p^{p_0} p^{-1/k}\,dp$$

to obtain

$$\frac{V^2}{2} = C^{1/k}\frac{k}{k-1}\left[p^{(k-1)/k}\right]_p^{p_0} = C^{1/k}\frac{k}{k-1}\left[p_0^{(k-1)/k} - p^{(k-1)/k}\right]$$

$$\frac{V^2}{2} = C^{1/k}\frac{k}{k-1}p^{(k-1)/k}\left[\left(\frac{p_0}{p}\right)^{(k-1)/k} - 1\right]$$

Since $C^{1/k} = p^{1/k}/\rho$, then

$$\frac{V^2}{2} = \frac{k}{k-1}\frac{p^{1/k}}{\rho}p^{(k-1)/k}\left[\left(\frac{p_0}{p}\right)^{(k-1)/k} - 1\right] = \frac{k}{k-1}\frac{p}{\rho}\left[\left(\frac{p_0}{p}\right)^{(k-1)/k} - 1\right]$$

Since we are seeking an expression for the stagnation pressure, we can rewrite this equation as

$$\left(\frac{p_0}{p}\right)^{(k-1)/k} = 1 + \frac{k-1}{k}\frac{\rho}{p}\frac{V^2}{2}$$

and

$$\frac{p_0}{p} = \left[1 + \frac{k-1}{k}\frac{\rho V^2}{2p}\right]^{k/(k-1)}$$

For an ideal gas $p = \rho RT$, and hence

$$\frac{p_0}{p} = \left[1 + \frac{k-1}{2}\frac{V^2}{kRT}\right]^{k/(k-1)}$$

Also, for an ideal gas the sonic speed is $c = \sqrt{kRT}$ and thus

$$\frac{p_0}{p} = \left[1 + \frac{k-1}{2}\frac{V^2}{c^2}\right]^{k/(k-1)}$$

$$\frac{p_0}{p} = \left[1 + \frac{k-1}{2}M^2\right]^{k/(k-1)} \tag{9.16a}$$

Equation 9.16a enables us to calculate the isentropic stagnation pressure at any point in a flow field of an ideal gas, provided that we know the static pressure and the Mach number at that point.

Since the process of decelerating the flow at a point to determine the corresponding isentropic stagnation properties was an isentropic process, we can readily obtain expressions for other isentropic stagnation properties by applying the relation

$$\frac{p}{\rho^k} = \text{constant}$$

between end points of the process. Thus

$$\frac{p_0}{p} = \left(\frac{\rho_0}{\rho}\right)^k$$

and

$$\frac{\rho_0}{\rho} = \left(\frac{p_0}{p}\right)^{1/k}$$

From the ideal gas equation of state, $p = \rho RT$. Then

$$\frac{T_0}{T} = \frac{p_0}{p}\frac{\rho}{\rho_0} = \frac{p_0}{p}\left(\frac{p_0}{p}\right)^{-1/k} = \left(\frac{p_0}{p}\right)^{(k-1)/k}$$

Utilizing Eq. 9.16a, we can summarize the equations for determining the isentropic stagnation properties of an ideal gas as follows:

$$\frac{p_0}{p} = \left[1 + \frac{k-1}{2}M^2\right]^{k/(k-1)} \qquad (9.16a)$$

$$\frac{T_0}{T} = 1 + \frac{k-1}{2}M^2 \qquad (9.16b)$$

$$\frac{\rho_0}{\rho} = \left[1 + \frac{k-1}{2}M^2\right]^{1/(k-1)} \qquad (9.16c)$$

From Eqs. 9.16 the ratio of each isentropic stagnation property to the corresponding static property at any point in a flow field for an ideal gas can be found if the local Mach number is known. In fact, these ratios could be calculated once and for all and tabulated. The calculation procedure is illustrated in Example Problem 9.4; ratios of local isentropic stagnation properties to the corresponding static properties for an ideal gas are tabulated in Table B.1 of Appendix B.

The Mach number range of validity of the assumption of incompressible flow is investigated in Example Problem 9.5.

Example 9.4

Air flows steadily through the duct shown in the sketch, from 350 kPa (absolute), 60 C and 183 m/sec at the initial state to a Mach number of 1.3

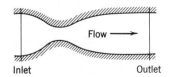

Inlet   Flow ⟶   Outlet

at the outlet, where the local isentropic stagnation conditions are known to be 385 kPa (absolute) and 350 K. Compute values of the isentropic stagnation pressure and temperature at the inlet and of the static pressure and temperature at the duct outlet. Locate the inlet and outlet static state points on a Ts diagram, and indicate the stagnation processes.

Example Problem 9.4

GIVEN:

Steady flow of air through a duct as shown in the sketch.

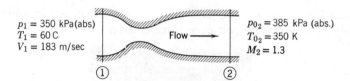

$p_1 = 350$ kPa(abs)
$T_1 = 60\,C$
$V_1 = 183$ m/sec

Flow $\longrightarrow$

$p_{0_2} = 385$ kPa (abs.)
$T_{0_2} = 350$ K
$M_2 = 1.3$

① ②

FIND:

(a) p_{0_1} (b) T_{0_1} (c) p_2 (d) T_2
(e) Locate state points ① and ② on a Ts diagram and indicate the stagnation processes.

SOLUTION:

To evaluate local isentropic stagnation conditions at section ①, we must calculate the Mach number, $M_1 = V_1/c_1$. For an ideal gas, $c = \sqrt{kRT}$. Then

$$c_1 = \sqrt{kRT_1} = \left[1.4 \times \frac{287 \text{ N} \cdot \text{m}}{\text{kg} \cdot \text{K}} \times (273 + 60) \text{ K} \times \frac{\text{kg} \cdot \text{m}}{\text{N} \cdot \text{sec}^2}\right]^{1/2} = 366 \text{ m/sec}$$

and

$$M_1 = \frac{V_1}{c_1} = \frac{183}{366} = 0.5$$

Local isentropic stagnation properties may be evaluated from Eqs. 9.16. Thus

$$\frac{p_{0_1}}{p_1} = \left[1 + \frac{k-1}{2} M_1^2\right]^{k/(k-1)} = [1 + 0.2(0.5)^2]^{3.5} = 1.186$$

and

$$p_{0_1} = 1.186 p_1 = (1.186)(350 \text{ kPa}) = 415 \text{ kPa (abs.)} \qquad \longleftarrow p_{0_1}$$

$$\frac{T_{0_1}}{T_1} = 1 + \frac{k-1}{2} M_1^2 = 1.05$$

and

$$T_{0_1} = 1.05T_1 = (1.05)(333 \text{ K}) = 350 \text{ K}$$ T_{0_1}

At section ②, Eqs. 9.16 may be applied again. Thus

$$\frac{p_{0_2}}{p_2} = \left[1 + \frac{k-1}{2} M_2^2\right]^{k/(k-1)} = [1 + 0.2(1.3)^2]^{3.5} = 2.77$$

and

$$p_2 = \frac{p_{0_2}}{2.77} = \frac{385 \text{ kPa}}{2.77} = 139 \text{ kPa (abs.)}$$ p_2

$$\frac{T_{0_2}}{T_2} = 1 + \frac{k-1}{2} M_2^2 = 1.338$$

and

$$T_2 = \frac{T_{0_2}}{1.338} = \frac{350 \text{ K}}{1.338} = 262 \text{ K}$$ T_2

The entropy change must be evaluated to locate state ② with respect to state ①. Using the $T ds$ equation

$$T ds = dh - v dp$$

or

$$ds = \frac{dh}{T} - \frac{v dp}{T} = c_p \frac{dT}{T} - R \frac{dp}{p}$$

Integrating gives

$$s_2 - s_1 = \int_{T_1}^{T_2} c_p \frac{dT}{T} - \int_{p_1}^{p_2} R \frac{dp}{p} = c_p \ln\frac{T_2}{T_1} - R \ln\frac{p_2}{p_1}$$

$$s_2 - s_1 = 1.00 \frac{\text{kJ}}{\text{kg} \cdot \text{K}} \times \ln\left(\frac{262}{333}\right) - 0.287 \frac{\text{kJ}}{\text{kg} \cdot \text{K}} \times \ln\left(\frac{1.39}{3.50}\right) = 0.0252 \text{ kJ/kg} \cdot \text{K}$$

Therefore, state ② lies to the right of state ① on the Ts plane as shown in the sketch below:

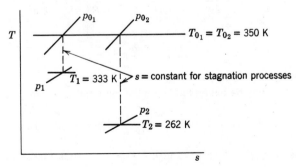

reference state: local isentropic stagnation properties/9-3 507

$$\left\{\begin{array}{l}\text{The process the fluid follows between states ① and ② is not specified. However,} \\ \text{it need not be known. A unique, isentropic stagnation process is defined at each} \\ \text{state point. Note that } s_{0_2} - s_{0_1} = s_2 - s_1.\end{array}\right\}$$

Example 9.5

We have derived equations for the ratio, p_0/p, for both compressible and "incompressible" flow. By writing both equations in terms of Mach number, compare their behavior. Find the Mach number below which the two equations agree within 5 percent.

Example Problem 9.5

GIVEN:

The incompressible and compressible forms of the equations for stagnation pressure, p_0.

Incompressible
$$p_0 = p + \tfrac{1}{2}\rho V^2 \tag{6.15}$$

Compressible
$$\frac{p_0}{p} = \left[1 + \frac{k-1}{2}M^2\right]^{k/(k-1)} \tag{9.16a}$$

FIND:

(a) Behavior of both equations as a function of Mach number.
(b) Mach number below which they agree within 5 percent.

SOLUTION:

First, let us write Eq. 6.15 in terms of Mach number. Using the ideal gas equation of state, and $c^2 = kRT$,

$$\frac{p_0}{p} = 1 + \frac{\rho V^2}{2p} = 1 + \frac{V^2}{2RT} = 1 + \frac{kV^2}{2kRT} = 1 + \frac{k}{2}\frac{V^2}{c^2}$$

Thus

$$\frac{p_0}{p} = 1 + \frac{k}{2}M^2 \tag{1}$$

for "incompressible" flow.

Equation 9.16a may be expanded using the binomial theorem,

$$(1 + x)^n = 1 + nx + \frac{n(n-1)}{2!}x^2 + \cdots, \ |x| < 1$$

For Eq. 9.16a, $x = [(k-1)/2]M^2$, and $n = k/(k-1)$. Thus the series converges for $[(k - 1)/2]M^2 < 1$, and

$$\frac{p_0}{p} = 1 + \left(\frac{k}{k-1}\right)\left[\frac{k-1}{2}M^2\right] + \left(\frac{k}{k-1}\right)\left(\frac{k}{k-1} - 1\right)\frac{1}{2!}\left[\frac{k-1}{2}M^2\right]^2$$

$$+ \left(\frac{k}{k-1}\right)\left(\frac{k}{k-1} - 1\right)\left(\frac{k}{k-1} - 2\right)\frac{1}{3!}\left[\frac{k-1}{2}M^2\right]^3 + \cdots$$

$$\frac{p_0}{p} = 1 + \frac{k}{2}M^2 + \frac{k}{8}M^4 + \frac{k(2-k)}{48}M^6 + \cdots$$

$$\frac{p_0}{p} = 1 + \frac{k}{2}M^2\left[1 + \frac{1}{4}M^2 + \frac{(2-k)}{24}M^4 + \cdots\right] \qquad (2)$$

for compressible flow.

In the limit, as $M \to 0$, the term in brackets in Eq. 2 approaches 1.0. Thus, for flow at low Mach number, the incompressible and compressible equations give the same result. The variation of p_0/p with Mach number is shown in the sketch below. As Mach number is increased, the compressible equation gives larger values for the ratio, p_0/p.

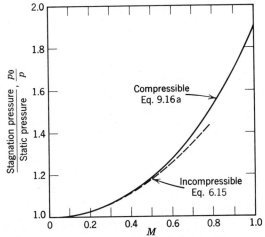

Equations 1 and 2 may be compared quantitatively most simply by writing

$$\frac{p_0}{p} - 1 = \frac{k}{2}M^2 \qquad \text{(``incompressible'')}$$

$$\frac{p_0}{p} - 1 = \frac{k}{2}M^2\left[1 + \frac{1}{4}M^2 + \frac{(2-k)}{24}M^4 + \cdots\right] \qquad \text{(compressible)}$$

The term in brackets is approximately equal to 1.02 at $M = 0.3$, and to 1.04 at $M = 0.4$. Thus, for calculations of engineering accuracy, flow may be considered incompressible if $M < 0.3$. The two agree with 5 percent for $M \gtrsim 0.45$.

9-3.2 CRITICAL CONDITIONS

Stagnation conditions are extremely useful as reference conditions for the thermodynamic properties; this is not true for velocity, since $V = 0$ by definition at stagnation. A more useful reference value for velocity is the critical speed, that is, the speed at a Mach number of unity. Even if there is no point in a given flow field where the Mach number is equal to unity, such a hypothetical condition is still useful as a reference condition.

Using an asterisk to denote conditions at $M = 1$, then by definition

$$V^* \equiv c^* \tag{9.17}$$

At critical conditions, Eqs. 9.16 for stagnation properties become (for $k = 1.4$)

$$\frac{p_0^*}{p^*} = \left[1 + \frac{k-1}{2} \right]^{k/(k-1)} = 1.893$$

$$\frac{T_0^*}{T^*} = 1 + \frac{k-1}{2} = 1.200$$

$$\frac{\rho_0^*}{\rho^*} = \left[1 + \frac{k-1}{2} \right]^{1/(k-1)} = 1.577$$

The critical speed may be written in terms of either the critical temperature, T^*, or the critical stagnation temperature, T_0^*.

For an ideal gas, $c^* = \sqrt{kRT^*}$, and thus $V^* = \sqrt{kRT^*}$. Since

$$T^* = \frac{T_0^*}{1 + (k-1)/2} = \frac{2}{k+1} T_0^*$$

then

$$V^* = c^* = \sqrt{\frac{2k}{k+1} RT_0^*} \tag{9.18}$$

We shall employ both the stagnation conditions and the critical conditions as reference conditions in the next chapter when we consider a variety of one-dimensional compressible flows.

summary objectives

After completing study of Chapter 9, you should be able to do the following:

1. Define:

Mach number	Mach angle
subsonic flow	zone of action
supersonic flow	zone of silence

transonic flow   local isentropic stagnation properties
hypersonic flow   critical conditions
Mach cone

2. For an ideal gas, write expressions for
   (a) The change in internal energy and enthalpy.
   (b) The Tds equations.
   (c) The relation between pressure and density for an isentropic process.

3. Derive an equation for the speed of sound in a medium and show that for an ideal gas, $c = \sqrt{kRT}$.

4. Write (and derive) expressions for the local isentropic stagnation properties (i.e. temperature, pressure, density) for the flow of an ideal gas.

5. Solve those problems at the end of the chapter that relate to the material you have studied.

problems

9.1 Five kilograms of air in a closed system expands reversibly with constant entropy from 300 kPa (absolute), 60 C, to 150 kPa (absolute). Calculate the air temperature after the expansion.

9.2 Ten pounds of air is cooled in a closed tank from 500 to 100 F. The initial pressure is 400 psia. Compute the changes in entropy, internal energy, and enthalpy.

9.3 A turbine operating on air has inlet conditions of 200 kPa (absolute), 95 C, and an exhaust pressure of 100 kPa (absolute). If the expansion is adiabatic and $\Delta KE = 0$, is an exhaust temperature of 5 C possible? Is it possible if $\Delta KE \neq 0$?

9.4 Air is compressed irreversibly from 100 kPa (absolute), 5 C, to 200 kPa (absolute), 115 C. Calculate the change in entropy of the air.

9.5 A fluid passing through a steady-flow system can be heated reversibly from state ① to state ② in either of two ways. In one case, $T = a + bs$; in the other case, $T = c + es^2$, where a, b, c, and e are constants. In which case is the heat transfer greater?

9.6 Is an adiabatic expansion of air from 300 kPa (absolute), 60 C, to 150 kPa (absolute), 27 C, possible? Justify your answer.

9.7 In a closed system, a gas undergoes a cycle made up of the following processes: 1–2 reversible isothermal compression, 2–3 reversible constant volume heating, 3–4 reversible constant pressure expansion, and 4–1 reversible adiabatic expansion.
   (a) Sketch pv and Ts diagrams.
   (b) State whether each of the following quantities is positive, zero, negative, or indeterminate in sign:

$$\oint \delta W, \quad \oint \delta Q, \quad \oint dS, \quad \oint dU, \quad \oint dH$$

9.8 An ideal gas is heated at constant volume from state ① to state ②; expanded isothermally to state ③, expanded adiabatically to state ④, which is at the same pressure as state ①, and then restored to state ① by a constant-pressure process. All four processes are reversible.

(a) Sketch pv and Ts diagrams of the cycle.

(b) State whether each of the following quantities is greater than zero, less than zero, equal to zero, or of indeterminate sign:

$$\oint \delta Q, \quad \oint \delta W, \quad \oint du, \quad \oint ds$$

9.9 An aircraft is flying at a speed of 960 km/hr through air at a pressure of 82 kPa and temperature of 0 C. Calculate the Mach number of the craft.

9.10 Air at a temperature of 20 C is flowing with a speed of 150 m/sec. A bullet is fired into the air stream with a speed of 800 m/sec. (The direction of the bullet is opposite to that of the air.) Calculate:

(a) The Mach number of the air flow.

(b) The Mach number of the bullet if it were fired in still air.

(c) The Mach number of the bullet with respect to the moving air.

9.11 What is the speed of sound in carbon dioxide at 150 C? What is the acoustic speed in water at 20 C?

9.12 A photograph of a bullet shows a Mach angle of 28°. Determine the speed of the bullet for standard air.

9.13 Air at 25 C is flowing at a Mach number of 1.9. Determine the air speed and the Mach angle.

9.14 A projectile is fired into a gas in which the pressure is 50 psia and the density is 0.27 lbm/ft³. It is observed experimentally that a Mach cone emanates from the projectile with a total angle of 20°. What is the speed of the projectile with respect to the gas?

9.15 A supersonic aircraft flies at an altitude of 10,000 ft at a speed of 3000 ft/sec on a standard day. How long after passing directly above a ground observer will it be before the sound of the aircraft is heard?

9.16 The Concorde supersonic transport cruises at $M = 2.2$ at an altitude of 17 km on a standard day. How long after the aircraft passes directly above a ground observer will it be before the sound of the aircraft is heard?

9.17 An airplane is flying at a speed of 180 m/sec at an altitude of 500 m, where the temperature is 20 C. The plane climbs to 15 km, where the temperature is -56 C and levels off at a speed of 320 m/sec. Calculate the Mach number of flight in both cases.

9.18 The pressure at the nose of an aircraft in flight was found to be 117 kPa (the velocity of air relative to the craft was zero at this point). Estimate the Mach number, speed and altitude of the craft, if the pressure and temperature of the undisturbed air were 27.6 kPa and -50 C, respectively.

9.19 A body moves through standard air at 200 m/sec. What is the pressure at a point on the body where the velocity of the air relative to the body is zero? Assume (a) compressible flow, and (b) incompressible flow.

9.20 Air flows in a duct where the static temperature is 50 F and the static pressure is 10 psia. Calculate the stagnation pressure if the air speed is (a) 200 ft/sec, and (b) 2000 ft/sec.

9.21 A plane is flying at an altitude where the air temperature is 10 C. At a point on the plane where the relative air velocity is zero the temperature is found to be 49 C. Determine the Mach number and speed of the plane.

9.22 Consider flow of standard air with a speed of 600 m/sec. What is the stagnation pressure? The stagnation enthalpy? The stagnation temperature?

9.23 Air flows steadily through a section (① denotes inlet and ② denotes exit) of an insulated constant-area duct. Properties change along the duct as a result of friction.

   (a) Beginning with the control volume form of the first law of thermodynamics, show that the equation can be reduced to

   $$h_1 + \frac{V_1^2}{2} = h_2 + \frac{V_2^2}{2} = \text{constant}$$

   (b) Denoting the constant by h_0 (the stagnation enthalpy), show that for adiabatic flow of an ideal gas with friction

   $$\frac{T_0}{T} = 1 + \frac{k-1}{2} M^2$$

   (c) For this flow does $T_{0_1} = T_{0_2}$? $p_{0_1} = p_{0_2}$?

9.24 Consider the steady adiabatic flow of air through a long, straight pipe with cross-sectional area 0.05 m². At the inlet (section ①) the air is at an absolute pressure of 200 kPa, 60 C, and has a velocity of 146 m/sec. At a section downstream the air is at 95.6 kPa (absolute) and has a velocity of 280 m/sec. Determine p_{0_1}, p_{0_2}, T_{0_1}, and T_{0_2}.

9.25 Air flows steadily through a constant-area duct. At section ①, the air is at 60 psia, 600 R, with velocity 500 ft/sec. As a result of heat transfer and friction, the air at section ② downstream is at 40 psia, 800 R. Determine p_{0_1}, p_{0_2}, T_{0_1}, and T_{0_2}.

Chapter 10

Steady One-Dimensional Compressible Flow

Fluid properties in compressible flow are affected by area change, friction, and heat transfer. In this chapter each of these effects is considered separately for steady, one-dimensional, compressible flow.

Isentropic flow, in which area is the independent variable (friction and heat transfer are neglected), is considered first for a general fluid. Then the isentropic flow of an ideal gas and applications to nozzles are considered in greater detail.

Following isentropic flow, adiabatic flow in a constant area duct with friction and frictionless flow in a constant area duct with heat transfer are considered. A discussion of normal shocks concludes the chapter.

10-1 BASIC EQUATIONS FOR ISENTROPIC FLOW

Consider the steady, one-dimensional, isentropic flow of any compressible fluid through a channel of arbitrary cross section; a portion of such a duct is shown in Fig. 10.1. In developing the governing equations for this flow, we shall apply the basic equations, derived in Chapter 4, to the finite, fixed control volume of Fig. 10.1. The properties at sections ① and ② are labeled with the appropriate subscripts; R_x is the x component of the surface force acting on the control volume.

515

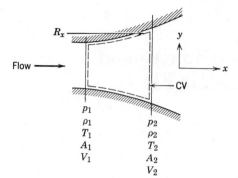

Fig. 10.1 Control volume for analysis of a general isentropic flow.

a. Continuity Equation

Basic equation:

$$0 = \overbrace{\frac{\partial}{\partial t} \int_{CV} \rho \, d\mathrm{V}}^{=0(1)} + \int_{CS} \rho \vec{V} \cdot d\vec{A} \tag{4.13}$$

Assumptions: (1) Steady flow
(2) One-dimensional flow

Then

$$0 = \{-|\rho_1 V_1 A_1|\} + \{|\rho_2 V_2 A_2|\}$$

Using scalar magnitudes and dropping absolute value signs gives the familiar form

$$\rho_1 V_1 A_1 = \rho_2 V_2 A_2 = \rho V A = \dot{m} = \text{constant} \tag{10.1a}$$

b. Momentum Equation

Basic equation:

$$F_{S_x} + \overbrace{F_{B_x}}^{=0(3)} = \overbrace{\frac{\partial}{\partial t} \int_{CV} V_x \rho \, d\mathrm{V}}^{=0(1)} + \int_{CS} V_x \rho \vec{V} \cdot d\vec{A} \tag{4.19a}$$

Assumptions: (3) $F_{B_x} = 0$

The surface force will be due to pressure forces at the surfaces ① and ②, and to the distributed pressure force, R_x, along the channel walls. Substituting gives

$$R_x + p_1 A_1 - p_2 A_2 = V_1\{-|\rho_1 V_1 A_1|\} + V_2\{|\rho_2 V_2 A_2|\}$$

Using scalar magnitudes and dropping absolute value signs, we obtain

$$R_x + p_1 A_1 - p_2 A_2 = \dot{m} V_2 - \dot{m} V_1 \qquad (10.1b)$$

c. First Law of Thermodynamics

Basic equation:

$$\overset{=0(4)}{\cancel{\dot{Q}}} + \overset{=0(5)}{\cancel{\dot{W_s}}} + \overset{=0(6)}{\cancel{\dot{W}_{\text{shear}}}} + \overset{0(6)}{\cancel{\dot{W}_{\text{other}}}} = \overset{=0(1)}{\cancel{\frac{\partial}{\partial t} \int_{\text{CV}} e\rho \, d\Psi}} + \int_{\text{CS}} (e + pv)\rho \vec{V} \cdot d\vec{A} \qquad (4.59)$$

where

$$e = u + \frac{V^2}{2} + \overset{\simeq 0(7)}{\cancel{gz}}$$

Assumptions: (4) $\dot{Q} = 0$ (isentropic, i.e. frictionless adiabatic flow)
(5) $\dot{W}_s = 0$
(6) $\dot{W}_{\text{shear}} = \dot{W}_{\text{other}} = 0$
(7) Effects of gravity are negligible

Under the assumptions, the first law reduces to

$$0 = \left(u_1 + p_1 v_1 + \frac{V_1^2}{2} \right) \{ -|\rho_1 V_1 A_1| \} + \left(u_2 + p_2 v_2 + \frac{V_2^2}{2} \right) \{ |\rho_2 V_2 A_2| \}$$

But we know from continuity that the mass flowrate terms in brackets are equal, so they may be canceled. We may also substitute $h \equiv u + pv$, to obtain

$$h_1 + \frac{V_1^2}{2} = h_2 + \frac{V_2^2}{2} = h + \frac{V^2}{2} = \text{constant} \qquad (10.1c)$$

The combination $h + V^2/2$ occurs often in compressible flow problems. It is convenient to define the stagnation enthalpy, h_0,

$$h_0 \equiv h + \frac{V^2}{2}$$

Physically, the stagnation enthalpy is that enthalpy which would be reached if the fluid were decelerated adiabatically to zero velocity. We note that the stagnation enthalpy is constant throughout an adiabatic flow field.

d. Second Law of Thermodynamics

Basic equation:

$$\int_{CS} \frac{1}{T} \underset{=0(4)}{\cancel{\frac{\dot{Q}}{A}}} dA \leq \underset{=0(1)}{\cancel{\frac{\partial}{\partial t}}} \int_{CV} s\rho \, d\Psi + \int_{CS} s\rho \vec{V} \cdot d\vec{A} \tag{4.60}$$

Then for a reversible adiabatic process,

$$0 = s_1\{-|\rho_1 V_1 A_1|\} + s_2\{|\rho_2 V_2 A_2|\}$$

Since the mass flowrate terms { } are equal by continuity,

$$s_1 = s_2 = s = \text{constant} \tag{10.1d}$$

e. Equation of State

Equations of state are relations among intensive thermodynamic properties. These relations may be in the form of tables, charts, or algebraic equations. Since, for a pure substance, it is possible to specify any intensive thermodynamic property in terms of any other two intensive thermodynamic properties, we can write

$$h = h(s, p) \tag{10.1e}$$

and

$$\rho = \rho(s, p) \tag{10.1f}$$

as equations of state.

Before summarizing the simplified forms of the basic equations for steady, one-dimensional, isentropic flow of any compressible fluid, let us turn to a representation of an isentropic flow on an hs diagram. At some point in the

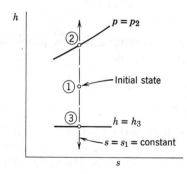

Fig. 10.2 Representation of isentropic flow in the hs plane.

isentropic flow field (call it state ①), the flow properties are h_1, s_1, p_1, V_1, and so on. Clearly, in isentropic flow, the flow may proceed to state ② (with properties h_2, $s_2 = s_1$, p_2, V_2, etc.) or to state ③ (with properties h_3, $s_3 = s_1$, p_3, V_3, etc.) as shown in Fig. 10.2.

How are the isentropic stagnation properties at states ①, ②, and ③ related? To answer this question, consider the results obtained from the first law analysis for isentropic flow. From Eq. 10.1c and the definition of stagnation enthalpy, we have

$$h_1 + \frac{V_1^2}{2} = h_2 + \frac{V_2^2}{2} = h + \frac{V^2}{2} = h_0 = \text{constant}$$

Thus all states in an isentropic flow have the same stagnation enthalpy. In addition, by definition, all states in an isentropic flow (including the stagnation state) have the same entropy. Thus all stagnation states in an isentropic flow have the same stagnation enthalpy and stagnation entropy. Since at the stagnation state the velocity is zero, it follows that the stagnation properties are constant for all points in an isentropic flow.

For isentropic flow, the first law, in the form

$$h + \frac{V^2}{2} = h_0 = \text{constant}$$

suggests another interpretation for the stagnation enthalpy. The stagnation enthalpy, h_0, represents the total energy per unit mass of flowing fluid. The kinetic energy per unit mass of the flow is then represented by the enthalpy difference, $h_0 - h$. This is illustrated graphically in Fig. 10.3.

Equations 10.1a–f are the simplified forms of the basic equations that describe steady, one-dimensional, isentropic flow of any compressible fluid.

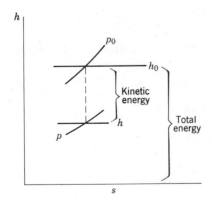

Fig. 10.3 Schematic *hs* diagram illustrating interpretation of energy per unit mass in a flow.

There are 6 independent equations. If all the properties of the flow are known at state ① then we have a total of seven unknowns (p_2, A_2, V_2, ρ_2, h_2, s_2, and R_x) in these six equations. Consequently, an isentropic flow can proceed to a variety of states from state ①. Each such state must have $s_2 = s_1$, satisfying Eq. 10.1d. This leaves, in effect, six unknowns and five equations. Thus the problem at this point is indeterminate. In order to determine conditions at state ②, one of the six unknowns must be specified.

What causes fluid properties to change in an isentropic flow? You undoubtedly recognize that it is the effect of area variation. In the next section we shall look at the effect of area variation on flow properties in isentropic flow.

10-2 EFFECT OF AREA VARIATION ON FLOW PROPERTIES IN ISENTROPIC FLOW

In considering the effect of area variation on flow properties in isentropic flow, we shall concern ourselves primarily with velocity and pressure. We wish to determine the effect of a change in the area, A, on the velocity, V, and the pressure, p; that is, for a change, dA, in the area, A, are dV and dp positive or negative?

To answer these questions, it is convenient to work with the differential forms of the governing equations. These were derived for the differential control volume of Fig. 9.3 (Section 9-3.1). The differential momentum equation for isentropic flow reduces to

$$\frac{dp}{\rho} + d\left(\frac{V^2}{2}\right) = 0 \tag{9.15c}$$

or

$$dp = -\rho V \, dV$$

Dividing by ρV^2, we obtain

$$\frac{dp}{\rho V^2} = -\frac{dV}{V} \tag{10.2}$$

A convenient differential form of the continuity equation can be obtained from Eq. 10.1a

$$\rho A V = \text{constant}$$

Taking the natural logarithm of both sides,

$$\ln \rho + \ln A + \ln V = \ln C$$

Differentiating,

$$\frac{d\rho}{\rho} + \frac{dA}{A} + \frac{dV}{V} = 0 \tag{10.3}$$

Solving Eq. 10.3 for dA/A gives

$$\frac{dA}{A} = -\frac{dV}{V} - \frac{d\rho}{\rho}$$

Substituting from Eq. 10.2,

$$\frac{dA}{A} = \frac{dp}{\rho V^2} - \frac{d\rho}{\rho}$$

or

$$\frac{dA}{A} = \frac{dp}{\rho V^2}\left[1 - \frac{V^2}{dp/d\rho}\right]$$

Now recall that for an isentropic process $dp/d\rho = \partial p/\partial \rho)_s = c^2$, so

$$\frac{dA}{A} = \frac{dp}{\rho V^2}\left[1 - \frac{V^2}{c^2}\right] = \frac{dp}{\rho V^2}[1 - M^2] \tag{10.4}$$

From Eq. 10.4 we see that for $M < 1$ an area change causes a pressure change of the same sign (positive dA means positive dp for $M < 1$); for $M > 1$ an area change causes a pressure change of opposite sign.

Substituting from Eq. 10.2 into Eq. 10.4, we obtain

$$\frac{dA}{A} = \frac{-dV}{V}[1 - M^2] \tag{10.5}$$

From Eq. 10.5 we see that for $M < 1$ an area change causes a velocity change of opposite sign (positive dA means negative dV for $M < 1$); for $M > 1$ an area change causes a velocity change of the same sign.

These results are summarized in Fig. 10.4. For subsonic flows ($M < 1$) flow acceleration in a *nozzle* requires a passage of diminishing cross section; that is, the area must decrease to cause a velocity increase. This produces a passage shaped like that shown in the upper left of Fig. 10.4, and this result is in accord with our experience. A subsonic *diffuser* requires that the passage area increase to cause a velocity decrease. Again this result agrees with our experience.

In supersonic flows ($M > 1$), the effects of area change are different. According to Eq. 10.5, a *supersonic nozzle* must be built with an area increase in the flow direction. A *supersonic diffuser* must be a converging channel. Although these predictions may be contrary to our experience, laboratory experiments show that they are valid. We can also recall seeing divergent nozzles designed to produce supersonic flow on missiles and launch vehicles.

What of the remaining case, $M = 1$? Further inspection of Eq. 10.5 shows that at $M = 1$, $dA/dV = 0$. The fact that $dA/dV = 0$ means that the passage

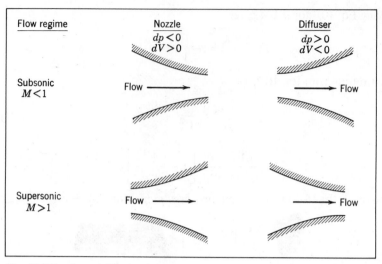

Fig. 10.4 Nozzle and diffuser shapes as a function of initial Mach number.

area must pass through a minimum or maximum at $M = 1$. Inspection of Fig. 10.4 shows that $M = 1$ can be reached only in a *throat* or section of minimum area.

To accelerate flow from rest to supersonic speed ($M > 1$) requires first a subsonic or converging nozzle. Under the proper conditions, the flow will be at $M = 1$ at the throat, where the area is a minimum. Further acceleration is possible if a supersonic (diverging) nozzle segment is added downstream from a throat. Isentropic flow in converging nozzles will be treated in Section 10-3.4, and isentropic flow in converging-diverging nozzles will be covered in Section 10-3.5.

To decelerate flow from supersonic ($M > 1$) to subsonic speed requires first a supersonic (converging) diffuser. In theory, the flow speed could be reduced isentropically to $M = 1$ at a throat where the area is a minimum, and further isentropic deceleration could take place in a diverging subsonic diffuser section. In practice, supersonic flow cannot be decelerated to exactly $M = 1$ at a throat because the sonic flow near the throat is unstable in a rising (adverse) pressure gradient. (Disturbances that are always present in a real subsonic flow propagate upstream, disturbing the sonic flow at the throat. Shock waves form and travel upstream where they are disgorged from the inlet of the supersonic diffuser.)

The throat area of a real supersonic diffuser must be slightly larger than that required to reduce the flow to $M = 1$. Under the proper downstream conditions, a weak normal shock forms in the diverging channel just downstream from the throat. Flow leaving the shock is subsonic, and decelerates in the diverging channel. Thus deceleration from supersonic to subsonic flow cannot occur isentropically in practice, since the weak normal shock causes an entropy increase. Normal shocks will be analyzed in Section 10-6.

For accelerating flows (favorable pressure gradients) the idealization of isentropic flow is generally a realistic model of the actual flow behavior. For decelerating flows, the idealization of isentropic flow may not be a realistic flow model because of the adverse pressure gradients and the attendant possibility of flow separation, as discussed for incompressible flow in Chapter 8.

10-3 ISENTROPIC FLOW OF AN IDEAL GAS

10-3.1 BASIC EQUATIONS

In Section 10-1 we applied the basic equations to a finite control volume for the steady, one-dimensional, isentropic flow of any compressible fluid. The only modification required in restricting our discussion to an ideal gas is in the equation of state. For an ideal gas the equation of state is $p = \rho RT$. In addition, for the isentropic flow of an ideal gas we have the process equation, $p/\rho^k = $ constant. Then, for the isentropic flow of an ideal gas we can summarize the basic equations as follows:

Continuity:	$\rho_1 V_1 A_1 = \rho_2 V_2 A_2 = \rho V A = \dot{m}$	(10.1a)
Momentum:	$R_x + p_1 A_1 - p_2 A_2 = \dot{m} V_2 - \dot{m} V_1$	(10.1b)
First law:	$h_1 + \dfrac{V_1^2}{2} = h_2 + \dfrac{V_2^2}{2} = h + \dfrac{V^2}{2}$	(10.1c)
Second law:	$s_1 = s_2 = s$	(10.1d)
Equation of state:	$p = \rho RT$	(9.1)
Process equation:	$p/\rho^k = $ constant	(9.11b)

These are the governing equations for steady, one-dimensional, isentropic flow of an ideal gas. If all the properties at state ① are known, then we have eight unknowns (ρ_2, A_2, V_2, p_2, h_2, s_2, T_2, and R_x) in these six equations. However, we have the known relationship between h and T for an ideal gas, $dh = c_p \, dT$. For an ideal gas with constant specific heats,

$$\Delta h = h_2 - h_1 = c_p \Delta T = c_p(T_2 - T_1) \tag{9.7b}$$

Thus, as in the general case (Section 10-1), the problem is indeterminate. One condition (other than s_2) must be specified at state ② before conditions at state ② can be completely determined.

10-3.2 REFERENCE CONDITIONS FOR ISENTROPIC FLOW OF AN IDEAL GAS

Expressions for local isentropic stagnation properties for an ideal gas were developed in Chapter 9 (Section 9-3.1). For completeness these expressions are repeated here.

Stagnation pressure:
$$\frac{p_0}{p} = \left[1 + \frac{k-1}{2} M^2\right]^{k/(k-1)} \tag{9.16a}$$

Stagnation temperature:
$$\frac{T_0}{T} = 1 + \frac{k-1}{2} M^2 \tag{9.16b}$$

Stagnation density:
$$\frac{\rho_0}{\rho} = \left[1 + \frac{k-1}{2} M^2\right]^{1/(k-1)} \tag{9.16c}$$

As shown in Section 10-1, the stagnation properties are constant throughout a steady, isentropic flow field.

The critical conditions, that is, the values of the flow properties at which the Mach number is unity, were introduced in Section 9-3.2. Since the stagnation properties are constant in an isentropic flow, then from Eqs. 9.16 we can write, for $k = 1.4$

$$\frac{p_0}{p^*} = \left[1 + \frac{k-1}{2}\right]^{k/(k-1)} = 1.893$$

$$\frac{T_0}{T^*} = 1 + \frac{k-1}{2} = 1.200$$

$$\frac{\rho_0}{\rho^*} = \left[1 + \frac{k-1}{2}\right]^{1/(k-1)} = 1.577$$

In addition, from Eq. 9.18 we have

$$V^* = c^* = \sqrt{\frac{2k}{k+1} R T_0}$$

In Section 10-2 we saw that it was necessary for a passage to have a section of minimum area (a throat) to accelerate a flow isentropically from rest to a Mach number greater than unity. Furthermore, in such a flow the Mach number is unity at the throat. If the area at which the Mach number is unity is designated by A^*, then it is possible to express the contour of a passage in terms of the area ratio, A/A^*.

Since for steady, one-dimensional flow, the continuity equation can be written

$$\rho A V = \text{constant} = \rho^* A^* V^*$$

Then

$$\frac{A}{A^*} = \frac{\rho^*}{\rho}\frac{V^*}{V} = \frac{\rho^*}{\rho}\frac{c^*}{Mc} = \frac{1}{M}\frac{\rho^*}{\rho}\sqrt{\frac{T^*}{T}}$$

$$\frac{A}{A^*} = \frac{1}{M}\frac{\rho^*}{\rho_0}\frac{\rho_0}{\rho}\sqrt{\frac{T^*/T_0}{T/T_0}}$$

$$\frac{A}{A^*} = \frac{1}{M}\frac{\left[1 + \dfrac{k-1}{2}M^2\right]^{1/(k-1)}}{\left[1 + \dfrac{k-1}{2}\right]^{1/(k-1)}}\left[\frac{1 + \dfrac{k-1}{2}M^2}{1 + \dfrac{k-1}{2}}\right]^{1/2}$$

$$\frac{A}{A^*} = \frac{1}{M}\left[\frac{1 + \dfrac{k-1}{2}M^2}{1 + \dfrac{k-1}{2}}\right]^{(k+1)/2(k-1)} \qquad (10.6)$$

From Eq. 10.6 we see that a choice of M gives a unique value of A/A^*. The variation of A/A^* with M is shown in Fig. 10.5. Note that the curve is double-valued; that is, for a given value of A/A^* other than unity, there are two possible values of Mach number. This is consistent with the results of Section 10-2 (see Fig. 10.4), where it was found that a converging-diverging passage with a section of minimum area is required to accelerate a flow from subsonic to supersonic speed.

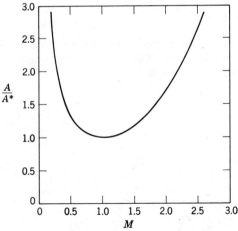

Fig. 10.5 Variation of A/A^* with Mach number in isentropic flow for $k = 1.4$.

Example 10.1

Air flows isentropically in a channel. At section ①, the Mach number is 0.3, the area is 0.001 m², and the absolute pressure and the temperature are 650 kPa and 62 C, respectively. At section ②, the Mach number is 0.8. Sketch the channel shape, plot a Ts diagram for the process, and evaluate the properties at section ②.

Example Problem 10.1

GIVEN:

Isentropic flow of air in a channel. At sections ① and ②, the following data are given: $M_1 = 0.3$, $T_1 = 62$ C, $p_1 = 650$ kPa (absolute), $A_1 = 0.001$ m², and $M_2 = 0.8$.

FIND:

(a) Sketch the channel shape.
(b) Plot a Ts diagram for the process.
(c) Properties at section ②.

SOLUTION:

To accelerate a subsonic flow requires a converging nozzle. The channel shape must be:

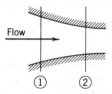

Flow

① ②

On the Ts plane, the process follows a line $s = $ constant. Stagnation conditions remain fixed for isentropic flow. Consequently, the stagnation temperature at section ② can be calculated (for air, $k = 1.4$) from

$$T_{0_2} = T_{0_1} = T_1 \left[1 + \frac{k-1}{2} M_1^2 \right]$$

$$= (62 + 273) \text{ K} \left[1 + \frac{1.4 - 1}{2} (0.3)^2 \right]$$

$$T_{0_2} = T_{0_1} = 341 \text{ K}$$

and

$$T_2 = \frac{T_{0_2}}{\left[1 + \frac{k-1}{2} M_2^2\right]} = \frac{341 \text{ K}}{[1 + 0.2(0.8)^2]} = 302 \text{ K or } 29 \text{ C} \qquad T_2$$

Also, for an ideal gas

$$c_2 = \sqrt{kRT_2} = \left[1.4 \times \frac{287 \text{ N} \cdot \text{m}}{\text{kg} \cdot \text{K}} \times 302 \text{ K} \times \frac{\text{kg} \cdot \text{m}}{\text{N} \cdot \text{sec}^2}\right]^{1/2} = 348 \text{ m/sec} \qquad c_2$$

From the definition of Mach number,

$$V_2 = M_2 c_2 = (0.8)348 \text{ m/sec} = 278 \text{ m/sec} \qquad V_2$$

Using the isentropic relation, $p/\rho^k = \text{constant}$,

$$\frac{p_2}{p_1} = \left(\frac{T_2}{T_1}\right)^{k/(k-1)} = \left(\frac{302 \text{ K}}{335 \text{ K}}\right)^{3.5} = 0.696$$

and

$$p_2 = 0.696 p_1 = (0.696)650 \text{ kPa (abs)} = 452 \text{ kPa (abs)} \qquad p_2$$

Also

$$\rho_2 = \frac{p_2}{RT_2} = \frac{4.52 \times 10^5 \text{ N}}{\text{m}^2} \times \frac{\text{kg} \cdot \text{K}}{287 \text{ N} \cdot \text{m}} \times \frac{1}{302 \text{ K}} = 5.21 \text{ kg/m}^3 \qquad \rho_2$$

From the continuity equation

$$\dot{m} = \rho_1 V_1 A_1 = \rho_2 V_2 A_2 = \text{constant}$$

so that

$$A_2 = A_1 \frac{\rho_1}{\rho_2} \frac{V_1}{V_2} = A_1 \left(\frac{T_1}{T_2}\right)^{1/(k-1)} \frac{M_1 c_1}{M_2 c_2} = A_1 \left(\frac{T_1}{T_2}\right)^{1/(k-1)} \frac{M_1}{M_2} \sqrt{\frac{kRT_1}{kRT_2}}$$

or

$$A_2 = A_1 \frac{M_1}{M_2}\left(\frac{T_1}{T_2}\right)^{(k+1)/2(k-1)} = 0.001 \text{ m}^2 \times \frac{0.3}{0.8}\left(\frac{335}{302}\right)^3 = 5.12 \times 10^{-4} \text{ m}^2 \qquad A_2$$

Thus $A_2 < A_1$, as expected. Finally,

$$p_{0_2} = p_2 \left[1 + \frac{k-1}{2} M_2^2\right]^{k/(k-1)} = 452 \text{ kPa}[1 + 0.2(0.8)^2]^{3.5}$$

$$p_{0_2} = 689 \text{ kPa (abs)} \qquad p_{0_2}$$

The stagnation pressure should be constant for isentropic flow. Checking gives

$$p_{0_1} = p_1 \left[1 + \frac{k-1}{2} M_1^2\right]^{k/(k-1)} = 650 \text{ kPa}[1 + 0.2(0.3)^2]^{3.5}$$

$$p_{0_1} = 692 \text{ kPa (abs)}$$

The discrepancy between the calculated values of p_{0_1} and p_{0_2} is due to use of temperature values rounded to three significant figures in the calculation of p_{0_2}. Thus we note again for isentropic flow that $T_{0_1} = T_{0_2}$ and $p_{0_1} = p_{0_2}$.

In the previous section we saw (Eqs. 9.16a, b, c and 10.6) that the properties at a point in a compressible flow of an ideal gas may be related to appropriate reference conditions by functions of the local Mach number. This makes it possible to tabulate or plot them as functions of Mach number, M, for a given value of k.

Table B.1 of Appendix B lists values of T/T_0, p/p_0, ρ/ρ_0, and A/A^* as functions of M for isentropic flow of an ideal gas with $k = 1.4$.[1] Use of tables can significantly reduce the labor of calculations.

Since the reference conditions remain constant in an isentropic flow, the ratio of properties at two points in a flow may readily be found from the tables. Use of the tables is illustrated in Example Problem 10.2.

Example 10.2

Air flows isentropically in a channel. At section ①, the Mach number is 0.3, the area is 0.001 m², and the absolute pressure and the temperature are 650 kPa and 62 C, respectively. Evaluate the properties at section ②, where the Mach number is 0.8, using the isentropic flow tables. (Note that these are the data of Example 10.1.)

Example Problem 10.2

GIVEN:

Isentropic flow of air in a channel. At sections ① and ②, the following data are given: $M_1 = 0.3$, $T_1 = 62$ C, $p_1 = 650$ kPa (absolute), $A_1 = 0.001$ m², and $M_2 = 0.8$.

FIND:

Properties at section ②, using isentropic flow tables.

SOLUTION:

The Ts diagram was sketched in Example Problem 10.1. From Table B.1, Appendix B, we find

M	T/T_0	p/p_0	ρ/ρ_0	A/A^*
0.3	0.9823	0.9395	0.9564	2.035
0.8	0.8865	0.6560	0.7400	1.038

** This section may be omitted without loss of continuity in the text material.

[1] Tables for other common values of k have been published. See, for example, J. H. Keenan and J. Kaye, *Gas Tables* (New York: John Wiley, 1948).

For isentropic flow, $T_{0_1} = T_{0_2} = T_0$. Thus

$$\frac{T_2}{T_1} = \frac{T_2}{T_0}\frac{T_0}{T_1} = \frac{(T/T_0)_2}{(T/T_0)_1} = \frac{0.8865}{0.9823} = 0.9025$$

$$T_2 = 0.9025 T_1 = 0.9025(273 + 62)\ \text{K} = 302\ \text{K} \qquad\qquad T_2$$

Also $p_{0_2} = p_{0_1} = p_0$, so

$$\frac{p_2}{p_1} = \frac{p_2}{p_0}\frac{p_0}{p_1} = \frac{(p/p_0)_2}{(p/p_0)_1} = \frac{0.6560}{0.9395} = 0.6982$$

$$p_2 = 0.6982 p_1 = 0.6982(650\ \text{kPa}) = 454\ \text{kPa (abs)} \qquad\qquad p_2$$

and

$$\rho_2 = \frac{p_2}{RT_2} = \frac{4.54 \times 10^5\ \text{N}}{\text{m}^2} \times \frac{\text{kg} \cdot \text{K}}{287\ \text{N} \cdot \text{m}} \times \frac{1}{302\ \text{K}} = 5.24\ \text{kg/m}^3 \qquad \rho_2$$

The stagnation properties are

$$T_{0_2} = T_{0_1} = \frac{T_1}{(T/T_0)_1} = \frac{(273 + 62)\ \text{K}}{0.9823} = 341\ \text{K} \qquad\qquad T_{0_2}$$

and

$$p_{0_2} = p_{0_1} = \frac{p_1}{(p/p_0)_1} = \frac{650\ \text{kPa}}{0.9395} = 692\ \text{kPa (abs)} \qquad\qquad p_{0_2}$$

The area may be computed using the ratio A/A^*. Thus since $A^* = \text{constant}$,

$$\frac{A_2}{A_1} = \frac{A_2}{A^*}\frac{A^*}{A_1} = \frac{(A/A^*)_2}{(A/A^*)_1} = \frac{1.038}{2.035} = 0.5101$$

$$A_2 = 0.5101 A_1 = 0.5101\ (0.001\ \text{m}^2) = 5.10 \times 10^{-4}\ \text{m}^2 \qquad\qquad A_2$$

The velocity at section ② may be calculated from $V_2 = M_2 c_2$.

10-3.4 ISENTROPIC FLOW IN A CONVERGING NOZZLE

In this section we are interested in investigating the operation of a converging nozzle under various back pressures. Flow through the converging nozzle shown in Fig. 10.6 is supplied from a large plenum chamber, where conditions are assumed to be stagnation conditions; the flow is induced by a vacuum pump downstream and is controlled by the valve shown.

The back pressure, p_b, to which the nozzle discharges is controlled by the valve. The upstream stagnation conditions (p_0, T_0, etc.) are maintained constant. The pressure in the exit plane of the nozzle is denoted as p_e. We are interested in investigating the effect of variations in back pressure on the pressure distribution through the nozzle, on the mass flowrate, and on the

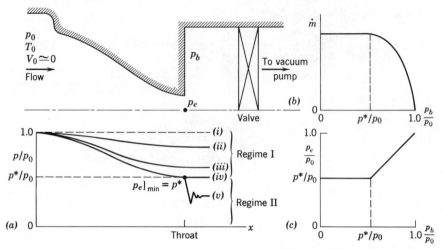

Fig. 10.6 Converging nozzle operating at various back pressures.

exit plane pressure. The results are illustrated graphically in Fig. 10.6. Let us look at each of the cases shown.

When the valve is closed, there is no flow through the nozzle. The pressure is p_0 throughout as shown by condition (i) in Fig. 10.6a.

If the back pressure, p_b, is now reduced to a value slightly less than p_0, there will be flow through the nozzle with a decrease in pressure in the direction of flow as shown by condition (ii). Flow at the exit plane will be subsonic with the exit plane pressure equal to the back pressure.

What happens as we continue to decrease the back pressure? The flow-rate will continue to increase and the exit plane pressure will continue to decrease as shown by condition (iii) in Fig. 10.6. Will these trends continue indefinitely as the back pressure is lowered?

Recall from our previous discussion that in a converging section the Mach number cannot increase beyond unity in isentropic flow. Thus with continued decrease of the back pressure, the flow at the exit plane of the nozzle will eventually reach a Mach number of unity. The corresponding pressure is the critical pressure, p^*. Condition (iv) illustrates the condition where M_e equals unity and p_b/p_0 equals p^*/p_0.

From Eq. 9.16a, with $M = 1$, the critical pressure ratio for an ideal gas is given by

$$\frac{p^*}{p_0} = \left(\frac{2}{k+1}\right)^{k/(k-1)}$$

For $k = 1.4$, $p^*/p_0 = 0.528$.

What happens when the back pressure is reduced further to a value below p^*, i.e. to condition (v)? Since the Mach number at the throat is unity $(V_e = c_e)$, information about conditions in the exhaust duct cannot be transmitted upstream. Consequently, reductions in p_b below a value equal to p^* have no effect on flow conditions in the nozzle; thus neither the pressure distribution through the nozzle, the nozzle exit pressure, nor the mass flowrate are affected by lowering p_b below p^*. When p_b is less than or equal to p^*, the nozzle is said to be *choked*.

For p_b less than p^*, the flow leaving the nozzle will expand to match the lower back pressure as shown for condition (v) in Fig. 10.6a. This unconfined expansion process is three-dimensional; the pressure distribution cannot be predicted by one-dimensional theory. Experiments show that a series of shocks form in the exit stream, resulting in an increase in entropy.

Flow through a converging nozzle may be divided into two regimes:

1. In Regime I, $1 \geq p_b/p_0 \geq p^*/p_0$. The flow to the throat is isentropic; $p_e = p_b$.
2. In Regime II, $p_b/p_0 < p^*/p_0$. The flow to the throat is isentropic, but a nonisentropic expansion occurs in the flow leaving the nozzle; $p_e = p^* > p_b$.

The flow processes corresponding to Regime II are shown on a Ts diagram in Fig. 10.7. Two problems involving converging nozzles are solved in Example Problems 10.3 and 10.4.

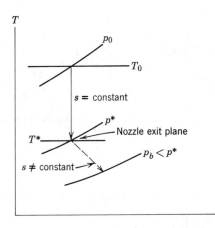

Fig. 10.7 Schematic Ts diagram for choked flow through a converging nozzle.

Although isentropic flow is an idealization, it is often a very good approximation for the actual behavior of nozzles. Since a nozzle is a device that accelerates a flow, the internal pressure gradient is favorable. This tends to keep the wall boundary layers thin and to minimize the effects of friction.

Example 10.3

A converging nozzle with a throat area of $0.001\ m^2$ is operated with air at a back pressure of 591 kPa (absolute). The nozzle is fed from a large plenum chamber where the absolute stagnation pressure and the temperature are 1.0 MPa and 60 C, respectively. The exit Mach number and mass flowrate are to be determined using (i) isentropic flow relations, and (ii) tables for isentropic flow.

Example Problem 10.3

GIVEN:

Air flow through a converging nozzle at the conditions shown:
Flow is isentropic.

FIND:

(a) M_e, and (b) $\dot{m}$.

$p_0 = 1.0$ MPa (abs.)
$T_0 = 333$ K

$p_b = 591$ kPa (abs.)

p_e

SOLUTION:

The first step is to check for choking. The pressure ratio is

$$\frac{p_b}{p_0} = \frac{5.91 \times 10^5}{1.0 \times 10^6} = 0.591 > 0.528$$

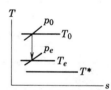

so the flow is not choked. Thus $p_b = p_e$, and the flow is isentropic, as sketched on the Ts diagram above.

(i) *Isentropic Flow Relations*

Since $p_0 = $ constant, the Mach number, M_e, may be found from the pressure ratio,

$$\frac{p_0}{p_e} = \left[1 + \frac{k-1}{2} M_e^2\right]^{k/(k-1)}$$

Solving for M_e, since $p_e = p_b$,

$$1 + \frac{k-1}{2} M_e^2 = \left(\frac{p_0}{p_b}\right)^{(k-1)/k}$$

and

$$M_e = \left\{\left[\left(\frac{p_0}{p_b}\right)^{(k-1)/k} - 1\right]\frac{2}{k-1}\right\}^{1/2} = \left\{\left[\left(\frac{1.0 \times 10^6}{5.91 \times 10^5}\right)^{0.286} - 1\right]\frac{2}{1.4-1}\right\}^{1/2} = 0.9 \qquad \overset{\longleftarrow}{M_e}$$

The mass flowrate will be given by

$$\dot{m} = \rho_e V_e A_e = \rho_e M_e c_e A_e$$

Thus we need the temperature, T_e, to find ρ_e and c_e. Since $T_0 = $ constant,

$$\frac{T_0}{T_e} = 1 + \frac{k-1}{2} M_e^2$$

or

$$T_e = \frac{T_0}{1 + \dfrac{k-1}{2} M_e^2} = \frac{(273 + 60)\text{ K}}{1 + 0.2(0.9)^2} = 287 \text{ K}$$

$$c_e = \sqrt{kRT_e} = \left[1.4 \times \frac{287 \text{ N} \cdot \text{m}}{\text{kg} \cdot \text{K}} \times 287 \text{ K} \times \frac{\text{kg} \cdot \text{m}}{\text{N} \cdot \text{sec}^2}\right]^{1/2} = 340 \text{ m/sec}$$

and

$$\rho_e = \frac{p_e}{RT_e} = 5.91 \times 10^5 \frac{\text{N}}{\text{m}^2} \times \frac{\text{kg} \cdot \text{K}}{287 \text{ N} \cdot \text{m}} \times \frac{1}{287 \text{ K}} = 7.18 \text{ kg/m}^3$$

Finally,

$$\dot{m} = \rho_e M_e c_e A_e = \frac{7.18 \text{ kg}}{\text{m}^3} \times 0.9 \times \frac{340 \text{ m}}{\text{sec}} \times 0.001 \text{ m}^2 = 2.20 \text{ kg/sec} \qquad \overset{\longleftarrow}{\dot{m}}$$

(ii) **Tables for isentropic Flow

The pressure ratio is

$$\frac{p_b}{p_0} = \frac{5.91 \times 10^5}{1.0 \times 10^6} = 0.591 > 0.528$$

so the flow is not choked. Thus $p_e = p_b$, and the flow is isentropic. From Table B.1, Appendix B, $p/p_0 = 0.591$ at $M = 0.90$. Thus

$$M_e = 0.90 \qquad \overset{\longleftarrow}{M_e}$$

The simplest procedure in using the tables is to express the desired result in terms of property ratios, which may then be found from the tables. Thus

$$\dot{m} = \rho_e V_e A_e = \rho_e M_e c_e A_e = \rho_e M_e \sqrt{kRT_e} A_e$$

$$\dot{m} = \rho_0 \frac{\rho_e}{\rho_0} M_e \sqrt{kRT_0} \sqrt{\frac{T_e}{T_0}} A_e = \frac{p_0}{RT_0} \frac{\rho_e}{\rho_0} M_e \sqrt{kRT_0} \sqrt{\frac{T_e}{T_0}} A_e$$

or finally,

$$\dot{m} = \frac{\rho_e}{\rho_0} \sqrt{\frac{T_e}{T_0}} p_0 M_e \sqrt{\frac{k}{RT_0}} A_e$$

Using values for $M_e = 0.9$ gives

$$\dot{m} = (0.6870)(0.8606)^{1/2}\left(1.0 \times 10^6 \frac{N}{m^2}\right)(0.9)$$

$$\times \left[1.4 \times \frac{kg \cdot K}{287\ N \cdot m} \times \frac{1}{333\ K} \times \frac{kg \cdot m}{N \cdot sec^2}\right]^{1/2}(0.001\ m^2)$$

$$\dot{m} = 2.20\ kg/sec \qquad\qquad \dot{m}$$

Example 10.4

Air flows isentropically through a converging nozzle. At a section where the nozzle area is 0.013 ft², the local pressure, temperature and Mach number are 60 psia, 40 F, and 0.52, respectively. The back pressure is 30 psia. The Mach number at the throat, the mass flowrate, and the throat area are to be determined, using (i) isentropic flow relations and (ii) tables for isentropic flow.

Example Problem 10.4

GIVEN:

Air flow through a converging nozzle at the conditions shown:

$M_1 = 0.52$
$T_1 = 40\ F$
$p_1 = 60\ psia$
$A_1 = 0.013\ ft^2$

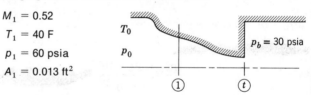

FIND:

(a) M_t, (b) $\dot{m}$, and (c) A_t

SOLUTION:

(i) Isentropic Flow Relations

We must first check for choking, to determine if flow is isentropic down to p_b. To check, we evaluate the stagnation conditions.

$$\frac{p_0}{p_1} = \left[1 + \frac{k-1}{2}M_1^2\right]^{k/(k-1)} = [1 + 0.2(0.52)^2]^{3.5} = 1.20$$

and

$$p_0 = 1.20\ p_1 = (1.2)60\ psia = 72.0\ psia$$

The back pressure ratio is

$$\frac{p_b}{p_0} = \frac{30.0}{72.0} = 0.417 < 0.528$$

so the flow is choked! For choked flow,

$$M_t = 1.0$$

The Ts diagram is

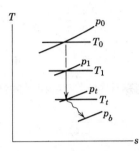

The mass flowrate may be found from conditions at section ①, using $\dot{m} = \rho_1 V_1 A_1$.

$$V_1 = M_1 c_1 = M_1 \sqrt{kRT_1}$$

$$= 0.52 \left[1.4 \times \frac{53.3}{\text{lbm} \cdot \text{R}} \frac{\text{ft} \cdot \text{lbf}}{\text{lbm} \cdot \text{R}} \times (460 + 40)\,\text{R} \times \frac{32.2}{\text{slug}} \frac{\text{lbm}}{\text{slug}} \times \frac{\text{slug} \cdot \text{ft}}{\text{lbf} \cdot \text{sec}^2} \right]^{1/2}$$

$$V_1 = 570 \text{ ft/sec}$$

$$\rho_1 = \frac{p_1}{RT_1} = \frac{60 \text{ lbf}}{\text{in.}^2} \times \frac{\text{lbm} \cdot \text{R}}{53.3 \text{ ft} \cdot \text{lbf}} \times \frac{1}{500 \text{ R}} \times \frac{144 \text{ in.}^2}{\text{ft}^2} = 0.324 \text{ lbm/ft}^3$$

$$\dot{m} = \rho_1 V_1 A_1 = \frac{0.324 \text{ lbm}}{\text{ft}^3} \times \frac{570 \text{ ft}}{\text{sec}} \times 0.013 \text{ ft}^2 = 2.40 \text{ lbm/sec}$$

The throat area may be computed by applying continuity between section ① and the throat; that is, $\dot{m} = \rho_1 V_1 A_1 = \rho_t V_t A_t$, so

$$A_t = A_1 \frac{\rho_1}{\rho_t} \frac{V_1}{V_t} = A_1 \frac{p_1}{RT_1} \frac{RT_t}{p_t} \frac{M_1 \sqrt{kRT_1}}{M_t \sqrt{kRT_t}} = A_1 \frac{p_1}{p_t} \frac{M_1}{M_t} \sqrt{\frac{T_t}{T_1}}$$

For isentropic flow, $T_0 = \text{constant}$, so

$$\frac{T_t}{T_1} = \frac{T_t}{T_0} \frac{T_0}{T_1} = \frac{1 + \dfrac{k-1}{2} M_1^2}{1 + \dfrac{k-1}{2} M_t^2} = \frac{1 + (0.2)(0.52)^2}{1.2} = 0.878$$

Also,

$$\frac{p_0}{p_t} = \left[1 + \frac{k-1}{2} M_t^2\right]^{k/(k-1)} = (1.2)^{3.5} = 1.89$$

so that

$$p_t = \frac{p_0}{1.89} = \frac{72.0 \text{ psia}}{1.89} = 38.1 \text{ psia}$$

Substituting gives

$$A_t = A_1 \frac{p_1}{p_t} \frac{M_1}{M_t} \sqrt{\frac{T_t}{T_1}} = 0.013 \text{ ft}^2 \times \frac{60.0 \text{ psia}}{38.1 \text{ psia}} \times \frac{0.52}{1.0} \sqrt{0.878} = 9.98 \times 10^{-3} \text{ ft}^2$$

$$A_t$$

(ii)   **Tables for Isentropic Flow*

We must first check for choking. From Table B.1, Appendix B, at $M_1 = 0.52$

$$\frac{p_1}{p_0} = 0.8317; \qquad p_0 = \frac{p_1}{0.8317} = \frac{60 \text{ psia}}{0.8317} = 72.1 \text{ psia}$$

From the table at $M = 1.0$, the minimum isentropic pressure ratio in a converging nozzle is

$$\frac{p}{p_0} = 0.5283$$

From the conditions given

$$\frac{p_b}{p_0} = \frac{30}{72.1} = 0.416 < 0.5283$$

so the flow is choked! For choked flow,

$$M_t = 1.0$$

$$M_t$$

The Ts diagram was given above. The mass flowrate calculation is the same as in the previous solution. From Table B.1, at $M_1 = 0.52$, $A_1/A^* = 1.303$. For choked flow, $A_t = A^*$. Thus

$$A_t = A^* = \frac{A_1}{1.303} = \frac{0.013 \text{ ft}^2}{1.303} = 9.98 \times 10^{-3} \text{ ft}^2$$

$$A_t$$

10-3.5   ISENTROPIC FLOW IN A CONVERGING-DIVERGING NOZZLE

Having considered isentropic flow in a converging nozzle, we turn now to isentropic flow in a converging-diverging nozzle. As in the previous case, the flow through the converging-diverging passage of Fig. 10.8 is induced by a

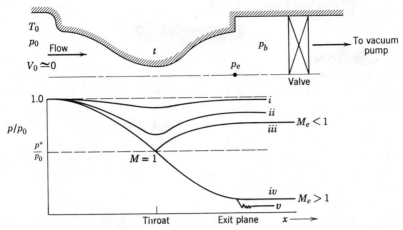

Fig. 10.8 Pressure distributions for isentropic flow in a converging-diverging nozzle.

vacuum pump downstream, and is controlled by the valve shown. The upstream stagnation conditions are assumed constant; the pressure in the exit plane of the nozzle is denoted as p_e; the nozzle discharges to the back pressure, p_b. We are interested in investigating the effect of variations in back pressure on the pressure distribution through the nozzle. The results are illustrated graphically in Fig. 10.8. Let us consider each of the cases shown.

With the valve initially closed, there is no flow through the nozzle; the pressure is constant at p_0. Opening the valve slightly (p_b slightly less than p_0) produces the pressure distribution curve (*i*). If the flowrate is low enough, at all points on this curve the flow will be subsonic and essentially incompressible. Under these conditions, the C–D nozzle will behave like a venturi, with flow accelerating in the converging portion, until a point of maximum velocity and minimum pressure is reached at the throat, then decelerating in the diverging portion to the nozzle exit.

As the valve is opened farther and the flowrate is increased, a more sharply defined pressure minimum occurs, as shown by curve (*ii*). Although compressibility effects become important, the flow is still subsonic everywhere, and deceleration takes place in the diverging section. Finally, as the valve is opened farther, curve (*iii*) results. At the section of minimum area the flow finally reaches $M = 1$ and the nozzle is choked; that is, the flowrate is the maximum possible for the given nozzle and stagnation conditions.

All of the flows giving rise to pressure distributions *i*, *ii*, and *iii* are isentropic; each curve is associated with a unique rate of mass flow. Finally when curve (*iii*) is reached, critical conditions are present at the throat. For this flowrate, the flow is choked, and

$$\dot{m} = \rho^* V^* A^*$$

where $A^* = A_t$.

In our discussion of the effect of area variation on isentropic flow we noted that a diverging section was required to accelerate a flow to supersonic speed from Mach one at a throat. At this point, then, we ask the question, "What back pressure, p_b, is necessary to accelerate the flow isentropically in the diverging portion of the nozzle?"

To accelerate flow in the diverging section requires a pressure decrease. This condition is illustrated by curve (iv) in Fig. 10.8. The flow will accelerate isentropically in the nozzle provided the exit pressure is set at the value p_{iv}. Thus we see that with a throat Mach number of unity there are two possible isentropic flow conditions in the converging-diverging nozzle. This is consistent with the results of Fig. 10.5, where we found two Mach numbers for a given value of A/A^* in isentropic flow.

Lowering the back pressure below that of condition (iv), say to condition (v), has no effect on the flow in the nozzle. The flow is isentropic from the plenum chamber to the nozzle exit (same as condition (iv)) and then undergoes a three-dimensional irreversible expansion to the lower back pressure. A nozzle operating under these conditions is said to be *underexpanded*, since an additional expansion takes place outside the nozzle.

A converging-diverging nozzle is generally intended to produce supersonic flow at the exit plane. If the back pressure is set at p_{iv}, flow will be isentropic through the nozzle, and supersonic at the nozzle exit. Nozzles operating at $p_b = p_{iv}$ (corresponding to curve (iv) in Fig. 10.8) are said to be at *design conditions*.

Flow leaving a C–D nozzle is supersonic when the back pressure is at or below the design value. The exit Mach number is fixed once the area ratio, A_e/A^*, is specified. All other exit plane properties (for isentropic flow) are uniquely related to the stagnation properties by the fixed exit plane Mach number.

The assumption of isentropic flow for a real nozzle at design conditions is a reasonable one. However, the one-dimensional flow model is inadequate for the design of relatively short nozzles to produce uniform supersonic exit flow.

Rocket-propelled vehicles use C–D nozzles to accelerate the exhaust gases to the maximum possible velocity to produce high thrust. A propulsion nozzle is subject to varying ambient discharge conditions during flight through the atmosphere, so it is impossible to attain the maximum theoretical thrust over the complete operating range. Because only a single supersonic Mach number can be obtained for a given area ratio, nozzles for supersonic wind tunnels are often built with interchangeable sections or variable geometry.

You have undoubtedly noticed that nothing has been said about the operation of converging-diverging nozzles when the back pressure is in the range $p_{iii} > p_b > p_{iv}$. The reason is that for such cases the flow cannot expand isentropically to p_b. Under these conditions a shock (which may be treated as an irreversible discontinuity involving entropy increase) occurs somewhere within the flow. Following a discussion of normal shocks in Section 10-6, we shall return to complete the discussion of converging-diverging nozzle flows. Nozzles operating with $p_{iii} > p_b > p_{iv}$ are said to be *overexpanded*, because the pressure at some point in the nozzle is less than the back pressure. Obviously, an overexpanded nozzle could be made to operate at a new design condition by cutting off a portion of the diverging section.

One other comment should be made at this point. Real compressible fluid flows are affected by friction, heating or cooling, and the possible presence (in supersonic flow) of shock waves. We have treated isentropic flow first because it is a useful idealized model for many real flow processes and because it gives us valuable insight into the behavior of fluids in compressible flow. In the next two sections we take up the effects of friction and heat transfer separately to gain insight into the effect of each factor on flow behavior. Following this we return to a discussion of the normal shock and complete our study of converging-diverging nozzle flows. In later courses it will be possible to explore real flows and the results of combining several of these effects.

Example 10.5

Air flows isentropically in a converging-diverging nozzle, with an exit area of 0.001 m². The nozzle is fed from a large plenum where the stagnation temperature and the absolute stagnation pressure are 350 K and 1.0 MPa, respectively. The exit pressure is 954 kPa (abs), and the Mach number at the throat is 0.68. Flow conditions at the throat and the exit Mach number are to be determined.

Example Problem 10.5

GIVEN:

Isentropic flow of air in C–D nozzle as shown:

$T_0 = 350$ K

$p_0 = 1.0$ MPa (abs)

$p_b = 954$ kPa (abs)

$M_t = 0.68 \qquad A_e = 0.001$ m²

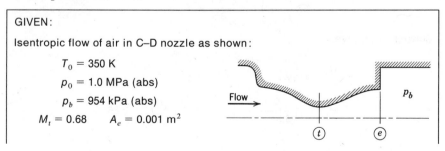

FIND:

(a) Properties at nozzle throat.
(b) M_e.

SOLUTION:

(i) *Isentropic Flow Relations*

The stagnation temperature is constant for isentropic flow. Thus, since

$$\frac{T_0}{T} = 1 + \frac{k-1}{2} M^2$$

$$T_t = \frac{T_0}{1 + \frac{k-1}{2} M_t^2} = \frac{350 \text{ K}}{1 + 0.2(0.68)^2} = 320 \text{ K} \qquad \longleftarrow \quad T_t$$

Also, since p_0 is constant for isentropic flow, then

$$p_t = p_0 \left(\frac{T_t}{T_0}\right)^{k/(k-1)} = p_0 \left[\frac{1}{1 + \frac{k-1}{2} M_t^2}\right]^{k/(k-1)}$$

$$p_t = 1.0 \times 10^6 \text{ Pa} \left[\frac{1}{1 + 0.2(0.68)^2}\right]^{3.5} = 734 \text{ kPa (abs)} \qquad \longleftarrow \quad p_t$$

$$\rho_t = \frac{p_t}{RT_t} = \frac{7.34 \times 10^5 \text{ N}}{\text{m}^2} \times \frac{\text{kg} \cdot \text{K}}{287 \text{ N} \cdot \text{m}} \times \frac{1}{320 \text{ K}} = 7.99 \text{ kg/m}^3 \qquad \longleftarrow \quad \rho_t$$

and

$$V_t = M_t c_t = M_t \sqrt{kRT_t}$$

$$V_t = 0.68 \left[\frac{1.4 \times 287 \text{ N} \cdot \text{m}}{\text{kg} \cdot \text{K}} \times \frac{320 \text{ K}}{1} \times \frac{\text{kg} \cdot \text{m}}{\text{N} \cdot \text{sec}^2}\right]^{1/2} = 244 \text{ m/sec} \qquad \longleftarrow \quad V_t$$

Since $M_t < 1$, flow at the exit must be subsonic. Therefore, $p_e = p_b$. The stagnation properties are constant, so

$$\frac{p_0}{p_e} = \left[1 + \frac{k-1}{2} M_e^2\right]^{k/(k-1)}$$

Solving for M_e gives

$$M_e = \left\{\left[\left(\frac{p_0}{p_e}\right)^{(k-1)/k} - 1\right] \frac{2}{k-1}\right\}^{1/2}$$

$$M_e = \left\{\left[\left(\frac{1.0 \times 10^6}{9.54 \times 10^5}\right)^{0.286} - 1\right] (5)\right\}^{1/2} = 0.26 \qquad \longleftarrow \quad M_e$$

The *Ts* diagram for this flow is

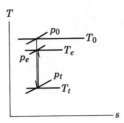

(ii)   **Tables for Isentropic Flow

The stagnation properties are constant for isentropic flow. From Table B.1, Appendix B, at $M = 0.68$,

$$\frac{T}{T_0} = 0.9154; \qquad T_t = 0.9154T_0 = (0.9154)(350 \text{ K}) = 320 \text{ K}$$

$$\frac{p}{p_0} = 0.7338; \qquad p_t = 0.7338p_0 = (0.7338)(1.0 \times 10^6 \text{ Pa}) = 734 \text{ kPa (abs)} \qquad p_t$$

$$\frac{\rho}{\rho_0} = 0.8016; \qquad \rho_t = 0.8016\rho_0 = 0.8016\frac{p_0}{RT_0}$$

$$\rho_t = 0.8016 \times \frac{1.0 \times 10^6 \text{ N}}{\text{m}^2} \times \frac{\text{kg} \cdot \text{K}}{287 \text{ N} \cdot \text{m}} \times \frac{1}{350 \text{ K}} = 7.98 \text{ kg/m}^3 \qquad \rho_t$$

and

$$\frac{A}{A^*} = 1.110; \qquad A_t = 1.110A^*$$

but at this point A^* is not known.

At the exit, $p_e = 954$ kPa (abs). Thus $p_e/p_0 = 0.954$, and from Table B.1,

$$M_e = 0.26 \qquad\qquad M_e$$

Since A_e is known, we can compute A^*. From Table B.1, at $M = 0.26$, $A/A^* = 2.317$. Thus

$$A^* = \frac{A_e}{2.317} = \frac{0.001 \text{ m}^2}{2.317} = 4.32 \times 10^{-4} \text{ m}^2$$

and

$$A_t = 1.110A^* = (1.110)(4.32 \times 10^{-4} \text{ m}^2) = 4.80 \times 10^{-4} \text{ m}^2 \qquad A_t$$

{Note that the solution for A_t using the tables was relatively effortless. To find A_t using the isentropic relations, we could have applied continuity between the throat and exit planes. Since this would have required calculation of all properties at the exit, it would have been a lengthy process.}

Example 10.6

The nozzle of Example 10.5 has a design back pressure of 72.8 kPa (abs) but is operated at a back pressure of 50.0 kPa (abs). Flow within the nozzle may be assumed isentropic. Determine the exit Mach number and mass flowrate. Use (i) isentropic flow relations, and (ii) tables for isentropic flow.

Example Problem 10.6

GIVEN:

Air flow through C–D nozzle as shown:

$$T_0 = 350 \text{ K}$$
$$p_0 = 1.0 \text{ MPa (abs)}$$
$$p_e \text{ (design)} = 72.8 \text{ kPa (abs)}$$
$$p_b = 50.0 \text{ kPa (abs)}$$
$$A_e = 0.001 \text{ m}^2$$

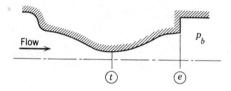

FIND:

(a) M_e, and (b) $\dot{m}$.

SOLUTION:

The operating back pressure is *below* the design value. Consequently, the nozzle is underexpanded, and the Ts diagram and pressure distribution will be as shown below:

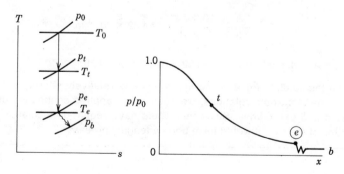

Flow *within* the nozzle will be isentropic, but the irreversible expansion from p_e to p_b will cause an entropy increase. $p_e = p_e$ (design) = 72.8 kPa (abs).

(*i*)   *Isentropic Flow Relations*

Since stagnation properties are constant for isentropic flow, the exit Mach number can be computed from the pressure ratio. Thus

$$\frac{p_0}{p_e} = \left[1 + \frac{k-1}{2} M_e^2\right]^{k/(k-1)}$$

or

$$M_e = \left\{\left[\left(\frac{p_0}{p_e}\right)^{(k-1)/k} - 1\right]\frac{2}{k-1}\right\}^{1/2} = \left\{\left[\left(\frac{1.0 \times 10^6}{7.28 \times 10^4}\right)^{0.286} - 1\right]\frac{2}{0.4}\right\}^{1/2} = 2.36 \quad M_e$$

The mass flowrate is given by

$$\dot{m} = \rho_e V_e A_e = \rho_e M_e c_e A_e = \rho_e M_e \sqrt{kRT_e} A_e = \frac{p_e}{RT_e} M_e \sqrt{kRT_e} A_e$$

or

$$\dot{m} = p_e M_e \sqrt{\frac{k}{RT_e}} A_e$$

Since T_0 is constant,

$$\frac{T_0}{T_e} = 1 + \frac{k-1}{2} M_e^2; \qquad T_e = \frac{T_0}{1 + \dfrac{k-1}{2}M_e^3} = \frac{350 \text{ K}}{1 + 0.2(2.36)^2} = 166 \text{ K}$$

Then

$$\dot{m} = p_e M_e \sqrt{\frac{k}{RT_e}} A_e$$

$$= \frac{7.28 \times 10^4 \text{ N}}{\text{m}^2} \times 2.36 \left[1.4 \times \frac{\text{kg} \cdot \text{K}}{287 \text{ N} \cdot \text{m}} \times \frac{1}{166 \text{ K}} \times \frac{\text{kg} \cdot \text{m}}{\text{N} \cdot \text{sec}^2}\right]^{1/2} 0.001 \text{ m}^2$$

$$\dot{m} = 0.931 \text{ kg/sec} \qquad\qquad\qquad \dot{m}$$

(*ii*)   **Tables for Isentropic Flow*

From Table B.1, Appendix B, at $p/p_0 = p_e/p_0 = 0.0728$,

$$M_e = 2.36 \qquad\qquad\qquad M_e$$

Also $\dfrac{T_e}{T_0} = 0.4731;$ $T_e = 0.4731 T_0 = 0.4731(350 \text{ K}) = 166 \text{ K}$

$\left\{\begin{array}{l}\text{This temperature checks the value obtained above. The mass flowrate calcula-}\\\text{tion would also be as above.}\end{array}\right\}$

10-4 ADIABATIC FLOW IN A CONSTANT AREA DUCT WITH FRICTION

To obtain an overall view of the problem of frictional adiabatic flow, apply the basic equations to the steady uniform flow of an ideal gas with constant specific heats through the finite control volume shown in Fig. 10.9.

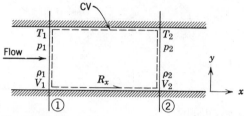

Fig. 10.9 Control volume used for integral analysis of frictional adiabatic flow.

10-4.1 BASIC EQUATIONS

a. Continuity Equation

Basic equation:

$$0 = \overset{=0(1)}{\cancel{\frac{\partial}{\partial t} \int_{CV} \rho \, d\Psi}} + \int_{CS} \rho \vec{V} \cdot d\vec{A} \tag{4.13}$$

Assumptions: (1) Steady flow
(2) Uniform flow at each section

Then

$$0 = \{-|\rho_1 V_1 A_1|\} + \{|\rho_2 V_2 A_2|\}$$

The area is constant, so

$$\rho_1 V_1 = \rho_2 V_2 \equiv G \tag{10.7a}$$

b. Momentum Equation

Basic equation:

$$F_{S_x} + \overset{=0(3)}{\cancel{F_{B_x}}} = \overset{=0(1)}{\cancel{\frac{\partial}{\partial t} \int_{CV} V_x \rho \, d\Psi}} + \int_{CS} V_x \rho \vec{V} \cdot d\vec{A} \tag{4.19a}$$

Assumption: (3) $F_{B_x} = 0$

The surface force is due to pressure forces at sections ① and ②, and to the friction force, R_x, of the duct wall *on* the flow. Substituting, recognizing that $A_2 = A_1 = A$, then

$$R_x + p_1 A - p_2 A = V_1\{-|\rho_1 V_1 A|\} + V_2\{|\rho_2 V_2 A|\}$$

Using scalar magnitudes and dropping absolute value signs, we obtain

$$R_x + p_1 A - p_2 A = \dot{m} V_2 - \dot{m} V_1 \qquad (10.7b)$$

c. First Law of Thermodynamics

Basic equation:

$$\overset{=0(4)}{\cancel{\dot{Q}}} + \overset{=0(5)}{\cancel{\dot{W}_s}} + \overset{=0(6)}{\cancel{\dot{W}_{shear}}} + \overset{=0(6)}{\cancel{\dot{W}_{other}}} = \overset{=0(1)}{\cancel{\frac{\partial}{\partial t} \int_{CV} e\rho\, d\mathbf{V}}} + \int_{CS} (e + pv)\rho \vec{V} \cdot d\vec{A} \quad (4.59)$$

where

$$e = u + \frac{V^2}{2} + \overset{\approx 0(7)}{\cancel{gz}}$$

Assumptions: (4) $\dot{Q} = 0$ (adiabatic flow)
(5) $\dot{W}_s = 0$
(6) $\dot{W}_{shear} = \dot{W}_{other} = 0$
(7) Effects of gravity are negligible

With these restrictions, the equation becomes

$$0 = \left(u_1 + \frac{V_1^2}{2} + p_1 v_1\right)\{-|\rho_1 V_1 A|\} + \left(u_2 + \frac{V_2^2}{2} + p_2 v_2\right)\{|\rho_2 V_2 A|\}$$

Since the mass flowrate terms in { } are identical by continuity,

$$u_1 + \frac{V_1^2}{2} + p_1 v_1 = u_2 + \frac{V_2^2}{2} + p_2 v_2$$

or

$$h_1 + \frac{V_1^2}{2} = h_2 + \frac{V_2^2}{2} \qquad (10.7c)$$

We could also write

$$h_{0_1} = h_{0_2}$$

which is a physical consequence of our assumption of adiabatic flow.

d. Second Law of Thermodynamics

Basic equation:

$$\int_{cs} \frac{1}{T} \cancelto{0(4)}{\frac{\dot{Q}}{A}} dA \le \cancelto{0(1)}{\frac{\partial}{\partial t}} \int_{cv} s\rho \, d\mathbb{V} + \int_{cs} s\rho \vec{V} \cdot d\vec{A} \tag{4.60}$$

Then because the flow is frictional and, hence, irreversible,

$$0 < s_1\{-|\rho_1 V_1 A|\} + s_2\{|\rho_2 V_2 A|\} = \dot{m}(s_2 - s_1)$$

The control volume form of the second law then tells us that $s_2 - s_1 > 0$.

This fact is of little help in calculating the actual entropy change between any two points in an adiabatic frictional flow. To calculate the entropy change, we rely on the $T\,ds$ equations. Since

$$T\,ds = dh - v\,dp$$

for an ideal gas we can write

$$ds = c_p \frac{dT}{T} - R \frac{dp}{p}$$

For constant specific heats the equation can be integrated to give

$$s_2 - s_1 = c_p \ln\frac{T_2}{T_1} - R \ln\frac{p_2}{p_1} \tag{10.7d}$$

e. Equations of State

For an ideal gas, the equation of state is given by

$$p = \rho R T \tag{10.7e}$$

Equations 10.7a–e are the governing equations for the steady, one-dimensional, adiabatic, frictional flow of an ideal gas in a constant area duct. If all the properties at state ① are known, then we have seven unknowns ($T_2, p_2, \rho_2, V_2, h_2, s_2$, and R_x) in these five equations. However, we have the known relationship between h and T for an ideal gas, $dh = c_p\,dT$. For an ideal gas with constant specific heats,

$$\Delta h = h_2 - h_1 = c_p \Delta T = c_p(T_2 - T_1) \tag{10.7f}$$

We thus have the situation of six equations and seven unknowns.

If all conditions at state ① are known, how many possible states ② are there? The mathematics of the situation (six equations and seven unknowns) indicate that there are an infinite number of possible states ②.

With an infinite number of possible states ② for a given state ①, what is to be expected if all possible states ② are plotted on a Ts diagram? It follows that the locus of all possible states ② reachable from state ① is a continuous curve passing through state ①.

How might we determine this curve? Perhaps the simplest way is to assume different values of T_2. For an assumed value of T_2 we could then calculate the corresponding values of all other properties at state ② and also R_x.

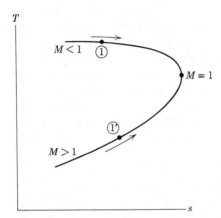

Fig. 10.10 Schematic Ts diagram for frictional adiabatic (Fanno line) flow in a constant area duct.

10-4.2   THE FANNO LINE

The results of these calculations are shown qualitatively on the Ts plane of Fig. 10.10. The locus of all possible downstream states is referred to as the *Fanno line*.[2] In carrying out the calculations, we discover some interesting features of Fanno line flow. At the point of maximum entropy, the Mach number is unity. On the upper branch of the curve the Mach number is always less than unity, and it increases monotonically as we proceed to the right along the curve. At every point on the lower portion of the curve the Mach number is greater than unity; the Mach number decreases monotonically as we move to the right along the curve.

For any initial state in a Fanno line flow any point on the Fanno line represents a mathematically possible downstream state. Indeed, we determined the locus of all possible downstream states by assuming values of T_2 and calculating the corresponding properties. While the Fanno line represents all mathematically possible downstream states, are they all physically attainable downstream states? A moment's reflection will indicate they are

---

[2]   A convenient way to remember this name is to think of *F*rictional *A*diabatic flow.

not. Why not? The second law requires that the entropy must increase. Consequently, we must always move to the right along the Fanno line of Fig. 10.10. Indeed, it is the effect of friction that causes the flow properties to change from the initial state. Referring again to Fig. 10.10, we see that for an initially subsonic flow (state ①), the effect of friction is to increase the Mach number toward unity. For a flow that is initially supersonic (state ①'), the effect of friction is to decrease the Mach number toward unity.

In developing the simplified form of the first law for Fanno line flow, we found that the stagnation enthalpy remains constant for the flow. Consequently, when the fluid is an ideal gas with constant specific heats, the stagnation temperature must also remain constant. What happens to the stagnation pressure? Friction causes the local isentropic stagnation pressure to decrease for all Fanno line flows as shown in Fig. 10.11. Recall that the second law requires $s_2 - s_1 > 0$. Since the entropy must increase in the direction of flow, the flow process must proceed to the right on the Ts diagram. In Fig. 10.11, a path from state ① to state ② is shown on the subsonic portion of the curve. The corresponding values of local isentropic stagnation pressure, p_{0_1} and p_{0_2}, show clearly that $p_{0_2} < p_{0_1}$. An identical result is obtained for flow on the supersonic branch of the curve from state ①' to state ②'. Again $p_{0_{2'}} < p_{0_{1'}}$. Thus p_0 decreases for any Fanno line flow.

At this point it may be beneficial to summarize the effects of friction on flow properties in Fanno line flow. The summary is presented in Table 10.1.

In deducing the effect of friction on flow properties for Fanno line flow, we have used the shape of the Fanno line on the Ts diagram and the basic governing equations (Eqs. 10.7a–f). You are encouraged to follow through the logic indicated in the right-hand column of the table.

We have noted that the entropy must increase in the direction of flow; it is the effect of friction that causes the change in flow properties along the Fanno line curve. From Fig. 10.10 we see that there is a maximum entropy point, corresponding to $M = 1$ for each Fanno line. The maximum entropy point is reached by increasing the amount of friction (through addition of duct length), just enough to produce a Mach number of unity (choked flow) at the exit. How do we compute this critical length of duct?

To compute the critical duct length, we must analyze the flow in detail, accounting for the effect of friction. This analysis requires that we begin with a differential control volume, develop expressions in terms of Mach number, and integrate along the duct to the section where $M = 1$. This analysis is completed in Section 10-4.3, where tables for Fanno line flow are developed. The algebra required for the detailed analysis tends to obscure the physics of the flow; the general trends in properties caused by friction can be demonstrated using finite control volumes and the basic governing equations. This approach is illustrated in Example Problem 10.7.

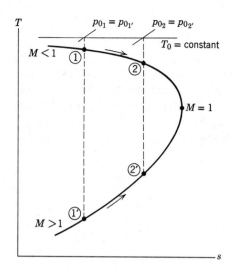

Fig. 10.11 Schematic of Fanno line flow on Ts plane, showing reduction in local isentropic stagnation pressure caused by friction.

Table 10.1 **Summary of Effects of Friction on Properties in Fanno Line Flow**

Property	Subsonic $M < 1$	Supersonic $M > 1$	Obtained from:
Stagnation temperature, T_0	Constant	Constant	Energy equation
Entropy, s	Increases	Increases	Second law
Stagnation pressure, p_0	Decreases	Decreases	$T_0 = $ constant; s increases
Temperature, T	Decreases	Increases	Shape of Fanno line
Velocity, V	Increases	Decreases	Energy equation, and trend of T
Mach number, M	Increases	Decreases	Trends of V, T, and definition of M
Density, ρ	Decreases	Increases	Continuity equation, and effect on V
Pressure, p	Decreases	Increases	Equation of state, and effects on ρ, T

adiabatic flow in a constant area duct with friction/10-4 *549*

Example 10.7

Air flow is induced in an insulated tube of 7.16 mm diameter by a vacuum pump. The air is drawn from a room where the absolute pressure is 101 kPa and the temperature is 23 C, through a smoothly contoured, converging nozzle. At section ①, where the nozzle joins the constant area tube, the absolute static pressure is 98.5 kPa. At section ②, located some distance downstream in the constant area tube, the air temperature is 14 C. Determine the mass flowrate, the local isentropic stagnation pressure at section ②, and the friction force on the duct wall between sections ① and ②.

Example Problem 10.7

GIVEN:

Air flow in insulated tube.

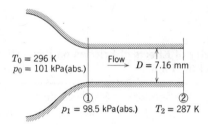

$T_0 = 296$ K
$p_0 = 101$ kPa(abs.)

Flow $D = 7.16$ mm

① $p_1 = 98.5$ kPa(abs.) ② $T_2 = 287$ K

FIND:

(a) $\dot{m}$.
(b) Stagnation pressure at section ②.
(c) Force on duct wall.

SOLUTION:

The mass flowrate can be obtained from properties at section ①. For isentropic flow through the converging nozzle, local isentropic stagnation properties remain constant. Thus

$$\frac{p_{01}}{p_1} = \left(1 + \frac{k-1}{2} M_1^2\right)^{k/(k-1)}$$

and

$$M_1 = \left\{\frac{2}{k-1}\left[\left(\frac{p_{0_1}}{p_1}\right)^{(k-1)/k} - 1\right]\right\}^{1/2} = \left\{\frac{2}{0.4}\left[\left(\frac{1.01 \times 10^5}{9.85 \times 10^4}\right)^{0.286} - 1\right]\right\}^{1/2} = 0.190$$

$$T_1 = \frac{T_{01}}{1 + \frac{k-1}{2}M_1^2} = \frac{296 \text{ K}}{1 + 0.2(0.190)^2} = 294 \text{ K}$$

For an ideal gas,

$$\rho_1 = \frac{p_1}{RT_1} = \frac{9.85 \times 10^4 \text{ N}}{\text{m}^2} \times \frac{\text{kg} \cdot \text{K}}{287 \text{ N} \cdot \text{m}} \times \frac{1}{294 \text{ K}} = 1.17 \text{ kg/m}^3$$

$$V_1 = M_1 c_1 = M_1 \sqrt{kRT_1} = (0.19) \left[1.4 \times \frac{287 \text{ N} \cdot \text{m}}{\text{kg} \cdot \text{K}} \times 294 \text{ K} \times \frac{\text{kg} \cdot \text{m}}{\text{N} \cdot \text{sec}^2} \right]^{1/2} = 65.3 \text{ m/sec}$$

$$A_1 = A = \frac{\pi D^2}{4} = \frac{\pi}{4} (7.16 \times 10^{-3})^2 \text{ m}^2 = 4.03 \times 10^{-5} \text{ m}^2$$

From continuity

$$\dot{m} = \rho_1 V_1 A_1 = \frac{1.17 \text{ kg}}{\text{m}^3} \times \frac{65.3 \text{ m}}{\text{sec}} \times 4.03 \times 10^{-5} \text{ m}^2$$

$$\dot{m} = 3.08 \times 10^{-3} \text{ kg/sec} \qquad\qquad\qquad \dot{m}$$

Flow is adiabatic and frictional and hence T_0 is constant, so

$$T_{0_2} = T_{0_1} = 296 \text{ K} \qquad\qquad\qquad T_{0_2}$$

Then

$$\frac{T_{0_2}}{T_2} = 1 + \frac{k-1}{2} M_2^2$$

Solving for M_2 gives

$$M_2 = \left[\frac{2}{k-1} \left(\frac{T_{0_2}}{T_2} - 1 \right) \right]^{1/2} = \left[\frac{2}{0.4} \left(\frac{296}{287} - 1 \right) \right]^{1/2} = 0.396 \qquad M_2$$

$$V_2 = M_2 c_2 = M_2 \sqrt{kRT_2} = (0.396) \left[1.4 \times \frac{287 \text{ N} \cdot \text{m}}{\text{kg} \cdot \text{K}} \times 287 \text{ K} \times \frac{\text{kg} \cdot \text{m}}{\text{N} \cdot \text{sec}^2} \right]^{1/2}$$

$$V_2 = 134 \text{ m/sec} \qquad\qquad\qquad V_2$$

From continuity, $\rho_1 V_1 = \rho_2 V_2$, so

$$\rho_2 = \rho_1 \frac{V_1}{V_2} = \frac{1.17 \text{ kg}}{\text{m}^3} \times \frac{65.3}{134} = 0.570 \text{ kg/m}^3 \qquad\qquad \rho_2$$

and

$$p_2 = \rho_2 R T_2 = \frac{0.570 \text{ kg}}{\text{m}^3} \times \frac{287 \text{ N} \cdot \text{m}}{\text{kg} \cdot \text{K}} \times 287 \text{ K} = 47.0 \text{ kPa (abs)} \qquad p_2$$

The local isentropic stagnation pressure is

$$p_{0_2} = p_2 \left(1 + \frac{k-1}{2} M_2^2 \right)^{k/(k-1)} = 4.70 \times 10^4 \text{ Pa}[1 + 0.2(0.396)^2]^{3.5}$$

$$p_{0_2} = 52.4 \text{ kPa (abs)} \qquad\qquad\qquad p_{0_2}$$

The friction force may be found by applying the momentum equation to the control volume shown below:

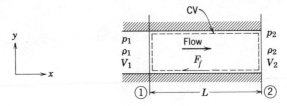

Basic equation:

$$F_{S_x} + \overset{=0(1)}{\cancel{F_{B_x}}} = \overset{=0(2)}{\cancel{\frac{\partial}{\partial t}\int_{CV} V_x \rho \, d\forall}} + \int_{CS} V_x \rho \vec{V} \cdot d\vec{A} \qquad (4.19a)$$

Assumptions: (1) $F_{B_x} = 0$
(2) Steady flow
(3) Uniform flow at each section

Then

$$-F_f + p_1 A - p_2 A = V_1\{-|\rho_1 V_1 A|\} + V_2\{|\rho_2 V_2 A|\} = \dot{m}(V_2 - V_1)$$

and

$$-F_f = (p_2 - p_1)A + \dot{m}(V_2 - V_1)$$

$$-F_f = \frac{(4.70 - 9.85)10^4}{m^2} \frac{N}{} \times 4.03 \times 10^{-5} \text{ m}^2$$

$$+ \; 3.08 \times 10^{-3} \frac{\text{kg}}{\text{sec}} (134 - 65.3) \frac{\text{m}}{\text{sec}} \times \frac{\text{N} \cdot \text{sec}^2}{\text{kg} \cdot \text{m}}$$

or

$$F_f = 1.86 \text{ N}$$

This is the force exerted on the control volume by the duct wall. The force of the *fluid* on the *duct* is

$$K_x = F_f = 1.86 \text{ N} \qquad \text{(to the right)} \qquad K_x$$

**10-4.3 TABLES FOR COMPUTATION OF FANNO LINE FLOW OF AN IDEAL GAS

The primary independent variable in Fanno line flow is the friction force, F_f. Knowledge of the total friction force between any two points in a Fanno line flow would enable us to predict downstream conditions when conditions

** This section may be omitted without loss of continuity in the text material.

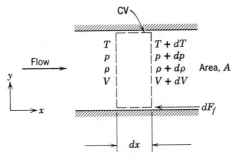

Fig. 10.12 Differential control volume used for analysis of Fanno line flow.

upstream were known. The total friction force is the integral of the wall shear stress over the duct surface area. Since the wall shear stress varies along the duct, we must develop a differential equation and then integrate to find property variations. To set up the differential equation, we use the differential control volume shown in Fig. 10.12 for our analysis.

a. Continuity Equation

Basic equation:

$$0 = \frac{\partial}{\partial t} \int_{CV} \rho \, d\Psi + \int_{CS} \rho \vec{V} \cdot d\vec{A} \tag{4.13}$$

$$= 0(1)$$

Assumptions: (1) Steady flow
(2) Uniform flow at each section

$$0 = \{ -|\rho V A| \} + \{ |(\rho + d\rho)(V + dV)A| \}$$

or

$$\rho V A = (\rho + d\rho)(V + dV)A$$

Simplifying and canceling A gives

$$0 = -\rho V + \rho V + \rho \, dV + V \, d\rho + d\rho \, dV$$

$$\simeq 0$$

which reduces to

$$\rho \, dV + V \, d\rho = 0 \tag{10.8a}$$

since products of differentials are negligible.

b. Momentum Equation

Basic equation:

$$\underset{\substack{\downarrow \\ =0(3)}}{F_{S_x}} + \underset{\substack{\downarrow \\ =0(1)}}{F_{B_x}} = \frac{\partial}{\partial t}\int_{CV} V_x \rho \, d\mathbf{V} + \int_{CS} V_x \rho \vec{V} \cdot d\vec{A} \qquad (4.19a)$$

Assumption: (3) $F_{B_x} = 0$

The momentum equation becomes

$$-dF_f + pA - (p + dp)A = V\{-|\rho VA|\} + (V + dV)\{|(\rho + d\rho)(V + dV)A|\}$$

which simplifies, using continuity, to give

$$-\frac{dF_f}{A} - dp = \rho V \, dV \qquad (10.8b)$$

c. First Law of Thermodynamics

Basic equation:

$$\underset{\substack{\downarrow \\ =0(4)}}{\dot{Q}} + \underset{\substack{\downarrow \\ =0(5)}}{\dot{W}_s} + \underset{\substack{\downarrow \\ =0(6)}}{\dot{W}_{shear}} + \underset{\substack{\downarrow \\ =0(6)}}{\dot{W}_{other}} = \underset{\substack{\downarrow \\ =0(1)}}{\frac{\partial}{\partial t}\int_{CV}} e\rho \, d\mathbf{V} + \int_{CS} (e + pv)\rho\vec{V} \cdot d\vec{A} \qquad (4.59)$$

where

$$e = u + \frac{V^2}{2} + \underset{\substack{\downarrow \\ \simeq 0(7)}}{gz}$$

Assumptions: (4) Adiabatic flow, $\dot{Q} = 0$
  (5) $\dot{W}_s = 0$
  (6) $\dot{W}_{shear} = \dot{W}_{other} = 0$
  (7) Effects of gravity are negligible

Under these restrictions, we obtain

$$0 = \left(u + \frac{V^2}{2} + pv\right)\{-|\rho VA|\} + \left[u + du + \frac{V^2}{2} + d\left(\frac{V^2}{2}\right) + pv + d(pv)\right]$$
$$\times \{|(\rho + d\rho)(V + dV)A|\}$$

Noting from continuity that the flowrate terms { } are equal and substituting $h = u + pv$, then

$$dh + d\left(\frac{V^2}{2}\right) = 0 \tag{10.8c}$$

To complete our formulation, we must relate the friction force, dF_f, to the flow variables at each cross section. We note that $dF_f = \tau_w \, dA_w = \tau_w P_w \, dx$, where P_w is the wetted perimeter of the duct. To obtain an expression for τ_w in terms of flow variables at each cross section, we assume that changes in flow variables with x are gradual, and use the correlations developed in Chapter 8 for fully-developed, incompressible duct flow. (This assumption, when checked against experimental data, shows surprisingly good agreement.for subsonic flows; data for supersonic flow are sparse.)

Ducts of other than circular shape can be included in our analysis by introducing the hydraulic diameter

$$D_h = \frac{4A}{P_w} \tag{8.39}$$

(Recall that the factor of 4 was included in Eq. 8.39 so that D_h would reduce to D, the diameter, for circular ducts.) In terms of the hydraulic diameter, Eq. 8.19 can be written

$$\tau_w = -\frac{D_h}{4} \frac{dp}{dx} \tag{10.9}$$

Combining Eqs. 8.26 and 8.28, we can obtain an expression for the pressure gradient,

$$\frac{dp}{dx} = -\frac{\Delta p}{L} = \frac{-f}{D_h} \frac{\rho V^2}{2} \tag{10.10}$$

Combining Eqs. 10.9 and 10.10, τ_w may be written

$$\tau_w = f \frac{\rho V^2}{8}$$

so that

$$dF_f = \tau_w P_w \, dx = f \frac{\rho V^2}{8} \frac{4A}{D_h} dx$$

or

$$dF_f = \frac{fA}{D_h} \frac{\rho V^2}{2} dx \tag{10.11}$$

Substituting this result into the momentum equation (Eq. 10.8b), we obtain

$$-\frac{f}{D_h} \frac{\rho V^2}{2} dx - dp = \rho V \, dV$$

or, after dividing by p,

$$\frac{dp}{p} = -\frac{f}{D_h}\frac{\rho V^2}{2p}dx - \frac{\rho V\,dV}{p}$$

Noting that $p/\rho = RT = c^2/k$, and $V\,dV = d(V^2/2)$, we obtain

$$\frac{dp}{p} = -\frac{f}{D_h}\frac{kM^2}{2}dx - \frac{k}{c^2}d\left(\frac{V^2}{2}\right)$$

and finally,

$$\frac{dp}{p} = -\frac{f}{D_h}\frac{kM^2}{2}dx - \frac{kM^2}{2}\frac{d(V^2)}{V^2} \tag{10.12}$$

In order to obtain an equation relating M and x, we need to eliminate dp/p and $d(V^2)/V^2$ from Eq. 10.12.

From the definition of Mach number, $M = V/c$, then $V^2 = M^2c^2 = M^2kRT$ and

$$\frac{d(V^2)}{V^2} = \frac{dT}{T} + \frac{d(M^2)}{M^2} \tag{10.13a}$$

From the continuity equation, $d\rho/\rho = -dV/V$ and

$$\frac{d\rho}{\rho} = -\frac{1}{2}\frac{d(V^2)}{V^2}$$

From the ideal gas equation of state, $p = \rho RT$, we can write

$$\frac{dp}{p} = \frac{d\rho}{\rho} + \frac{dT}{T}$$

Combining these three equations, we obtain

$$\frac{dp}{p} = \frac{1}{2}\frac{dT}{T} - \frac{1}{2}\frac{d(M^2)}{M^2} \tag{10.13b}$$

Substituting Eqs. 10.13 into Eq. 10.12 gives

$$\frac{1}{2}\frac{dT}{T} - \frac{1}{2}\frac{d(M^2)}{M^2} = -\frac{f}{D_h}\frac{kM^2}{2}dx - k\frac{M^2}{2}\frac{dT}{T} - \frac{kM^2}{2}\frac{d(M^2)}{M^2}$$

This equation can be simplified to

$$\left(\frac{1 + kM^2}{2}\right)\frac{dT}{T} = -\frac{f}{D_h}\frac{kM^2}{2}dx + \left(\frac{1 - kM^2}{2}\right)\frac{d(M^2)}{M^2} \tag{10.14}$$

We have been successful in reducing the number of variables somewhat.

However, in order to obtain an equation relating M and x, we must obtain an expression for dT/T in terms of M.

Such an expression can be most readily obtained from the stagnation temperature equation

$$\frac{T_0}{T} = 1 + \frac{k-1}{2} M^2 \tag{9.16b}$$

Since the stagnation temperature is constant for Fanno line flow,

$$T\left(1 + \frac{k-1}{2} M^2\right) = \text{constant}$$

and

$$\frac{dT}{T} + \frac{M^2 \dfrac{(k-1)}{2}}{\left(1 + \dfrac{k-1}{2} M^2\right)} \frac{d(M^2)}{M^2} = 0$$

Substituting for dT/T into Eq. 10.14,

$$\frac{M^2 \dfrac{(k-1)}{2} \left(\dfrac{1 + kM^2}{2}\right)}{\left(1 + \dfrac{k-1}{2} M^2\right)} \frac{d(M^2)}{M^2} = \frac{f}{D_h} \frac{kM^2}{2} dx - \left(\frac{1 - kM^2}{2}\right) \frac{d(M^2)}{M^2}$$

Combining terms,

$$\frac{(1 - M^2)}{\left(1 + \dfrac{k-1}{2} M^2\right)} \frac{d(M^2)}{kM^4} = \frac{f}{D_h} dx \tag{10.15}$$

We have succeeded in obtaining a differential equation that relates changes in M with x. All we have to do now is integrate the equation to find M as a function of x.

Integration of Eq. 10.15 between states ① and ② would produce a complicated function of both M_1 and M_2. The function would have to be evaluated numerically for each new combination of M_1 and M_2 encountered in a problem. Calculations can be simplified considerably by use of the critical conditions (conditions where, by definition, $M = 1$). All Fanno line flows tend toward $M = 1$, so the integration is carried out between a section where the Mach number is M and the section where sonic conditions occur (the critical conditions). The Mach number will reach unity when the maximum possible length of duct is used as shown schematically in Fig. 10.13.

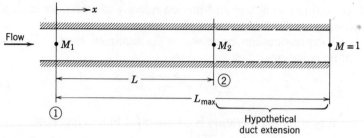

Fig. 10.13 Coordinates and notation used for analysis of Fanno line flow.

The task then is to perform the integration

$$\int_M^1 \frac{(1 - M^2)}{kM^4 \left(1 + \dfrac{k-1}{2} M^2\right)} d(M^2) = \int_0^{L_{max}} \frac{f}{D_h} dx \qquad (10.16)$$

The left side may be evaluated using integration by parts. On the right side the friction factor, f, may vary with x, since the Reynolds number will vary along the duct. Note, however, that since the ρV product is constant along the duct (from continuity) the variation in Reynolds number is caused solely by variations in fluid viscosity.

If we define a mean friction factor, $\bar{f}$, over the duct length as

$$\bar{f} = \frac{1}{L_{max}} \int_0^{L_{max}} f \, dx$$

then integration of Eq. 10.16 leads to

$$\frac{1 - M^2}{kM^2} + \frac{k+1}{2k} \ln\left[\frac{(k+1)M^2}{2\left(1 + \dfrac{k-1}{2} M^2\right)}\right] = \frac{\bar{f} L_{max}}{D_h} \qquad (10.17)$$

Equation 10.17 gives the maximum value of $\bar{f} L/D_h$ corresponding to any given initial Mach number. For any given Mach number we can compute the corresponding value of $\bar{f} L_{max}/D_h$. These values are tabulated in Table B.2 of Appendix B.

Since the parameter $\bar{f} L_{max}/D_h$ is a function only of M, the duct length, L, required for the flow Mach number to change from an initial value, M_1, to a final value, M_2 (as illustrated in Fig. 10.13), may be found from

$$\frac{\bar{f} L}{D_h} = \left(\frac{\bar{f} L_{max}}{D_h}\right)_{M_1} - \left(\frac{\bar{f} L_{max}}{D_h}\right)_{M_2}$$

The critical conditions are appropriate reference conditions to use in tabulating fluid properties as a function of local Mach number. Thus, for example, since T_0 is constant, we can write

$$\frac{T}{T^*} = \frac{T/T_0}{T^*/T_0} = \frac{1}{1 + \dfrac{k-1}{2}M^2} \bigg/ \frac{1}{1 + \dfrac{k-1}{2}M^2} = \frac{\left(\dfrac{k+1}{2}\right)}{\left(1 + \dfrac{k-1}{2}M^2\right)}$$

(10.18a)

Similarly,

$$\frac{V}{V^*} = \frac{M\sqrt{kRT}}{\sqrt{kRT^*}} = M\sqrt{\frac{T}{T^*}} = \left[\frac{\left(\dfrac{k+1}{2}\right)M^2}{1 + \dfrac{k-1}{2}M^2}\right]^{1/2}$$

(10.18b)

From continuity

$$\frac{\rho}{\rho^*} = \frac{V^*}{V} = \left[\frac{1 + \dfrac{k-1}{2}M^2}{\left(\dfrac{k+1}{2}\right)M^2}\right]^{1/2}$$

(10.18c)

From the ideal gas equation of state

$$\frac{p}{p^*} = \frac{\rho}{\rho^*}\frac{T}{T^*} = \frac{1}{M}\left[\frac{\left(\dfrac{k+1}{2}\right)}{1 + \dfrac{k-1}{2}M^2}\right]^{1/2}$$

(10.18d)

The ratio of local stagnation pressure to the reference stagnation pressure is given by

$$\frac{p_0}{p_0^*} = \frac{p_0}{p}\frac{p}{p^*}\frac{p^*}{p_0^*}$$

$$= \left(1 + \frac{k-1}{2}M^2\right)^{k/(k-1)}\frac{1}{M}\left[\frac{\left(\dfrac{k+1}{2}\right)}{1 + \dfrac{k-1}{2}M^2}\right]^{1/2}\frac{1}{\left(1 + \dfrac{k-1}{2}\right)^{k/(k-1)}}$$

or

$$\frac{p_0}{p_0^*} = \frac{1}{M}\left[\left(\frac{2}{k+1}\right)\left(1 + \frac{k-1}{2}M^2\right)\right]^{(k+1)/2(k-1)}$$

(10.18e)

The ratios in Eqs. 10.18 are tabulated as functions of Mach number in Table B.2 of Appendix B.

Example 10.8

Air flow is induced in a smooth insulated tube of 7.16 mm diameter by a vacuum pump. The air is drawn from a room where the absolute pressure is 760 mm Hg and the temperature is 23 C through a smoothly contoured, converging nozzle. At section ①, where the nozzle joins the constant area tube, the static gage pressure is −18.9 mm Hg. At section ②, located some distance downstream in the constant area tube, the static gage pressure is −412 mm Hg. The duct walls are smooth; the average friction factor, $\bar{f}$, may be taken as the value at section ①. Determine the length of duct required for choking from section ①, the Mach number at section ②, and the duct length, L_{12}, between sections ① and ②.

Example Problem 10.8

GIVEN:

Air flow (with friction) in an insulated constant area tube.

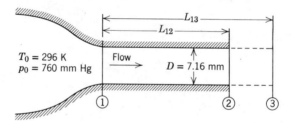

$T_0 = 296$ K
$p_0 = 760$ mm Hg

$D = 7.16$ mm

Gage pressures: $p_1 = -18.9$ mm Hg, $p_2 = -412$ mm Hg, and $M_3 = 1.0$

FIND:

(a)L_{13}, (b) M_2, and (c) L_{12}.

SOLUTION:

Flow in the constant area tube is frictional and adiabatic, that is, a Fanno line flow. To find the friction factor, we need to know the flow conditions at section ①. If it is assumed that flow in the nozzle is isentropic, local properties at the nozzle exit may be computed using isentropic relations. Thus

$$\frac{p_{0_1}}{p_1} = \left(1 + \frac{k-1}{2}M_1^2\right)^{k/(k-1)}$$

Solving for M_1,

$$M_1 = \left\{ \frac{2}{k-1} \left[\left(\frac{p_{0_1}}{p_1} \right)^{(k-1)/k} - 1 \right] \right\}^{1/2} = \left\{ \frac{2}{0.4} \left[\left(\frac{760}{760 - 18.9} \right)^{0.286} - 1 \right] \right\}^{1/2} = 0.190$$

$$T_1 = \frac{T_{0_1}}{1 + \frac{k-1}{2} M_1^2} = \frac{296 \text{ K}}{1 + 0.2(0.190)^2} = 294 \text{ K}$$

$$V_1 = M_1 c_1 = M_1 \sqrt{kRT_1} = 0.190 \left[1.4 \times \frac{287 \text{ N} \cdot \text{m}}{\text{kg} \cdot \text{K}} \times 294 \text{ K} \times \frac{\text{kg} \cdot \text{m}}{\text{N} \cdot \text{sec}^2} \right]^{1/2}$$

$$V_1 = 65.3 \text{ m/sec}$$

$$p_1 = g \rho_{\text{Hg}} h_1 = g SG \rho_{\text{H}_2\text{O}} h_1$$

$$= \frac{9.81 \text{ m}}{\text{sec}^2} \times 13.6 \times \frac{999 \text{ kg}}{\text{m}^3} \times (760 - 18.9)10^{-3} \text{ m} \times \frac{\text{N} \cdot \text{sec}^2}{\text{kg} \cdot \text{m}}$$

$$p_1 = 98.8 \text{ kPa (abs)}$$

$$\rho_1 = \frac{p_1}{RT_1} = \frac{9.88 \times 10^4 \text{ N}}{\text{m}^2} \times \frac{\text{kg} \cdot \text{K}}{287 \text{ N} \cdot \text{m}} \times \frac{1}{294 \text{ K}} = 1.17 \text{ kg/m}^3$$

At $T = 294$ K (21 C), $\mu = 1.9 \times 10^{-5}$ kg/m·sec from Fig. A.2, Appendix A. Thus

$$Re_1 = \frac{\rho_1 V_1 D_1}{\mu_1} = \frac{1.17 \text{ kg}}{\text{m}^3} \times \frac{65.3 \text{ m}}{\text{sec}} \times 0.00716 \text{ m} \times \frac{\text{m} \cdot \text{sec}}{1.9 \times 10^{-5} \text{ kg}} = 2.86 \times 10^4$$

From Fig. 8.12, for smooth pipe, $f = 0.0235$. From Table B.2 at $M_1 = 0.19$, $p/p^* = 5.745$, and $\bar{f}L_{\text{max}}/D_h = 16.38$. Thus, assuming $\bar{f} = f_1$,

$$L_{13} = (L_{\text{max}})_1 = \left(\frac{\bar{f}L_{\text{max}}}{D_h} \right)_1 \frac{D_h}{f_1} = 16.38 \times 0.00716 \text{ m} \times \frac{1}{0.0235} = 4.99 \text{ m} \qquad \underline{L_{13}}$$

Since p^* is constant for Fanno line flow, conditions at section ② can be determined from the pressure ratio, $(p/p^*)_2$. Thus

$$\left(\frac{p}{p^*} \right)_2 = \frac{p_2}{p^*} = \frac{p_2}{p_1} \frac{p_1}{p^*} = \frac{p_2}{p_1} \left(\frac{p}{p^*} \right)_1 = \left(\frac{760 - 412}{760 - 18.9} \right) 5.745 = 2.698$$

From Table B.2, at $(p/p^*)_2 = 2.698$, $M_2 \simeq 0.40$ $\qquad \underline{M_2}$

At $M_2 = 0.40$, $\bar{f}L_{\text{max}}/D_h = 2.309$ (Table B.2). Thus

$$L_{23} = (L_{\text{max}})_2 = \left(\frac{\bar{f}L_{\text{max}}}{D_h} \right)_2 \frac{D_h}{f} = 2.309 \times 0.00716 \text{ m} \times \frac{1}{0.0235} = 0.704 \text{ m}$$

Finally,

$$L_{12} = L_{13} - L_{23} = (4.99 - 0.704) \text{ m} = 4.29 \text{ m} \qquad \underline{L_{12}}$$

This is the same physical system as that analyzed in Example Problem 10.7. Use of the tables simplifies the calculations and makes it possible to determine the duct length.

10-5 FRICTIONLESS FLOW IN A CONSTANT AREA DUCT WITH HEAT TRANSFER

To explore the effects of heat transfer on a compressible flow, let us apply the basic equations to the steady, one-dimensional, frictionless flow of an ideal gas with constant specific heats through the finite control volume shown in Fig. 10.14.

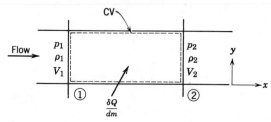

Fig. 10.14 Control volume used for integral analysis of frictionless flow with heat transfer.

10-5.1 BASIC EQUATIONS

a. Continuity Equation

Basic equation:

$$0 = \overset{=0(1)}{\cancel{\frac{\partial}{\partial t} \int_{CV} \rho \, d\Psi}} + \int_{CS} \rho \vec{V} \cdot d\vec{A} \tag{4.13}$$

Assumptions: (1) Steady flow
(2) Uniform flow at each section

$$0 = \{-|\rho_1 V_1 A_1|\} + \{|\rho_2 V_2 A_2|\}$$

The area is constant, so

$$\rho_1 V_1 = \rho_2 V_2 = G \tag{10.19a}$$

b. Momentum Equation

Basic equation:

$$F_{S_x} + \overset{=0(3)}{\cancel{F_{B_x}}} = \overset{=0(1)}{\cancel{\frac{\partial}{\partial t} \int_{CV} V_x \rho \, d\Psi}} + \int_{CS} V_x \rho \vec{V} \cdot d\vec{A} \tag{4.19a}$$

Assumption: (3) $F_{B_x} = 0$

Since there is no friction between the duct walls and the flow, and $A_2 = A_1 = A$, then

$$p_1 A - p_2 A = V_1\{-|\rho_1 V_1 A|\} + V_2\{|\rho_2 V_2 A|\}$$

This can be written as

$$p_1 A - p_2 A = \dot{m} V_2 - \dot{m} V_1 \qquad (10.19b)$$

or

$$p_1 + \rho_1 V_1^2 = p_2 + \rho_2 V_2^2$$

c. First Law of Thermodynamics

Basic equation:

$$= 0(4) \quad = 0(5) \quad = 0(5) \quad = 0(1)$$

$$\dot{Q} + \cancel{\dot{W}_s} + \cancel{\dot{W}_{shear}} + \cancel{\dot{W}_{other}} = \frac{\partial}{\partial t} \int_{CV} e\rho \, d\Psi + \int_{CS} (e + pv)\rho \vec{V} \cdot d\vec{A} \quad (4.59)$$

where

$$\simeq 0(6)$$

$$e = u + \frac{V^2}{2} + \cancel{gz}$$

Assumptions: (4) $\dot{W}_s = 0$
(5) $\dot{W}_{shear} = \dot{W}_{other} = 0$
(6) Effects of gravity are negligible

With these restrictions

$$\dot{Q} = \left(u_1 + \frac{V_1^2}{2} + p_1 v_1 \right) \{-|\rho_1 V_1 A|\} + \left(u_2 + \frac{V_2^2}{2} + p_2 v_2 \right) \{|\rho_2 V_2 A|\}$$

or

$$\dot{Q} = \dot{m} \left(h_2 + \frac{V_2^2}{2} - h_1 - \frac{V_1^2}{2} \right)$$

But

$$\frac{\delta Q}{dm} = \frac{1}{\dot{m}} \dot{Q}$$

so

$$\frac{\delta Q}{dm} + h_1 + \frac{V_1^2}{2} = h_2 + \frac{V_2^2}{2} \qquad (10.19c)$$

or

$$\frac{\delta Q}{dm} = h_{0_2} - h_{0_1}$$

We see that heat transfer causes the stagnation enthalpy, and hence the stagnation temperature to change.

d. Second Law of Thermodynamics

Basic equation:

$$\int_{cs} \frac{1}{T} \frac{\dot{Q}}{A} \, dA \le \overset{=0(1)}{\cancel{\frac{\partial}{\partial t} \int_{cv}}} s\rho \, d\Psi + \int_{cs} s\rho \vec{V} \cdot d\vec{A} \qquad (4.60)$$

$$\int_{cs} \frac{1}{T} \frac{\dot{Q}}{A} \, dA \le \dot{m}(s_2 - s_1)$$

The flow is frictionless, and the heat transfer may be carried out over an arbitrarily small difference in temperature. Since under these conditions the process could be considered reversible, the equality in Eq. 4.60 would hold. However, even with heat transfer uniform over the control surface, we would need to know the temperature distribution over the area to evaluate the integral on the left side of Eq. 4.60. Consequently, the control volume form of the second law does not enable us to readily calculate the actual entropy change between any two points in the flow. The rate of heat transfer, $\dot{Q}$, may be positive (heat addition to the flow) or negative (heat rejection from the flow). Therefore, the entropy change in a frictionless flow with heat transfer may be either positive or negative.

To calculate the entropy change, we rely on the $T\,ds$ equations. Since

$$T\,ds = dh - v\,dp$$

for an ideal gas we can write

$$ds = c_p \frac{dT}{T} - R \frac{dp}{p}$$

For constant specific heats the equation can be integrated to give

$$s_2 - s_1 = c_p \ln \frac{T_2}{T_1} - R \ln \frac{p_2}{p_1} \qquad (10.19d)$$

e. Equations of State

For an ideal gas, the equation of state is given by

$$p = \rho R T \qquad (10.19e)$$

Equations 10.19a–e are the governing equations for the steady, one-dimensional, frictionless flow of an ideal gas in a constant area duct with heat transfer. If all properties at state $①$ are known, then we have seven unknowns $(\rho_2, V_2, p_2, h_2, s_2, T_2,$ and $\delta Q/dm)$ in these five equations. However, we have the known relationship between h and T for an ideal gas, $dh = c_p dT$. For an ideal gas with constant specific heats,

$$\Delta h = h_2 - h_1 = c_p \Delta T = c_p(T_2 - T_1) \tag{10.19f}$$

We thus have the situation of six equations and seven unknowns.

10-5.2 THE RAYLEIGH LINE

If all conditions at state $①$ are known, how many possible states $②$ are these? The mathematics of the situation (six equations and seven unknowns) indicate that there are an infinite number of possible states $②$.

With an infinite number of possible states $②$ for a given state $①$, what is to be expected if all possible states $②$ are plotted on a Ts diagram? It follows that the locus of all possible states $②$, reachable from state $①$, is a continuous curve passing through state $①$.

How can we determine this curve? Perhaps the simplest way is to assume different values of T_2. For an assumed value of T_2 we could then calculate the corresponding values of all other properties at state $②$ and also $\delta Q/dm$.

The results of these calculations are shown qualitatively on the Ts plane in Fig. 10.15. The locus of all possible downstream states is referred to as the Rayleigh line. In carrying out the calculations, we would discover some interesting features of Rayleigh line flow. At the point of maximum temperature (point a of Fig. 10.15), the value of the Mach number for an ideal gas is $1/\sqrt{k}$. At the point of maximum entropy, the Mach number is unity. On the upper branch of the curve the Mach number is always less than unity, and it increases monotonically as we proceed to the right along the curve. At every point on the lower portion of the curve the Mach number is greater than unity, and it decreases monotonically as we move to the right along the curve. Regardless of the initial Mach number, with heat addition the flow state proceeds to the right and with heat rejection the flow state proceeds to the left along the Rayleigh line.

For any initial state in a Rayleigh line flow, any point on the Rayleigh line represents a mathematically possible downstream state. Indeed, we determined the locus of all possible downstream states by assuming values of T_2 and proceeding to calculate the corresponding properties. Although the Rayleigh line represents all mathematically possible states, are they all physically attainable downstream states? A moment's reflection will indicate they are.

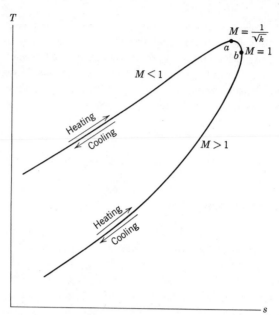

Fig. 10.15 Schematic Ts diagram for frictionless flow in a constant area duct with heat transfer (Rayleigh line flow).

Since we are considering a flow with heat transfer, the second law does not impose any restrictions on the sign of the entropy change.

The effects of heat transfer on properties in the steady, frictionless compressible flow of an ideal gas may be found from the basic equations, Eqs. 10.19a–f, and the Rayleigh line, Fig. 10.15. These effects are summarized in Table 10.2; the basis of the indicated trends is discussed in the next few paragraphs.

The direction of entropy change is always determined by the heat transfer; entropy increases with heating and decreases with cooling. Similarly, the first law, Eq. 10.19c, shows that heating increases the stagnation enthalpy and cooling decreases it; since $\Delta h_0 = c_p \Delta T_0$, the same effect is found on the stagnation temperature.

The effect on temperature of heating and cooling may be deduced from the shape of the Rayleigh line in Fig. 10.15. We see that for $M < 1/\sqrt{k}$ (for air, $1/\sqrt{k} = 0.85$) or for $M > 1$, heating causes T to increase, and in the same regions cooling causes T to decrease. However, we also see the unexpected result that for $1/\sqrt{k} < M < 1$, *heat addition* causes the stream temperature to *decrease*, and *heat rejection* causes the stream temperature to *increase*!

Table 10.2 Summary of Heat Transfer Effects on Fluid Properties

Property	Heating $M < 1$	Heating $M > 1$	Cooling $M < 1$	Cooling $M > 1$	Obtained from:
Entropy, s	Increase	Increase	Decrease	Decrease	Second law
Stagnation temperature, T_0	Increase	Increase	Decrease	Decrease	First law, and $\Delta h_0 = c_p \Delta T_0$
Temperature, T $\left(M < \dfrac{1}{\sqrt{k}} \right)$	Increase	Increase	Decrease	Decrease	Shape of Rayleigh line
$\left(\dfrac{1}{\sqrt{k}} < M < 1 \right)$	Decrease		Increase		
Mach number, M	Increase	Decrease	Decrease	Increase	Trend on Rayleigh line
Pressure, p	Decrease	Increase	Increase	Decrease	Trend on Rayleigh line
Velocity, V	Increase	Decrease	Decrease	Increase	Momentum equation, and effect on p
Density, ρ	Decrease	Increase	Increase	Decrease	Continuity, and effect on V
Stagnation pressure, p_0	Decrease	Decrease	Increase	Increase	Fig. 10.16

For subsonic flow, the Mach number increases monotonically with heating, until the value $M = 1$ is reached. For a given set of inlet conditions, all possible downstream states lie on a single Rayleigh line. Therefore, the point $M = 1$ determines the maximum possible heat addition without choking. If the flow is initially supersonic, heating will reduce the Mach number. Again the maximum possible heat addition without choking is that which reduces the Mach number to $M = 1.0$.

The effect of heat transfer on static pressure is obtained from the shapes of the Rayleigh line and of constant pressure lines on the Ts plane (see Fig. 10.16). For $M < 1$, pressure falls with heating and for $M > 1$, pressure increases with heating, as shown by the shapes of the constant pressure lines. Once the pressure variation has been found, the effect on velocity may be found from the momentum equation

$$p_1 A - p_2 A = \dot{m}V_2 - \dot{m}V_1 \tag{10.19b}$$

or

$$p + \left(\frac{\dot{m}}{A}\right) V = \text{constant}$$

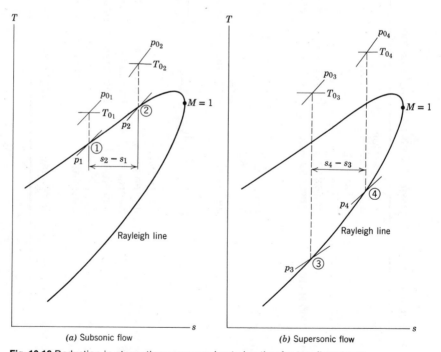

(a) Subsonic flow   (b) Supersonic flow

Fig. 10.16 Reduction in stagnation pressure due to heating for two flow cases.

Thus, since $\dot{m}/A =$ constant, trends in p and V must be opposite. From the continuity equation the trend in ρ is opposite that in V.

The local isentropic stagnation pressure always decreases with heating. This is illustrated schematically in Fig. 10.16. A reduction in stagnation pressure has obvious practical implications for heating processes, such as combustion chambers. Adding the same amount of energy per unit mass (same change in T_0) causes a larger change in p_0 for supersonic flow; because the heating occurs at a lower temperature in supersonic flow, the entropy increase is larger.

Example 10.9
Air flows with negligible friction through a duct with cross-sectional area 0.25 ft². At section ①, the flow properties are $T_1 = 600$ R, $p_1 = 20$ psia, and $V_1 = 360$ ft/sec. At section ②, the pressure is 10 psia. The flow is heated between sections ① and ②. Determine the properties at section ②, the energy added in Btu/lbm, and the entropy change. Finally, plot the process on a Ts diagram.

Example Problem 10.9

GIVEN:

Frictionless flow of air in duct shown:

$T_1 = 600$ R

$p_1 = 20$ psia $\qquad p_2 = 10$ psia

$V_1 = 360$ ft/sec $\qquad A_1 = A_2 = A = 0.25$ ft²

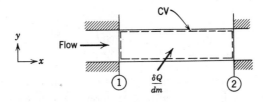

FIND:

(a) Properties at section ②. $\qquad$ (c) $s_2 - s_1$.
(b) $\delta Q/dm$. $\qquad$ (d) Sketch on Ts diagram.

SOLUTION:

Apply the x component of the momentum equation, using the coordinates and control volume shown.

Basic equation:

$$\overset{=0(1)}{\cancel{}} \quad \overset{=0(2)}{\cancel{}}$$

$$F_{S_x} + \cancel{F_{B_x}} = \frac{\partial}{\partial t} \int_{\text{CV}} V_x \rho \, d\forall + \int_{\text{CS}} V_x \rho \vec{V} \cdot d\vec{A} \tag{4.19a}$$

Assumptions: (1) $F_{B_x} = 0$
(2) Steady flow
(3) Uniform flow at each section

Then

$$p_1 A - p_2 A = V_1\{-|\rho_1 V_1 A|\} + V_2\{|\rho_2 V_2 A|\} = \dot{m}(V_2 - V_1)$$

or

$$p_1 - p_2 = \frac{\dot{m}}{A}(V_2 - V_1) = \rho_1 V_1 (V_2 - V_1)$$

Solving for V_2 gives

$$V_2 = \frac{p_1 - p_2}{\rho_1 V_1} + V_1$$

For an ideal gas,

$$\rho_1 = \frac{p_1}{RT_1} = \frac{20 \text{ lbf}}{\text{in.}^2} \times \frac{144 \text{ in.}^2}{\text{ft}^2} \times \frac{\text{lbm} \cdot \text{R}}{53.3 \text{ ft} \cdot \text{lbf}} \times \frac{1}{600 \text{ R}} = 0.090 \text{ lbm/ft}^3$$

$$V_2 = \frac{(20 - 10) \text{ lbf}}{\text{in.}^2} \times \frac{144 \text{ in.}^2}{\text{ft}^2} \times \frac{\text{ft}^3}{0.090 \text{ lbm}} \times \frac{\text{sec}}{360 \text{ ft}} \times \frac{32.2 \text{ lbm}}{\text{slug}}$$

$$\times \frac{\text{slug} \cdot \text{ft}}{\text{lbf} \cdot \text{sec}^2} + \frac{360 \text{ ft}}{\text{sec}}$$

$$V_2 = 1790 \text{ ft/sec} \qquad\qquad V_2$$

From continuity, $G = \rho_1 V_1 = \rho_2 V_2$, so

$$\rho_2 = \rho_1 \frac{V_1}{V_2} = 0.090 \frac{\text{lbm}}{\text{ft}^3}\left(\frac{360}{1790}\right) = 0.0181 \text{ lbm/ft}^3 \qquad\qquad \rho_2$$

Solving for T_2,

$$T_2 = \frac{p_2}{\rho_2 R} = \frac{10 \text{ lbf}}{\text{in.}^2} \times \frac{144 \text{ in.}^2}{\text{ft}^2} \times \frac{\text{ft}^3}{0.0181 \text{ lbm}} \times \frac{\text{lbm} \cdot \text{R}}{53.3 \text{ ft} \cdot \text{lbf}} = 1490 \text{ R} \qquad T_2$$

The local isentropic stagnation temperature is given by

$$T_{0_2} = T_2\left(1 + \frac{k-1}{2}M_2^2\right)$$

$$c_2 = \sqrt{kRT_2} = 1890 \text{ ft/sec}; \qquad M_2 = \frac{V_2}{c_2} = \frac{1790}{1890} = 0.947$$

$$T_{0_2} = 1490 \text{ R}[1 + 0.2(0.947)^2] = 1760 \text{ R} \qquad\qquad T_{0_2}$$

and

$$p_{0_2} = p_2 \left(\frac{T_{0_2}}{T_2}\right)^{k/(k-1)} = 10 \text{ psia} \left(\frac{1760}{1490}\right)^{3.5} = 17.8 \text{ psia} \qquad \longleftarrow \; p_{0_2}$$

The heat transfer is determined from the energy equation.

Basic equation:

$$\dot{Q} + \overset{=\,0(4)}{\cancel{\dot{W}_s}} + \overset{=\,0(5)}{\cancel{\dot{W}_{shear}}} + \overset{=\,0(5)}{\cancel{\dot{W}_{other}}} = \overset{=\,0(2)}{\cancel{\frac{\partial}{\partial t}\int_{CV} e\rho \, d\forall}} + \int_{CS} (e + pv)\rho\vec{V}\cdot d\vec{A} \qquad (4.59)$$

where

$$e = u + \frac{V^2}{2} + \overset{\simeq\,0(6)}{\cancel{gz}}$$

Assumptions: (4) $\dot{W}_s = 0$
(5) $\dot{W}_{shear} = \dot{W}_{other} = 0$
(6) Neglect changes in z

Then

$$\dot{Q} = \left(u_1 + p_1 v_1 + \frac{V_1^2}{2}\right)\{-|\rho_1 V_1 A|\} + \left(u_2 + p_2 v_2 + \frac{V_2^2}{2}\right)\{|\rho_2 V_2 A|\}$$

$$\dot{Q} = \dot{m}\left(h_2 + \frac{V_2^2}{2} - h_1 - \frac{V_1^2}{2}\right) = \dot{m}(h_{0_2} - h_{0_1}) = \dot{m}c_p(T_{0_2} - T_{0_1})$$

and

$$\frac{\delta Q}{dm} = \frac{1}{\dot{m}}\dot{Q} = c_p(T_{0_2} - T_{0_1})$$

$$T_{0_1} = T_1\left(1 + \frac{k-1}{2}M_1^2\right)$$

$$c_1 = \sqrt{kRT_1} = 1200 \text{ ft/sec}; \qquad M_1 = \frac{V_1}{c_1} = \frac{360}{1200} = 0.3$$

$$T_{0_1} = (600 \text{ R})[1 + 0.2(0.3)^2] = 611 \text{ R}$$

so

$$\frac{\delta Q}{dm} = 0.240 \frac{\text{Btu}}{\text{lbm}\cdot\text{R}}(1760 - 611)\text{ R} = 276 \text{ Btu/lbm} \qquad \longleftarrow \; \delta Q/dm$$

Using the $T \, ds$ equation, $T \, ds = dh - v \, dp$, we obtain for an ideal gas with constant specific heats,

$$s_2 - s_1 = c_p \ln\frac{T_2}{T_1} - R \ln\frac{p_2}{p_1} = c_p \ln\frac{T_2}{T_1} - (c_p - c_v) \ln\frac{p_2}{p_1}$$

Then

$$s_2 - s_1 = \frac{0.240 \ \text{Btu}}{\text{lbm} \cdot \text{R}} \times \ln\left(\frac{1490}{600}\right) - \frac{(0.240 - 0.171) \ \text{Btu}}{\text{lbm} \cdot \text{R}} \times \ln\left(\frac{10}{20}\right)$$

$$s_2 - s_1 = 0.266 \ \text{Btu/lbm} \cdot \text{R} \qquad\qquad\qquad s_2 - s_1$$

The process follows a Rayleigh line:

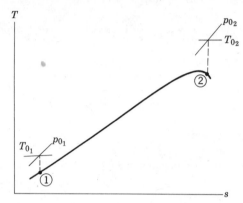

To complete our analysis, let us examine the change in p_0 by comparing p_{0_2} with p_{0_1}.

$$p_{0_1} = p_1 \left(\frac{T_{0_1}}{T_1}\right)^{k/(k-1)} = 20.0 \ \text{psia} \left(\frac{611}{600}\right)^{3.5} = 21.3 \ \text{psia} \qquad\qquad p_{0_1}$$

Comparing the two values, we see that p_{0_2} is less than p_{0_1}.

In general, the stagnation pressure is decreased by heating and increased by cooling in Rayleigh line flow.

**10-5.3 TABLES FOR COMPUTATION OF RAYLEIGH LINE FLOW OF AN IDEAL GAS

In Section 10-5.1 we wrote the basic equations for Rayleigh line flow between two arbitrary states ① and ② in the flow. To facilitate the solution of problems, it is convenient to tabulate dimensionless properties in terms of local Mach number as we did for Fanno line flow. The reference state is again taken as the critical condition, that is, the state at which the Mach number is unity; the properties at the critical condition are denoted by (*).

** This section may be omitted without loss of continuity in the text material.

The dimensionless properties (such as T/T^* and p/p^*) may be obtained by writing the basic equations between a point in the flow where the properties are M, T, p, and so on (unsubscripted) and the critical state ($M = 1$ and properties denoted, e.g. as T^* and p^*).

The pressure ratio p/p^* may be obtained from the momentum equation

$$pA - p^*A = \dot{m}V^* - \dot{m}V \tag{10.19b}$$

or

$$p + \rho V^2 = p^* + \rho^* V^{*2}$$

Substituting $\rho = p/RT$ and factoring out pressures,

$$p\left[1 + \frac{V^2}{RT}\right] = p^*\left[1 + \frac{V^{*2}}{RT^*}\right]$$

Noting that $V^2/RT = k(V^2/kRT) = kM^2$, then

$$p[1 + kM^2] = p^*[1 + k]$$

and finally,

$$\frac{p}{p^*} = \frac{1 + k}{1 + kM^2} \tag{10.20a}$$

From the ideal gas equation of state

$$\frac{T}{T^*} = \frac{p}{p^*}\frac{\rho^*}{\rho}$$

From the continuity equation

$$\frac{\rho^*}{\rho} = \frac{V}{V^*} = M\frac{c}{c^*} = M\sqrt{\frac{T}{T^*}}$$

Then substituting for ρ^*/ρ, we obtain

$$\frac{T}{T^*} = \frac{p}{p^*}M\sqrt{\frac{T}{T^*}}$$

Squaring, and substituting from Eq. 10.20a,

$$\frac{T}{T^*} = \left[\frac{p}{p^*}M\right]^2 = \left[M\left(\frac{1 + k}{1 + kM^2}\right)\right]^2 \tag{10.20b}$$

Also from continuity, using Eq. 10.20b,

$$\frac{\rho^*}{\rho} = \frac{V}{V^*} = \frac{M^2(1 + k)}{1 + kM^2} \tag{10.20c}$$

The dimensionless stagnation temperature, T_0/T_0^*, can be determined from

$$\frac{T_0}{T_0^*} = \frac{T_0}{T}\frac{T}{T^*}\frac{T^*}{T_0^*} = \left(1 + \frac{k-1}{2}M^2\right)\left[M\left(\frac{1+k}{1+kM^2}\right)\right]^2 \frac{1}{\left(1 + \frac{k-1}{2}\right)}$$

$$\frac{T_0}{T_0^*} = \frac{2(k+1)M^2\left(1 + \frac{k-1}{2}M^2\right)}{(1 + kM^2)^2} \qquad (10.20d)$$

Similarly,

$$\frac{p_0}{p_0^*} = \frac{p_0}{p}\frac{p}{p^*}\frac{p^*}{p_0} = \left(1 + \frac{k-1}{2}M^2\right)^{k/(k-1)}\left(\frac{1+k}{1+kM^2}\right)\frac{1}{\left(1 + \frac{k-1}{2}\right)^{k/(k-1)}}$$

$$\frac{p_0}{p_0^*} = \frac{1+k}{1+kM^2}\left[\left(\frac{2}{k+1}\right)\left(1 + \frac{k-1}{2}M^2\right)\right]^{k/(k-1)} \qquad (10.20e)$$

Since the ratios in Eqs. 10.20a–e are functions of Mach number only, they may be computed once and presented in tabular form. These ratios are tabulated as functions of Mach number in Table B.3 of Appendix B.

Example 10.10

Air flows with negligible friction in a constant area duct. At section 1, the flow properties are $T_1 = 60$ C, $p_1 = 135$ kPa (abs), and $V_1 = 732$ m/sec. Heat is added to the flow between section ① and section ②, where the Mach number is 1.2. Determine the flow properties at section ②, the heat transfer per unit mass, the entropy change, and sketch the process on a Ts diagram. Use tables.

Example Problem 10.10

GIVEN:

Frictionless flow of air as shown:

$T_1 = 333$ K $M_2 = 1.2$
$p_1 = 135$ kPa (abs)
$V_1 = 732$ m/sec

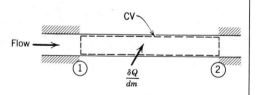

FIND:

(a) Properties at section ②.
(b) $\delta Q/dm$.
(c) $s_2 - s_1$.
(d) Sketch on a Ts diagram.

SOLUTION:

To obtain property ratios from tables, we need both Mach numbers.

$$c_1 = \sqrt{kRT_1} = \left[1.4 \times \frac{287 \text{ N} \cdot \text{m}}{\text{kg} \cdot \text{K}} \times 333 \text{ K} \times \frac{\text{kg} \cdot \text{m}}{\text{N} \cdot \text{sec}^2} \right]^{1/2} = 366 \text{ m/sec}$$

$$M_1 = \frac{V_1}{c_1} = \frac{732 \text{ m}}{\text{sec}} \times \frac{\text{sec}}{366 \text{ m}} = 2.00$$

From Table B.3, Appendix B,

M	T_0/T_0^*	p_0/p_0^*	T/T^*	p/p^*	V/V^*
2.00	0.7934	1.503	0.5289	0.3636	1.455
1.20	0.9787	1.019	0.9119	0.7958	1.146

Using these data and recognizing that critical properties are constant,

$$\frac{T_2}{T_1} = \frac{T_2/T^*}{T_1/T^*} = \frac{0.9119}{0.5289} = 1.72; \qquad T_2 = 1.72T_1 = (1.72)\ 333 \text{ K} = 573 \text{ K} \qquad \longleftarrow T_2$$

$$\frac{p_2}{p_1} = \frac{p_2/p^*}{p_1/p^*} = \frac{0.7958}{0.3636} = 2.19; \qquad p_2 = 2.19p_1 = (2.19)\ 135 \text{ kPa} = 296 \text{ kPa (abs)} \qquad \longleftarrow p_2$$

$$\frac{V_2}{V_1} = \frac{V_2/V^*}{V_1/V^*} = \frac{1.146}{1.455} = 0.788; \qquad V_2 = 0.788V_1 = \frac{(0.788)\ 732 \text{ m}}{\text{sec}} = 577 \text{ m/sec} \qquad \longleftarrow V_2$$

$$\rho_2 = \frac{p_2}{RT_2} = \frac{2.96 \times 10^5 \text{ N}}{\text{m}^2} \times \frac{\text{kg} \cdot \text{K}}{287 \text{ N} \cdot \text{m}} \times \frac{1}{573 \text{ K}} = 1.80 \text{ kg/m}^3 \qquad \longleftarrow \rho_2$$

The heat transfer may be determined from the energy equation, which reduces to (see Example Problem 10.9)

$$\frac{\delta Q}{dm} = h_{0_2} - h_{0_1} = c_p(T_{0_2} - T_{0_1})$$

From the isentropic flow tables (Table B.1), at $M = 2.0$,

$$\frac{T}{T_0} = \frac{T_1}{T_{0_1}} = 0.5556; \qquad T_{0_1} = \frac{T_1}{0.5556} = \frac{333 \text{ K}}{0.5556} = 599 \text{ K}$$

and at $M = 1.2$

$$\frac{T}{T_0} = \frac{T_2}{T_{0_2}} = 0.7764; \qquad T_{0_2} = \frac{T_2}{0.7764} = \frac{573 \text{ K}}{0.7764} = 738 \text{ K} \qquad \longleftarrow T_{0_2}$$

Substituting gives

$$\frac{\delta Q}{dm} = c_p(T_{0_2} - T_{0_1}) = 1.00\,\frac{kJ}{kg \cdot K}\,(738 - 599)\,K = 139\ kJ/kg$$

$\delta Q/dm$

The entropy change may be found from the $T\,ds$ equation, $T\,ds = dh - v\,dp$. For an ideal gas with constant specific heats,

$$s_2 - s_1 = c_p \ln\frac{T_2}{T_1} - R\ln\frac{p_2}{p_1}$$

$$= \frac{1.00}{kg \cdot K}\,\frac{kJ}{kg \cdot K}\ln\left(\frac{573}{333}\right) - \frac{287\ N \cdot m}{kg \cdot K}\ln\left(\frac{2.96 \times 10^5}{1.35 \times 10^5}\right)\frac{kJ}{1000\ N \cdot m}$$

$$s_2 - s_1 = 0.317\ kJ/kg \cdot K$$

$s_2 - s_1$

Finally, let us check the effect on p_0. From Table B.1, at $M = 2.0$,

$$\frac{p}{p_0} = \frac{p_1}{p_{0_1}} = 0.1278; \qquad p_{0_1} = \frac{p_1}{0.1278} = \frac{135\ kPa}{0.1278} = 1.06\ MPa\ (abs)$$

and at $M = 1.2$,

$$\frac{p}{p_0} = \frac{p_2}{p_{0_2}} = 0.4124; \qquad p_{0_2} = \frac{p_2}{0.4124} = \frac{196\ kPa}{0.4124} = 718\ kPa\ (abs)$$

p_{0_2}

Thus $p_{0_2} < p_{0_1}$, as expected for a heating process.

The process follows the supersonic branch of a Rayleigh line:

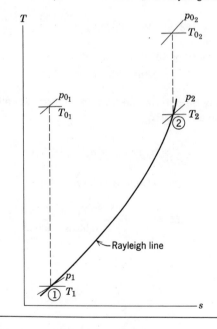

10-6 NORMAL SHOCKS

We have previously mentioned normal shocks in the section on nozzle flow. In practice, these irreversible discontinuities can occur in any supersonic flow field, in either internal flow or external flow.[3] Knowledge of property changes across shocks and of shock behavior is important in understanding the design of supersonic diffusers, for example, for inlets on high performance aircraft, and supersonic wind tunnels. Accordingly, the purpose of the present section is to analyze the normal shock process.

Before applying the basic equations to normal shocks, it is important that we have in mind a clear physical picture of the shock itself. Although it is physically impossible to have discontinuities in fluid properties, the normal shock is nearly discontinuous. The thickness of a shock is in the order of 0.2 microns (10^{-5} in.), or roughly 4 times the mean free path of the gas molecules. Across this small distance, large changes in pressure, temperature, and other properties occur. Local fluid accelerations can reach tens of millions of "G's"! These considerations justify treating the normal shock as an abrupt discontinuity; we are interested in changes occurring across the shock rather than in the details of its structure.

10-6.1 BASIC EQUATIONS

To begin our analysis, let us apply the basic equations to the thin control volume shown in Fig. 10.17, where for generality we have depicted a shock standing in a passage of arbitrary shape.

a. Continuity Equation

Basic equation:

$$0 = \overset{=0(1)}{\cancel{\frac{\partial}{\partial t} \int_{cv} \rho\, d\Psi}} + \int_{cs} \rho \vec{V} \cdot d\vec{A} \qquad (4.13)$$

Assumptions: (1) Steady flow
(2) Uniform flow at each section
(3) $A_1 = A_2 = A$, because the shock is so extraordinarily thin

$$0 = \{-|\rho_1 V_1 A|\} + \{|\rho_2 V_2 A|\}$$

Writing the result in terms of scalar magnitudes, we obtain

$$\rho_1 V_1 = \rho_2 V_2 = G \qquad (10.21a)$$

---

[3] The Shell film *Schlieren* shows several examples of shock formation in external flow.

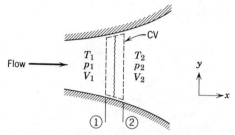

Fig. 10.17 Control volume used for analysis of normal shock.

b. Momentum Equation

Basic equation:

$$= 0(5) = 0(1)$$

$$F_{S_x} + F_{B_x} = \frac{\partial}{\partial t} \int_{CV} V_x \rho \, d\mathbf{V} + \int_{CS} V_x \rho \vec{V} \cdot d\vec{A} \qquad (4.19a)$$

Assumptions: (4) Negligible friction at duct walls because the shock is so thin

(5) $F_{B_x} = 0$

Under these conditions,

$$F_{S_x} = p_1 A - p_2 A = V_1\{-|\rho_1 V_1 A|\} + V_2\{|\rho_2 V_2 A|\}$$

Using scalar magnitudes and dropping absolute value signs, we obtain

$$p_1 A - p_2 A = \dot{m} V_2 - \dot{m} V_1 \qquad (10.21b)$$

or

$$p_1 + \rho_1 V_1^2 = p_2 + \rho_2 V_2^2$$

c. First Law of Thermodynamics

Basic equation:

$$= 0(6) = 0(7) = 0(8) \quad = 0(8) \quad = 0(1)$$

$$\dot{Q} + \dot{W}_s + \dot{W}_{shear} + \dot{W}_{other} = \frac{\partial}{\partial t} \int_{CV} e\rho \, d\mathbf{V} + \int_{CS} (e + pv)\rho \vec{V} \cdot d\vec{A} \qquad (4.59)$$

where

$$\approx 0(9)$$

$$e = u + \frac{V^2}{2} + gz$$

Assumptions: (6) $\dot{Q} = 0$ (adiabatic flow)
  (7) $\dot{W}_s = 0$
  (8) $\dot{W}_{shear} = \dot{W}_{other} = 0$
  (9) Effects of gravity are negligible

Then

$$0 = \left(u_1 + p_1 v_1 + \frac{V_1^2}{2}\right)\{-|\rho_1 V_1 A|\} + \left(u_2 + p_2 v_2 + \frac{V_2^2}{2}\right)\{|\rho_2 V_2 A|\}$$

However, from continuity, the mass flowrate terms in brackets are equal. We may also substitute $h \equiv u + pv$, to obtain

$$h_1 + \frac{V_1^2}{2} = h_2 + \frac{V_2^2}{2} \tag{10.21c}$$

or, in terms of stagnation enthalpy,

$$h_{0_1} = h_{0_2}$$

Physically, we should expect the total energy of the flow to remain constant, since there is no energy addition.

d. Second Law of Thermodynamics

Basic equation:

$$\overset{=0(6)}{\overbrace{}} \qquad \overset{=0(1)}{\overbrace{}}$$

$$\int_{CS} \frac{1}{T}\frac{\dot{Q}}{A}\, dA \leq \frac{\partial}{\partial t}\int_{CV} s\rho\, dV + \int_{CS} s\rho \vec{V} \cdot d\vec{A} \tag{4.60}$$

Then

$$0 \leq s_1\{-|\rho_1 V_1 A|\} + s_2\{|\rho_2 V_2 A|\}$$

Flow through the normal shock is irreversible because of the almost discontinuous property changes across the shock. Consequently, the inequality in the above equation holds. The control volume form of the second law then tells us that $s_2 - s_1 > 0$.

This fact is of little help in calculating the actual entropy change across the shock. To calculate the entropy change, we rely on the $T\, ds$ equations. Since

$$T\, ds = dh - v\, dp$$

then for an ideal gas we can write

$$ds = c_p \frac{dT}{T} - R\frac{dp}{p}$$

For constant specific heats this equation can be integrated to give

$$s_2 - s_1 = c_p \ln \frac{T_2}{T_1} - R \ln \frac{p_2}{p_1} \qquad (10.21d)$$

e. Equation of State

For an ideal gas, the equation of state is given by

$$p = \rho R T \qquad (10.21e)$$

Equations 10.21 a–e are the governing equations for the flow of an ideal gas through a normal shock. If all the properties at state ① (immediately upstream of the shock) are known, then we have six unknowns $(T_2, p_2, \rho_2, V_2, h_2, s_2)$ in these five equations. However, we have the known relationship between h and T for an ideal gas, $dh = c_p\, dT$. For an ideal gas with constant specific heats,

$$\Delta h = h_2 - h_1 = c_p \Delta T = c_p(T_2 - T_1) \qquad (10.21f)$$

We thus have the situation of six equations and six unknowns.

Then if all conditions at state ① (immediately ahead of the shock) are known, how many possible states ② (immediately behind the shock) are there? The mathematics of the situation (six equations and six unknowns) indicates that there is a unique state ② for a given state ①.

We can obtain a physical picture of the flow through a normal shock by employing some of the notions developed in the consideration of Fanno line and Rayleigh line flows. For convenience let us first rewrite the governing equations for a normal shock.

$$\rho_1 V_1 = \rho_2 V_2 = G \qquad (10.21a)$$

$$p_1 A - p_2 A = \dot{m} V_2 - \dot{m} V_1 \qquad (10.21b)$$

$$h_1 + \frac{V_1^2}{2} = h_2 + \frac{V_2^2}{2} \qquad (10.21c)$$

$$s_2 - s_1 = c_p \ln \frac{T_2}{T_1} - R \ln \frac{p_2}{p_1} \qquad (10.21d)$$

$$p = \rho R T \qquad (10.21e)$$

$$h_2 - h_1 = c_p(T_2 - T_1) \qquad (10.21f)$$

Flow through a normal shock must satisfy Eqs. 10.21a–f. Since all conditions at state ① are known, we can locate state ① on a Ts diagram. If we were to draw a Fanno line curve through state ①, we would have a locus of mathematical states that satisfy Eqs. 10.21a, c, d, e, and f. (The Fanno line curve does not satisfy Eq. 10.21b.) Drawing a Rayleigh line curve through state ① gives

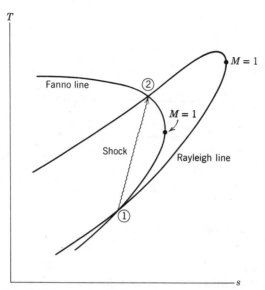

Fig. 10.18 Intersection of Fanno line and Rayleigh line as a solution of the normal shock equations.

a locus of mathematical states that satisfy Eqs. 10.21a, b, d, e, and f. (The Rayleigh line curve does not satisfy Eq. 10.21c.) These curves are shown in Fig. 10.18.

The normal shock must satisfy all six of Eqs. 10.21a–f. Consequently, for a given state ①, the end state (state ②) of the normal shock must lie on both the Fanno line and the Rayleigh line passing through state ①. Hence the intersection of the two lines at state ② represents the conditions downstream from the shock, corresponding to the given upstream conditions at state ①. In Fig. 10.18 the flow through the shock has been indicated as occurring from state ① to state ②. This is the only possible direction of the shock process as dictated by the second law ($s_2 > s_1$).

From Fig. 10.18 we note also that the flow through a normal shock involves a change from supersonic to subsonic speeds. Normal shocks can occur only in a flow that is initially supersonic.

As an aid in summarizing the effects of a normal shock on the flow properties, a schematic of the normal shock process is illustrated on the Ts plane of Fig. 10.19. This figure, together with the governing basic equations, is the basis for Table 10.3. You are encouraged to follow through the logic indicated in the table.

In theory the solution of six equations in six unknowns poses no difficulty, but in practice the algebra may become involved. It makes sense to recast

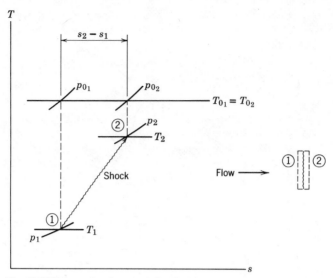

Fig. 10.19 Schematic of normal shock process on the Ts plane.

Table 10.3   Summary of Property Changes Across a Normal Shock

Property	Effect	Obtained from
Stagnation temperature, T_0	Constant	Energy equation
Entropy, s	Increase	Second law
Stagnation pressure, p_0	Decrease	Ts diagram
Temperature, T	Increase	Ts diagram
Velocity, V	Decrease	Energy equation, and effect on T
Density, ρ	Increase	Continuity equation, and effect on V
Pressure, p	Increase	Momentum equation, and effect on V
Mach number, M	Decrease	$M = V/c$, and effect on V and T

the equations in a form suitable for tabulating property ratios across the shock. This we shall do in the next section. To illustrate the direct application of the control volume equations to a normal shock, consider Example Problem 10.11, in which the problem is overspecified (one property downstream of the shock is given).

Example 10.11

A normal shock stands in a duct. The fluid is air, which may be considered an ideal gas. Properties upstream from the shock are $T_1 = 5$ C, $p_1 = 65.0$ kPa (abs), and $V_1 = 668$ m/sec. The temperature at section ②, downstream from the shock, is $T_2 = 469$ K. Determine the property values at section ② and compare them with the upstream values. Sketch the process on a Ts diagram.

Example Problem 10.11

GIVEN:

Normal shock in a duct as shown:

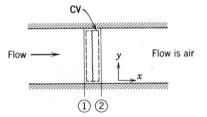

$T_1 = 5$ C $T_2 = 469$ K

$p_1 = 65.0$ kPa (abs)

$V_1 = 668$ m

FIND:

(a) Properties at section ②, (b) Ts diagram.

SOLUTION:

Let us first compute the remaining properties at section ①. For an ideal gas

$$\rho_1 = \frac{p_1}{RT_1} = \frac{6.5 \times 10^4 \text{ N}}{\text{m}^2} \times \frac{\text{kg} \cdot \text{K}}{287 \text{ N} \cdot \text{m}} \times \frac{1}{278 \text{ K}} = 0.815 \text{ kg/m}^3$$

$$c_1 = \sqrt{kRT_1} = \left[1.4 \times \frac{287 \text{ N} \cdot \text{m}}{\text{kg} \cdot \text{K}} \times 278 \text{ K} \times \frac{\text{kg} \cdot \text{m}}{\text{N} \cdot \text{sec}^2} \right]^{1/2} = 334 \text{ m/sec}$$

$$M_1 = \frac{V_1}{c_1} = \frac{668}{334} = 2.00$$

$$T_{0_1} = T_1 \left(1 + \frac{k-1}{2} M_1^2 \right) = 278 \text{ K}[1 + 0.2(2.0)^2] = 500 \text{ K}$$

$$p_{0_1} = p_1 \left(1 + \frac{k-1}{2} M_1^2 \right)^{k/(k-1)} = 65.0 \text{ kPa}[1 + 0.2(2.0)^2]^{3.5} = 509 \text{ kPa (abs)}$$

Also, V_2 may be evaluated by applying the energy equation to the control volume shown.

Basic equation:

$$\overset{=0(1)}{\dot{Q}} + \overset{=0(2)}{\dot{W}_s} + \overset{=0(3)}{\dot{W}_{shear}} + \overset{=0(3)}{\dot{W}_{other}} = \overset{=0(4)}{\frac{\partial}{\partial t}\int_{CV} e\rho\, d\forall} + \int_{CS}(e + pv)\rho\vec{V}\cdot d\vec{A} \qquad (4.59)$$

where

$$e = u + \frac{V^2}{2} + \overset{\simeq 0(6)}{gz}$$

Assumptions: (1) $\dot{Q} = 0$ (5) Uniform flow at each section
              (2) $\dot{W}_s = 0$ (6) Neglect gravity term
              (3) $\dot{W}_{shear} = \dot{W}_{other} = 0$ (7) $A_1 = A_2 = A$
              (4) Steady flow

Then

$$0 = \left(u_1 + p_1v_1 + \frac{V_1^2}{2}\right)\{-|\rho_1 V_1 A|\} + \left(u_2 + p_2v_2 + \frac{V_2^2}{2}\right)\{|\rho_2 V_2 A|\}$$

$$0 = \dot{m}\left(u_2 + p_2v_2 + \frac{V_2^2}{2} - u_1 - p_1v_1 - \frac{V_1^2}{2}\right)$$

or

$$h_1 + \frac{V_1^2}{2} = h_2 + \frac{V_2^2}{2}$$

Solving for V_2 gives

$$V_2 = [V_1^2 + 2(h_1 - h_2)]^{1/2} = [V_1^2 + 2c_p(T_1 - T_2)]^{1/2}$$

$$= \left[\frac{(668)^2 \; m^2}{sec^2} + 2 \times \frac{1000 \; N\cdot m \; (278 - 469) \; K}{kg\cdot K} \times \frac{kg\cdot m}{N\cdot sec^2}\right]^{1/2}$$

$$V_2 = 253 \; m/sec \xleftarrow{\hspace{3cm}} V_2$$

Continuity reduces to $G = \rho_1 V_1 = \rho_2 V_2$, so

$$\rho_2 = \rho_1\frac{V_1}{V_2} = 0.815\;\frac{kg}{m^3}\left(\frac{668}{253}\right) = 2.15 \; kg/m^3 \xleftarrow{\hspace{1cm}} \rho_2$$

The pressure can be obtained two ways:
(1) From the ideal gas equation of state

$$p_2 = \rho_2 RT_2 = \frac{2.15 \; kg}{m^3} \times \frac{287 \; N\cdot m}{kg\cdot K} \times 469 \; K = 289 \; kPa \; (abs) \xleftarrow{\hspace{1cm}} p_2$$

(2) From the momentum equation

Basic equation:

$$= 0(8) = 0(4)$$

$$F_{S_x} + F_{B_x} = \frac{\partial}{\partial t} \int_{CV} V_x \rho \, d\forall + \int_{CS} V_x \rho \vec{V} \cdot d\vec{A} \qquad (4.19a)$$

Assumption: (8) $F_{B_x} = 0$

Then

$$p_1 A - p_2 A = V_1\{-|\rho_1 V_1 A|\} + V_2\{|\rho_2 V_2 A|\} = \dot{m}(V_2 - V_1)$$

or

$$p_1 - p_2 = \rho_1 V_1 (V_2 - V_1)$$

Solving gives

$$p_2 = p_1 - \rho_1 V_1 (V_2 - V_1)$$

$$p_2 = \frac{6.5 \times 10^4 \, \text{N}}{\text{m}^2} - \frac{0.815 \, \text{kg}}{\text{m}^3} \times \frac{668 \, \text{m}}{\text{sec}} \frac{(253 - 668) \, \text{m}}{\text{sec}} \times \frac{\text{N} \cdot \text{sec}^2}{\text{kg} \cdot \text{m}} = 291 \, \text{kPa (abs)}$$

(The two calculated pressure values are in close, but not exact, agreement due to round-off errors.)

For adiabatic flow, T_0 = constant. Thus

$$T_{0_2} = T_{0_1} = 500 \text{ K} \qquad\qquad\qquad T_{0_2}$$

The local isentropic stagnation pressure at section ② is

$$p_{0_2} = p_2 \left(\frac{T_{0_2}}{T_2}\right)^{k/(k-1)} = 289 \text{ kPa} \left(\frac{500}{469}\right)^{3.5} = 362 \text{ kPa (abs)} \qquad p_{0_2}$$

Comparing, we see that

$$T_{0_2} = T_{0_1}$$
$$p_{0_2} < p_{0_1}$$
$$T_2 > T_1$$
$$p_2 > p_1$$

and

$$V_2 < V_1$$

in accord with Table 10.3.

The entropy change may be computed from the $T \, ds$ equation

$$T \, ds = dh - v \, dp$$

For an ideal gas

$$ds = c_p \frac{dT}{T} - R \frac{dp}{p}$$

Integrating, for constant specific heats,

$$s_2 - s_1 = c_p \ln\frac{T_2}{T_1} - R \ln\frac{p_2}{p_1}$$

Since $s_{0_2} - s_{0_1} = s_2 - s_1$, the entropy change is easiest to evaluate at stagnation conditions, because $T_{0_2} = T_{0_1}$. At stagnation conditions, we have

$$s_2 - s_1 = s_{0_2} - s_{0_1} = c_p \ln \frac{\cancelto{0}{T_{0_2}}}{T_{0_1}} - R \ln \frac{p_{0_2}}{p_{0_1}} = -\frac{287 \text{ N} \cdot \text{m}}{\text{kg} \cdot \text{K}} \ln\left(\frac{3.62 \times 10^5}{5.09 \times 10^5}\right)$$

$$s_2 - s_1 = 0.0978 \text{ kJ/kg} \cdot \text{K}$$

Finally, the Ts diagram may be sketched:

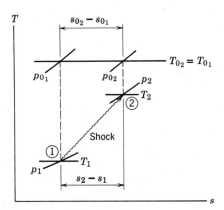

{Note once again that $s_{0_2} - s_{0_1} = s_2 - s_1$, since $s_{0_1} = s_1$ and $s_{0_2} = s_2$.}

**10-6.2 TABLES FOR COMPUTATION OF NORMAL SHOCKS IN AN IDEAL GAS

Consideration of the basic equations for the flow through a normal shock has indicated that for given conditions ahead of the shock there is a unique downstream state. Therefore, it is possible to develop expressions for the property ratios across the shock in terms of the Mach number, M_1, ahead of the shock; these results can be tabulated as functions of M_1.

To obtain the results, we proceed in three steps. First, we obtain property ratios (e.g. T_2/T_1 and p_2/p_1) in terms of M_1 and M_2. Then we develop a relation between M_1 and M_2. Finally, we use this relation to obtain expressions for the property ratios in terms of the upstream Mach number, M_1.

** This section may be omitted without loss of continuity in the text material.

The temperature ratio can be expressed as

$$\frac{T_2}{T_1} = \frac{T_2}{T_{0_2}} \frac{T_{0_2}}{T_{0_1}} \frac{T_{0_1}}{T_1}$$

Since the stagnation temperature is constant, we have

$$\frac{T_2}{T_1} = \frac{1 + \dfrac{k-1}{2} M_1^2}{1 + \dfrac{k-1}{2} M_2^2} \qquad (10.22a)$$

A velocity ratio may be obtained by using

$$\frac{V_2}{V_1} = \frac{M_2 c_2}{M_1 c_1} = \frac{M_2}{M_1} \frac{\sqrt{kRT_2}}{\sqrt{kRT_1}} = \frac{M_2}{M_1} \sqrt{\frac{T_2}{T_1}}$$

or

$$\frac{V_2}{V_1} = \frac{M_2}{M_1} \left[\frac{1 + \dfrac{k-1}{2} M_1^2}{1 + \dfrac{k-1}{2} M_2^2} \right]^{1/2} \qquad (10.22b)$$

A ratio of densities may be obtained from the continuity equation

$$\rho_1 V_1 = \rho_2 V_2 \qquad (10.21a)$$

Substituting from Eq. 10.22b gives

$$\frac{\rho_2}{\rho_1} = \frac{V_1}{V_2} = \frac{M_1}{M_2} \left[\frac{1 + \dfrac{k-1}{2} M_2^2}{1 + \dfrac{k-1}{2} M_1^2} \right]^{1/2} \qquad (10.22c)$$

Finally, we can obtain a pressure ratio from the momentum equation

$$p_1 A - p_2 A = \dot{m} V_2 - \dot{m} V_1 \qquad (10.21b)$$

or

$$p_1 + \rho_1 V_1^2 = p_2 + \rho_2 V_2^2$$

Substituting $\rho = p/RT$ and factoring out pressures,

$$p_1 \left[1 + \frac{V_1^2}{RT_1} \right] = p_2 \left[1 + \frac{V_2^2}{RT_2} \right]$$

Since

$$\frac{V^2}{RT} = k\frac{V^2}{kRT} = kM^2$$

then

$$p_1[1 + kM_1^2] = p_2[1 + kM_2^2]$$

and finally,

$$\frac{p_2}{p_1} = \frac{1 + kM_1^2}{1 + kM_2^2} \qquad (10.22d)$$

In order to solve for M_2 in terms of M_1, we need to obtain another expression for one of the property ratios given by Eqs. 10.22a–d.

From the ideal gas equation of state the temperature ratio may be written

$$\frac{T_2}{T_1} = \frac{p_2/\rho_2 R}{p_1/\rho_1 R} = \frac{p_2}{p_1}\frac{\rho_1}{\rho_2}$$

Substituting from Eqs. 10.22c and 10.22d,

$$\frac{T_2}{T_1} = \left[\frac{1 + kM_1^2}{1 + kM_2^2}\right]\frac{M_2}{M_1}\left[\frac{1 + \dfrac{k-1}{2}M_1^2}{1 + \dfrac{k-1}{2}M_2^2}\right]^{1/2} \qquad (10.23)$$

Equations 10.22a and 10.23 are two equations for T_2/T_1. We can combine them and solve for M_2 in terms of M_1. Combining and canceling gives

$$\left[\frac{1 + \dfrac{k-1}{2}M_1^2}{1 + \dfrac{k-1}{2}M_2^2}\right]^{1/2} = \frac{M_2}{M_1}\left[\frac{1 + kM_1^2}{1 + kM_2^2}\right]$$

Squaring, we obtain

$$\frac{1 + \dfrac{k-1}{2}M_1^2}{1 + \dfrac{k-1}{2}M_2^2} = \frac{M_2^2}{M_1^2}\left[\frac{1 + 2kM_1^2 + k^2M_1^4}{1 + 2kM_2^2 + k^2M_2^4}\right]$$

which may be solved explicitly for M_2^2. Two solutions are obtained:

$$M_2^2 = M_1^2 \qquad (10.24a)$$

and

$$M_2^2 = \frac{M_1^2 + \dfrac{2}{k-1}}{\dfrac{2k}{k-1} M_1^2 - 1} \tag{10.24b}$$

Obviously, the first of these is trivial. The second expresses the unique dependence of M_2 on M_1. This relation is tabulated in Appendix B, Table B.4. Now, having a relationship between M_2 and M_1, we can solve for the property ratios across a shock, knowing the upstream Mach number. Knowing M_1, M_2 is obtained from Eq. 10.24b. The property ratios can subsequently be determined from Eqs. 10.22a–d. These property ratios are tabulated as functions of M_1 in Table B.4 of Appendix B.

Since the stagnation temperature remains constant, the stagnation temperature ratio across the shock is unity. The ratio of stagnation pressures is evaluated as

$$\frac{p_{0_2}}{p_{0_1}} = \frac{p_{0_2}}{p_2} \frac{p_2}{p_1} \frac{p_1}{p_{0_1}} = \frac{p_2}{p_1} \left[\frac{1 + \dfrac{k-1}{2} M_2^2}{1 + \dfrac{k-1}{2} M_1^2} \right]^{k/(k-1)} \tag{10.25}$$

Combining Eqs. 10.22d and 10.24b, we obtain (after considerable algebra)

$$\frac{p_2}{p_1} = \frac{1 + kM_1^2}{1 + kM_2^2} = \frac{2k}{k+1} M_1^2 - \frac{k-1}{k+1} \tag{10.26}$$

Using Eqs. 10.24b and 10.26, Eq. 10.25 becomes

$$\frac{p_{0_2}}{p_{0_1}} = \frac{\left[\dfrac{k+1}{2} M_1^2 \right]^{k/(k-1)}}{\left[\dfrac{2k}{k+1} M_1^2 - \dfrac{k-1}{k+1} \right]^{1/(k-1)}} \tag{10.27}$$

The stagnation pressure ratio as a function of upstream Mach number is included in Table B.4 of Appendix B.

Use of the tables in the solution of problems involving a normal shock is illustrated in Example Problem 10.12.

Example 10.12

A normal shock stands in a duct. The fluid is air, which can be considered an ideal gas. Properties upstream from the shock are $T_1 = 5$ C, $p_1 = 65.0$ kPa (abs), and $V_1 = 668$ m/sec. Determine the properties downstream and $s_2 - s_1$. Include a Ts plot.

(Note that these upstream conditions are the same as those given in Example Problem 10.11. However, with the tables available, we do not need to know any properties downstream to complete a solution.)

Example Problem 10.12

GIVEN:

Normal shock in a duct as shown:

$T_1 = 278$ K

$p_1 = 65.0$ kPa (abs)

$V_1 = 668$ m/sec

FIND:

(a) Properties at section ②.
(b) $s_2 - s_1$.
(c) Ts plot.

SOLUTION:

To use the shock tables, we need to know M_1. For an ideal gas,

$$c_1 = \sqrt{kRT_1} = \left[1.4 \times \frac{287 \text{ N} \cdot \text{m}}{\text{kg} \cdot \text{K}} \times 278 \text{ K} \times \frac{\text{kg} \cdot \text{m}}{\text{N} \cdot \text{sec}^2} \right]^{1/2} = 334 \text{ m/sec}$$

$$M_1 = \frac{V_1}{c_1} = \frac{668}{334} = 2.0$$

Normal shock property ratios are given in Table B.4, Appendix B. At $M_1 = 2.0$,

M_1	M_2	p_{0_2}/p_{0_1}	T_2/T_1	p_2/p_1	ρ_2/ρ_1
2.00	0.5774	0.7209	1.688	4.500	2.667

From these data

$$T_2 = 1.688T_1 = (1.688)278 \text{ K} = 469 \text{ K} \qquad T_2$$

$$p_2 = 4.500p_1 = (4.500)65.0 \text{ kPa (abs)} = 293 \text{ kPa (abs)} \qquad p_2$$

For an ideal gas,

$$\rho_2 = \frac{p_2}{RT_2} = \frac{2.93 \times 10^5 \text{ N}}{\text{m}^2} \times \frac{\text{kg} \cdot \text{K}}{287 \text{ N} \cdot \text{m}} \times \frac{1}{469 \text{ K}} = 2.18 \text{ kg/m}^3 \qquad \underleftarrow{\rho_2}$$

and

$$V_2 = M_2 c_2 = M_2 \sqrt{kRT_2} = (0.5774) \left[1.4 \times \frac{287 \text{ N} \cdot \text{m}}{\text{kg} \cdot \text{K}} \times \frac{469 \text{ K}}{} \times \frac{\text{kg} \cdot \text{m}}{\text{N} \cdot \text{sec}^2} \right]^{1/2}$$

$$V_2 = 251 \text{ m/sec} \qquad \underleftarrow{V_2}$$

Local isentropic stagnation properties at section ① may be evaluated using tables for isentropic flow. From Table B.1, Appendix B, at $M = 2.0$,

$$\frac{T}{T_0} = \frac{T_1}{T_{0_1}} = 0.5556; \qquad T_{0_1} = \frac{T_1}{0.5556} = \frac{278 \text{ K}}{0.5556} = 500 \text{ K}$$

$$\frac{p}{p_0} = \frac{p_1}{p_{0_1}} = 0.1278; \qquad p_{0_1} = \frac{p_1}{0.1278} = \frac{65.0 \text{ kPa}}{0.1278} = 509 \text{ kPa (abs)}$$

The stagnation temperature is constant in adiabatic flow. Thus

$$T_{0_2} = T_{0_1} = 500 \text{ K} \qquad \underleftarrow{T_{0_2}}$$

Using the property ratios for a normal shock,

$$p_{0_2} = p_{0_1} \frac{p_{0_2}}{p_{0_1}} = 509 \text{ kPa} (0.7209) = 367 \text{ kPa (abs)} \qquad \underleftarrow{p_{0_2}}$$

The entropy change across the shock may be found from the $T ds$ equation

$$T ds = dh - v dp$$

For an ideal gas,

$$ds = c_p \frac{dT}{T} - R \frac{dp}{p}$$

Integrating for constant specific heats gives

$$s_2 - s_1 = c_p \ln \frac{T_2}{T_1} - R \ln \frac{p_2}{p_1}$$

But $s_{0_2} - s_{0_1} = s_2 - s_1$, so

$$s_{0_2} - s_{0_1} = s_2 - s_1 = c_p \ln \overset{0}{\cancel{\frac{T_{0_2}}{T_{0_1}}}} - R \ln \frac{p_{0_2}}{p_{0_1}} = -\frac{0.287}{\text{kg} \cdot \text{K}} \frac{\text{kJ}}{} \ln (0.7209)$$

$$s_2 - s_1 = 0.0939 \text{ kJ/kg} \cdot \text{K} \qquad \underleftarrow{s_2 - s_1}$$

The *Ts* diagram is

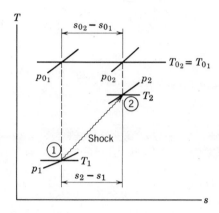

> Comparing the solutions presented in Example Problems 10.11 and 10.12, we see that using the tables simplifies the computations appreciably. In addition, the consistency of the calculated values is improved as a result of using tabulated values accurate to four significant figures. You must be careful when using the tables to check your work as you proceed. Each calculated result should be examined to be sure both its trend and magnitude are reasonable. This process is simplified and made almost automatic when a *Ts* diagram is drawn for each problem.

10-6.3 FLOW IN A CONVERGING-DIVERGING NOZZLE

Since we have considered normal shocks, we are now in a position to complete our discussion of flow in a converging-diverging nozzle operating under varying back pressures, discussed previously in Section 10-3.5. The pressure distribution through the nozzle for different back pressures is shown in Fig. 10.20.

Four different regimes of flow are possible. In Regime I the flow is subsonic throughout. The flowrate increases with decreasing back pressure. At condition (*iii*), which forms the dividing line between Regimes I and II, the flow at the throat is sonic, that is, $M_t = 1$.

As the back pressure is lowered below that of condition (*iii*), a normal shock appears downstream of the throat. There is a pressure rise across the shock. Since the flow is subsonic ($M < 1$) behind the shock, the flow decelerates, with an accompanying increase in pressure, through the diverging portion of the channel. As the back pressure is lowered further, the shock moves downstream until it appears at the exit plane of the nozzle (condition *vii*). In Regime II, as in Regime I, the exit flow is subsonic and consequently

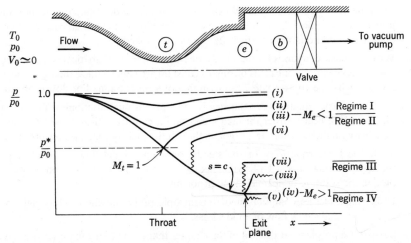

Fig. 10.20 Pressure distributions for flow in a converging-diverging nozzle as a function of back pressure.

$p_e = p_b$. Since the flow properties at the throat are constant for all conditions in Regime II, the flowrate in Regime II does not vary with back pressure.

In Regime III, as exemplified by condition (*viii*), the back pressure is higher than the exit pressure but not sufficiently high to sustain a normal shock in the exit plane. The flow adjusts to the back pressure through a series of oblique compression shocks that cannot be treated by one-dimensional theory.

As was previously noted in Section 10-3.5, condition (*iv*) represents the design condition[4]. In Regime IV the flow adjusts to the lower back pressure through a series of oblique expansion waves that cannot be treated by one-dimensional theory.

Summary Objectives

After completing study of Chapter 10, you should be able to do the following:

1. Write the basic equations for the steady, one-dimensional isentropic flow of (a) any compressible fluid, and (b) an ideal gas, through a channel of arbitrary cross section. Utilize these equations, together with the appropriate Ts plot, in the solution of isentropic flow problems.

2. Determine the effect of area change on fluid properties for isentropic flow. Sketch flow passages for the following: subsonic nozzle, subsonic diffuser, supersonic nozzle, and supersonic diffuser.

---

[4] Flow behavior in converging-diverging nozzles and applications to supersonic wind tunnels are shown in the film, *Channel Flow of a Compressible Fluid*, D. Coles, principal.

3. For flow in (i) a converging nozzle and (ii) a converging-diverging nozzle, plot the pressure distribution through the nozzle as a function of the nozzle back pressure. State the conditions under which the nozzle is choked.

4. Write the basic equations for the steady, one-dimensional, adiabatic flow of an ideal gas with constant specific heats through a constant area duct. Utilize these equations, together with the appropriate Ts plot, in the solution of Fanno line flow problems.

5. Write the basic equations for the steady, one-dimensional, frictionless flow of an ideal gas with heat transfer through a constant area duct. Utilize these equations, together with the appropriate Ts plot, in the solution of Rayleigh line flow problems.

6. Write the basic equations for the steady, one-dimensional flow of an ideal gas through a normal shock. Utilize these equations, together with the appropriate Ts plot, in the solution of normal shock problems.

**7. Utilize the tables for computation of (i) isentropic, (ii) Fanno line, (iii) Rayleigh line, and (iv) normal shock flow of an ideal gas.

8. Solve those problems at the end of the chapter that relate to the material you have studied.

problems

10.1 Steam flows steadily and isentropically through a nozzle. At section 1 the steam is at 700 F and 240 psia, and the flow velocity is 600 ft/sec. The steam is to be accelerated to the maximum possible velocity such that the exit stream quality is 1.0. Determine the corresponding exit velocity.

10.2 Steam flows steadily and isentropically through a nozzle. At section ① the steam is at 700 F and 240 psia, and the flow velocity is 600 ft/sec. The steam is to be accelerated to the maximum possible velocity such that the exit stream quality is 0.9. Determine the corresponding exit velocity.

10.3 Steam flows steadily and isentropically through a nozzle. At section ① the steam is at 350 C and an absolute pressure of 1.0 MPa, and the flow velocity is 200 m/sec. The steam is to be accelerated to the maximum possible velocity such that the exit stream quality is 1.0. Determine the corresponding exit velocity.

10.4 Steam flows steadily and isentropically through a nozzle. At section ① the steam is at 350 C and an absolute pressure of 1.0 MPa, and the flow velocity is 200 m/sec. The steam is to be accelerated to the maximum possible velocity such that the exit stream quality is 0.9. Determine the corresponding exit velocity.

10.5 Steam flows steadily and isentropically through a nozzle. At the upstream section where the flow velocity is negligible, the temperature and pressure are 900 F and 900 psia, respectively. At section ① where the nozzle diameter

** This objective applies to sections that may be omitted without loss of continuity in the text material.

is 0.188 in., the steam pressure is 600 psia. Determine the velocity and Mach number at section ① and the mass flowrate of steam. Sketch the passage shape.

10.6 Steam flows steadily and isentropically through a nozzle. At the upstream section where the flow velocity is negligible, the temperature and pressure are 880 F and 875 psia, respectively. At section ① where the nozzle diameter is 0.50 in., the steam pressure is 290 psia. Determine the velocity and Mach number at section ① and the mass flowrate of steam. Sketch the passage shape.

10.7 Steam flows steadily and isentropically through a nozzle. At the upstream section where the flow velocity is negligible, the temperature and absolute pressure are 475 C and 6.0 MPa, respectively. At section ① where the nozzle diameter is 6 mm, the absolute pressure of the steam is 4.0 MPa. Determine the velocity and Mach number at section ① and the mass flowrate of steam. Sketch the passage shape.

10.8 Steam flows steadily and isentropically through a nozzle. At the upstream section where the flow velocity is negligible, the temperature and absolute pressure are 475 C and 6.0 MPa, respectively. At section ① where the nozzle diameter is 12 mm, the absolute pressure of the steam is 2.0 MPa. Determine the velocity and Mach number at section ① and the mass flowrate of steam. Sketch the passage shape.

10.9 A nozzle is designed to expand air isentropically to atmospheric pressure from a large tank in which properties are held constant at 5 C and 304 kPa (abs). The desired flowrate is 1 kg/sec. Determine the exit area of the nozzle. Sketch a plot of Mach number and pressure as a function of distance along the nozzle.

10.10 At a section upstream of the throat in a converging-diverging nozzle the flow is at a pressure of 30 psia, and a temperature of 90 F; its velocity is 575 ft/sec. For the isentropic flow of air determine the Mach number at the point where the pressure is 12 psia.

10.11 Air flows steadily and isentropically into an aircraft inlet at a rate of 100 kg/sec. At the section where the area is 0.464 m², $M = 3$, $T = -60$ C and the absolute pressure is 15.0 kPa. Determine the velocity and cross-sectional area downstream where $T = 138$ C. Sketch the flow passage.

10.12 Air flows steadily and isentropically through a passage. At section ① where the cross-sectional area is 0.02 m², the air is at 40.0 kPa (abs), 60 C, and the Mach number is 2.0. At a section ② downstream, the velocity is 519 m/sec. Calculate the Mach number at section ②. Sketch the shape of the passage between sections ① and ②.

10.13 Air at an absolute pressure of 60.0 kPa and 27 C enters a passage at 486 m/sec. The cross-sectional area at the entrance is 0.02 m². At section ②, further downstream, the pressure is 78.8 kPa (abs). Assuming isentropic flow, calculate the Mach number at section ②.

10.14   A supersonic diffuser decelerates air isentropically from a Mach number of 3 to a Mach number of 1.4. If the static pressure at the diffuser inlet is 30.0 kPa (abs), calculate the static pressure rise in the diffuser, and the ratio of inlet to outlet area of the diffuser.

10.15   Air, with a stagnation temperature of 600 R, flows isentropically through a converging nozzle. At the point in the flow where the temperature is 571 R, the pressure is 100 psia. Determine the velocity and the stagnation pressure at the downstream position where $T = 532$ R.

10.16   Air flows isentropically through a converging nozzle into a receiver where the pressure is 33 psia. If the pressure is 50 psia, and the velocity is 500 ft/sec at the nozzle location where the Mach number is 0.4, determine the pressure, velocity, and Mach number at the nozzle throat.

10.17   Air flows isentropically through a converging nozzle into a receiver in which the absolute pressure is 240 kPa. The air enters the nozzle with negligible velocity at a pressure of 406 kPa (abs) and a temperature of 95 C. Determine the flowrate through the nozzle for a nozzle throat area of 0.01 m².

10.18   Air flows isentropically through a converging nozzle attached to a large tank where the absolute pressure is 171 kPa and the temperature is 27 C. At the inlet section the Mach number is 0.2. The nozzle discharges to the atmosphere; the discharge area is 0.015 m². Determine the magnitude and direction of the force that must be applied to hold the nozzle in place.

10.19   A stream of air flowing in a duct (area = 1 in.²) is at a pressure of 20 psia, has a Mach number of 0.6, and flows at a rate of 0.5 lbm/sec.
(a)   Determine the local isentropic stagnation temperature.
(b)   If the cross-sectional area of the passage were reduced downstream, determine the maximum percentage reduction of area allowable without reducing the flowrate (assume isentropic flow).
(c)   Determine the velocity and pressure at the minimum area location.

10.20   Air flowing isentropically through a converging nozzle discharges to the atmosphere. At the section where the absolute pressure is 179 kPa, the temperature is 39 C and the air velocity is 177 m/sec. Determine the nozzle throat pressure.

10.21   Air flows from a large tank ($p = 650$ kPa (abs), $T = 550$ C) through a converging nozzle, with a throat area of 600 mm², and discharges to the atmosphere. Determine the mass rate of flow for isentropic flow through the nozzle.

10.22   A large tank supplies air to a convergent nozzle that discharges to atmospheric pressure. Assume the flow to be reversible and adiabatic.
(a)   For what range of tank pressures will the flow at the nozzle exit be sonic ($M = 1$)?
(b)   If the tank pressure is 600 kPa (abs) and the temperature is 600 K, what is the mass flowrate through the nozzle if the exit area is 1.29×10^{-3} m²?

10.23   Helium in a very large tank is maintained at 800 kPa (abs), 250 C. Helium leaves the tank steadily and isentropically through a converging nozzle that discharges to the atmosphere. The nozzle throat area is 0.002 m². Determine the helium density at the nozzle throat.

10.24   Carbon dioxide at 500 MPa (abs) and 150 C discharges isentropically from a large tank to the atmosphere through a converging nozzle whose throat area is 1.0 mm². Find the temperature at the exit, the pressure difference between nozzle exit and atmosphere and the mass flowrate.

10.25   Air, with a stagnation pressure of 650 kPa (abs) and a stagnation temperature of 350 K, is flowing isentropically through a converging nozzle. At the section in the nozzle where the area is 2.6×10^{-3} m² the Mach number is 0.5. The nozzle discharges to a back pressure of 270 kPa (abs). Determine the exit area of the nozzle.

10.26   A converging nozzle discharges helium to atmosphere. Conditions at section ① in the nozzle result in a pressure of 350 kPa (abs), a temperature of 20 C, and a velocity of 201 m/sec. The mass flowrate is to be 0.15 kg/sec. Assume the flow is frictionless and adiabatic and determine the Mach number at the nozzle exit, and the exit area of the nozzle.

10.27   A converging nozzle is bolted to the side of a large tank. Air inside the tank is maintained at a constant pressure of 50 psia and a temperature of 100 F. The inlet area of the nozzle is 10 in.² and the exit area is 1 in.² The nozzle discharges to the atmosphere. For isentropic flow in the nozzle determine the total force on the bolts and indicate whether the bolts are in tension or compression.

10.28   Consider the isentropic flow of helium through a converging-diverging wind tunnel nozzle. The stagnation pressure at the nozzle entrance is 700 kPa (abs) and the stagnation temperature is 60 C. At a section downstream of the throat the pressure is 528 kPa (abs) and the area is 1.2×10^{-3} m². Determine the Mach number, the temperature, the stagnation pressure at this section, and the mass flowrate.

10.29   A converging-diverging nozzle is attached to a very large tank of air in which the pressure is 20 psia and the temperature is 40 F. The nozzle exhausts to the atmosphere where the pressure is 14.7 psia. The exit area of the nozzle is 2 in.² What is the flowrate through the nozzle? Assume the flow to be isentropic.

10.30   Air enters a converging-diverging nozzle with negligible velocity at an absolute pressure of 1.0 MPa and a temperature of 60 C. If the flow is isentropic and the exit temperature is − 11 C, what is the Mach number at exit?

10.31   Methane gas enters a converging-diverging nozzle with negligible velocity at 1.10 MPa (abs) and a temperature of 50 C. For a frictionless, adiabatic process, what is the Mach number at a section where the absolute pressure is 500 kPa?

10.32   A converging-diverging nozzle with a throat area of 2 in.² is connected to a large tank in which air is kept at a pressure of 80 psia at a temperature of 60 F. If the nozzle is to operate at design conditions (flow is isentropic) and the ambient pressure outside the nozzle is 12.9 psia, calculate the exit area of the nozzle, and the mass flowrate.

10.33   Air is to be expanded through a converging-diverging nozzle by a frictionless adiabatic process from a pressure of 1.10 MPa (abs) and a temperature of 115 C to a pressure of 141 kPa (abs). Determine the throat and exit areas for a well-designed shockless nozzle if the mass flowrate is 2 kg/sec.

10.34   Air, at a stagnation pressure of 7.20 MPa (abs) and a stagnation temperature of 1100 K flows isentropically through a converging-diverging nozzle having a throat area of 0.01 m^2. Determine the velocity at the downstream section where the Mach number is 4.0 and the mass flowrate.

10.35   A large tank supplies helium to a converging-diverging nozzle. The pressure in the tank remains constant at 8.00 MPa. The tank temperature remains constant at 1000 K. The flow throughout the nozzle is isentropic. The nozzle is designed to discharge to atmospheric pressure with an exit Mach number, $M = 3.5$. The exit area of the nozzle is 100 mm^2. What is the mass flowrate through the nozzle?

10.36   Air flows isentropically through a converging-diverging nozzle attached to a large tank in which the pressure is 100 psia and the temperature is 500 R. The nozzle is operating at design conditions for which the nozzle exit pressure, p_e, is equal to the surrounding atmospheric pressure, p_a. The exit area of the nozzle, $A_e = 4.0$ in.^{2}

(a)   Calculate the flowrate through the nozzle.

(b)   If the temperature of the air in the tank is increased to 2000 R (all pressures remaining the same), how will the flowrate be affected?

10.37   Nitrogen at a pressure and temperature of 371 kPa (abs) and 400 K enters a nozzle with negligible velocity. The exhaust jet is directed against a large flat plate that is perpendicular to the jet axis. The flow leaves the nozzle at atmospheric pressure. The exit area is 0.003 m^2. Find the force required to hold the plate.

10.38   At a point upstream of the throat in a converging-diverging nozzle, the air velocity is 172 m/sec; $p = 200$ kPa (abs), and $T = 22$ C. The flow is isentropic and is supersonic at the nozzle exit. If the nozzle throat area is 0.01 m^2, determine the flowrate.

10.39   Air is flowing isentropically in a converging-diverging nozzle. At the section in the converging portion where the area is 1250 mm^2, the pressure is 600 kPa (abs), the temperature is 22 C, and the Mach number is 0.50. Determine the area in the diverging section where the Mach number is 2.

10.40   Air flows steadily and isentropically through a converging-diverging nozzle. At the throat the air is at 140 kPa (abs), 60 C. The throat cross-sectional area is 0.05 m^2. At a certain section in the diverging part of the nozzle, the pressure is 70.0 kPa (abs). Calculate the velocity and the area at this section.

10.41   A small, solid fuel rocket motor is tested on a thrust stand. The chamber pressure and temperature are 600 psia and 6000 R, respectively. The propulsion nozzle is designed to expand the exhaust gases isentropically to a back pressure of 10.0 psia. The nozzle exit area is 0.60 ft^2. The gases may be treated as an ideal gas with $k = 1.2$ and $R = 60.0$ ft·lbf/lbm·R. Determine the mass flowrate of propellant gases and the thrust produced by the motor, that is, the force exerted against the test stand.

10.42 A liquid rocket motor is fueled with hydrogen and oxygen. The chamber temperature and absolute pressure are 3300 K and 6.90 MPa, respectively. The nozzle is to be designed to expand the exhaust gases isentropically to a design back pressure corresponding to an altitude of 10 km on a standard day. The thrust produced by the motor is to be 100 kN at the design conditions. Treat the exhaust gases as water vapor and assume ideal gas behavior. Determine the propellant mass flowrate needed to produce the desired thrust, the nozzle exit area, and the area ratio, A_e/A_t.

10.43 Air is flowing through a constant area duct. At section ① the temperature, $T_1 = 550$ K, the pressure, $p_1 = 973$ kPa (abs), and the Mach number, $M_1 = 0.20$. At section ② in the duct the following information is given: $T_2 = 1100$ K, $p_2 = 910$ kPa (abs), $M_2 = 0.30$.
(a) Is the flow isentropic? Justify your answer.
(b) Calculate the stagnation enthalpy, stagnation temperature, and stagnation pressure at sections ① and ②.

10.44 Air flows steadily through a constant area duct. At section ①, the air is at 60 psia, 600 R, with a velocity of 500 ft/sec. As a result of heat transfer and friction, the air at section ② downstream is at 40 psia, 800 R. Calculate the heat transfer per pound of air between sections ① and ②, and the stagnation pressure at section ②.

10.45 Air flows adiabatically with friction through a duct of 0.3 m square cross section. At one section, the pressure, temperature, and velocity are 60.0 kPa (abs), 50 C, and 180 m/sec, respectively. At the exit from the duct, the pressure and temperature are, respectively, 31.7 kPa (abs) and 19 C. Calculate the exit Mach number and the stagnation temperature midway between the first section and the exit.

10.46 Air flows steadily and adiabatically in a horizontal, 50 mm diameter pipe. At section ① the pressure is 340 kPa (abs) and the temperature is 53 C. At section ② (downstream) the temperature is 114 C and the velocity is 591 m/sec. Determine the velocity and Mach number at section ①.

10.47 Air is flowing through a well-insulated, constant area channel. At a section where the velocity is 144 m/sec, $T = 50$ C and $p = 600$ kPa (abs). Determine the temperature and stagnation pressure at the downstream section where the density is 3.05 kg/m³, and the entropy increase.

10.48 Air flows steadily and adiabatically from a large tank through a converging nozzle connected to a constant area duct. The nozzle itself may be considered frictionless. Air in the tank is at $p = 1.00$ MPa (abs), $T = 125$ C. The absolute pressure at the nozzle exit (duct inlet) is 784 kPa. Determine the pressure at the end of the duct length, L, if the temperature there is 65 C, and the entropy increase.

10.49 Consider compressible flow in a long straight duct. The inlet to the duct is from the atmosphere where $T = 25$ C. The flow is considered to be adiabatic and the pipe is sufficiently long that the flow is choked. Find the velocity and temperature at the pipe exit.

10.50 Air flows steadily and adiabatically through a constant area duct. The density at the inlet is 3.51 kg/m^3. If the exit Mach number is unity and the exit temperature and pressure are 227 C and 110 kPa (abs), respectively, determine the inlet Mach number and the entropy change from inlet to exit.

10.51 A converging-diverging nozzle supplies air to a well-insulated, constant area duct. At the inlet to the duct, $M = 2.0$, $p = 19.4$ psia, and $T = 278$ R. At the duct exit the Mach number is unity and the stagnation pressure is 90 psia. Determine the pressure and temperature at the duct exit and the entropy change.

10.52 Consider adiabatic flow in a constant area pipe with friction. At one section of the pipe, $p_0 = 100$ psia, $T_0 = 500$ R, and $M = 0.70$. If the cross-sectional area of the pipe is 1 ft^2 and the Mach number at the exit is $M_2 = 1$, find the friction force exerted on the fluid by the pipe.

10.53 A converging-diverging nozzle discharges air into an insulated pipe with area, $A = 650$ mm^2. At the pipe inlet, $p = 128$ kPa (abs), $T = 39$ C, and $M = 2.00$. For a shockless flow to a Mach number of unity at the pipe exit, calculate the exit temperature, the net force of the fluid on the pipe and the entropy change.

10.54 Air flows through a smooth well-insulated 4 in. diameter pipe at a rate of 600 lbm/min. At one section the air is at 100 psia, 80 F. Determine the minimum pressure and the maximum velocity that can occur in the pipe.

*10.55 For the conditions of Problem 10.48, find the length, L, of commerical steel pipe of 50 mm diameter between sections ① and ②.

*10.56 Solve Problem 10.49 for the duct length, L, if the duct diameter is 12 mm and the surface is smooth. At the duct inlet, $p = 94.1$ kPa (abs).

*10.57 The duct of Problem 10.51 has relative roughness, $e/D = 0.0002$, and has a hydraulic diameter of 0.5 ft. Determine the duct length, L.

*10.58 For the conditions of Problem 10.52, determine the duct length. Assume that the duct is circular and made from commerical steel. Plot the variations of pressure and Mach number versus distance along the duct.

*10.59 For the conditions of Problem 10.53, determine the duct length. Assume that the duct is circular and made from commerical steel. Plot the variations of pressure and Mach number versus distance along the duct.

*10.60 Air is flowing in an insulated duct with a velocity of 140 m/sec. The temperature and absolute pressure are 200 C and 2.00 MPa, respectively.
(a) Find the temperature in this duct where the pressure has dropped to 1.26 MPa (abs) as a result of friction.

* Problems marked with an asterisk require use of Tables.

(b)   If the duct ($e/D = 0.0003$) has a diameter of 150 mm, find the distance between the two points.

*10.61   Consider the flow described in Example Problem 10.8. Using the tables for Fanno line flow of an ideal gas, plot the static pressure, temperature and Mach number versus L/D measured from the tube inlet; continue until the choked state is reached.

*10.62   Using coordinates T/T_0 and $(s - s^*)/c_p$, where s^* is the entropy at the condition where $M = 1$, plot the Fanno line starting from the inlet conditions specified in Example Problem 10.8. Use tables for Fanno line flow, and proceed to the section where $M = 1$.

*10.63   Using coordinates T/T^* and $(s - s^*)/c_p$, where s^* is the entropy at the condition where $M = 1$, plot the Fanno line for air flow ($k = 1.4$) for Mach numbers in the range $0.1 < M < 3.0$.

*10.64   Beginning with the inlet conditions of Problem 10.53, and using coordinates T/T_0 and $(s - s^*)/c_p$, plot the supersonic and subsonic branches of the Fanno line for the flow.

*10.65   Consider the frictionless flow of air in a constant area duct. At section ①, $M_1 = 0.50$, $p_1 = 1.10$ MPa (abs), and $T_{0_1} = 333$ K. Through the effect of heat transfer, the Mach number at section ② is $M_2 = 0.90$ and the stagnation temperature, $T_{0_2} = 478$ K.
(a)   Determine the amount of heat transfer per unit mass to or from the fluid between sections ① and ②.
(b)   Determine the pressure difference, $p_1 - p_2$.

10.66   Air is flowing in a constant area duct without friction. At section ① in the duct, $p_1 = 97.3$ psia and $T_1 = 992$ R, and the velocity is $V_1 = 309$ ft/sec. Through the effect of heat transfer the velocity at section ② downstream is $V_2 = 652$ ft/sec. Determine the pressure, temperature, stagnation pressure, and stagnation temperature at section ②, and the heat transfer per unit mass between sections ① and ②.

10.67   At a position 3 m from the exit of a constant area duct (area $= 0.02$ m²), air is at an absolute pressure of 126 kPa and a temperature of 260 C. The air flows steadily through the duct (without friction) at the rate of 1.83 kg/sec. The air leaves the duct subsonically at atmospheric pressure. Determine the Mach number, temperature, and stagnation temperature at the exit of the duct and the heat transfer over the 3 m of duct length.

10.68   A constant area duct is fed with air from a converging-diverging nozzle. At the entrance to the duct the following properties are known: $p_{0_1} = 800$ kPa (abs), $T_{01} = 700$ K, and $M_1 = 3.0$. A short distance down the duct (at section ②) the density is 0.334 kg/m³. Assuming frictionless flow, determine the velocity, pressure and Mach number at section ②, and the heat transfer between the inlet and section ②.

---

*   Problems marked with an asterisk require use of Tables.

10.69 Air flows without friction through a short duct of constant area. At the duct entrance, $M_1 = 0.30$, $T_1 = 50$ C and $\rho_1 = 2.16$ kg/m^3. As a result of heating, the Mach number and density at the tube outlet are $M_2 = 0.60$ and $\rho_2 = 0.721$ kg/m^3. Determine the heat transfer per unit mass and the entropy change for the process.

10.70 Air flows at the rate of 1.42 kg/sec through a duct of 100 mm diameter. At the inlet section the temperature and absolute pressure are 52 C and 60.0 kPa, respectively. At the section downstream where the flow is choked, $T_2 = 45$ C. Determine the heat transfer per unit mass, the entropy change, and the change in stagnation pressure for the process, assuming frictionless flow.

10.71 Air flows without friction through a duct of 75 mm diameter. At the duct inlet, the Mach number, stagnation temperature and absolute pressure are $M_1 = 0.2$, $T_{0_1} = 278$ K, and $p_1 = 275$ kPa. Heat is added at the rate of 219 kJ/kg of flowing fluid, with the result that the temperature at the duct outlet is 489 K. Determine the stagnation temperature and Mach number at the duct outlet, and the entropy change for the process.

10.72 Air flows steadily and without friction in a duct of 0.5 ft^2 cross-sectional area. The inlet conditions are $M_1 = 0.30$, $T_{0_1} = 400$ R, and $p_{0_1} = 12.0$ psia. The exit stagnation temperature and pressure are 797 R and 9.43 psia, respectively. Determine the rate of heat transfer and the exit Mach number.

*10.73 Air flows without friction in a constant area duct. At section ①, $M_1 = 0.50$, $p_1 = 1.10$ MPa (abs), and $T_{0_1} = 335$ K. Through the effect of heat transfer, the Mach number is raised to $M_2 = 0.9$ at section ②. Determine the heat transfer per unit mass between sections ① and ②, and the pressure, p_2.

*10.74 Air flows without friction in a constant area duct. The properties at section ① are $T_1 = 992$ R and $p_1 = 97.3$ psia, and the velocity is $V_1 = 309$ ft/sec. Heat transfer causes the velocity to increase to 652 ft/sec at section ②. Determine the Mach number, temperature, pressure, stagnation temperature, and stagnation pressure at section ②.

*10.75 Air flows steadily and without friction at 1.83 kg/sec through a duct with cross-sectional area of 0.02 m^2. At the duct inlet, the temperature and absolute pressure are 260 C and 126 kPa, respectively. The exit flow discharges subsonically to atmospheric pressure. Determine the Mach number, temperature, and stagnation temperature at the duct outlet, and the heat transfer rate.

*10.76 A converging-diverging nozzle feeds a short duct of constant area. At the duct inlet, the local isentropic stagnation conditions are $T_{0_1} = 700$ K and $p_{0_1} = 800$ kPa (abs), and the Mach number is $M_1 = 3.0$. A short distance down the duct at section ②, the density is $\rho_2 = 0.334$ kg/m^3. Assuming frictionless flow in the duct, determine the velocity, pressure and Mach number at section ②, and the heat transfer between the inlet and section ②.

*10.77 Air flows without friction in a short section of constant area duct. At the duct inlet, $M_1 = 0.30$, $T_1 = 50$ C, and $\rho_1 = 2.16$ kg/m^3. At the duct outlet, $M_2 =$

* Problems marked with an asterisk are designed to be solved using Tables.

0.60. Determine the heat transfer per unit mass, the entropy change, and the change in stagnation pressure for the process.

*10.78 In the frictionless flow of air through a 100 mm diameter duct, 1.42 kg/sec enters at a temperature of 52 C and an absolute pressure of 60.0 kPa. Determine the amount of heat that must be added to choke the flow, and the fluid properties at the choked state.

*10.79 Air flows without friction through a duct of 75 mm diameter. The inlet Mach number, stagnation temperature, and absolute pressure are $M_1 = 0.20$, $T_{0_1} = 278$ K, and $p_1 = 275$ kPa, respectively. Heat is added at a rate equivalent to 219 kJ/kg of flowing air. Determine the stagnation temperature and Mach number at the duct outlet, and the changes in entropy and stagnation pressure for the process.

*10.80 Air flows steadily in a constant area, frictionless duct of 0.5 ft^2 cross-sectional area. The inlet conditions are $M_1 = 0.3$, $T_{0_1} = 400$ R, $p_{0_1} = 10$ psia, and the exit stagnation temperature, $T_{0_2} = 797$ R. Determine the amount of heat transfer, and the exit Mach number and pressure.

*10.81 Using coordinates T/T^* and $(s - s^*)/c_p$, where s^* is the entropy at the condition where $M = 1$, plot the Rayleigh line for air flow ($k = 1.4$) for Mach numbers in the range $0.4 < M < 3.0$.

*10.82 Beginning with the inlet conditions of Problem 10.53, and using coordinates T/T_{0_1} and $(s - s^*)/c_p$, plot the supersonic and subsonic branches of the Rayleigh line for the flow.

10.83 In long, constant area pipelines, such as those used for natural gas, the temperature may be considered constant. Assume gas leaves a pumping station at 50 psia and 70 F with a Mach number of 0.10 and density, $\rho = 0.255$ lbm/ft^3. At the section along the pipe where the pressure has dropped to 20 psia:
(a) Calculate the Mach number of the flow.
(b) Is heat added to or removed from the gas over the length between the pressure taps? Justify your answer.
(c) Sketch the process on a Ts diagram. Indicate (qualitatively) T_{0_1}, T_{0_2}, and p_{0_2}.

10.84 Natural gas (molecular mass, $M_m = 18$, $k = 1.3$) is to be pumped through a 36 in. i.d. pipe connecting two compressor stations 40 miles apart. At the upstream station the pressure is not to exceed 90 psig, and at the downstream station it is to be at least 10 psig. Calculate the maximum allowable rate of flow (ft^3/day at 70 F and 1 atm) assuming that there is sufficient heat transfer through the pipe to maintain the gas at 70 F.

10.85 An air stream with temperature, $T_1 = 0$ C, absolute pressure, $p_1 = 60.0$ kPa, and velocity, $V_1 = 497$ m/sec, undergoes a normal shock. The temperature downstream from the shock is 87 C. Determine the Mach number, velocity, and stagnation pressure downstream from the shock.

* Problems marked with an asterisk are designed to be solved using Tables.

10.86 Air with stagnation temperature, $T_{0_1} = 333$ K, and stagnation pressure, $p_{0_1} = 600$ kPa (abs), approaches a normal shock at Mach number, $M_1 = 2.0$. The velocity downstream from the shock is $V_2 = 204$ m/sec. Determine the static pressure downstream from the shock.

10.87 Air approaches a normal shock with velocity, $V_1 = 951$ m/sec. The stagnation temperature and absolute pressure of the air stream are $T_{0_1} = 700$ K and $p_1 = 125$ kPa, respectively; the Mach number is $M_1 = 3.0$. The absolute pressure downstream from the shock is 1.29 MPa. Determine the downstream velocity and temperature.

10.88 A normal shock stands in a constant area duct. Air approaches the shock with $T_{0_1} = 1000$ R and $p_{0_1} = 100$ psia, at a Mach number of 3.0. The Mach number downstream from the shock is known to be $M_2 = 0.475$. Determine the static pressure downstream from the shock.

10.89 Air undergoes a normal compression shock. The upstream temperature, absolute pressure, and velocity are $T_1 = 35$ C, $p_1 = 229$ kPa, and $V_1 = 704$ m/sec, respectively. The density downstream from the shock is $\rho_2 = 6.91$ kg/m³. Determine the temperature and stagnation pressure of the air stream leaving the shock.

10.90 An air stream approaches a normal shock at Mach number, $M_1 = 2.64$. The upstream stagnation pressure and density are $p_{0_1} = 3.00$ MPa (abs) and $\rho_1 = 1.65$ kg/m³, respectively. The ratio of static pressure to stagnation pressure immediately behind the shock is 0.843. Determine the downstream Mach number and temperature.

10.91 A normal shock occurs in air at a section where the flow velocity is 924 m/sec. The temperature and absolute pressure at this point are $T_1 = 10$ C and $p_1 = 35.0$ kPa. The absolute pressure downstream from the shock is $p_2 = 301$ kPa. Determine the velocity and Mach number downstream from the shock.

10.92 Air approaches a normal compression shock with temperature, absolute pressure, and velocity of $T_1 = 18$ C, $p_1 = 101$ kPa, and $V_1 = 766$ m/sec, respectively. The temperature immediately downstream from the shock is $T_2 = 551$ K. Determine the velocity immediately downstream from the shock and the pressure change across the shock. Calculate the corresponding pressure change for a frictionless, shockless deceleration between the same velocities.

10.93 Air flows steadily through a long, insulated constant area pipe. At section ①, the Mach number is 2.0, and the temperature and pressure are 140 F and 35.9 psia, respectively. At section ② downstream from a normal shock (the shock stands in the duct between sections ① and ②), the velocity is 1080 ft/sec. Determine the density and Mach number at section ②. Make a qualitative sketch of the pressure distribution along the pipe.

*10.94 An air stream with temperature, $T_1 = 0$ C, absolute pressure, $p_1 = 60.0$ kPa, and velocity, $V_1 = 497$ m/sec, undergoes a normal shock. Determine the Mach number, velocity, and stagnation pressure downstream from the shock.

*10.95 Air with stagnation temperature, $T_{0_1} = 333$ K, and stagnation pressure, $p_{0_1} = 600$ kPa (abs), approaches a normal shock at Mach number, $M_1 = 2.0$. Determine the static pressure downstream from the shock, and the decrease in stagnation pressure across the shock.

*10.96 Air approaches a normal shock with velocity, $V_1 = 951$ m/sec. The stagnation temperature and absolute pressure of the air stream are $T_{0_1} = 700$ K and $p_1 = 125$ kPa, respectively; the Mach number is $M_1 = 3.0$. Determine the velocity and temperature of the air leaving the shock, and the entropy change across the shock.

*10.97 A normal shock stands in a constant area duct. Air approaches the shock with $T_{0_1} = 1000$ R, $p_{0_1} = 100$ psia, at a Mach number of 3.0. Determine the static pressure downstream from the shock. Compare the downstream pressure with the value that would be reached by decelerating isentropically to the same subsonic Mach number.

*10.98 Air undergoes a normal compression shock. The upstream temperature, absolute pressure, and velocity are $T_1 = 35$ C, $p_1 = 229$ kPa, and $V_1 = 704$ m/sec, respectively. Determine the temperature and stagnation pressure of the air stream leaving the shock.

*10.99 An air stream approaches a normal shock at Mach number, $M_1 = 2.64$. The upstream stagnation pressure and density are $p_{0_1} = 3.00$ MPa (abs) and $\rho_1 = 1.65$ kg/m^3, respectively. Determine the downstream Mach number and temperature, and the entropy change cross the shock.

*10.100 A normal shock occurs in air at a section where the flow velocity is 924 m/sec. The temperature and absolute pressure at this point are $T_1 = 10$ C and $p_1 = 35.0$ kPa. Determine the velocity and Mach number downstream from the shock, and the change in stagnation pressure across the shock.

*10.101 Air approaches a normal compression shock with temperature, absolute pressure, and velocity of $T_1 = 18$ C, $p_1 = 101$ kPa, and $V_1 = 766$ m/sec, respectively. Determine the velocity immediately downstream from the shock, and the pressure change across the shock. Calculate the corresponding pressure change for a frictionless, shockless deceleration between the same velocities.

*10.102 Air flows steadily through a long insulated pipe of 2 in. diameter. At section ①, the Mach number is 2.0, and the temperature and pressure are 140 F and 35.9 psia, respectively. A normal shock occurs at section ② in the duct where $M_2 = 1.88$. At section ④, some distance downstream from the shock, the velocity is 1080 ft/sec. Determine the flow properties at sections ② and ③, immediately upstream and downstream from the shock, and the Mach

---

* Problems marked with an asterisk are designed to be solved using Tables.

problems *605*

number and pressure at section ④. Determine the length of the pipe and plot the pressure distribution along the pipe.

*10.103 A supersonic aircraft flies at $M = 2.7$ at an altitude of 20 km on a standard day. Air approaching a forward-facing total head (pitot) tube passes through a normal shock and then is brought isentropically to rest relative to the aircraft. Determine the final temperature of the air and the pressure sensed by the total head tube.

*10.104 The Concorde supersonic transport flies at $M = 2.2$ at an altitude of 20 km. Air is decelerated isentropically by the engine inlet system to a local Mach number of 1.3. The air passes through a normal shock and is decelerated further to $M = 0.4$ at the engine compressor section. Assume as a first approximation that this subsonic diffusion process is isentropic, and use standard atmosphere data for freestream conditions. Determine the temperature, pressure, and stagnation pressure of the air entering the engine compressor.

*10.105 A supersonic aircraft flies at an altitude of 22 km where the ambient pressure is 4.05 kPa on a standard day. A forward-facing stagnation tube senses a total pressure of 55.3 kPa. Determine the Mach number and speed of the aircraft.

*10.106 The subsonic portion of the inlet diffuser described in Problem 10.104 has an isentropic efficiency of 0.97; the actual static pressure rise is 97 percent of the value that could be obtained from an isentropic deceleration between the same initial and final velocities. The supersonic diffusion may be assumed to be isentropic. The entire flow is adiabatic. Determine the overall reduction in stagnation pressure for the inlet flow and the static pressure of the air stream entering the engine compressor at $M = 0.4$.

10.107 Air flows adiabatically from a reservoir, where the temperature and absolute pressure are 60 C and 600 kPa, through a converging-diverging nozzle. The design Mach number of the nozzle is 2.94. A normal shock occurs at the location in the nozzle where $M = 2.42$. (The Mach number immediately behind the shock is $M = 0.521$.) Assuming isentropic flow before and after the shock, determine the back pressure downstream from the nozzle, if the temperature there is 54.6 C. Sketch the pressure distribution.

10.108 A normal shock occurs in the diverging section of a converging-diverging nozzle at the location where the area is 4.0 in.^{2} and the local Mach number is 2.50. (The Mach number immediately downstream from the shock is 0.513.) The stagnation conditions for the upstream flow are $T_0 = 1000$ R and $p_0 = 100$ psia. The nozzle exit area is 6.0 in.^{2} and the exit temperature is 981 R. Assume that the flow is isentropic except across the shock. Determine the nozzle exit pressure, throat area, and mass flowrate.

10.109 A converging-diverging nozzle is designed to expand air isentropically to atmospheric pressure from a large tank where the temperature and absolute pressure are 150 C and 790 kPa. A normal shock stands in the diverging

---

* Problems marked with an asterisk are designed to be solved using Tables.

section at a location where the absolute pressure is 160 kPa and the cross-sectional area is 600 mm^2. The Mach number immediately behind the shock is 0.641. Determine the nozzle back pressure, exit area, and throat area.

10.110 Air flows through a converging-diverging nozzle designed to give an exit Mach number, $M = 2.80$. The upstream stagnation conditions are atmospheric; the back pressure is maintained by a vacuum pump. Determine the back pressure required to cause a normal shock to stand in the exit plane and the flow velocity after the shock. (The pressure ratio across a normal shock occurring at $M = 2.80$ is 8.98.)

*10.111 A converging-diverging nozzle with throat area, $A_t = 1.0$ in.^{2}, is attached to a large tank in which the pressure and temperature are maintained at 100 psia and 600 R. The nozzle exit area is 1.58 in.^{2}. Determine the exit Mach number at design conditions. Referring to Fig. 10.20, determine the back pressures corresponding to the boundaries of Regimes I, II, III, and IV. Sketch the corresponding plot for this nozzle.

*10.112 A converging-diverging nozzle with area ratio, $A_e/A_t = 4.0$, is designed to expand air isentropically to atmospheric pressure. Determine the exit Mach number at design conditions, and the required value of inlet stagnation pressure. Referring to Fig. 10.20, determine the back pressures that correspond to the boundaries of Regimes I, II, III, and IV. Sketch the plot of pressure ratio versus axial distance for this nozzle.

*10.113 Air flows adiabatically from a reservoir, where the temperature and absolute pressure are 60 C and 600 kPa, through a converging-diverging nozzle of area ratio, $A_e/A_t = 4.0$. A normal shock occurs at the location in the nozzle where $M = 2.42$. Assuming isentropic flow before and after the shock, determine the back pressure downstream from the nozzle. Sketch the pressure distribution.

*10.114 A normal shock occurs in the diverging section of a converging-diverging nozzle at the location where the area is 4.0 in.^{2} and the local Mach number is 2.50. The stagnation conditions for the upstream flow are $T_0 = 1000$ R and $p_0 = 100$ psia. The nozzle exit area is 6.0 in.^{2}. Assume that the flow is isentropic except across the shock. Determine the nozzle exit pressure, throat area, and mass flowrate.

*10.115 A converging-diverging nozzle is designed to expand air isentropically to atmospheric pressure from a large tank where the temperature and absolute pressure are 150 C and 790 kPa. A normal shock stands in the diverging section at a location where the pressure is 160 kPa (abs) and the cross-sectional area is 600 mm^2. Determine the nozzle back pressure, exit area, and throat area.

*10.116 A converging-diverging nozzle with a design pressure ratio of $p_e/p_0 = 0.1278$ is operated with a back pressure condition such that $p_b/p_0 = 0.830$, causing

*   Problems marked with an asterisk are designed to be solved using Tables.

a normal shock to stand in the diverging section. Determine the Mach number at which the shock occurs.

*10.117 Air flows through a converging-diverging nozzle with area ratio, $A_e/A_t = 3.5$. The upstream stagnation conditions are atmospheric; the back pressure is maintained by a vacuum pump. Determine the back pressure required to cause a normal shock to stand in the nozzle exit plane, and the flow velocity leaving the shock.

*10.118 Air flows through a converging-diverging nozzle with area ratio, $A_e/A_t = 3.5$. The upstream stagnation conditions are atmospheric; the back pressure is maintained by a vacuum system. Determine the range of back pressures for which a normal shock will occur within the nozzle, and the corresponding mass flowrate if $A_t = 500$ mm^2.

*10.119 Air flows through a converging-diverging nozzle with area ratio, $A_e/A_t = 1.87$. The upstream stagnation conditions are $T_{0_1} = 240$ F and $p_{0_1} = 100$ psia. The back pressure is maintained at 40 psia. Determine the Mach number and flow velocity in the nozzle exit plane.

*10.120 A converging-diverging nozzle with area ratio, $A_e/A_t = 1.633$ is designed to operate with atmospheric pressure at the exit plane. Determine the range(s) of stagnation pressures for which the nozzle will be free from normal shocks.

---

* Problems marked with an asterisk are designed to be solved using Tables.

Fluid Property Data

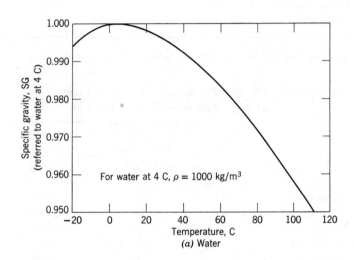

(a) Water

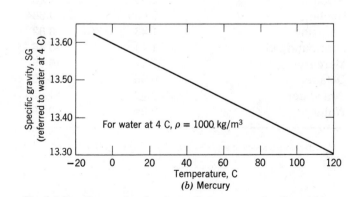

(b) Mercury

Fig. A.1 Specific gravity of water and mercury as functions of temperature (data from Ref. 1).

609

TABLE A.1 **Specific Gravities of Several Common Manometer Fluids at 20 C (data from Refs. 1, 2 and 3)**

Fluid	Specific Gravity*
E. V. Hill blue oil	0.797
Meriam red oil	0.827
Benzene	0.879
Dibutyl phthalate	1.04
Monochloronaphthalene	1.20
Carbon tetrachloride	1.595
Bromoethylbenzene (Meriam blue)	1.75
Tetrabromoethane	2.95
Mercury	13.55

* Specific gravity, SG $\equiv \rho/\rho_{H_2O}$ (at 4 C); ρ_{H_2O} (at 4 C) = 1000 kg/m^3 (1.94 slug/ft^3)

TABLE A.2 **Physical Properties of Common Liquids at 20 C (data from Refs. 1, 4 and 5)**

Liquid	Isentropic Bulk Modulus* (GN/m^2)	Specific Gravity (−)
Benzene	1.48	0.879
Carbon tetrachloride	1.36	1.595
Castor oil	2.11	0.969
Gasoline	—	0.72
Glycerin	4.59	1.26
Heptane	0.886	0.684
Kerosine	1.43	0.82
Lubricating oil	1.44	0.88
Mercury	28.5	13.55
Octane	0.963	0.702
Sea water	2.42	1.025
Water	2.24	0.998

* Calculated from speed of sound; 1 GN/m^2 = 10^9 N/m^2 (1 N/m^2 = 1.45 × 10^{-4} lbf/in.^{2})

SURFACE TENSION

The values of surface tension, σ, for most organic compounds are remarkably similar at room temperature: the typical range is 25 to 40 mN/m. Water is higher, at about 73 mN/m at 20 C. Liquid metals have values in the range between 300 and 600 mN/m; liquid mercury has a value of about 480 mN/m at 20 C. Surface tension decreases with temperature; the decrease is nearly

TABLE A.3   **Surface Tension of Common Liquids at 20 C (data from Refs. 1, 4–7)**

Liquid	Surface Tension, σ (mN/m)*	Contact Angle, θ (degrees)
(a) In contact with air		
Benzene	28.9	
Carbon tetrachloride	27.0	
Glycerin	63.0	
Hexane	18.4	
Kerosine	26.8	
Lube oil	25–35	
Mercury	484	140
Methanol	22.6	
Octane	21.8	
Water	72.8	~ 0
(b) In contact with water		
Benzene	35.0	
Carbon tetrachloride	45.0	
Hexane	51.1	
Mercury	375	140
Methanol	22.7	
Octane	50.8	

*   1 mN/m = 10^{-3} N/m

TABLE A.4 **Properties of the U.S. Standard Atmosphere (data from Ref. 8)**

Geometric Altitude (meters)	Temperature (K)	p/p_0 (—)	ρ/ρ_0 (—)
−500	291.4	1.061	1.049
0	288.2	1.000*	1.000†
500	284.9	0.9421	0.9529
1,000	281.7	0.8870	0.9075
1,500	278.4	0.8345	0.8638
2,000	275.2	0.7846	0.8217
2,500	271.9	0.7372	0.7812
3,000	268.7	0.6920	0.7423
3,500	265.4	0.6492	0.7048
4,000	262.2	0.6085	0.6689
4,500	258.9	0.5700	0.6343
5,000	255.7	0.5334	0.6012
6,000	249.2	0.4660	0.5389
7,000	242.7	0.4057	0.4817
8,000	236.2	0.3519	0.4292
9,000	229.7	0.3040	0.3813
10,000	223.3	0.2615	0.3376
11,000	216.8	0.2240	0.2978
12,000	216.7	0.1915	0.2546
13,000	216.7	0.1636	0.2176
14,000	216.7	0.1399	0.1860
15,000	216.7	0.1195	0.1590
16,000	216.7	0.1022	0.1359
17,000	216.7	0.08734	0.1162
18,000	216.7	0.07466	0.09930
19,000	216.7	0.06383	0.08489
20,000	216.7	0.05457	0.07258
22,000	218.6	0.03995	0.05266
24,000	220.6	0.02933	0.03832
26,000	222.5	0.02160	0.02797
28,000	224.5	0.01595	0.02047
30,000	226.5	0.01181	0.01503
40,000	250.4	0.002834	0.003262
50,000	270.7	0.0007874	0.0008383
60,000	255.8	0.0002217	0.0002497
70,000	219.7	0.00005448	0.00007146
80,000	180.7	0.00001023	0.00001632
90,000	180.7	0.000001622	0.000002588

* $p_0 = 1.01325 \times 10^5$ N/m² absolute (= 14.696 psia)
† $\rho_0 = 1.2250$ kg/m³ (= 0.002377 slug/ft³)

612

linear with absolute temperature. Surface tension at the critical temperature is zero.

Values of σ are usually reported for surfaces in contact with the pure vapor of the liquid being studied or with air. At low pressures both values are about the same.

THE PHYSICAL NATURE OF VISCOSITY

Viscosity is a measure of internal fluid friction, that is, resistance to deformation. The mechanism of gas viscosity is reasonably well understood, but the theory is poorly developed for liquids. We can gain some insight into the physical nature of viscous flow by discussing these mechanisms briefly.

The viscosity of a Newtonian fluid is fixed by the state of the material. Thus $\mu = \mu(T, p)$. Temperature is the more important variable, so let us consider it first. Excellent empirical equations for prediction of viscosity as a function of temperature are available.

a. EFFECT OF TEMPERATURE ON VISCOSITY

1. Gases

All gas molecules are in continuous random motion. When there is bulk motion due to flow, the bulk motion is superimposed on the random motions. It is then distributed throughout the fluid by molecular collisions. Analyses based on kinetic theory predict

$$\mu \propto \sqrt{T}$$

The kinetic theory prediction is in fair agreement with experimental trends, but the constant of proportionality and one or more correction factors must be determined. This limits practical application of this simple equation.

If two or more experimental datum points are available, the data may be correlated using the empirical Sutherland correlation

$$\mu = \frac{bT^{1/2}}{1 + S/T} \tag{A.1}$$

The constants b and S may be determined most simply by writing

$$\mu = \frac{bT^{3/2}}{S + T}$$

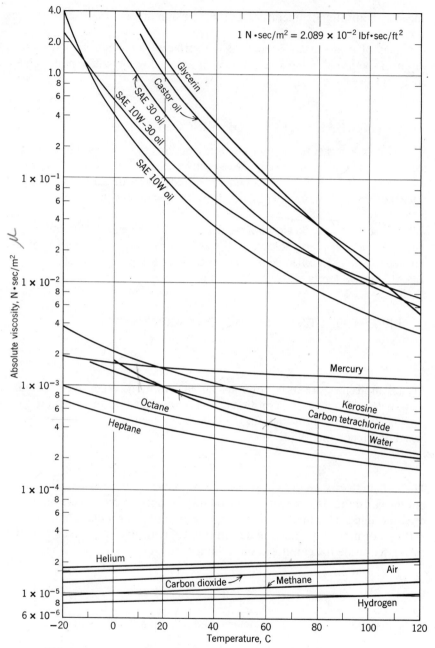

Fig. A.2 Dynamic (absolute) viscosity of common fluids as a function of temperature (data from Refs. 1, 5 and 9).

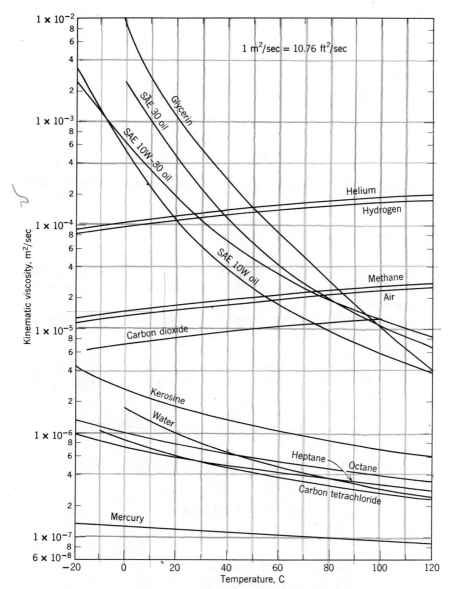

Fig. A.3 Kinematic viscosity of common fluids (at atmospheric pressure) as a function of temperature (data from Refs. 1, 5 and 9).

or

$$\frac{T^{3/2}}{\mu} = \left(\frac{1}{b}\right) T + \frac{S}{b}$$

(Compare this with $y = mx + c$.) From a plot of $T^{3/2}/\mu$ versus T, one obtains the slope $1/b$ and the intercept S/b. For air,

$$b = 1.458 \times 10^{-6} \frac{\text{kg}}{\text{m} \cdot \text{sec} \cdot \text{K}^{1/2}}$$

$$S = 110.4 \text{ K}$$

These values were used with Eq. A.1 to compute viscosities for the standard atmosphere in Ref. 8.

2. Liquids

Viscosities for liquids cannot be estimated well theoretically. The phenomenon of momentum transfer by molecular collisions seems overshadowed in liquids by the effects of the interacting force fields among the closely packed liquid molecules.

Liquid viscosities are affected drastically by temperature. This dependence on absolute temperature is well represented by the empirical equation

$$\mu = Ae^{B/T} \tag{A.2}$$

Equation A.2 requires two datum points to fit A and B. This may be done easily by plotting $\log \mu$ versus $1/T$ in the form

$$\log \mu = \log A + \frac{B}{T} \log e = \log A + \frac{0.434B}{T}$$

Log A may be identified as the intercept and $0.434B$ as the slope of the resulting plot.

b. EFFECT OF PRESSURE ON VISCOSITY

1. Gases

The viscosity of gases is essentially independent of pressure for values between a few hundredths of an atmosphere and a few atmospheres. However, viscosity at high pressures increases with pressure (or density).

2. Liquids

The viscosities of most liquids are not affected by moderate pressures, but large increases have been found at very high pressures. For example, the viscosity of water at 10,000 atm is twice the value at 1 atm. More complex compounds show a viscosity increase of several orders of magnitude over the same pressure range.

More information may be found in Reference 10.

LUBRICATING OILS

Engine and transmission lubricating oils are classified by viscosity according to standards established by the Society of Automotive Engineers (Ref. 11). The allowable ranges of viscosity for several grades are given in Table A.5.

Viscosity numbers with W (e.g. $20W$) are classified by viscosity at 0 F. Those without W are classified by viscosity at 210 F.

TABLE A.5 **Allowable Viscosity Ranges for SAE Lubricant Classifications (data from Ref. 11)**

| Lubricant Type | SAE Viscosity Number | Viscosity Range (centistokes)* | | | |
| | | At 0 F | | At 210 F | |
		Minimum	Maximum	Minimum	Maximum
Crankcase	5 W		1,200	3.9	
	10 W	1,200	2,400	3.9	
	20 W	2,400	9,600	3.9	
	20			5.7	9.6
	30			9.6	12.9
	40			12.9	16.8
	50			16.8	22.7
Transmission and axle	75		15,000		
	80	15,000	100,000		
	90			75	120
	140			120	200
	250			200	
Automatic transmission fluid	Type A	39†	43†	7	8.5

* 1 centistoke = 1 cSt = 10^{-6} m^2/sec (= 1.08×10^{-5} ft^2/sec)
† At 100 F.

TABLE A.6   Thermodynamic Properties of Common Gases at STP* (data from Refs. 8, 13, and 14)

Gas	Chemical Symbol	Molecular Mass, M_m	$R^†$ $\left(\dfrac{J}{kg \cdot K}\right)$	c_p $\left(\dfrac{J}{kg \cdot K}\right)$	c_v $\left(\dfrac{J}{kg \cdot K}\right)$	$k = \dfrac{c_p}{c_v}$ (—)	$R^†$ $\left(\dfrac{ft \cdot lbf}{lbm \cdot R}\right)$	c_p $\left(\dfrac{Btu}{lbm \cdot R}\right)$	c_v $\left(\dfrac{Btu}{lbm \cdot R}\right)$
Air	—	28.98	286.9	1,004	717.4	1.40	53.33	0.2399	0.1713
Carbon dioxide	CO_2	44.01	188.9	840.4	651.4	1.29	35.11	0.2007	0.1556
Carbon monoxide	CO	28.01	296.8	1,039	742.1	1.40	55.17	0.2481	0.1772
Helium	He	4.003	2,077	5,225	3,147	1.66	386.1	1.248	0.7517
Hydrogen	H_2	2.016	4,124	14,180	10,060	1.41	766.5	3.388	2.402
Methane	CH_4	16.04	518.3	2,190	1,672	1.31	96.32	0.5231	0.3993
Nitrogen	N_2	28.01	296.8	1,039	742.0	1.40	55.16	0.2481	0.1772
Oxygen	O_2	32.00	259.8	909.4	649.6	1.40	48.29	0.2172	0.1772
Steam‡	H_2O	18.02	461.4	~2,000	~1,540	~1.30	85.78	~0.478	~0.368

* STP = standard temperature and pressure, T = 15 C (59 F) and p = 101.325 kPa absolute (14.696 psia).
† $R \equiv R_u/M_m$; R_u = 8314.3 J/kgmol · K (1545.3 ft · lbf/lbmol · R); 1 Btu = 778.2 ft · lbf.
‡ Water vapor behaves as an ideal gas when, superheated by 55 C (100 F) or more.

618

Multigrade oils (e.g. $10W - 40$) are formulated to minimize viscosity variation with temperature. High polymer "viscosity index" improvers are used in blending these multigrade oils. Such additives are highly non-Newtonian; they may suffer permanent viscosity loss due to shearing.

Special charts used to estimate the viscosity of petroleum products as a function of temperature are available. The charts were used to develop the data for typical lubricating oils plotted in Figs. A.2 and A.3. For details, see Ref. 12.

references

1. **Handbook of Chemistry and Physics,** 54th ed. Cleveland, Ohio: Chemical Rubber Publishing Co., 1973–1974.

2. "Meriam Standard Indicating Fluids," Pamphlet No. 920GEN:430-1, The Meriam Instrument Co., 10920 Madison Avenue, Cleveland, Ohio 44102.

3. E. Vernon Hill, Inc., P.O. Box 14248, San Francisco, Calif. 94114.

4. **Handbook of Tables for Applied Engineering Science.** Cleveland, Ohio: Chemical Rubber Publishing Co., 1970.

5. Vargaftik, N. B., **Tables on the Thermophysical Properties of Liquids and Gases,** 2nd ed. Washington, D. C.: Hemisphere Publishing Corp., 1975.

6. Trefethen, L., "Surface Tension in Fluid Mechanics," in **Illustrated Experiments in Fluid Mechanics.** Cambridge, Mass.: The M.I.T. Press, 1972.

7. Streeter, V. L., ed. **Handbook of Fluid Dynamics.** New York: McGraw-Hill 1961.

8. **The U.S. Standard Atmosphere (1962).** Washington, D.C.: U.S. Government Printing Office, 1962. Pages 10, 14 and Table I, pp. 39–53.

9. Touloukian, Y. S., Saxena, S. C. and Hestermans, P., **Thermophysical Properties of Matter, the TRRC Data Series, Volume II—Viscosity.** New York: Plenum Publishing Corp., 1975.

10. Reid, R. C. and Sherwood, T. K., **The Properties of Gases and Liquids,** 2nd ed. New York: McGraw-Hill, 1966.

11. "Crankcase Oil Viscosity Classification, Recommended Practice SAE J300a," **SAE Handbook,** 1976 ed. Warrendale, Pa.: Society of Automotive Engineers, 1976.

12. ASTM Standard D 341-74, "Viscosity-Temperature Charts for Liquid Petroleum Products," American Society for Testing and Materials, 1916 Race Street, Philadelphia, Pa. 19103.

13. NASA, **Compressed Gas Handbook (Revised).** Washington, D.C.: National Aeronautics and Space Administration, SP-3045, 1970.

14. ASME, **Thermodynamic and Transport Properties of Steam.** New York: American Society of Mechanical Engineers, 1967.

Appendix B

Tables for Computations in Compressible Flow

TABLE B.1 **Isentropic Flow Functions**
        (one-dimensional flow, ideal gas, $k = 1.4$)

M	T/T_0	p/p_0	ρ/ρ_0	A/A^*
0.00	1.0000	1.0000	1.0000	∞
0.02	0.9999	0.9997	0.9998	28.94
0.04	0.9997	0.9989	0.9992	14.48
0.06	0.9993	0.9975	0.9982	9.666
0.08	0.9987	0.9955	0.9968	7.262
0.10	0.9980	0.9930	0.9950	5.822
0.12	0.9971	0.9900	0.9928	4.864
0.14	0.9961	0.9864	0.9903	4.182
0.16	0.9949	0.9823	0.9873	3.673
0.18	0.9936	0.9777	0.9840	3.278
0.20	0.9921	0.9725	0.9803	2.964
0.22	0.9904	0.9669	0.9762	2.708
0.24	0.9886	0.9607	0.9718	2.496
0.26	0.9867	0.9541	0.9670	2.317
0.28	0.9846	0.9470	0.9619	2.166
0.30	0.9823	0.9395	0.9564	2.035

TABLE B.1   (Continued)

M	T/T_0	p/p_0	ρ/ρ_0	A/A^*
0.30	0.9823	0.9395	0.9564	2.035
0.32	0.9799	0.9315	0.9506	1.922
0.34	0.9774	0.9231	0.9445	1.823
0.36	0.9747	0.9143	0.9380	1.736
0.38	0.9719	0.9052	0.9313	1.659
0.40	0.9690	0.8956	0.9243	1.590
0.42	0.9659	0.8857	0.9170	1.529
0.44	0.9627	0.8755	0.9094	1.474
0.46	0.9594	0.8650	0.9016	1.425
0.48	0.9560	0.8541	0.8935	1.380
0.50	0.9524	0.8430	0.8852	1.340
0.52	0.9487	0.8317	0.8766	1.303
0.54	0.9449	0.8201	0.8679	1.270
0.56	0.9410	0.8082	0.8589	1.240
0.58	0.9370	0.7962	0.8498	1.213
0.60	0.9328	0.7840	0.8405	1.188
0.62	0.9286	0.7716	0.8310	1.166
0.64	0.9243	0.7591	0.8213	1.145
0.66	0.9199	0.7465	0.8115	1.127
0.68	0.9154	0.7338	0.8016	1.110
0.70	0.9108	0.7209	0.7916	1.094
0.72	0.9061	0.7080	0.7814	1.081
0.74	0.9013	0.6951	0.7712	1.068
0.76	0.8964	0.6821	0.7609	1.057
0.78	0.8915	0.6691	0.7505	1.047
0.80	0.8865	0.6560	0.7400	1.038
0.82	0.8815	0.6430	0.7295	1.030
0.84	0.8763	0.6300	0.7189	1.024
0.86	0.8711	0.6170	0.7083	1.018
0.88	0.8659	0.6041	0.6977	1.013
0.90	0.8606	0.5913	0.6870	1.009
0.92	0.8552	0.5785	0.6764	1.006
0.94	0.8498	0.5658	0.6658	1.003
0.96	0.8444	0.5532	0.6551	1.001
0.98	0.8389	0.5407	0.6445	1.000
1.00	0.8333	0.5283	0.6339	1.000

TABLE B.1 (*Continued*)

M	T/T_0	p/p_0	ρ/ρ_0	A/A^*
1.00	0.8333	0.5283	0.6339	1.000
1.02	0.8278	0.5160	0.6234	1.000
1.04	0.8222	0.5039	0.6129	1.001
1.06	0.8165	0.4919	0.6024	1.003
1.08	0.8108	0.4801	0.5920	1.005
1.10	0.8052	0.4684	0.5817	1.008
1.12	0.7994	0.4568	0.5714	1.011
1.14	0.7937	0.4455	0.5612	1.015
1.16	0.7880	0.4343	0.5511	1.020
1.18	0.7822	0.4232	0.5411	1.025
1.20	0.7764	0.4124	0.5311	1.030
1.22	0.7706	0.4017	0.5213	1.037
1.24	0.7648	0.3912	0.5115	1.043
1.26	0.7590	0.3809	0.5019	1.050
1.28	0.7532	0.3708	0.4923	1.058
1.30	0.7474	0.3609	0.4829	1.066
1.32	0.7416	0.3512	0.4736	1.075
1.34	0.7358	0.3417	0.4644	1.084
1.36	0.7300	0.3323	0.4553	1.094
1.38	0.7242	0.3232	0.4463	1.104
1.40	0.7184	0.3142	0.4374	1.115
1.42	0.7126	0.3055	0.4287	1.126
1.44	0.7069	0.2969	0.4201	1.138
1.46	0.7011	0.2886	0.4116	1.150
1.48	0.6954	0.2804	0.4032	1.163
1.50	0.6897	0.2724	0.3950	1.176
1.52	0.6840	0.2646	0.3869	1.190
1.54	0.6783	0.2570	0.3789	1.204
1.56	0.6726	0.2496	0.3711	1.219
1.58	0.6670	0.2423	0.3633	1.234
1.60	0.6614	0.2353	0.3557	1.250
1.62	0.6558	0.2284	0.3483	1.267
1.64	0.6502	0.2217	0.3409	1.284
1.66	0.6447	0.2152	0.3337	1.301
1.68	0.6392	0.2088	0.3266	1.319
1.70	0.6337	0.2026	0.3197	1.338

TABLE B.1 (*Continued*)

M	T/T_0	p/p_0	ρ/ρ_0	A/A^*
1.70	0.6337	0.2026	0.3197	1.338
1.72	0.6283	0.1966	0.3129	1.357
1.74	0.6229	0.1907	0.3062	1.376
1.76	0.6175	0.1850	0.2996	1.397
1.78	0.6121	0.1794	0.2931	1.418
1.80	0.6068	0.1740	0.2868	1.439
1.82	0.6015	0.1688	0.2806	1.461
1.84	0.5963	0.1637	0.2745	1.484
1.86	0.5911	0.1587	0.2686	1.507
1.88	0.5859	0.1539	0.2627	1.531
1.90	0.5807	0.1492	0.2570	1.555
1.92	0.5756	0.1447	0.2514	1.580
1.94	0.5705	0.1403	0.2459	1.606
1.96	0.5655	0.1360	0.2405	1.633
1.98	0.5605	0.1318	0.2352	1.660
2.00	0.5556	0.1278	0.2301	1.688
2.02	0.5506	0.1239	0.2250	1.716
2.04	0.5458	0.1201	0.2200	1.745
2.06	0.5409	0.1164	0.2152	1.775
2.08	0.5361	0.1128	0.2105	1.806
2.10	0.5314	0.1094	0.2058	1.837
2.12	0.5266	0.1060	0.2013	1.869
2.14	0.5219	0.1027	0.1968	1.902
2.16	0.5173	0.09956	0.1925	1.935
2.18	0.5127	0.09650	0.1882	1.970
2.20	0.5081	0.09352	0.1841	2.005
2.22	0.5036	0.09064	0.1800	2.041
2.24	0.4991	0.08784	0.1760	2.078
2.26	0.4947	0.08514	0.1721	2.115
2.28	0.4903	0.08252	0.1683	2.154
2.30	0.4859	0.07997	0.1646	2.193
2.32	0.4816	0.07751	0.1610	2.233
2.34	0.4773	0.07513	0.1574	2.274
2.36	0.4731	0.07281	0.1539	2.316
2.38	0.4689	0.07057	0.1505	2.359
2.40	0.4647	0.06840	0.1472	2.403

M	T/T_0	p/p_0	ρ/ρ_0	A/A^*
2.40	0.4647	0.06840	0.1472	2.403
2.42	0.4606	0.06630	0.1440	2.448
2.44	0.4565	0.06426	0.1408	2.494
2.46	0.4524	0.06229	0.1377	2.540
2.48	0.4484	0.06038	0.1347	2.588
2.50	0.4444	0.05853	0.1317	2.637
2.52	0.4405	0.05674	0.1288	2.687
2.54	0.4366	0.05500	0.1260	2.737
2.56	0.4328	0.05332	0.1232	2.789
2.58	0.4289	0.05169	0.1205	2.842
2.60	0.4252	0.05012	0.1179	2.896
2.62	0.4214	0.04859	0.1153	2.951
2.64	0.4177	0.04711	0.1128	3.007
2.66	0.4141	0.04568	0.1103	3.065
2.68	0.4104	0.04429	0.1079	3.123
2.70	0.4068	0.04295	0.1056	3.183
2.72	0.4033	0.04166	0.1033	3.244
2.74	0.3998	0.04039	0.1010	3.306
2.76	0.3963	0.03917	0.09885	3.370
2.78	0.3928	0.03800	0.09671	3.434
2.80	0.3894	0.03685	0.09462	3.500
2.82	0.3860	0.03574	0.09259	3.567
2.84	0.3827	0.03467	0.09059	3.636
2.86	0.3794	0.03363	0.08865	3.706
2.88	0.3761	0.03262	0.08674	3.777
2.90	0.3729	0.03165	0.08489	3.850
2.92	0.3697	0.03071	0.08308	3.924
2.94	0.3665	0.02980	0.08130	3.999
2.96	0.3633	0.02891	0.07957	4.076
2.98	0.3602	0.02805	0.07788	4.155
3.00	0.3571	0.02722	0.07623	4.235
3.10	0.3422	0.02345	0.06852	4.657
3.20	0.3281	0.02023	0.06165	5.121
3.30	0.3147	0.01748	0.05554	5.629
3.40	0.3019	0.01512	0.05009	6.184
3.50	0.2899	0.01311	0.04523	6.790

TABLE B.1 (Continued)

M	T/T_0	p/p_0	ρ/ρ_0	A/A^*
3.50	0.2899	0.01311	0.04523	6.790
3.60	0.2784	0.01138	0.04089	7.450
3.70	0.2675	0.009903	0.03702	8.169
3.80	0.2572	0.008629	0.03355	8.951
3.90	0.2474	0.007532	0.03044	9.799
4.00	0.2381	0.006586	0.02766	10.72
4.10	0.2293	0.005769	0.02516	11.71
4.20	0.2208	0.005062	0.02292	12.79
4.30	0.2129	0.004449	0.02090	13.95
4.40	0.2053	0.003918	0.01909	15.21
4.50	0.1980	0.003455	0.01745	16.56
4.60	0.1911	0.003053	0.01597	18.02
4.70	0.1846	0.002701	0.01463	19.58
4.80	0.1783	0.002394	0.01343	21.26
4.90	0.1724	0.002126	0.01233	23.07
5.00	0.1667	0.001890	0.01134	25.00

TABLE B.2 Fanno Line Functions
(one-dimensional flow, ideal gas, $k = 1.4$)

M	p_0/p_0^*	T/T^*	p/p^*	V/V^*	$\bar{f}L_{max}/D_h$
0.00	∞	1.200	∞	0.0000	∞
0.02	28.94	1.200	54.77	0.02191	1778
0.04	14.48	1.200	27.38	0.04381	440.5
0.06	9.666	1.199	18.25	0.06570	193.0
0.08	7.262	1.199	13.68	0.08758	106.7
0.10	5.822	1.198	10.94	0.1094	66.92
0.12	4.864	1.197	9.116	0.1313	45.41
0.14	4.182	1.195	7.809	0.1531	32.51
0.16	3.673	1.194	6.829	0.1748	24.20
0.18	3.278	1.192	6.066	0.1965	18.54
0.20	2.964	1.191	5.456	0.2182	14.53
0.22	2.708	1.189	4.955	0.2398	11.60
0.24	2.496	1.186	4.538	0.2614	9.387
0.26	2.317	1.184	4.185	0.2829	7.688
0.28	2.166	1.182	3.882	0.3044	6.357
0.30	2.035	1.179	3.619	0.3257	5.299
0.32	1.922	1.176	3.389	0.3470	4.447
0.34	1.823	1.173	3.185	0.3682	3.752
0.36	1.736	1.170	3.004	0.3894	3.180
0.38	1.659	1.166	2.842	0.4104	2.706
0.40	1.590	1.163	2.696	0.4313	2.309
0.42	1.529	1.159	2.563	0.4522	1.974
0.44	1.474	1.155	2.443	0.4729	1.692
0.46	1.425	1.151	2.333	0.4936	1.451
0.48	1.380	1.147	2.231	0.5141	1.245
0.50	1.340	1.143	2.138	0.5345	1.069
0.52	1.303	1.138	2.052	0.5548	0.9174
0.54	1.270	1.134	1.972	0.5750	0.7866
0.56	1.240	1.129	1.898	0.5951	0.6736
0.58	1.213	1.124	1.828	0.6150	0.5757
0.60	1.188	1.119	1.763	0.6348	0.4908

TABLE B.2   (Continued)

M	p_0/p_0^*	T/T^*	p/p^*	V/V^*	$\bar{f}L_{max}/D_h$
0.60	1.188	1.119	1.763	0.6348	0.4908
0.62	1.166	1.114	1.703	0.6545	0.4172
0.64	1.145	1.109	1.646	0.6740	0.3533
0.66	1.127	1.104	1.592	0.6934	0.2979
0.68	1.110	1.098	1.541	0.7127	0.2498
0.70	1.094	1.093	1.493	0.7318	0.2081
0.72	1.081	1.087	1.448	0.7508	0.1722
0.74	1.068	1.082	1.405	0.7696	0.1411
0.76	1.057	1.076	1.365	0.7883	0.1145
0.78	1.047	1.070	1.326	0.8068	0.09167
0.80	1.038	1.064	1.289	0.8251	0.07229
0.82	1.030	1.058	1.254	0.8433	0.05593
0.84	1.024	1.052	1.221	0.8614	0.04226
0.86	1.018	1.045	1.189	0.8793	0.03097
0.88	1.013	1.039	1.158	0.8970	0.02180
0.90	1.0089	1.033	1.129	0.9146	0.01451
0.92	1.0056	1.026	1.101	0.9320	0.008913
0.94	1.0031	1.020	1.074	0.9493	0.004815
0.96	1.0014	1.013	1.049	0.9663	0.002057
0.98	1.0003	1.007	1.024	0.9832	0.0004947
1.00	1.0000	1.000	1.000	1.000	0.0000
1.02	1.0003	0.9933	0.9771	1.017	0.0004587
1.04	1.0013	0.9866	0.9551	1.033	0.001769
1.06	1.0029	0.9798	0.9338	1.049	0.003838
1.08	1.0051	0.9730	0.9134	1.065	0.006585
1.10	1.0079	0.9662	0.8936	1.081	0.009935
1.12	1.011	0.9593	0.8745	1.097	0.01382
1.14	1.015	0.9524	0.8561	1.113	0.01819
1.16	1.020	0.9455	0.8383	1.128	0.02298
1.18	1.025	0.9386	0.8210	1.143	0.02814
1.20	1.030	0.9317	0.8044	1.158	0.03364
1.22	1.037	0.9247	0.7882	1.173	0.03942
1.24	1.043	0.9178	0.7726	1.188	0.04547
1.26	1.050	0.9108	0.7574	1.203	0.05174
1.28	1.058	0.9038	0.7427	1.217	0.05820
1.30	1.066	0.8969	0.7285	1.231	0.06483

TABLE B.2 (*Continued*)

M	p_0/p_0^*	T/T^*	p/p^*	V/V^*	$\bar{f}L_{max}/D_h$
1.30	1.066	0.8969	0.7285	1.231	0.06483
1.32	1.075	0.8899	0.7147	1.245	0.07161
1.34	1.084	0.8829	0.7012	1.259	0.07850
1.36	1.094	0.8760	0.6882	1.273	0.08550
1.38	1.104	0.8690	0.6755	1.286	0.09259
1.40	1.115	0.8621	0.6632	1.300	0.09974
1.42	1.126	0.8551	0.6512	1.313	0.1069
1.44	1.138	0.8482	0.6396	1.326	0.1142
1.46	1.150	0.8413	0.6282	1.339	0.1215
1.48	1.163	0.8345	0.6172	1.352	0.1288
1.50	1.176	0.8276	0.6065	1.365	0.1361
1.52	1.190	0.8208	0.5960	1.377	0.1434
1.54	1.204	0.8139	0.5858	1.389	0.1506
1.56	1.219	0.8072	0.5759	1.402	0.1579
1.58	1.234	0.8004	0.5662	1.414	0.1651
1.60	1.250	0.7937	0.5568	1.425	0.1724
1.62	1.267	0.7870	0.5476	1.437	0.1795
1.64	1.284	0.7803	0.5386	1.449	0.1867
1.66	1.301	0.7736	0.5299	1.460	0.1938
1.68	1.319	0.7670	0.5213	1.471	0.2008
1.70	1.338	0.7605	0.5130	1.482	0.2078
1.72	1.357	0.7539	0.5048	1.494	0.2147
1.74	1.376	0.7474	0.4969	1.504	0.2216
1.76	1.397	0.7410	0.4891	1.515	0.2284
1.78	1.418	0.7345	0.4815	1.526	0.2352
1.80	1.439	0.7282	0.4741	1.536	0.2419
1.82	1.461	0.7218	0.4668	1.546	0.2485
1.84	1.484	0.7155	0.4597	1.556	0.2551
1.86	1.507	0.7093	0.4528	1.566	0.2616
1.88	1.531	0.7030	0.4460	1.576	0.2680
1.90	1.555	0.6969	0.4394	1.586	0.2743
1.92	1.580	0.6907	0.4329	1.596	0.2806
1.94	1.606	0.6847	0.4265	1.605	0.2868
1.96	1.633	0.6786	0.4203	1.615	0.2930
1.98	1.660	0.6726	0.4142	1.624	0.2990
2.00	1.688	0.6667	0.4083	1.633	0.3050

TABLE B.2 (*Continued*)

M	p_0/p_0^*	T/T^*	p/p^*	V/V^*	$\bar{f}L_{max}/D_h$
2.00	1.688	0.6667	0.4083	1.633	0.3050
2.02	1.716	0.6608	0.4024	1.642	0.3109
2.04	1.745	0.6549	0.3967	1.651	0.3168
2.06	1.775	0.6491	0.3911	1.660	0.3225
2.08	1.806	0.6433	0.3856	1.668	0.3282
2.10	1.837	0.6376	0.3802	1.677	0.3339
2.12	1.869	0.6320	0.3750	1.685	0.3394
2.14	1.902	0.6263	0.3698	1.694	0.3449
2.16	1.935	0.6208	0.3648	1.702	0.3503
2.18	1.970	0.6152	0.3598	1.710	0.3556
2.20	2.005	0.6098	0.3549	1.718	0.3609
2.22	2.041	0.6043	0.3502	1.726	0.3661
2.24	2.078	0.5990	0.3455	1.734	0.3712
2.26	2.115	0.5936	0.3409	1.741	0.3763
2.28	2.154	0.5883	0.3364	1.749	0.3813
2.30	2.193	0.5831	0.3320	1.756	0.3862
2.32	2.233	0.5779	0.3277	1.764	0.3911
2.34	2.274	0.5728	0.3234	1.771	0.3959
2.36	2.316	0.5677	0.3193	1.778	0.4006
2.38	2.359	0.5626	0.3152	1.785	0.4053
2.40	2.403	0.5576	0.3111	1.792	0.4099
2.42	2.448	0.5527	0.3072	1.799	0.4144
2.44	2.494	0.5478	0.3033	1.806	0.4189
2.46	2.540	0.5429	0.2995	1.813	0.4233
2.48	2.588	0.5381	0.2958	1.819	0.4277
2.50	2.637	0.5333	0.2921	1.826	0.4320
2.52	2.687	0.5286	0.2885	1.832	0.4362
2.54	2.737	0.5239	0.2850	1.839	0.4404
2.56	2.789	0.5193	0.2815	1.845	0.4445
2.58	2.842	0.5147	0.2781	1.851	0.4486
2.60	2.896	0.5102	0.2747	1.857	0.4526
2.62	2.951	0.5057	0.2714	1.863	0.4565
2.64	3.007	0.5013	0.2682	1.869	0.4604
2.66	3.065	0.4969	0.2650	1.875	0.4643
2.68	3.123	0.4925	0.2619	1.881	0.4681
2.70	3.183	0.4882	0.2588	1.887	0.4718

M	p_0/p_0^*	T/T^*	p/p^*	V/V^*	$\bar{f}L_{max}/D_h$
2.70	3.183	0.4882	0.2588	1.887	0.4718
2.72	3.244	0.4839	0.2558	1.892	0.4755
2.74	3.306	0.4797	0.2528	1.898	0.4792
2.76	3.370	0.4755	0.2499	1.903	0.4827
2.78	3.434	0.4714	0.2470	1.909	0.4863
2.80	3.500	0.4673	0.2441	1.914	0.4898
2.82	3.567	0.4632	0.2414	1.919	0.4932
2.84	3.636	0.4592	0.2386	1.925	0.4966
2.86	3.706	0.4553	0.2359	1.930	0.5000
2.88	3.777	0.4513	0.2333	1.935	0.5033
2.90	3.850	0.4474	0.2307	1.940	0.5065
2.92	3.924	0.4436	0.2281	1.945	0.5097
2.94	3.999	0.4398	0.2256	1.950	0.5129
2.96	4.076	0.4360	0.2231	1.954	0.5160
2.98	4.155	0.4323	0.2206	1.959	0.5191
3.00	4.235	0.4286	0.2182	1.964	0.5222
3.50	6.79	0.3478	0.1685	2.064	0.5864
4.00	10.72	0.2857	0.1336	2.138	0.6331
4.50	16.56	0.2376	0.1083	2.194	0.6676
5.00	25.00	0.2000	0.08944	2.236	0.6938

TABLE B.3 Rayleigh Line Functions
(one-dimensional flow, ideal gas, $k = 1.4$)

M	T_0/T_0^*	p_0/p_0^*	T/T^*	p/p^*	V/V^*
0.00	0.0000	1.268	0.0000	2.400	0.0000
0.02	0.001918	1.268	0.002301	2.399	0.0009595
0.04	0.007648	1.267	0.009175	2.395	0.003831
0.06	0.01712	1.265	0.02053	2.388	0.008597
0.08	0.03021	1.262	0.03621	2.379	0.01522
0.10	0.04678	1.259	0.05602	2.367	0.02367
0.12	0.06661	1.255	0.07970	2.353	0.03388
0.14	0.08947	1.251	0.1070	2.336	0.04578
0.16	0.1151	1.246	0.1374	2.317	0.05931
0.18	0.1432	1.241	0.1708	2.296	0.07438
0.20	0.1736	1.235	0.2066	2.273	0.09091
0.22	0.2057	1.228	0.2445	2.248	0.1088
0.24	0.2395	1.221	0.2841	2.221	0.1279
0.26	0.2745	1.214	0.3250	2.193	0.1482
0.28	0.3104	1.206	0.3667	2.163	0.1696
0.30	0.3469	1.199	0.4089	2.131	0.1918
0.32	0.3837	1.190	0.4512	2.099	0.2149
0.34	0.4206	1.182	0.4933	2.066	0.2388
0.36	0.4572	1.174	0.5348	2.031	0.2633
0.38	0.4935	1.165	0.5755	1.996	0.2883
0.40	0.5290	1.157	0.6152	1.961	0.3137
0.42	0.5638	1.148	0.6535	1.925	0.3395
0.44	0.5975	1.139	0.6903	1.888	0.3656
0.46	0.6301	1.131	0.7254	1.852	0.3918
0.48	0.6614	1.122	0.7587	1.815	0.4181
0.50	0.6914	1.114	0.7901	1.778	0.4445
0.52	0.7199	1.106	0.8196	1.741	0.4708
0.54	0.7470	1.098	0.8470	1.704	0.4970
0.56	0.7725	1.090	0.8723	1.668	0.5230
0.58	0.7965	1.083	0.8955	1.632	0.5489
0.60	0.8189	1.075	0.9167	1.596	0.5745
0.62	0.8398	1.068	0.9359	1.560	0.5998
0.64	0.8592	1.061	0.9530	1.525	0.6248
0.66	0.8771	1.055	0.9682	1.491	0.6494
0.68	0.8935	1.049	0.9814	1.457	0.6737
0.70	0.9085	1.043	0.9929	1.424	0.6975

TABLE B.3 (Continued)

M	T_0/T_0^*	p_0/p_0^*	T/T^*	p/p^*	V/V^*
0.70	0.9085	1.043	0.9929	1.424	0.6975
0.72	0.9221	1.038	1.003	1.391	0.7209
0.74	0.9344	1.033	1.011	1.359	0.7439
0.76	0.9455	1.028	1.017	1.327	0.7665
0.78	0.9553	1.023	1.022	1.296	0.7885
0.80	0.9639	1.019	1.025	1.266	0.8101
0.82	0.9715	1.016	1.028	1.236	0.8313
0.84	0.9781	1.012	1.029	1.207	0.8519
0.86	0.9836	1.010	1.028	1.179	0.8721
0.88	0.9883	1.007	1.027	1.152	0.8918
0.90	0.9921	1.005	1.025	1.125	0.9110
0.92	0.9951	1.003	1.021	1.098	0.9297
0.94	0.9973	1.002	1.017	1.073	0.9480
0.96	0.9988	1.001	1.012	1.048	0.9658
0.98	0.9997	1.000	1.006	1.024	0.9831
1.00	1.000	1.000	1.000	1.000	1.000
1.02	0.9997	1.000	0.9930	0.9770	1.016
1.04	0.9990	1.001	0.9855	0.9546	1.032
1.06	0.9977	1.002	0.9776	0.9328	1.048
1.08	0.9960	1.003	0.9691	0.9115	1.063
1.10	0.9939	1.005	0.9603	0.8909	1.078
1.12	0.9915	1.007	0.9512	0.8708	1.092
1.14	0.9887	1.010	0.9417	0.8512	1.106
1.16	0.9856	1.012	0.9320	0.8322	1.120
1.18	0.9823	1.016	0.9220	0.8137	1.133
1.20	0.9787	1.019	0.9119	0.7958	1.146
1.22	0.9749	1.023	0.9015	0.7783	1.158
1.24	0.9709	1.028	0.8911	0.7613	1.171
1.26	0.9668	1.033	0.8805	0.7447	1.182
1.28	0.9624	1.038	0.8699	0.7287	1.194
1.30	0.9580	1.044	0.8592	0.7130	1.205
1.32	0.9534	1.050	0.8484	0.6978	1.216
1.34	0.9487	1.056	0.8377	0.6830	1.226
1.36	0.9440	1.063	0.8270	0.6686	1.237
1.38	0.9392	1.070	0.8161	0.6546	1.247
1.40	0.9343	1.078	0.8054	0.6410	1.256

TABLE B.3 (*Continued*)

M	T_0/T_0^*	p_0/p_0^*	T/T^*	p/p^*	V/V^*
1.40	0.9343	1.078	0.8054	0.6410	1.256
1.42	0.9293	1.086	0.7947	0.6278	1.266
1.44	0.9243	1.094	0.7841	0.6149	1.275
1.46	0.9193	1.103	0.7735	0.6024	1.284
1.48	0.9143	1.112	0.7629	0.5902	1.293
1.50	0.9093	1.122	0.7525	0.5783	1.301
1.52	0.9042	1.132	0.7422	0.5668	1.310
1.54	0.8992	1.142	0.7319	0.5555	1.318
1.56	0.8942	1.153	0.7217	0.5446	1.325
1.58	0.8892	1.164	0.7117	0.5339	1.333
1.60	0.8842	1.176	0.7017	0.5236	1.340
1.62	0.8792	1.188	0.6919	0.5135	1.348
1.64	0.8743	1.200	0.6822	0.5036	1.355
1.66	0.8694	1.213	0.6726	0.4941	1.361
1.68	0.8645	1.226	0.6631	0.4847	1.368
1.70	0.8597	1.240	0.6538	0.4756	1.375
1.72	0.8549	1.255	0.6446	0.4668	1.381
1.74	0.8502	1.269	0.6355	0.4581	1.387
1.76	0.8455	1.284	0.6265	0.4497	1.393
1.78	0.8409	1.300	0.6177	0.4415	1.399
1.80	0.8363	1.316	0.6089	0.4335	1.405
1.82	0.8317	1.332	0.6004	0.4257	1.410
1.84	0.8273	1.349	0.5919	0.4181	1.416
1.86	0.8228	1.367	0.5836	0.4107	1.421
1.88	0.8185	1.385	0.5754	0.4035	1.426
1.90	0.8141	1.403	0.5673	0.3964	1.431
1.92	0.8099	1.422	0.5594	0.3896	1.436
1.94	0.8057	1.442	0.5516	0.3828	1.441
1.96	0.8015	1.462	0.5439	0.3763	1.446
1.98	0.7974	1.482	0.5364	0.3699	1.450
2.00	0.7934	1.503	0.5289	0.3636	1.455
2.02	0.7894	1.525	0.5216	0.3575	1.459
2.04	0.7855	1.547	0.5144	0.3516	1.463
2.06	0.7816	1.569	0.5074	0.3458	1.467
2.08	0.7778	1.592	0.5004	0.3401	1.471
2.10	0.7741	1.616	0.4936	0.3345	1.475

M	T_0/T_0^*	p_0/p_0^*	T/T^*	p/p^*	V/V^*
2.10	0.7741	1.616	0.4936	0.3345	1.475
2.12	0.7704	1.640	0.4868	0.3291	1.479
2.14	0.7667	1.665	0.4802	0.3238	1.483
2.16	0.7631	1.691	0.4737	0.3186	1.487
2.18	0.7596	1.717	0.4673	0.3136	1.490
2.20	0.7561	1.743	0.4611	0.3086	1.494
2.22	0.7527	1.771	0.4549	0.3038	1.497
2.24	0.7493	1.799	0.4488	0.2991	1.501
2.26	0.7460	1.827	0.4429	0.2945	1.504
2.28	0.7428	1.856	0.4370	0.2899	1.507
2.30	0.7395	1.886	0.4312	0.2855	1.510
2.32	0.7364	1.917	0.4256	0.2812	1.513
2.34	0.7333	1.948	0.4200	0.2770	1.517
2.36	0.7302	1.979	0.4145	0.2728	1.520
2.38	0.7272	2.012	0.4091	0.2688	1.522
2.40	0.7242	2.045	0.4038	0.2648	1.525
2.42	0.7213	2.079	0.3986	0.2609	1.528
2.44	0.7184	2.114	0.3935	0.2571	1.531
2.46	0.7156	2.149	0.3885	0.2534	1.533
2.48	0.7128	2.185	0.3836	0.2497	1.536
2.50	0.7101	2.222	0.3787	0.2462	1.539
2.52	0.7074	2.259	0.3739	0.2427	1.541
2.54	0.7047	2.298	0.3692	0.2392	1.543
2.56	0.7021	2.337	0.3646	0.2359	1.546
2.58	0.6995	2.377	0.3601	0.2326	1.548
2.60	0.6970	2.418	0.3556	0.2294	1.551
2.62	0.6945	2.459	0.3512	0.2262	1.553
2.64	0.6921	2.502	0.3469	0.2231	1.555
2.66	0.6896	2.545	0.3427	0.2201	1.557
2.68	0.6873	2.589	0.3385	0.2171	1.559
2.70	0.6849	2.634	0.3344	0.2142	1.561
2.72	0.6826	2.680	0.3304	0.2113	1.563
2.74	0.6804	2.727	0.3264	0.2085	1.565
2.76	0.6782	2.775	0.3225	0.2058	1.567
2.78	0.6760	2.824	0.3186	0.2031	1.569
2.80	0.6738	2.873	0.3149	0.2004	1.571

TABLE B.3   (*Continued*)

M	T_0/T_0^*	p_0/p_0^*	T/T^*	p/p^*	V/V^*
2.80	0.6738	2.873	0.3149	0.2004	1.571
2.82	0.6717	2.924	0.3111	0.1978	1.573
2.84	0.6696	2.975	0.3075	0.1953	1.575
2.86	0.6675	3.028	0.3039	0.1927	1.577
2.88	0.6655	3.081	0.3004	0.1903	1.578
2.90	0.6635	3.136	0.2969	0.1879	1.580
2.92	0.6615	3.191	0.2934	0.1855	1.582
2.94	0.6596	3.248	0.2901	0.1832	1.583
2.96	0.6577	3.306	0.2868	0.1809	1.585
2.98	0.6558	3.365	0.2835	0.1787	1.587
3.00	0.6540	3.424	0.2803	0.1765	1.588
3.50	0.6158	5.328	0.2142	0.1322	1.620
4.00	0.5891	8.227	0.1683	0.1026	1.641
4.50	0.5698	12.50	0.1354	0.08177	1.656
5.00	0.5556	18.63	0.1111	0.06667	1.667

TABLE B.4   **Normal Shock Functions**
            (one-dimensional flow, ideal gas, $k = 1.4$)

M_1	M_2	p_{0_2}/p_{0_1}	T_2/T_1	p_2/p_1	ρ_2/ρ_1
1.00	1.000	1.000	1.000	1.000	1.000
1.02	0.9805	1.000	1.013	1.047	1.033
1.04	0.9620	0.9999	1.026	1.095	1.067
1.06	0.9444	0.9998	1.039	1.144	1.101
1.08	0.9277	0.9994	1.052	1.194	1.135
1.10	0.9118	0.9989	1.065	1.245	1.169
1.12	0.8966	0.9982	1.078	1.297	1.203
1.14	0.8820	0.9973	1.090	1.350	1.238
1.16	0.8682	0.9961	1.103	1.403	1.272
1.18	0.8549	0.9946	1.115	1.458	1.307
1.20	0.8422	0.9928	1.128	1.513	1.342
1.22	0.8300	0.9907	1.141	1.570	1.376
1.24	0.8183	0.9884	1.153	1.627	1.411
1.26	0.8071	0.9857	1.166	1.686	1.446
1.28	0.7963	0.9827	1.178	1.745	1.481
1.30	0.7860	0.9794	1.191	1.805	1.516
1.32	0.7760	0.9757	1.204	1.866	1.551
1.34	0.7664	0.9718	1.216	1.928	1.585
1.36	0.7572	0.9676	1.229	1.991	1.620
1.38	0.7483	0.9630	1.242	2.055	1.655
1.40	0.7397	0.9582	1.255	2.120	1.690
1.42	0.7314	0.9531	1.268	2.186	1.724
1.44	0.7235	0.9477	1.281	2.253	1.759
1.46	0.7157	0.9420	1.294	2.320	1.793
1.48	0.7083	0.9360	1.307	2.389	1.828
1.50	0.7011	0.9298	1.320	2.458	1.862
1.52	0.6941	0.9233	1.334	2.529	1.896
1.54	0.6874	0.9166	1.347	2.600	1.930
1.56	0.6809	0.9097	1.361	2.673	1.964
1.58	0.6746	0.9026	1.374	2.746	1.998
1.60	0.6684	0.8952	1.388	2.820	2.032

M_1	M_2	p_{0_2}/p_{0_1}	T_2/T_1	p_2/p_1	ρ_2/ρ_1
1.60	0.6684	0.8952	1.388	2.820	2.032
1.62	0.6625	0.8876	1.402	2.895	2.065
1.64	0.6568	0.8799	1.416	2.971	2.099
1.66	0.6512	0.8720	1.430	3.048	2.132
1.68	0.6458	0.8640	1.444	3.126	2.165
1.70	0.6406	0.8557	1.458	3.205	2.198
1.72	0.6355	0.8474	1.473	3.285	2.230
1.74	0.6305	0.8389	1.487	3.366	2.263
1.76	0.6257	0.8302	1.502	3.447	2.295
1.78	0.6210	0.8215	1.517	3.530	2.327
1.80	0.6165	0.8127	1.532	3.613	2.359
1.82	0.6121	0.8038	1.547	3.698	2.391
1.84	0.6078	0.7947	1.562	3.783	2.422
1.86	0.6036	0.7857	1.577	3.870	2.454
1.88	0.5996	0.7766	1.592	3.957	2.485
1.90	0.5956	0.7674	1.608	4.045	2.516
1.92	0.5918	0.7581	1.624	4.134	2.546
1.94	0.5880	0.7488	1.639	4.224	2.577
1.96	0.5844	0.7395	1.655	4.315	2.607
1.98	0.5808	0.7302	1.671	4.407	2.637
2.00	0.5774	0.7209	1.687	4.500	2.667
2.02	0.5740	0.7115	1.704	4.594	2.696
2.04	0.5707	0.7022	1.720	4.689	2.725
2.06	0.5675	0.6928	1.737	4.784	2.755
2.08	0.5643	0.6835	1.754	4.881	2.783
2.10	0.5613	0.6742	1.770	4.978	2.812
2.12	0.5583	0.6649	1.787	5.077	2.840
2.14	0.5554	0.6557	1.805	5.176	2.868
2.16	0.5525	0.6464	1.822	5.277	2.896
2.18	0.5498	0.6373	1.839	5.378	2.924
2.20	0.5471	0.6281	1.857	5.480	2.951
2.22	0.5444	0.6191	1.875	5.583	2.978
2.24	0.5418	0.6100	1.892	5.687	3.005
2.26	0.5393	0.6011	1.910	5.792	3.032
2.28	0.5368	0.5921	1.929	5.898	3.058
2.30	0.5344	0.5833	1.947	6.005	3.085

M_1	M_2	p_{0_2}/p_{0_1}	T_2/T_1	p_2/p_1	ρ_2/ρ_1
2.30	0.5344	0.5833	1.947	6.005	3.085
2.32	0.5321	0.5745	1.965	6.113	3.110
2.34	0.5297	0.5658	1.984	6.222	3.136
2.36	0.5275	0.5572	2.002	6.331	3.162
2.38	0.5253	0.5486	2.021	6.442	3.187
2.40	0.5231	0.5402	2.040	6.553	3.212
2.42	0.5210	0.5318	2.059	6.666	3.237
2.44	0.5189	0.5234	2.079	6.779	3.261
2.46	0.5169	0.5152	2.098	6.894	3.285
2.48	0.5149	0.5071	2.118	7.009	3.310
2.50	0.5130	0.4990	2.137	7.125	3.333
2.52	0.5111	0.4910	2.157	7.242	3.357
2.54	0.5092	0.4832	2.177	7.360	3.380
2.56	0.5074	0.4754	2.198	7.479	3.403
2.58	0.5056	0.4677	2.218	7.599	3.426
2.60	0.5039	0.4601	2.238	7.720	3.449
2.62	0.5022	0.4526	2.259	7.842	3.471
2.64	0.5005	0.4452	2.280	7.965	3.494
2.66	0.4988	0.4379	2.301	8.088	3.516
2.68	0.4972	0.4307	2.322	8.213	3.537
2.70	0.4956	0.4236	2.343	8.338	3.559
2.72	0.4941	0.4166	2.364	8.465	3.580
2.74	0.4926	0.4097	2.386	8.592	3.601
2.76	0.4911	0.4028	2.407	8.721	3.622
2.78	0.4897	0.3961	2.429	8.850	3.643
2.80	0.4882	0.3895	2.451	8.980	3.664
2.82	0.4868	0.3829	2.473	9.111	3.684
2.84	0.4854	0.3765	2.496	9.243	3.704
2.86	0.4840	0.3701	2.518	9.376	3.724
2.88	0.4827	0.3639	2.540	9.510	3.743
2.90	0.4814	0.3577	2.563	9.645	3.763
2.92	0.4801	0.3517	2.586	9.781	3.782
2.94	0.4788	0.3457	2.609	9.918	3.801
2.96	0.4776	0.3398	2.632	10.06	3.820
2.98	0.4764	0.3340	2.656	10.19	3.839
3.00	0.4752	0.3283	2.679	10.33	3.857

TABLE B.4 (*Continued*)

M_1	M_2	p_{0_2}/p_{0_1}	T_2/T_1	p_2/p_1	ρ_2/ρ_1
3.00	0.4752	0.3283	2.679	10.33	3.857
3.10	0.4695	0.3012	2.799	11.05	3.947
3.20	0.4644	0.2762	2.922	11.78	4.031
3.30	0.4596	0.2533	3.049	12.54	4.112
3.40	0.4552	0.2322	3.180	13.32	4.188
3.50	0.4512	0.2130	3.315	14.13	4.261
3.60	0.4474	0.1953	3.454	14.95	4.330
3.70	0.4440	0.1792	3.596	15.81	4.395
3.80	0.4407	0.1645	3.743	16.68	4.457
3.90	0.4377	0.1510	3.893	17.58	4.516
4.00	0.4350	0.1388	4.047	18.50	4.571
4.10	0.4324	0.1276	4.205	19.45	4.624
4.20	0.4299	0.1173	4.367	20.41	4.675
4.30	0.4277	0.1080	4.532	21.41	4.723
4.40	0.4255	0.09948	4.702	22.42	4.768
4.50	0.4236	0.09170	4.875	23.46	4.812
4.60	0.4217	0.08459	5.052	24.52	4.853
4.70	0.4199	0.07809	5.233	25.61	4.893
4.80	0.4183	0.07214	5.418	26.71	4.930
4.90	0.4167	0.06670	5.607	27.85	4.966
5.00	0.4152	0.06172	5.800	29.00	5.000

Films and Film Loops for Fluid Mechanics

Listed below by supplier are titles of 16 mm sound films and Super 8 silent film "loops" (silent films enclosed in plastic cartridges designed for use in the Technicolor Instant Movie Projector).

1. Encyclopaedia Britannica Educational Corporation
    425 North Michigan Avenue
    Chicago, Illinois 60611
    (Films available for rental or purchase.)

a. 16 mm sound films[1] (length as noted)
    Aerodynamic Generation of Sound
        (44 min, principals: M. J. Lighthill, J. E. Ffowcs-Williams)
    Boundary-Layer Control (25 min, principal: D. C. Hazen)
    Cavitation (31 min, principal: P. Eisenberg)
    Channel Flow of a Compressible Fluid (29 min, principal: D. E. Coles)
    Deformation of Continuous Media (38 min, principal: J. L. Lumley)
    Eulerian and Lagrangian Descriptions in Fluid Mechanics
        (27 min, principal: J. L. Lumley)
    Flow Instabilities (27 min, principal: E. L. Mollo-Christensen)
    Flow Visualization (31 min, principal: S. J. Kline)
    The Fluid Dynamics of Drag[2]
        (4 parts, 118 min, principal: A. H. Shapiro)

---

[1] Detailed summaries of these films have been prepared by the National Committee for Fluid Mechanics Films. See *Illustrated Experiments in Fluid Mechanics* (Cambridge, Mass.: The M.I.T. Press, 1972).
[2] The contents of this film are summarized and illustrated in *Shape and Flow, The Fluid Dynamics of Drag* (New York: Anchor Books, 1961).

Fundamentals of Boundary Layers
   (24 min, principal: F. H. Abernathy)
Low-Reynolds-Number Flows (33 min, principal: Sir G. I. Taylor)
Magnetohydrodynamics (27 min, principal: J. A. Shercliff)
Pressure Fields and Fluid Acceleration
   (30 min, principal: A. H. Shapiro)
Rarefied Gas Dynamics
   (33 min, principals: F. C. Hurlbut, F. S. Sherman)
Rheological Behavior of Fluids (22 min, principal: H. Markovitz)
Rotating Flows (29 min, principal: D. Fultz)
Secondary Flow (30 min, principal: E. S. Taylor)
Stratified Flow (26 min, principal: R. R. Long)
Surface Tension in Fluid Mechanics
   (29 min, principal: L. M. Trefethen)
Turbulence (29 min, principal: R. W. Stewart)
Vorticity (2 parts, 44 min, principal: A. H. Shapiro)
Waves in Fluids (33 min, principal: A. E. Bryson)

b. Supper 8 film loops[3] (all are 3–5 min in length)
   S-FM001   *Some Regimes of Boundary-Layer Transition*
   S-FM002   *Structure of the Turbulent Boundary Layer*
   S-FM003   *Shear Deformation of Viscous Fluids*
   S-FM004   *Separated Flows—Part I*
   S-FM005   *Separated Flows—Part II*
   S-FM006   *Boundary-Layer Formation*
   S-FM007   *Propagating Stall in Airfoil Cascade*
   S-FM008   *The Occurrence of Turbulence*
   S-FM009   *Aerodynamic Heating as Shown by Temperature-Sensitive Paints*
   S-FM010   *Generation of Circulation and Lift for an Airfoil*
   S-FM011   *The Magnus Effect*
   S-FM012   *Flow Separation and Vortex Shedding*
   S-FM013   *The Bathtub Vortex*
   S-FM014A *Visualization of Vorticity with Vorticity Meter—Part I*
   S-FM014B *Visualization of Vorticity with Vorticity Meter—Part II*
   S-FM015   *Incompressible Flow through Area Contractions and Expansions*
   S-FM016   *Flow from a Reservoir to a Duct*
   S-FM017   *Flow Patterns in Venturis, Nozzles, and Orifices*
   S-FM018   *Secondary Flow in a Teacup*
   S-FM019   *Secondary Flow in a Bend*

[3] For summaries, see *Illustrated Experiments in Fluid Mechanics* (footnote 1).

S-FM020   *The Horseshoe Vortex*

S-FM021   *Techniques of Visualization for Low Speed Flows—Part I*

S-FM022   *Techniques of Visualization for Low Speed Flows—Part II*

S-FM023   *Tollmien–Schlichting Waves*

S-FM024   *Wing Tip Vortex*

S-FM025   *Low Speed Jets: Stability and Mixing*

S-FM026   *Tornadoes in Nature and the Laboratory*

S-FM027   *Interaction of Oblique Shock with Flat-Plate Boundary Layer*

S-FM028   *Transonic Flow Past a Symmetric Airfoil*

S-FM029   *Supersonic Zones on Airfoils in Subsonic Flow*

S-FM030   *Shock and Boundary-Layer Interaction on Transonic Airfoil*

S-FM031   *Instabilities in Circular Couette Flow*

S-FM032   *Examples of Turbulent Flow between Concentric Rotating Cylinders*

S-FM033   *Stagnation Pressure*

S-FM034   *The Coanda Effect*

S-FM035   *Radial Flow between Parallel Disks*

S-FM036   *Venturi Passage*

S-FM037   *Streamline Curvature and Normal Pressure Gradient*

S-FM038   *Streamwise Pressure Gradient in Inviscid Flow*

S-FM039   *Interpretation of Flow Using Wall Tufts*

S-FM043   *Buoyancy-Induced Waves in Rotating Fluid*

S-FM044   *Elastoid-Inertia Oscillations in a Rotating Fluid*

S-FM045   *Velocities near an Airfoil*

S-FM046   *Current-Induced Instabilities of a Mercury Jet*

S-FM047   *Pathlines, Streaklines, Streamlines, and Timelines in Steady Flow*

S-FM048   *Pathlines, Streaklines, Streamlines, and Timelines in Unsteady Flow*

S-FM049   *Flow Regimes in Subsonic Diffusers*

S-FM050   *Flow over an Upstream-Facing Step*

S-FM051   *Simple Supersonic Inlet (Axially Symmetric Geometry)*

S-FM052   *Supersonic Conical-Spike Inlet*

S-FM054   *Leading-Edge Separation Bubble in Two-Dimensional Flow*

S-FM055   *Bow Waves in Hypersonic Flow*

S-FM056   *Three-Dimensional Boundary-Layer Separation*

S-FM057   *Effect of Axial Jet on Afterbody Separation*

S-FM058	*Effect of Jet Blowing over Airfoil Flap*
S-FM059	*Leading-Edge Vortices on Delta Wing in Subsonic Flow*
S-FM060	*Breakdown of Leading-Edge Vortices on Delta Wing in Subsonic Flow*
S-FM061	*Ablation of Ice Models in a Water Tunnel*
S-FM062	*Interactions between Oblique Shocks and Expansion Waves*
S-FM063	*Slot Blowing to Suppress Shock-Induced Separation*
S-FM065	*Wide-Angle Diffuser with Suction*
S-FM066	*Thin Bodies of Revolution at Incidence*
S-FM067	*Flow through Right-Angle Bends*
S-FM068	*Flow through Ported Chambers*
S-FM069	*Flow through Tee-Elbow*
S-FM070	*The Sink Vortex*
S-FM071	*Flow near Tip of Lifting Wing*
S-FM072	*Examples of Surface Tension*
S-FM073	*Surface Tension and Contact Angles*
S-FM074	*Formation of Bubbles*
S-FM075	*Surface Tension and Curved Surfaces*
S-FM076	*Breakup of Liquid into Drops*
S-FM077	*Motions Caused by Composition Gradients along Liquid Surfaces*
S-FM078	*Motions Caused by Electrical and Chemical Effects on Liquid Surfaces*
S-FM079	*Motions Caused by Temperature Gradients along Liquid Surfaces*
S-FM080	*Hele-Shaw Analog to Potential Flows, Part I: Sources and Sinks in Uniform Flow*
S-FM081	*Hele-Shaw Analog to Potential Flows, Part II: Sources and Sinks*
S-FM082	*Water Jet Instability in Electric Field*
S-FM083	*Induced $\vec{J} \times \vec{B}$ Forces in Solids and Liquids*
S-FM084	*An MHD Pump*
S-FM086	*Suppression of Vorticity by MHD Forces*
S-FM087	*The Hartmann Layer*
S-FM088	*Laminar Boundary Layers*
S-FM089	*Turbulent Boundary Layers*
S-FM090	*Supersonic Flow past Diamond Airfoil*
S-FM091	*Modes of Sloshing in Tanks*
S-FM092	*Stages of Boundary-Layer Instability and Transition*
S-FM093	*Supersonic Spike Inlet with Variable Geometry*
S-FM097	*Flow through Fans and Propellers*

S-FM098   *Aerodynamic Heating and Ablation of Missile Shapes*
S-FM099   *Passage of Shock Waves over Bodies*
S-FM100   *Passage of Shock Waves through Constrictions*
S-FM101   *Reflections of Shock Waves*
S-FM102   *Passage of a Shock Wave through a Circular Orifice*
S-FM103   *Effect of Knudsen Number on Flow Past a Blunt Body*
S-FM104   *Effect of Knudsen Number on a Jet*
S-FM105   *Ripple-Tank Radiation Patterns of Source, Dipole, and Quadrupole*
S-FM107   *Deformation in Fluids Illustrated by a Rectilinear Shear Flow*
S-FM108   *Small-Amplitude Waves*
S-FM109   *Source Moving at Speeds below and above Wave Speeds*
S-FM112   *Examples of Low-Reynolds-Number Flows*
S-FM113   *Hydrodynamic Lubrication*
S-FM114   *Sedimentation at Low Reynolds Number*
S-FM115   *Kinematic Reversibility of Low-Reynolds-Number Flows*
S-FM116   *Swimming Propulsion at Low Reynolds Number*
S-FM117   *Subsonic Flow Patterns and Pressure Distributions for an Airfoil*
S-FM118   *Laminar-Flow Versus Conventional Airfoils*
S-FM119   *Reduction of Airfoil Friction Drag by Suction*
S-FM120   *Some Methods for Increasing Lift Coefficient*
S-FM122   *Nonlinear Shear Stress Behavior in Steady Flows*
S-FM123   *Normal-Stress Effects in Viscoelastic Fluids*
S-FM124   *Memory Effects in Viscoelastic Fluids*
S-FM125   *Examples of Cavitation*
S-FM126   *Cavitation on Hydrofoils*
S-FM127   *Cavitation Bubble Dynamics*
S-FM128   *Cavity Flows*
S-FM129   *Compressible Flow through Convergent-Divergent Nozzle*
S-FM130   *Starting of Supersonic Wind Tunnel with Variable-Throat Diffuser*
S-FM134   *Laminar and Turbulent Pipe Flow*
S-FM135   *Averages and Transport in Turbulence*
S-FM136   *Structure of Turbulence*
S-FM137   *Effects of Density Stratification on Turbulence*
S-FM138   *Taylor Columns in Rotating Flows (at Low Rossby Number)*
S-FM139   *Small-Amplitude Gravity Waves in an Open Channel*
S-FM140   *Flattening and Steepening of Large-Amplitude Gravity Waves*

appendix c/films and film loops for fluid mechanics   *645*

S-FM141	*The Hydraulic Surge Wave*
S-FM142	*The Hydraulic Jump*
S-FM143	*Free-Surface Flow over a Towed Obstacle*
S-FM144	*Flow of a Two-Liquid Stratified Fluid past an Obstacle*
S-FM145	*Flow of a Continuously-Stratified Fluid past an Obstacle*
S-FM146	*Examples of Flow Instability (Part I)*
S-FM147	*Examples of Flow Instability (Part II)*
S-FM148	*Experimental Study of a Flow Instability*

2. University of Iowa
The Audiovisual Center
Iowa City, Iowa 52240
(Films available for rental or purchase.)

The following six films were prepared as a series, in the order listed. They can be viewed individually without serious loss of continuity.

Introduction to the Study of Fluid Motion (24 min, principal: H. Rouse) This orientation film shows a variety of familar flow phenomena. Use of scale models for empirical study of complex phenomena is illustrated and the significance of the Euler, Froude, Mach, and Reynolds numbers as similarity parameters is shown using several sequences of model and prototype flows.

Fundamental Principles of Flow (23 min, principal: H. Rouse) The basic concepts and physical relationships needed to analyze fluid motions are developed in this film. The continuity, momentum, and energy equations are derived and used to analyze a jet propulsion device.

Fluid Motion in a Gravitational Field (23 min, principal: H. Rouse) Buoyancy effects and free-surface flows are illustrated in this film. The Froude number is shown to be a fundamental parameter for flows with a free surface. Wave motions are shown for open-channel and density-stratified flows.

Characteristics of Laminar and Turbulent Flow (26 min, principal: H. Rouse) Dye, smoke, suspended particles, and hydrogen bubbles are used to visualize laminar and turbulent flows. Instabilities that lead to turbulence are shown; production and decay of turbulence and mixing are described.

Form Drag, Lift, and Propulsion (24 min, principal: H. Rouse) The effects of boundary-layer separation on flow patterns and pressure distributions are shown for several body shapes. The basic characteristics of lifting shapes, including effects of aspect ratio, are discussed, and the results are applied to analysis of the performance of propellers and torque converters.

Effects of Fluid Compressibility (17 min, principal: H. Rouse)　The hydraulic analogy between open-channel liquid flow and compressible gas flow is used to show representative wave patterns. Schlieren optical flow visualization is used in a supersonic wind tunnel to show patterns of flow past several bodies at subsonic and supersonic speeds.

3.　Ohio State University
　　Film Distribution Supervisor
　　Motion Picture Division
　　1885 Neil Avenue
　　Columbus, Ohio 44223
　　(The following film contains spectacular original footage from this disaster, which occurred in 1940.)
　　Tacoma Narrows Bridge Collapse

4.　University of Minnesota
　　Saint Anthony Falls Hydraulics Laboratory
　　Mississipi River and 3rd Avuenue, SE
　　Minneapolis, Minn. 55414
　　Some Phenomena of Open-Channel Flow (33 min, silent)　Many features of open-channel flow are demonstrated. Applications of the specific energy and pressure-momentum curves are illustrated.
　　Fluid Mechanics—The Boundary Layer (30 min, sound)　The film contains a series of demonstrations of physical principles. It is planned to be used as a summary after regular classroom discussion of boundary-layer phenomena.

5.　Shell Film Library
　　1433 Sadlier Circle West Drive
　　Indianapolis, Ind. 46239
　　(Films available for free loan; borrower pays only return postage.)
　　Approaching the Speed of Sound　(~ 20 min)　This film includes a good discussion of the cirtical Mach number for high subsonic speed flight, including a description of the effects of profile thickness and wing sweepback. Shock formation at the critical Mach number is shown with color Schlieren visualization. The Mach cone is introduced with an animated sequence that illustrates it very clearly.
　　Beyond the Speed of Sound (~ 20 min)　The film includes a good discussion of drag in both transonic and supersonic flow.
　　High Speed Flight (20 min)　This film combines *Approaching the Speed of Sound*, *Transonic Flight*, and *Beyond the Speed of Sound* in simplified form. It retains most of the important features of the individual films.

How an Airplane Flies (2 reels, 50 min)   This four-part film treats Weight and Lift, Thrust and Drag, Balance and Stability, and The Controls and Their Effect. Footage of actual aircraft in flight, wind tunnel models, and animation are used to develop a complete treatment of aircraft flight. An outstanding film.

Schlieren (~ 20 min)   The principles of optical visualization techniques for compressible flows are discussed. The color Schlieren is used to analyze and discuss flow fields about a variety of shapes. This is an excellent introductory film.

Transonic Flight (~ 20 min)   This film deals specifically with transonic flight, where subsonic, critical, and supersonic flow are present simultaneously. Early problems with transonic flight are reviewed, and some popular misconceptions, for example, the "sound barrier" are dispelled. The "area rule" is explained.

6. American Institute of Aeronautics and Astronautics
   Director of Public Information
   1290 Avenue of the Americas
   New York, N.Y. 10019
   (Film available for free loan; borrower pays only return postage.)
   America's Wings (29 min) Individuals who made significant contributions to development of aircraft for high speed flight are interviewed. These developments are explained in their own words. This is an effective film for a relatively sophisticated audience.

Review of Vector Concepts and Operations

This appendix is included to serve as a quick review of vector concepts and manipulations. Applications are illustrated by giving derivations or proofs of some results used in the text.

D-1 SYMBOLS

Scalars are represented by a, b, and so on. Vector quantities are denoted by $\vec{a}$, $\vec{b}$, etc. The unit vectors in the x, y, and z coordinate directions are given by $\hat{i}$, $\hat{j}$, and $\hat{k}$, respectively. A vector is written in terms of its components as

$$\vec{a} = a_x\hat{i} + a_y\hat{j} + a_z\hat{k} \tag{D.1}$$

In cylindrical coordinates, the unit vectors in the r, θ, and z coordinate directions are denoted by $\hat{i}_r$, $\hat{i}_\theta$, and $\hat{i}_z$, respectively.

A unit vector in the direction of $\vec{a}$ may be obtained by writing

$$\hat{i}_a = \frac{\vec{a}}{|\vec{a}|}$$

where $|\vec{a}|$ is the magnitude of $\vec{a}$. From Fig. D.1, we can obtain an expression for $|\vec{a}|$ in terms of components, a_x, a_y, and a_z.

From trigonometry,

$$c^2 = (a_x)^2 + (a_z)^2$$

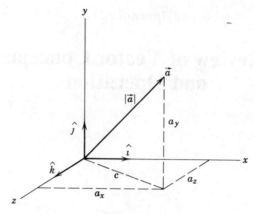

Fig. D.1 Magnitude of vector $\vec{a}$.

and

$$[|\vec{a}|]^2 = c^2 + (a_y)^2$$

or

$$|\vec{a}| = \sqrt{(a_x)^2 + (a_y)^2 + (a_z)^2} \qquad \text{(D.2)}$$

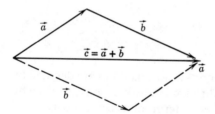

Fig. D.2 Addition of vectors by parallelogram rule.

Vectors are added by accounting for both magnitude and direction, using the parallelogram rule, Fig. D.2. Thus

$$\vec{c} = \vec{a} + \vec{b}$$

and

$$\vec{c} = c_x \hat{i} + c_y \hat{j} + c_z \hat{k}$$
$$= a_x \hat{i} + a_y \hat{j} + a_z \hat{k} + b_x \hat{i} + b_y \hat{j} + b_z \hat{k}$$
$$\vec{c} = (a_x + b_x)\hat{i} + (a_y + b_y)\hat{j} + (a_z + b_z)\hat{k}$$

Hence

$$c_x = a_x + b_x \qquad c_y = a_y + b_y \qquad \text{and} \qquad c_z = a_z + b_z \qquad \text{(D.3)}$$

If a vector, $\vec{d}$, equals zero, then $d_x = d_y = d_z = 0$.

D-2 PRODUCTS OF VECTORS

D-2.1 THE DOT PRODUCT (OR SCALAR PRODUCT)

The dot product of two vectors, $\vec{a}$ and $\vec{b}$, is written $\vec{a} \cdot \vec{b}$, and defined as

$$\vec{a} \cdot \vec{b} = |\vec{a}|\,|\vec{b}| \cos \theta_{ab} \tag{D.4}$$

where θ_{ab} is the angle between the vectors. Physically, it represents the product of $|\vec{a}|$ with the component of $|\vec{b}|$ in the direction of $\vec{a}$. The dot product is a scalar; if $\theta_{ab} < \pi/2$, its magnitude is positive, whereas for $\theta_{ab} > \pi/2$, it is negative. If $\theta_{ab} = \pi/2$, $\vec{a} \cdot \vec{b} = 0$.

The dot products of unit vectors in the xyz coordinate system are:

$$\hat{\imath} \cdot \hat{\imath} = 1 \qquad \hat{\jmath} \cdot \hat{\imath} = 0 \qquad \hat{k} \cdot \hat{\imath} = 0$$
$$\hat{\imath} \cdot \hat{\jmath} = 0 \qquad \hat{\jmath} \cdot \hat{\jmath} = 1 \qquad \hat{k} \cdot \hat{\jmath} = 0$$
$$\hat{\imath} \cdot \hat{k} = 0 \qquad \hat{\jmath} \cdot \hat{k} = 0 \qquad \hat{k} \cdot \hat{k} = 1$$

Then

$$\vec{a} \cdot \vec{b} = (a_x\hat{\imath} + a_y\hat{\jmath} + a_z\hat{k}) \cdot (b_x\hat{\imath} + b_y\hat{\jmath} + b_z\hat{k})$$
$$\vec{a} \cdot \vec{b} = a_xb_x + a_yb_y + a_zb_z \tag{D.5}$$

Using Eq. D.5, we can show:

1. The dot product is *commutative*, that is, $\vec{a} \cdot \vec{b} = \vec{b} \cdot \vec{a}$. This follows from Eq. D.5 because $a_xb_x = b_xa_x$, and so on.

2. The dot product is *distributive*, that is, $\vec{a} \cdot (\vec{b} + \vec{c}) = \vec{a} \cdot \vec{b} + \vec{a} \cdot \vec{c}$. This follows from Eqs. D.3 and D.5, since

$$\vec{a} \cdot (\vec{b} + \vec{c}) = (a_x\hat{\imath} + \cdots) \cdot [(b_x + c_x)\hat{\imath} + \cdots]$$
$$= a_x(b_x + c_x) + \cdots$$
$$\vec{a} \cdot (\vec{b} + \vec{c}) = a_xb_x + a_xc_x + \cdots = \vec{a} \cdot \vec{b} + \vec{a} \cdot \vec{c}$$

3. The dot product is *not associative*, that is, $\vec{a}(\vec{b} \cdot \vec{c}) \neq (\vec{a} \cdot \vec{b})\vec{c}$. This follows, because $\vec{a}(\vec{b} \cdot \vec{c}) = \alpha\vec{a}$, since $\vec{b} \cdot \vec{c}$ is a scalar. Likewise $(\vec{a} \cdot \vec{b})\vec{c} = \beta\vec{c}$. Since vectors have both magnitude and direction, $\alpha\vec{a} \neq \beta\vec{c}$.

Applications of the dot product may be illustrated using work or flowrate as examples. The work done by a force, $\vec{F}$, moving through a distance, $d\vec{s}$, is a scalar, equal to the component of the force in the direction of $d\vec{s}$ times the distance moved. Thus

$$\delta W = |\vec{F}| \cos \theta_{Fs}|d\vec{s}| = \vec{F} \cdot d\vec{s}$$

The volumetric flowrate is given by the component of velocity, $\vec{V}$, normal to an element of area, dA, times the area of the element. Taking $d\vec{A}$ as the vector normal to dA gives

$$dQ = |\vec{V}| \cos \theta_{V\,dA}|d\vec{A}| = \vec{V} \cdot d\vec{A}$$

D-2.2 CROSS PRODUCT (OR VECTOR PRODUCT)

The cross product of two vectors, $\vec{a}$ and $\vec{b}$, is written $\vec{a} \times \vec{b}$. It is a vector that has the properties:

1. $|\vec{a} \times \vec{b}| = |\vec{a}|\,|\vec{b}| \sin \theta_{ab}$. (If $\vec{a}$ and $\vec{b}$ are parallel, then $\sin \theta_{ab} = 0$ and $\vec{a} \times \vec{b} = 0$.)
2. $\vec{a} \times \vec{b}$ is a vector perpendicular to *both* $\vec{a}$ and $\vec{b}$.
3. The *sense* of $\vec{a} \times \vec{b}$ is given by the right-hand rule, that is, as $\vec{a}$ is rotated into $\vec{b}$, then $\vec{a} \times \vec{b}$ points in the direction of the right thumb. This is shown in Fig. D.3.

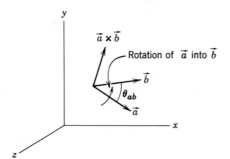

Fig. D.3 Cross product of vectors.

Applying these rules, we obtain the cross products among unit vectors:

$$\begin{array}{lll}
\hat{\imath} \times \hat{\imath} = 0 & \hat{\jmath} \times \hat{\imath} = -\hat{k} & \hat{k} \times \hat{\imath} = \hat{\jmath} \\
\hat{\imath} \times \hat{\jmath} = \hat{k} & \hat{\jmath} \times \hat{\jmath} = 0 & \hat{k} \times \hat{\jmath} = -\hat{\imath} \\
\hat{\imath} \times \hat{k} = -\hat{\jmath} & \hat{\jmath} \times \hat{k} = \hat{\imath} & \hat{k} \times \hat{k} = 0
\end{array}$$

The cross product, $\vec{a} \times \vec{b}$, is conveniently evaluated from the determinant

$$\vec{a} \times \vec{b} = \begin{vmatrix} \hat{\imath} & \hat{\jmath} & \hat{k} \\ a_x & a_y & a_z \\ b_x & b_y & b_z \end{vmatrix} \qquad \text{(D.6a)}$$

Expanding in minors of the first row,

$$\vec{a} \times \vec{b} = \hat{\imath}(a_y b_z - a_z b_y) + (-\hat{\jmath})(a_x b_z - a_z b_x) + \hat{k}(a_x b_y - a_y b_x)$$

or

$$\vec{a} \times \vec{b} = \hat{\imath}(a_y b_z - a_z b_y) + \hat{\jmath}(a_z b_x - a_x b_z) + \hat{k}(a_x b_y - a_y b_x) \qquad \text{(D.6b)}$$

Using Eqs. D.6, we can show:

1. The cross product is *not commutative*, that is, $\vec{a} \times \vec{b} \neq \vec{b} \times \vec{a}$. Interchanging the order of two rows or columns changes the sign of a determinant. Thus $\vec{a} \times \vec{b} = -\vec{b} \times \vec{a}$.
2. The cross product is *distributive*, that is, $\vec{a} \times (\vec{b} + \vec{c}) = \vec{a} \times \vec{b} + \vec{a} \times \vec{c}$. This follows by direct substitution of the components of $(\vec{b} + \vec{c})$ into Eq. D.6a.
3. The cross product is *not associative*, that is, $\vec{a} \times (\vec{b} \times \vec{c}) \neq (\vec{a} \times \vec{b}) \times \vec{c}$. This may be shown by choosing $\vec{b} = \vec{c}$. In the left side, then $\vec{a} \times (\vec{b} \times \vec{c}) = \vec{a} \times \vec{0} = \vec{0}$. But $\vec{a} \times \vec{b} \neq \vec{0}$, and the right side, $(\vec{a} \times \vec{b}) \times \vec{c} \neq \vec{0}$. Thus $\vec{a} \times (\vec{b} \times \vec{c}) \neq (\vec{a} \times \vec{b}) \times \vec{c}$.

One application of the cross product is in formulation of the torque or moment applied to a body. Physically, the torque is $|\vec{r}|$ times the component of $\vec{F}$ perpendicular to $\vec{r}$. Thus for the body shown in Fig. D.4.

$$|\vec{T}| = |\vec{r}| \, |\vec{F}| \sin \theta_{rF}$$

or

$$\vec{T} = \vec{r} \times \vec{F}$$

according to the right-hand rule.

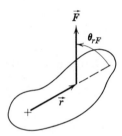

Fig. D.4 Formulation of torque.

D-3 DIFFERENTIATION OF VECTORS

The conventional definition of a derivative may be used to differentiate a vector. Consider a vector $\vec{a} = \vec{a}(t)$. Then in rectangular coordinates $a_x = a_x(t)$, etc., and

$$\frac{d\vec{a}}{dt} = \lim_{\Delta t \to 0} \frac{\vec{a}(t + \Delta t) - \vec{a}(t)}{\Delta t}$$

$$\frac{d\vec{a}}{dt} = \lim_{\Delta t \to 0} \frac{[a_x(t + \Delta t) - a_x(t)]\hat{\imath} + \cdots}{\Delta t}$$

The limiting process applies to each term, so

$$\frac{d\vec{a}}{dt} = \lim_{\Delta t \to 0} \frac{a_x(t + \Delta t) - a_x(t)}{\Delta t} \hat{\imath} + \lim_{\Delta t \to 0} \frac{a_y(t + \Delta t) - a_y(t)}{\Delta t} \hat{\jmath} + \cdots$$

appendix d/review of vector concepts and operations *653*

or

$$\frac{d\vec{a}}{dt} = \frac{da_x}{dt}\,\hat{\imath} + \frac{da_y}{dt}\,\hat{\jmath} + \frac{da_z}{dt}\,\hat{k} \qquad (D.7)$$

Similarly, if $\vec{a} = \vec{a}(x, y, z)$, that is, $a_x = a_x(x, y, z)$, etc., then

$$\frac{\partial \vec{a}}{\partial x} = \lim_{\Delta x \to 0} \frac{\vec{a}(x + \Delta x, y, z) - \vec{a}(x, y, z)}{\Delta x}$$

$$\frac{\partial \vec{a}}{\partial x} = \frac{\partial a_x}{\partial x}\,\hat{\imath} + \frac{\partial a_y}{\partial x}\,\hat{\jmath} + \frac{\partial a_z}{\partial x}\,\hat{k} \qquad (D.8)$$

In cylindrical coordinates the unit vectors, $\hat{\imath}_r$ and $\hat{\imath}_\theta$ are functions of θ. It may be shown using the method of Example 4.12 that

$$\frac{\partial \hat{\imath}_r}{\partial \theta} = \hat{\imath}_\theta \qquad \text{and} \qquad \frac{\partial \hat{\imath}_\theta}{\partial \theta} = -\hat{\imath}_r$$

Consequently, in cylindrical coordinates, if $\vec{a} = \vec{a}(r, \theta, z)$, that is, $a_r = a_r(r, \theta, z)$, etc., then

$$\frac{\partial \vec{a}}{\partial \theta} = \frac{\partial a_r}{\partial \theta}\,\hat{\imath}_r + a_r\frac{\partial \hat{\imath}_r}{\partial \theta} + \cdots = \frac{\partial a_r}{\partial \theta}\,\hat{\imath}_r + a_r\hat{\imath}_\theta + \cdots$$

This fact is used in applying the operator, ∇, in the next section.

D-4 THE VECTOR OPERATOR, ∇

D-4.1 DEFINITION OF ∇

Certain operations occur so frequently in mathematical formulations that a shorthand notation is desirable. The vector operator del, ∇, has been defined as

$$\nabla \equiv \hat{\imath}\frac{\partial}{\partial x} + \hat{\jmath}\frac{\partial}{\partial y} + \hat{k}\frac{\partial}{\partial z} \qquad \text{(Cartesian coordinates)}$$

$$\nabla \equiv \hat{\imath}_r\frac{\partial}{\partial r} + \hat{\imath}_\theta\frac{1}{r}\frac{\partial}{\partial \theta} + \hat{\imath}_z\frac{\partial}{\partial z} \qquad \text{(cylindrical coordinates)}$$

$\qquad (D.9)$

The terms in the definition of ∇ are written $\hat{\imath}(\partial/\partial x)$, and so on, to emphasize that ∇ is an operator. Thus the partial derivatives, such as $\partial/\partial x$, *operate* on any function that follows them.

Since ∇ is a vector operator, three possible "products" can be formed with other functions. These are discussed in the next three sections.

D-4.2 THE GRADIENT

When ∇ operates on a differentiable scalar function (or field), the result is called the gradient. Thus, if $\phi(x, y, z)$ is a scalar field, then

$$\nabla\phi = \text{gradient } \phi = \text{grad } \phi = \hat{i}\frac{\partial\phi}{\partial x} + \hat{j}\frac{\partial\phi}{\partial y} + \hat{k}\frac{\partial\phi}{\partial z} \qquad \text{(D.10)}$$

Note that ϕ is a scalar field, but grad ϕ is a vector field. The gradient, $\nabla\phi$, is a vector perpendicular to a surface of constant ϕ. Its magnitude and direction are those of the greatest space rate of increase of ϕ.

The pressure gradient, ∇p, is important in fluid mechanics. ∇p for a pressure field, $p = p(x, y, z)$, is evaluated and discussed in Section 3-1.

D-4.3 THE DIVERGENCE

When the dot product is formed between ∇ and a vector function (or field), the result is called the *divergence*. In Cartesian coordinates, if $\vec{V} = \vec{V}(x, y, z)$, then

$$\nabla \cdot \vec{V} = \text{divergence } \vec{V} = \text{div } \vec{V} = \left(\hat{i}\frac{\partial}{\partial x} + \hat{j}\frac{\partial}{\partial y} + \hat{k}\frac{\partial}{\partial z}\right) \cdot (V_x\hat{i} + V_y\hat{j} + V_z\hat{k})$$

$$= \hat{i}\cdot\frac{\partial V_x}{\partial x}\hat{i} + \hat{i}\cdot V_x\overset{=0}{\cancel{\frac{\partial\hat{i}}{\partial x}}} + \hat{j}\cdot\frac{\partial V_y}{\partial y}\hat{j} + \hat{j}\cdot V_y\overset{=0}{\cancel{\frac{\partial\hat{j}}{\partial y}}} + \hat{k}\cdot\frac{\partial V_z}{\partial x}\hat{k} + \hat{k}\cdot V_z\overset{=0}{\cancel{\frac{\partial\hat{k}}{\partial z}}}$$

$$\nabla \cdot \vec{V} = \frac{\partial V_x}{\partial x} + \frac{\partial V_y}{\partial y} + \frac{\partial V_z}{\partial z} \qquad \text{(Cartesian coordinates)} \qquad \text{(D.11)}$$

In cylindrical coordinates, if $\vec{V} = \vec{V}(r, \theta, z)$, then

$$\nabla \cdot \vec{V} = \left(\hat{i}_r\frac{\partial}{\partial r} + \hat{i}_\theta\frac{1}{r}\frac{\partial}{\partial\theta} + \hat{i}_z\frac{\partial}{\partial z}\right) \cdot (\hat{i}_r V_r + \hat{i}_\theta V_\theta + \hat{i}_z V_z)$$

$$= \hat{i}_r\cdot\overset{=0}{\cancel{\frac{\partial\hat{i}_r}{\partial r}}}V_r + \hat{i}_r\cdot\hat{i}_r\frac{\partial V_r}{\partial r} + \hat{i}_\theta\cdot\frac{1}{r}\frac{\partial\hat{i}_r}{\partial\theta}V_r + \hat{i}_\theta\cdot\hat{i}_\theta\frac{1}{r}\frac{\partial V_\theta}{\partial\theta}$$

$$+ \hat{i}_z\cdot\overset{=0}{\cancel{\frac{\partial\hat{i}_z}{\partial z}}}V_z + \hat{i}_z\cdot\hat{i}_z\frac{\partial V_z}{\partial z}$$

$$\nabla \cdot \vec{V} = \frac{\partial V_r}{\partial r} + \hat{i}_\theta\cdot\frac{1}{r}\frac{\partial\hat{i}_r}{\partial\theta}V_r + \frac{1}{r}\frac{\partial V_\theta}{\partial\theta} + \frac{\partial V_z}{\partial z}$$

But $\partial \hat{\imath}_r / \partial \theta = \hat{\imath}_\theta$, as shown in Example Problem 4.12. Thus

$$\nabla \cdot \vec{V} = \frac{\partial V_r}{\partial r} + \frac{V_r}{r} + \frac{1}{r} \frac{\partial V_\theta}{\partial \theta} + \frac{\partial V_z}{\partial z}$$

But

$$\frac{\partial V_r}{\partial r} + \frac{V_r}{r} = \frac{1}{r} \frac{\partial}{\partial r} (rV_r)$$

so finally,

$$\nabla \cdot \vec{V} = \frac{1}{r} \frac{\partial}{\partial r} (rV_r) + \frac{1}{r} \frac{\partial V_\theta}{\partial \theta} + \frac{\partial V_z}{\partial z} \qquad \text{(cylindrical coordinates)} \quad \text{(D.12)}$$

Here $\nabla \cdot \vec{V}$, where $\vec{V}$ is the velocity field (Section 2-2), represents the net volume rate of fluid flow from an infinitesimal control volume, per unit volume. For an incompressible flow, $\nabla \cdot \vec{V} = 0$, that is, the net rate of flux of fluid from any infinitesimal control volume must vanish.

D-4.4 THE CURL

When the cross product is formed between ∇ and a vector field, the result is called the *curl*. If $\vec{V} = \vec{V}(x, y, z)$, then in Cartesian coordinates

$$\nabla \times \vec{V} = \text{curl } \vec{V} = \begin{vmatrix} \hat{\imath} & \hat{\jmath} & \hat{k} \\ \dfrac{\partial}{\partial x} & \dfrac{\partial}{\partial y} & \dfrac{\partial}{\partial z} \\ V_x & V_y & V_z \end{vmatrix}$$

$$\nabla \times \vec{V} = \hat{\imath} \left(\frac{\partial V_z}{\partial y} - \frac{\partial V_y}{\partial z} \right) + \hat{\jmath} \left(\frac{\partial V_x}{\partial z} - \frac{\partial V_z}{\partial x} \right) + \hat{k} \left(\frac{\partial V_y}{\partial x} - \frac{\partial V_x}{\partial y} \right) \quad \text{(D.13)}$$

For cylindrical coordinates, $\vec{V} = \vec{V}(r, \theta, z) = \hat{\imath}_r V_r + \hat{\imath}_\theta V_\theta + \hat{\imath}_z V_z$. Since the unit vectors change with angle, θ, we must proceed with caution. Thus

$$\nabla \times \vec{V} = \left(\hat{\imath}_r \frac{\partial}{\partial r} + \hat{\imath}_\theta \frac{1}{r} \frac{\partial}{\partial \theta} + \hat{\imath}_z \frac{\partial}{\partial z} \right) \times (\hat{\imath}_r V_r + \hat{\imath}_\theta V_\theta + \hat{\imath}_z V_z)$$

$$= \hat{\imath}_r \frac{\partial}{\partial r} \times \hat{\imath}_r V_r + \hat{\imath}_r \frac{\partial}{\partial r} \times \hat{\imath}_\theta V_\theta + \hat{\imath}_r \frac{\partial}{\partial r} \times \hat{\imath}_z V_z$$

$$+ \hat{\imath}_\theta \frac{1}{r} \frac{\partial}{\partial \theta} \times \hat{\imath}_r V_r + \hat{\imath}_\theta \frac{1}{r} \frac{\partial}{\partial \theta} \times \hat{\imath}_\theta V_\theta + \hat{\imath}_\theta \frac{1}{r} \frac{\partial}{\partial \theta} \times \hat{\imath}_z V_z$$

$$+ \hat{\imath}_z \frac{\partial}{\partial z} \times \hat{\imath}_r V_r + \hat{\imath}_z \frac{\partial}{\partial z} \times \hat{\imath}_\theta V_\theta + \hat{\imath}_z \frac{\partial}{\partial z} \times \hat{\imath}_z V_z$$

$$\nabla \times \vec{V} = \hat{\imath}_r \times \hat{\imath}_r \frac{\partial V_r}{\partial r} \overset{=0}{\diagup} + \hat{\imath}_r \times V_r \frac{\partial \hat{\imath}_r}{\partial r} \overset{=0}{\diagup} + \hat{\imath}_r \times \hat{\imath}_\theta \frac{\partial V_\theta}{\partial r} + \hat{\imath}_r \times V_\theta \frac{\partial \hat{\imath}_\theta}{\partial r} \overset{=0}{\diagup}$$

$$+ \hat{\imath}_r \times \hat{\imath}_z \frac{\partial V_z}{\partial r} + \hat{\imath}_r \times V_z \frac{\partial \hat{\imath}_z}{\partial r} \overset{=0}{\diagup}$$

$$+ \hat{\imath}_\theta \times \hat{\imath}_r \frac{1}{r}\frac{\partial V_r}{\partial \theta} + \hat{\imath}_\theta \times V_r \frac{1}{r}\frac{\partial \hat{\imath}_r}{\partial \theta} \overset{=\hat{\imath}_\theta}{\diagup} + \hat{\imath}_\theta \times \hat{\imath}_\theta \frac{1}{r}\frac{\partial V_\theta}{\partial \theta} \overset{=0}{\diagup}$$

$$+ \hat{\imath}_\theta \times V_\theta \frac{1}{r}\frac{\partial \hat{\imath}_\theta}{\partial \theta} \overset{=-\hat{\imath}_r}{\diagup} + \hat{\imath}_\theta \times \hat{\imath}_z \frac{1}{r}\frac{\partial V_z}{\partial \theta} + \hat{\imath}_\theta \times V_z \frac{1}{r}\frac{\partial \hat{\imath}_z}{\partial \theta} \overset{=0}{\diagup}$$

$$+ \hat{\imath}_z \times \hat{\imath}_r \frac{\partial V_r}{\partial z} + \hat{\imath}_z \times V_r \frac{\partial \hat{\imath}_r}{\partial z} \overset{=0}{\diagup} + \hat{\imath}_z \times \hat{\imath}_\theta \frac{\partial V_\theta}{\partial z} + \hat{\imath}_z \times V_\theta \frac{\partial \hat{\imath}_\theta}{\partial z} \overset{=0}{\diagup}$$

$$+ \hat{\imath}_z \times \hat{\imath}_z \frac{\partial V_z}{\partial z} \overset{=0}{\diagup} + \hat{\imath}_z \times V_z \frac{\partial \hat{\imath}_z}{\partial z} \overset{=0}{\diagup}$$

or

$$\nabla \times \vec{V} = \hat{\imath}_z \frac{\partial V_\theta}{\partial r} - \hat{\imath}_\theta \frac{\partial V_z}{\partial r} - \hat{\imath}_z \frac{1}{r}\frac{\partial V_r}{\partial \theta} + \hat{\imath}_z \frac{V_\theta}{r} + \hat{\imath}_r \frac{1}{r}\frac{\partial V_z}{\partial \theta} + \hat{\imath}_\theta \frac{\partial V_r}{\partial z} - \hat{\imath}_r \frac{\partial V_\theta}{\partial z}$$

and finally,

$$\nabla \times \vec{V} = \hat{\imath}_r \left(\frac{1}{r}\frac{\partial V_z}{\partial \theta} - \frac{\partial V_\theta}{\partial z} \right) + \hat{\imath}_\theta \left(\frac{\partial V_r}{\partial z} - \frac{\partial V_z}{\partial r} \right) + \hat{\imath}_z \frac{1}{r}\left(\frac{\partial}{\partial r}(rV_\theta) - \frac{\partial V_r}{\partial \theta} \right) \quad \text{(D.14)}$$

Equation D.13 was derived in Section 5-5 by evaluating the rates of rotation of three pairs of mutually perpendicular line segments in a fluid particle. In Section 5-5, we showed that

$$\nabla \times \vec{V} = 2\vec{\omega}$$

where $\vec{V}$ is the velocity field and $\vec{\omega}$ is the rotation vector for a fluid particle. Thus the curl of a velocity field is related to the rotation of the field.

D-4.5   THE LAPLACIAN

Another operator that occurs frequently is obtained by evaluating the dot product $\nabla \cdot \nabla$. This operator is called the Laplacian, and is given the symbol ∇^2. Thus, in Cartesian coordinates.

$$\nabla^2 = \nabla \cdot \nabla = \frac{\partial^2}{\partial x^2} + \frac{\partial^2}{\partial y^2} + \frac{\partial^2}{\partial z^2} \tag{D.15}$$

In cylindrical coordinates,

$$\nabla^2 = \nabla \cdot \nabla = \left(\hat{\imath}_r \frac{\partial}{\partial r} + \hat{\imath}_\theta \frac{1}{r} \frac{\partial}{\partial \theta} + \hat{\imath}_z \frac{\partial}{\partial z} \right) \cdot \left(\hat{\imath}_r \frac{\partial}{\partial r} + \hat{\imath}_\theta \frac{1}{r} \frac{\partial}{\partial \theta} + \hat{\imath}_z \frac{\partial}{\partial z} \right)$$

$$= \frac{\partial^2}{\partial r^2} + \hat{\imath}_\theta \cdot \frac{1}{r} \left(\frac{\overset{= \hat{\imath}_\theta}{\partial \hat{\imath}_r}}{\partial \theta} \frac{\partial}{\partial r} + \hat{\imath}_r \frac{\partial^2}{\partial \theta \, \partial r} + \frac{\overset{= -\hat{\imath}_r}{\partial \hat{\imath}_\theta}}{\partial \theta} \frac{1}{r} \frac{\partial}{\partial \theta} + \hat{\imath}_\theta \frac{1}{r} \frac{\partial^2}{\partial \theta^2} \right) + \frac{\partial^2}{\partial z^2}$$

$$\nabla^2 = \frac{\partial^2}{\partial r^2} + \frac{1}{r} \frac{\partial}{\partial r} + \frac{1}{r^2} \frac{\partial^2}{\partial \theta^2} + \frac{\partial^2}{\partial z^2}$$

But

$$\frac{\partial^2}{\partial r^2} + \frac{1}{r} \frac{\partial}{\partial r} = \frac{1}{r} \frac{\partial}{\partial r} \left(r \frac{\partial}{\partial r} \right)$$

so

$$\nabla^2 = \frac{1}{r} \frac{\partial}{\partial r} \left(r \frac{\partial}{\partial r} \right) + \frac{1}{r^2} \frac{\partial^2}{\partial \theta^2} + \frac{\partial^2}{\partial z^2} \tag{D.16}$$

in cylindrical coordinates.

As pointed out in Sections 5-3 and 5-6.2, both the stream function, ψ, and velocity potential, ϕ, satisfy Laplace's equation,

$$\nabla^2 \psi = \nabla^2 \phi = 0$$

for two-dimensional, irrotational, incompressible flow. Laplace's equation also describes many other physical phenomena, including steady state heat conduction, for which $\nabla^2 T = 0$.

D-5   VECTOR IDENTITIES

D-5.1 $\nabla \times \nabla\phi = 0$

In Section 5-6.1, we noted that $\nabla \times \nabla\phi \equiv 0$, and used that relation to define the velocity potential, ϕ. This relation may be verified by expanding it into components. Thus, in Cartesian coordinates,

$$\nabla \times \nabla\phi = \nabla \times \left(\hat{\imath}\frac{\partial\phi}{\partial x} + \hat{\jmath}\frac{\partial\phi}{\partial y} + \hat{k}\frac{\partial\phi}{\partial z} \right)$$

$$= \begin{vmatrix} \hat{\imath} & \hat{\jmath} & \hat{k} \\ \dfrac{\partial}{\partial x} & \dfrac{\partial}{\partial y} & \dfrac{\partial}{\partial z} \\ \dfrac{\partial\phi}{\partial x} & \dfrac{\partial\phi}{\partial y} & \dfrac{\partial\phi}{\partial z} \end{vmatrix}$$

$$\nabla \times \nabla\phi = \hat{\imath}\left(\frac{\partial^2\phi}{\partial y\,\partial z} - \frac{\partial^2\phi}{\partial z\,\partial y} \right) + \hat{\jmath}\left(\frac{\partial^2\phi}{\partial z\,\partial x} - \frac{\partial^2\phi}{\partial x\,\partial z} \right) + \hat{k}\left(\frac{\partial^2\phi}{\partial x\,\partial y} - \frac{\partial^2\phi}{\partial y\,\partial x} \right)$$

If $\phi = \phi(x, y, z)$ is a continuous, differentiable function, then the order of repeated differentiation is not important, that is,

$$\frac{\partial^2\phi}{\partial x\,\partial y} = \frac{\partial^2\phi}{\partial y\,\partial x}, \qquad \frac{\partial^2\phi}{\partial x\,\partial z} = \frac{\partial^2\phi}{\partial z\,\partial x} \qquad \text{and} \qquad \frac{\partial^2\phi}{\partial y\,\partial z} = \frac{\partial^2\phi}{\partial z\,\partial y}$$

Consequently, $\nabla \times \nabla\phi \equiv 0$.

Although it is a more lengthy process, this identity may also be proved in cylindrical coordinates.

D-5.2 $(\vec{V} \cdot \nabla)\vec{V} = \frac{1}{2}\nabla(\vec{V} \cdot \vec{V}) - \vec{V} \times (\nabla \times \vec{V})$

This relation was used in Section 6-5.2 in developing the Bernoulli integral in rectangular coordinates. It may also be proved by expanding both sides into components. Thus, in Cartesian coordinates,

$$\vec{V} \cdot \nabla = V_x\frac{\partial}{\partial x} + V_y\frac{\partial}{\partial y} + V_z\frac{\partial}{\partial z}$$

and

$$(\vec{V} \cdot \nabla)\vec{V} = \hat{\imath}\left(V_x\frac{\partial V_x}{\partial x} + V_y\frac{\partial V_x}{\partial y} + V_z\frac{\partial V_x}{\partial z} \right)$$

$$+ \hat{\jmath}\left(V_x\frac{\partial V_y}{\partial x} + V_y\frac{\partial V_y}{\partial y} + V_z\frac{\partial V_y}{\partial z} \right)$$

$$+ \hat{k}\left(V_x\frac{\partial V_z}{\partial x} + V_y\frac{\partial V_z}{\partial y} + V_z\frac{\partial V_z}{\partial z} \right) \tag{D.17}$$

Also

$$\vec{V} \cdot \vec{V} = V_x^2 + V_y^2 + V_z^2$$

so

$$\frac{1}{2}\nabla(\vec{V}\cdot\vec{V}) = \hat{\imath}\left(V_x\frac{\partial V_x}{\partial x} + V_y\frac{\partial V_y}{\partial x} + V_z\frac{\partial V_z}{\partial x}\right)$$

$$+\,\hat{\jmath}\left(V_x\frac{\partial V_x}{\partial y} + V_y\frac{\partial V_y}{\partial y} + V_z\frac{\partial V_z}{\partial y}\right)$$

$$+\,\hat{k}\left(V_x\frac{\partial V_x}{\partial z} + V_y\frac{\partial V_y}{\partial z} + V_z\frac{\partial V_z}{\partial z}\right) \qquad \text{(D.18a)}$$

and

$$-\vec{V}\times(\nabla\times\vec{V}) = -\begin{vmatrix} \hat{\imath} & \hat{\jmath} & \hat{k} \\ V_x & V_y & V_z \\ \left(\dfrac{\partial V_z}{\partial y}-\dfrac{\partial V_y}{\partial z}\right) & \left(\dfrac{\partial V_x}{\partial z}-\dfrac{\partial V_z}{\partial x}\right) & \left(\dfrac{\partial V_y}{\partial x}-\dfrac{\partial V_x}{\partial y}\right) \end{vmatrix}$$

$$=\hat{\imath}\left(V_z\frac{\partial V_x}{\partial z} - V_z\frac{\partial V_z}{\partial x} - V_y\frac{\partial V_y}{\partial x} + V_y\frac{\partial V_x}{\partial y}\right)$$

$$+\,\hat{\jmath}\left(-V_z\frac{\partial V_z}{\partial y} + V_z\frac{\partial V_y}{\partial z} + V_x\frac{\partial V_y}{\partial x} - V_x\frac{\partial V_x}{\partial y}\right)$$

$$+\,\hat{k}\left(-V_x\frac{\partial V_x}{\partial z} + V_x\frac{\partial V_z}{\partial x} + V_y\frac{\partial V_z}{\partial y} - V_y\frac{\partial V_y}{\partial z}\right) \qquad \text{(D.18b)}$$

Adding Eqs. D.18a and D.18b,

$$\frac{1}{2}\nabla(\vec{V}\cdot\vec{V}) - \vec{V}\times(\nabla\times\vec{V}) = \hat{\imath}\left(V_x\frac{\partial V_x}{\partial x} + V_y\frac{\partial V_x}{\partial y} + V_z\frac{\partial V_x}{\partial z}\right)$$

$$+\,\hat{\jmath}\left(V_x\frac{\partial V_y}{\partial x} + V_y\frac{\partial V_y}{\partial y} + V_z\frac{\partial V_y}{\partial z}\right)$$

$$+\,\hat{k}\left(V_x\frac{\partial V_z}{\partial x} + V_y\frac{\partial V_z}{\partial y} + V_z\frac{\partial V_z}{\partial z}\right) \qquad \text{(D.19)}$$

Comparing Eqs. D.17 and D.19 shows that

$$(\vec{V}\cdot\nabla)\vec{V} = \tfrac{1}{2}\nabla(\vec{V}\cdot\vec{V}) - \vec{V}\times(\nabla\times\vec{V})$$

This identity could also be proved in cylindrical coordinates, but the proof would be extremely lengthy!

Appendix E

SI Units, Prefixes and Conversion Factors

TABLE E.1 **SI Units and Prefixes***

SI Units	Quantity	Unit	SI Symbol	Formula
SI base units:	Length	meter	m	—
	Mass	kilogram	kg	—
	Time	second	sec	—
	Temperature	kelvin	K	—
SI supplementary unit:	Plane angle	radian	rad	—
SI derived units:	Energy	joule	J	$N \cdot m$
	Force	newton	N	$kg \cdot m/sec^2$
	Power	watt	W	J/sec
	Pressure	pascal	Pa	N/m^2
	Work	joule	J	$N \cdot m$

SI prefixes	Multiplication Factor	Prefix	SI Symbol
	$1\ 000\ 000\ 000\ 000 = 10^{12}$	tera	T
	$1\ 000\ 000\ 000 = 10^{9}$	giga	G
	$1\ 000\ 000 = 10^{6}$	mega	M
	$1\ 000 = 10^{3}$	kilo	k
	$0.01 = 10^{-2}$	centi[†]	c
	$0.001 = 10^{-3}$	milli	m
	$0.000\ 001 = 10^{-6}$	micro	μ
	$0.000\ 000\ 001 = 10^{-9}$	nano	n
	$0.000\ 000\ 000\ 001 = 10^{-12}$	pico	p

* Source: ASTM Metric Practice Guide E 380–74, Nov. 1974 (American National Standard Z210.1).
† To be avoided where possible.

TABLE E.2 Conversion Factors and Definitions

Fundamental Conversion Factors:	English Unit	Exact SI Value	Approximate SI Value
Length	1 in.	0.0254 m	—
Mass	1 lbm	0.453 592 37 kg	0.4536 kg
Temperature	1 F	5/9 K	—

Definitions:

Acceleration of gravity: $g = 9.8066$ m/sec^2 ($= 32.174$ ft/sec^2)

Energy: Btu (British thermal unit) $\equiv$ amount of energy required to raise the temperature of 1 lbm of water 1 F (1 Btu = 778.2 ft · lbf)

kilocalorie $\equiv$ amount of energy required to raise 1 kg of water 1 K (1 kcal = 4187 J)

Length: 1 mile = 5280 ft; 1 nautical mile = 6076.1 ft

Power: 1 horsepower $\equiv$ 550 ft · lbf/sec

Pressure: 1 bar $\equiv 10^5$ Pa

Temperature: degree Fahrenheit $t_F = \frac{9}{5}t_C + 32$ (where t_C is degrees Celsius)

degree Rankine $t_R = t_F + 459.67$

Kelvin $t_K = t_C + 273.15$ (exact)

Viscosity: 1 Poise $\equiv 0.1$ kg/m · sec

1 Stoke $\equiv 0.0001$ m^2/sec

Volume: 1 cubic foot = 7.48 gal

Useful Conversion Factors:

1 lbf = 4.448 N

1 lbf/in.^{2} = 6895 Pa

1 Btu = 1055 J

1 hp = 746 W = 2545 Btu/hr

1 kW = 3413 Btu/hr

1 quart = 0.000946 m^3 = 0.946 liter

1 kcal = 3.968 Btu

UNIT CONVERSIONS

The data needed to solve problems are not always available in consistent units. Thus it is often necessary to convert from one system of units to another.

In principle, all derived units can be expressed in terms of basic units. Then only conversion factors for basic units would be required.

In practice, many engineering quantities are expressed in terms of defined units, for example, the horsepower, British thermal unit (Btu), quart, or nautical mile. Definitions for such quantities are necessary, and additional conversion factors are useful in calculations.

Basic SI units and necessary conversion factors, plus a few definitions and convenient conversion factors are given in Table E.2.

Answers to Even-Numbered Problems

CHAPTER 1

1.6 $T \simeq 269$ F, $p = 41$ psia, Work $= 5390$ Btu

1.8 (a) $R = 2V_0^2 \sin \theta \cos \theta / g$ (b) $\theta = \pi/4$

1.10 $t = 3$ W/gk

1.14 (a) $KE/m = 5 \times 10^3$ ft·lbf/slug (b) $KE/\Psi = 31.7$ ft·lbf/ft^3

1.16 $F = 7.36$ mN

1.18 $W = 59.9$ lbf, $\Psi = 0.964$ ft^3

1.20 $SG = 1.22 \times 10^{-3}$

1.22 $a = -2, b = 1, c = -3$

CHAPTER 2

2.6 Slope $= 1/4$

2.8 (a) $y = x^{-a/2}$, $y = z/3$ (b) $y = x^{-a/2}$, $y = \left[\dfrac{(bz - 3)}{(3b - 3)} \right]^{a/b}$

2.12 $\vec{F}_B = 16.8\hat{j} + 88\hat{k}$ N

2.14 $\vec{F}_B = 144\hat{i} + 192\hat{j}$ N

2.20 $\tau_{A-A} = -158$ psi, $\sigma_{A-A} = -51.0$ psi

2.22 $\tau_{yx} = -2.64$ N/m^2

2.24 $\mu = 8.07 \times 10^{-4}$ N·sec/m²

2.26 $|\vec{F}| = 34.8$ lbf

2.28 $F = 1.01$ N

2.30 $U = 9.84$ m/sec

2.34 $\vec{V}_{max} = 5.07 \times 10^{-2}$ m/sec

CHAPTER 3

3.2 $D = 0.254$ m, $p = 154$ kPa (gage)

3.4 $m = 14.7$ kg, $t = 11.9$ mm

3.6 $\Delta p = 0.04$ psi

3.8 $p = 1.18$ psig

3.10 $l = 0.316$ m

3.12 $h = 1.11$ in.

3.14 $\theta = 12.5°$

3.16 $T = 323$ K

3.18 $p = 13.0$ psi

3.20 $d\rho/\rho = 0.451\%$

3.22 $\Delta\rho/\rho_0 = 4.33\%$, $\Delta p/p_i = 2.12\%$

3.24 $dp/dz = -12.1$ N/m³, $dp/dz = -7.25$ N/m³

3.26 $p_{\rho=c} = 57.5$ kPa, $p_{ad} = 60.2$ kPa

3.28 $T_{z=0} = 100$ C, $T_{z=1000\,m} = 96.6$ C, $T_{z=2000\,m} = 93.2$ C

3.30 $\Delta h = 75.6$ mm H₂O

3.32 $\rho_s = 5.09$ kg/m²

3.34 $p_{abs} = 927$ kPa, $p_{gage} = 824$ kPa

3.36 $|\vec{F}| = 71.2$ kN

3.38 $|\vec{F}_R| = 61,700$ lbf

3.40 $|\vec{F}| = \rho g w b \left(h + \dfrac{b}{2}\right)$, $a = \left(h + \dfrac{b}{2}\right) + \dfrac{1}{12}b^2 \Big/ \left(h + \dfrac{b}{2}\right)$

3.42 $|\vec{F}| = 1940$ lbf

3.44 $|\vec{R}| = 52.6$ kN

3.46 $|\vec{F}| = 1800$ lbf

3.48 $D = 2.60$ m

3.50 $|\vec{F}| = 33.3$ kN, $d_b = 7.28$ mm

3.52 $|\vec{F}| = 63,800$ lbf

3.54 $F_A/b = 174$ lbf/ft

3.56 $|\vec{M}| = 6340$ ft·lbf

3.60 $F_{R_y} = -73.9$ kN, $x' = 1.06$ m

3.62 $F_{R_y} = 3.68$ kN, $|\vec{M}| = 1.05$ kN·m

3.64 $\vec{M}_0 = 3.19\hat{k}$ kN·m

3.66 $F_{R_y} = -17,100$ lbf, $x' = 2.14$ ft

3.68 $F_{R_x} = 78.4$ kN, $F_{R_y} = 56.0$ kN, $y' = 0.833$ m, $\theta = 35.5°$

3.70 $F_{R_x} = -12,460$ lbf, $z' = 1.93$ ft

3.72 $M = 302$ kg

3.74 $F_v = 356$ kN

3.76 $h = 17.7$ cm

3.78 $D = 27.4$ m, $\mathrm{V_{sw}/V} = 0.507$

3.80 $D = 82.7$ m, $M = 637$ kg

3.82 $\theta = 23.8°$

3.84 Slope $= 0.22$

3.86 $\alpha = 13.3°$

3.88 $F_{w_{rear}} = 107$ N, $F_{w_{front}} = 54.1$ N, $F_{w_{bottom}} = 156$ N

3.90 $\partial p/\partial r = \rho \omega^2 r$

3.92 $\theta = 14.3°$, toward center of curve

CHAPTER 4

4.2 $t_2 - t_1 = 16.7$ sec

4.4 $\vec{T}_{avg} = -0.325\hat{k}$ mN·m

4.6 $W = 469$ kJ

4.8 $\delta Q = 9.57$ kJ

4.10 $p_f = 300$ psia, $T_f \simeq 947$ F, $W = -1.65 \times 10^5$ ft·lbf, $Q = 2130$ Btu

4.12 (a) $-awy\,dz$, (b) 0, (c) $-bw\,dy$, (d) $-abwy\,dy\hat{j} - b^2w\,dy\hat{k}$,
(e) $-25w\hat{j} - 25w\hat{k}$ m³/sec²

4.14 $\int \vec{V} \cdot d\vec{A} = 0, \int \vec{V}(\vec{V} \cdot d\vec{A}) = 0$

4.16 (a) $u_{max}\pi R^2/2$, (b) $u_{max}^2\pi R^2/3$

4.18 $\vec{V}_3 = 4.04\hat{\imath} - 2.33\hat{\jmath}$ m/sec

4.20 $D_2 = 55.6$ mm

4.22 $\bar{V} = V_{max}/3$

4.24 $U_{max} = 7.50$ m/sec

4.26 $\partial\rho_0/\partial t = 2.50 \times 10^{-3}$ slug/ft$^3\cdot$sec

4.28 $V_3 = 1.00$ m/sec (level falling)

4.30 $y = 0.134$ m

4.32 $t = 6\mathbb{V}_0/5Q_0$

4.34 $r = (2g)^{1/4}(A/\pi s)^{1/2}h^{1/4}$; $\mathbb{V} = 2A(2g)^{1/2}n^{3/2}s^{1/2}/3$

4.36 Ratio = 4/3

4.38 Ratio = 4/3

4.40 $R_x = 199$ N

4.42 $M = 1150\,(1 - \cos\theta)$ kg

4.44 $T = 30.0$ N, $\vec{F} = 30.0\hat{\imath} - 51.9\hat{\jmath}$ N

4.46 $K_x = 0.0230$ lbf

4.48 $F = 97.0$ (down)

4.50 $F = 213$ lbf, $W = 203$ lbf

4.52 $K_x = 2.80$ kN

4.54 $R_x = -1.81$ kN (tension)

4.56 $K_x = 5.11$ kN

4.60 Drag = 0.558 N

4.62 $V_2 = 6.60$ m/sec, $p_2 - p_1 = 84.2$ kPa

4.64 $\Delta p/\frac{1}{2}\rho V_1^2 = 2[1 - (1 - Q_3/Q_1)^2]$

4.66 $\vec{K} = 822\hat{\imath} - 220\hat{\jmath}$ N

4.68 $\vec{K} = 570\hat{\imath} - 329\hat{\jmath}$ lbf

4.70 $\vec{K} = 1.73\hat{\imath}$ kN

4.72 $V = 95.4$ ft/sec (relative to boat)

4.74 $a = -1.19$ m/sec^2 (sled decelerates)

4.76 $t = 24.7$ sec

4.78 $a = 2\rho V^2 A/M\,(1 + 2\rho V At/M)^{-2}$, $t = M/2\rho V A$

4.80 $U = 51.5$ m/sec

4.82 $U = 16.5$ m/sec

4.84 $U = 83.9$ ft/sec

4.86 $a = -14.7$ m/sec^2 (deceleration)

4.88 $U = 344$ m/sec

4.90 $t = 12.5$ sec, $\theta_{max} = 29.0°$

4.92 $V = 734$ m/sec

4.94 $\partial M_{cv}/\partial t = -0.165$ kg/sec, $dP_{xcv}/dt/K_x = -4.62 \times 10^{-4}$ percent

4.96 $U/V = 1 - 1/(1 + 2\rho V At/M_0)^{1/2}$

4.98 $p_0 = 18.2$ kN/m^2 (gage), $Q_0 = 3.97 \times 10^{-4}$ m^3/sec

4.100 $K_{x_1} = 46.0$ kN, $K_{x_2} = -44.4$ kN, $\vec{M} = -920\hat{k}$ kN $\cdot$ m (clockwise)

4.102 $\vec{T}_{out} = R(1 - \cos\theta)\rho V A(V - U)\hat{k}$, $U = R\omega$, $\dot{W}_{out} = \vec{\omega} \cdot \vec{T}_{out}$, $U/V = 1/2$

4.104 $Q_{des} = 4140$ gpm, $T_{cv} = 2280$ ft $\cdot$ lbf, $\dot{W}_{in} = 759$ hp (at design conditions with $\beta_r = 90°$)

4.106 $\dot{W}_{in} = 80.0$ kW

4.108 $\dot{W}_s = 108$ kW

4.110 $\dot{W}_s$, actual $= 3.41$ kW

4.112 $\dot{W}_s = 42.9$ kW

4.114 $\cos\theta = (1 - V_2/V_1)[(1 + V_2/V_1)/(1 + (V_2/V_1)^3)]^{1/2}/\sqrt{k}$

4.116 $\eta = 57.1$ percent

4.118 $Q = 0.0166$ m^3/sec, $z_{max} = 61.4$ m, $K_x = -561$ N

4.120 $\Delta mef/\dot{m} = -1.88$ N $\cdot$ m/kg, $\Delta T = 4.49 \times 10^{-4}$ K

CHAPTER 5

5.2 $r = a$; $\theta = 0$, π; $r = a$, $\theta = 0$, π

5.4   Yes, no, no, no, no

5.6   Yes, yes, no

5.8 $A + E + J = 0$, the rest are arbitrary

5.10 $V_\theta = -\Lambda \sin\theta/r^2 + f(r)$

5.12 $\vec{V} = (-U\cos\theta + Q/2\pi r)\hat{i}_r + U\sin\theta\hat{i}_\theta$; $(r, \theta) = (Q/2\pi U, 0)$; $\psi_{stag} = 0$

5.14   2-D., yes, $\psi = -y^3z - 2z^2$

5.16 $\vec{V} = -8A\hat{i} + 6A\hat{j}$, $Q = 4$ m^3/m $\cdot$ sec

5.18   Yes, yes

5.20 Yes, yes

5.22 $\vec{\omega} = \omega\hat{\imath}_z$, $\vec{\xi} = 2\omega\hat{\imath}_z$

5.24 $g(r) = c/r$

5.26 $\vec{V} = a\hat{\imath} - b\hat{\jmath}$; $(x, y) = (100, -100)$ m

5.28 $\psi = b(x^2 - y^2)/2 - 2axy$; $\vec{a}_p = (4a^2 + b^2)\hat{\imath}$

5.30 3-D.; no; $\vec{a}_p = 27\hat{\imath} + 9\hat{\jmath} + 64\hat{k}$ m/sec^2

5.32 $f_1 = x_0 e^{At}$, $f_2 = y_0 e^{-At}$; 0.693 sec, 1.39 sec; $(\hat{\imath} + \hat{\jmath})$ m/sec^2, $(2\hat{\imath} + \frac{1}{2}\hat{\jmath})$ m/sec^2

5.34 $a_r = -5.56$ ft/sec^2, $a_\theta = 1.36$ ft/sec^2; $a_r = -22.2$ ft/sec^2, $a_\theta = 0$

5.36 $a_r = \dfrac{U^2}{a}\left(\dfrac{a}{r}\right)^3\left[1 - \left(\dfrac{a}{r}\right)^2\right]$, $a_\theta = 0$; $a_r = \dfrac{-4U^2\sin^2\theta}{a}$,

$a_\theta = \dfrac{4U^2}{a}\sin\theta\cos\theta$; $a_{\theta\,\text{max}} = \pm\dfrac{2U^2}{a}$

5.38 $\vec{a}_p = 44.5\hat{\imath}$ ft/sec^2; $\vec{a}_p = 103\hat{\imath}$ ft/sec^2

5.40 $a_x = 4[20 + 2\sin(\omega t)]^2 + 1.2\cos(\omega t)$ m/sec^2

5.42 $\dfrac{DT}{Dt} = \left[\dfrac{U}{L}\sin\left(\dfrac{2\pi t}{\tau}\right) - \dfrac{2\pi}{\tau}\cos\left(\dfrac{2\pi t}{\tau}\right)\right]\alpha e^{-x/L}$ deg/sec

CHAPTER 6

6.2 $\vec{a}_p = 2\hat{\imath} + 2\hat{\jmath}$ ft/sec^2, $\nabla p = -(4\hat{\imath} + 68.4\hat{\jmath})$ lbf/ft^2/ft

6.4 $|\vec{a}_p| = 2.24$ m/sec^2, $\theta = 63.4°$

6.6 $\nabla p = -(3\hat{\imath} + 9\hat{\jmath})$ kN/m^2/m

6.8 $\nabla p = \rho\left[\hat{\imath}_r\dfrac{(Q^2 + k^2)}{4\pi^2 r^3} - \hat{\imath}_z g\right]$

6.10 $dy/dx = -ax/g$, $d = 8$ in., $b = 12$ in.

6.12 $\omega_{\text{max}} \le \sqrt{2gh}/R$

6.14 $\partial p/\partial r = \rho\omega^2 r$, $p_2 - p_1 = 150$ kN/m^2

6.16 $\partial p/\partial r = \rho k^2/4\pi^2 r^3$, $p_2 - p_1 = 37.5$ kN/m^2

6.18 $p_1 - p_2 = 2.43$ lbf/ft^2

6.20 $p_1 = 227$ kPa

6.22 $V = 21.5$ ft/sec, $Q = 0.469$ ft^3/sec

6.24 $h = 16.3$ m

6.26 $Q = 1.25 \times 10^{-2}$ m^3/sec, $p_c = 71.6$ kPa

6.28 $p - p_{\text{atm}} = \dfrac{\rho q^2}{8h^2}(R^2 - r^2)$, $F_z = 3.06$ mN

6.30 $F = 83.3$ kN

6.32 $p_i - p_\infty = 112$ kPa

6.34 $V = 27.5$ m/sec

6.36 $Q_1 = 0.176$ ft^3/sec, $u_4 - u_3 = 1.8$ Btu/slug, $Q_2 = 0.353$ ft^3/sec

6.38 $p_2 - p_1 = -18.8$ kN/m^2

6.40 $\psi = \dfrac{A}{3}x^2y^3 + c$, $\vec{\omega} = -\dfrac{B}{2}(y^3 + 3x^2y)\hat{k}$

6.42 $p_{\text{gage}} = 3\rho V^2 R^2/8b^2$

CHAPTER 7

7.2 $V = 1370$ mi/hr, $V = 1570$ mi/hr

7.4 $We = 109$

7.6 $F/\mu V D$

7.8 $\mathcal{P}/\rho D^5 \omega^3 = f(Q/D^3\omega)$

7.10 3 dimensionless parameters, $\sigma/\rho D V^2$, $\mu/\rho V D$, d/D

7.12 $V = \sqrt{gD}f(\lambda/D)$

7.14 $V \sim (\sigma/\rho\lambda)^{1/2}$

7.16 $\mathcal{P}/p\omega D^3 = f(\mu\omega/p, c/D)$

7.18 $T/R^3\mu\omega = f(h/R)$

7.20 $V = 9.58$ m/sec

7.22 $V_m = 2.24$ m/sec, $v_m = 1.12 \times 10^{-8}$ m^2/sec

7.24 $V_m = 39.2$ m/sec, $V_p = 39.2$ m/sec

7.26 $H_m = 145$ ft·lbf/slug, $Q_m = 5.92$ ft^3/sec, $D_m = 0.491$ ft

7.28 $fd/V = g(\rho V d/\mu)$, $V_1/V_2 = 1/2$, $f_1/f_2 = 1/4$

7.30 $\omega_2 = \omega_1/8$, $H_2 = H_1/16$

CHAPTER 8

8.2   7.5 m $< L <$ 12.0 m

8.4 $Q/b = 2u_{\text{max}}h/3$

8.6 $\tau_{yx} = y\,\partial p/\partial x$, $\tau_{max} = -0.00835$ lbf/ft^2

8.8 $\mu = 0.0695$ N$\cdot$sec/m^2

8.10 $u_{max} = 1.0$ ft/sec at $y = 0.01$ ft, $V/w = 0.0522$ ft^3/ft

8.12 $u = U_0$ at $y = 0$, $\tau = 0$ at $y = h$, $u = \rho g(y^2/2 - hy)/\mu + U_0$

8.14   For water: $Q = 7.13 \times 10^{-6}$ m^3/sec, $\partial p/\partial x = -146$ N/m$^2\cdot$m;
       For oil: $Q = 4.10 \times 10^{-4}$ m^3/sec, $\partial p/\partial x = -436$ kN/m$^2\cdot$m

8.16 $u = -R^2\,\partial p/\partial x\,[1 - (r/R)^2]/4\mu$

8.18 $u = \bar{V}$ when $r = 0.757$ R

8.20 $\tau_{rx} = r\,\partial p/\partial x/2$

8.22 $\beta = 2\left(\dfrac{U}{\bar{V}}\right)^2 \displaystyle\int_0^1 \left(\dfrac{u}{U}\right)^2 \left(\dfrac{r}{R}\right) d\left(\dfrac{r}{R}\right)$, $\beta_{lam} = 4/3$, $\beta_{turb} = 1.02$

8.24 $p_1 = 345$ kPa (gage)

8.26 $Q = 2.45 \times 10^{-3}$ m^3/sec, $f = 0.025$

8.28 $\Delta p \sim \bar{V}^{7/4}$, $\Delta p \sim 1/D^{19/4} \sim 1/D^5$

8.30 $f = 0.042$

8.32 $Q = 1.09 \times 10^{-3}$ m^3/sec

8.34 $AR \simeq 2.7$, $2\phi \simeq 6°$, $Q = 0.172$ m^3/sec

8.36 $z_1 - z_2 = 9.80$ m

8.38 $p_1 = 1.40$ MPa (gage), $T_2 - T_1 = 0.215$ K

8.40 $z_1 - z_2 = 51.8$ m

8.42 $z_1 - z_5 = 8.13$ m

8.44 $L = 212$ m

8.46 $Q = 108$ gpm, $V_4 = 124$ ft/sec, $\dot{W}_s = 13.4$ hp

8.48 $p_A = -25.5$ kPa (gage), $Q = 0.0223$ m^3/sec

8.50   Flow from ② to ① $Q = 0.0134$ m^3/sec

8.52 $\Delta t \simeq 168$ sec

8.54 $D \geq 14$ mm

8.56 $D = 6$ in. (nominal) pipe

8.58 $Q_1 \simeq 498$ gpm, $Q_2 \simeq 470$ gpm, $Q_3 \simeq 532$ gpm, $p_A \simeq 73.5$ psig,
       $p_1 \simeq 53.1$ psig, $p_2 \simeq 47.3$ psig, $p_3 \simeq 60.8$ psig

8.62 $Q \simeq 10{,}000$ ft^3/min

8.64 0.684 m $< x < 0.912$ m

8.66 $A = U$, $B = \pi/2\delta$, $C = 0$

8.70 δ^*/δ: Linear, 1/2; Parabolic, 1/3; Cubic, 3/8; Sinusoidal, 0.363

8.72 $\delta^*/x = 1.74/\sqrt{\text{Re}_x}$

8.74 $\dot{m}_{ab} = 0.0133$ kg/sec, $F_x = -0.133$ N (to the left)

8.76 $\tau = 1.57\mu U \cos(\pi y/2\delta)/\delta$, $\tau_w = 8.67 \times 10^{-4}$ lbf/ft^2, $C_f = 0.00358$

8.78 $\delta_{\text{lam}} = 5.48$ mm, $\tau_w = 0.101$ N/m^2, $\delta_{\text{turb}} = 23.7$ mm, $\tau_w = 0.502$ N/m^2

8.80 $\delta/x = 4.64/\sqrt{\text{Re}_x}$, $C_f = 0.647/\sqrt{\text{Re}_x}$

8.82 $\delta/x = 0.398/(\text{Re}_x)^{1/5}$, $C_f = 0.0567/(\text{Re}_x)^{1/5}$

8.84   For $\phi = 0$, $p_2 < p_{2i}$, for $\phi > 0$ (no separation), $p_2 < p_{2i}$

8.86 $F_D = 0.767$ N

8.88 $\mathscr{P} = 642$ W

8.90 $F_D = 92.3$ kN

8.92 $D = 6.90$ m

8.94 $V = 265$ mph, 252 mph

8.96 $C_D = 24/\text{Re} = 61.9$, $\rho_s = 3720$ kg/m^3, $V = 0.731$ m/sec

8.98 $\mathscr{P} = 4.80$ kW

8.100 $T = 11.9$ N·m

8.102 $\Delta\mathscr{P} = 4.56$ kW

8.104 $A_p = 7.03$ m^2

8.106 $V = 251$ kt

8.108 $M = 7260$ kg, $T = 2.67$ kN

8.110 $F_L = 0.701$ oz, $R = 82.8$ ft

8.112 $\dot{m} = 0.0592$ g/sec

8.114 $D_t = 225$ mm, $\dot{m}_{\text{min}} = 8.48$ kg/sec

8.118 $\dot{m} = 2.10$ kg/sec, $\Delta h = 185$ mm Hg

CHAPTER 9

9.2 $S_2 - S_1 = -0.923$ Btu/R, $U_2 - U_1 = -684$ Btu,
      $H_2 - H_1 = -960$ Btu

9.4 $s_2 - s_1 = 134$ J/kg·K

9.6 $s_2 - s_1 = -18.9$ J/kg·K, adiabatic process is not possible

9.8 $\oint \delta W < 0, \oint \delta Q > 0$

9.10 $M_a = 0.437, M_b = 2.33, M_{rb} = 2.77$

9.12 $V = 725$ m/sec

9.14 $V = 6320$ ft/sec

9.16 $t = 53.5$ sec

9.18 $M = 1.60, V = 479$ m/sec, $z = 10$ km

9.20 $p_0 = 10.2$ psia, $p_0 = 57.5$ psia

9.22 $p_0 = 546$ kPa (abs.), $T_0 = 466$ K, $h_0 - h = 178$ kJ/kg

9.24 $T_{0_1} = 344$ K, $p_{0_1} = 223$ kPa (abs.), $T_{0_2} = 344$ K, $p_{0_2} = 145$ kPa (abs.)

CHAPTER 10

10.2 $V = 4210$ ft/sec

10.4 $V = 1280$ m/sec

10.6 $V = 2620$ ft/sec, $\dot{m} = 1.76$ lbm/sec, $M = 1.36$

10.8 $V = 797$ m/sec, $\dot{m} = 0.706$ kg/sec, $M = 1.35$

10.10 $M_2 = 1.35$

10.12 $M_2 = 1.20$

10.14 $p_2 - p_1 = 315$ kPa, $A_1/A_2 = 3.79$

10.16 $p_t = 33.0$ psia, $M_t = 0.90, V_t = 1060$ ft/sec

10.18 $R_x = 1560$ N (to the left)

10.20 $p_t = 112$ kPa (abs.)

10.22 $p_0 \geqslant 191$ kPa (abs.), $\dot{m} = 1.28$ kg/sec

10.24 $p_t = 2.64$ MPa (gage), $T_t = 369$ K, $\dot{m} = 0.0114$ kg/sec

10.26 $M_t = 1.0, A_t = 450$ mm^2

10.28 $p_{01} = 700$ kPa (abs.), $M_1 = 0.6, T_1 = 298$ K, $\dot{m} = 0.622$ kg/sec

10.30 $M_1 = 1.16$

10.32 $\dot{m} = 3.74$ lbm/sec, $A_1 = 2.99$ in.^{2}

10.34 $V_1 = 1300$ m/sec, $\dot{m} = 87.4$ kg/sec

10.36 $\dot{m} = 6.05$ lbm/sec, $\dot{m}$ decreases by factor of 2

10.38 $\dot{m} = 5.44$ kg/sec

10.40 $V_1 = 504$ m/sec, $A_1 = 0.0596$ m^2

10.42 $\dot{m} = 33.0$ kg/sec, $A_e = 0.158$ m^2, $A_e/A_t = 18.0$

10.44 $\delta Q/dm = 63.0$ Btu/lbm, $p_{02} = 56.6$ psia

10.46 $V_1 = 688$ m/sec, $M_1 = 1.90$

10.48 $p_2 = 477$ kPa (abs.), $s_2 - s_1 = 49.5$ J/kg·K

10.50 $M_1 = 0.20$, $s_2 - s_1 = 311$ J/kg·K

10.52 $F_f = 822$ lbf (opposes motion)

10.54 $V_{max} = 1040$ ft/sec, $p_{min} = 18.5$ psia

10.56 $L_{12} = 2.88$ m

10.58 $L_{12} = 18.8$ ft

10.60 $T_2 = 459$ K, $L_{12} = 34.5$ m

10.66 $p_2 = 91.2$ psia, $T_2 = 1960$ R, $T_{02} = 2000$ R,
$\delta Q/dm = 240$ Btu/lbm, $p_{02} = 97.1$ psia

10.68 $V_2 = 866$ m/sec, $p_2 = 46.4$ kPa (abs.), $M_2 = 1.96$, $\delta Q/dm = 156$ kJ/kg

10.70 $s_2 - s_1 = 0.0532$ kJ/kg·K, $\delta Q/dm = 18$ kJ/kg,
$p_{01} - p_{02} = 2.0$ kPa

10.72 $\dot{Q} = 1050$ Btu/sec, $M_2 = 0.50$

10.74 $M_2 = 0.30$, $T_2 = 1964$ R, $p_2 = 91.2$ psia, $T_{02} = 1998$ R,
$p_{02} = 97.1$ psia

10.76 $M_2 = 1.96$, $p_2 = 46.7$ kPa (abs.), $V_2 = 865$ m/sec, $\delta Q/dm = 162$ kJ/kg

10.78 $T_{02} = 382$ K, $p_{02} = 87.7$ kPa (abs.), $T_2 = 318$ K,
$p_2 = 46.3$ kPa (abs.), $\rho_2 = 0.507$ kg/m^3, $V_2 = 356$ m/sec,
$\delta Q/dm = 17$ kJ/kg

10.80 $\dot{Q} = 898$ Btu/sec, $M_2 = 0.50$, $p_2 = 7.84$ psia

10.84 $Q = 1.84 \times 10^8$ ft^3/day

10.86 $p_2 = 344$ kPa (abs.)

10.88 $p_2 = 28.1$ psia

10.90 $M_2 = 0.500$, $T_2 = 679$ K

10.92 $V_2 = 258$ m/sec, $p_2 - p_1 = 471$ kPa, $(p_2 - p_1)_{s=c} = 842$ kPa

10.94 $M_2 = 0.701$, $V_2 = 267$ m/sec, $p_{02} = 205$ kPa (abs.)

10.96 $V_2 = 247$ m/sec, $T_2 = 670$ K, $s_2 - s_1 = 0.315$ kJ/kg·K

10.98 $T_2 = 520$ K, $p_{02} = 1.29$ MPa (abs.)

10.100 $M_2 = 0.493$, $V_2 = 257$ m/sec, $p_{01} - p_{02} = 512$ kPa

10.102

Section	M	T_0 (R)	p_0 (psia)	T (R)	p (psia)	ρ (lbm/ft³)	V (ft/sec)
2	—	1080	255	633	39.2	0.157	2320
3	0.600	1080	198	1010	155	0.390	934
4	0.700	1080	—	—	131	—	—

$L_{14} = 5.93$ ft

10.104 $p_{03} = 57.9$ kPa (abs.), $T_3 = 414$ K, $p_3 = 51.9$ kPa (abs.)

10.106 $p_3 = 51.5$ kPa (abs.), $p_{01} - p_{03} = 1.60$ kPa

10.108 $\dot{m} = 2.55$ lbm/sec, $A_t = 1.52$ in.², $p_3 = 46.7$ psia

10.110 $p_2 = 33.4$ kPa (abs.), $V_2 = 162$ m/sec

10.112 $M_e = 2.94$, $p_0 = 3.39$ MPa (abs.), $p_{b_1} = 3.35$ MPa (abs.), $p_{b_2} = 1.00$ MPa (abs.), $p_{b_3} = 101$ kPa (abs.)

10.114 $A_t = 1.52$ in.², $\dot{m} = 2.55$ lbm/sec, $p_3 = 46.7$ psia

10.116 $M_1 = 1.50$

10.118 33.4 kPa $< p_b <$ 99.6 kPa (abs.), $\dot{m} = 0.121$ kg/sec

10.120 $p_{atm} < p_0 < 112$ kPa (abs.), or $p_0 > 743$ kPa (abs.)

Index

Absolute pressure, 66
Absolute viscosity, 35
Acceleration, convective, 249
  gravitational, 15
  local, 249 *accel CV 156*
  particle in a velocity field,
    248, 262
Adiabatic process, 488
Adverse pressure gradient, 43
Aging of pipes, 366
Angle of attack, 437
Angular momentum, *see* Moment
  of momentum
Archimedes' principle, 87
Aspect ratio, airfoil, 440
  rectangular duct, 397
Atmosphere, standard, 65, 612
Average velocity, 330

Barometer, 67
Barotropic fluid, 63
Barrel, petroleum industry, 470
Basic equation of fluid statics, 53
Basic equations for control volume,
    conservation of mass, 125
  first law of thermodynamics, 191
  moment of momentum, for
    inertial control volume, 172
  for rotating control volume,
    183
  Newton's second law (linear
    momentum), for control
    volume with arbitrary
    acceleration, 166
  for control volume with
    rectilinear acceleration, 156

for nonaccelerating control
    volume, 134, 150
  second law of thermodynamics,
    196
Basic laws for system, conservation
    of mass, 114
  first law of thermodynamics, 115
  moment of momentum, 114
  Newton's second law (linear
    momentum), 114
  second law of thermodynamics,
    115
Basic pressure-height relation, 57
Bearing, journal, 341
Bernoulli equation, 149, 273
  irrotational flow, 292
  unsteady flow, 293
Body force, 28
Boundary layer, 40, 398
  displacement thickness, 400
  effect of pressure gradient on, 418
  flat plate, 409
  laminar, 411
  momentum integral equation for,
    408
  separation, 418
  thickness, 400
  transition, 399
  turbulent, 413
Boundary-layer control, 442
Buckingham Pi theorem, 308
Bulk modulus, 63, 96, 610
Buoyancy force, 86

Capillary effect, 313
Capillary viscometer, 350

677

Cavitation number, 322
Choking, 531, 537, 548
Chord, 437
Coanda effect, 207
Components of a vector, 649
Compressible flow, 38, 45, 485
Concentric-cylinder viscometer, 50
Conical diffuser, 372
Conservation of energy, *see* First
    law of thermodynamics
  of mass, 114, 125
Constitutive equations, 4
Contact angle, 611
Continuity, *see* Conservation,
    of mass
Continuity equation, differential
    form, 226, 228
Continuum, 21
Contraction coefficient, 451
Control surface, 7
Control volume, 7
Convective acceleration, 249
Converging nozzle, 529
Converging-diverging nozzle, 536,
    592
Conversion factors, 662
Critical conditions, 510
Critical pressure ratio, 530
Critical Reynolds number, 366, 431
Critical speed, 510
Cross product, 652
Curl, 239, 656
Cylinder, flow around, 41
  inviscid flow around, 41, 299

Deformation, angular, 236
  linear, 236
  rate of, 34, 243
Del operator, cylindrical
    coordinates, 229, 654
  rectangular coordinates, 226, 654

Density, 22
Density field, 23
Derivative, substantial, 249
Design conditions, nozzle, 538
Diffuser, 372, 421, 521
Dimension, 13
  of flow field, 24
Dimensional homogeneity, 13
Displacement thickness, 400
Divergence, 655
Doppler effect, 499
Dot product, 651
Downwash, 441
Drag, 42, 424
  friction, 425
  pressure, 42, 426
Drag coefficient, 425
  airfoil, 438
  cylinder, 431
  flat plate normal to flow, 426
  flat plate parallel to flow, 426
  selected objects, 428
  sphere, 429
  streamlined strut, 435
Dynamic pressure, 284
Dynamic similarity, 318
Dyne, 14

Effective viscosity, 51
Elbow flowmeter, 447
Energy equation, *see* First law
    of thermodynamics
Enthalpy, 487
Entrance length, 330
Entropy, 115, 488
Equation of state, 486
Equivalent length, 368
  bends, 374
  fittings and valves, 376
  miter bends, 375
Ergometer, 481

Euler equations, 266
   along streamline, 271
   normal to streamline, 272
Eulerian method of description, 11
Euler number, 317
Euler turbine equation, 174
Extensive property, 116
External flow, 331

Fanno line flow, 544
   basic equations for, 544
   choking length, 557
   effects on properties, 548, 549
   maximum duct length, 557
   tables for computation of, 552,
      627
   Ts diagram, 547
Field representation, 23
Films and film loops, 641
First law of thermodynamics,
      115, 191
Fittings, losses in, 375
Flap, 442
Flat plate, flow over, 39, 409
Flow coefficient, 451
   flow nozzle, 455
   orifice plate, 453
   venturi meter, 456
Flow measurement, 451
Flowmeter, 449
   mechanical, 460
Flow nozzle, 453
Fluid, 1
Fluidic device, 207
Fluid particle, 23
Fluid properties, 609
Force, body, 28
   buoyancy, 86
   pressure, 54
   shear, 2
   surface, 28

Force, hydrostatic,
   on curved submerged surfaces, 76
   on plane submerged surfaces, 68
Forced vortex, 299
Francis turbine, 217
Free surface, 58
Free vortex, 261, 299
Friction drag, 402
Friction factor, 363, 364
   fully-rough flow regime
      correlation, 467
   smooth pipe correlation, 467
Frictionless flow, 38, 266
Froude number, 317
Fully-developed flow, 330
Fully-rough flow regime, 366

Gage pressure, 66
Gas constant, 486, 618
   universal, 486
Geometric similarity, 318
Gibbs equation, 290
Gradient, 55, 655
Gravity, acceleration of, 15

Head, 174, 361
Head loss, 359
   major, 361
   minor, 367
   total, 360
Head loss coefficient, 481
Head loss in, diffusers, 372
   enlargements and contractions,
      371
   exits, 373
   gradual contractions, 371
   miter bends, 375
   nozzles, 371
   pipe bends, 373
   pipe entrances, 368
   pipes, 363

sudden area changes, 371
valves and fittings, 375
Hydraulic accumulator, 219
Hydraulic diameter, 397
Hydraulic jump, 221
Hydrometer, 110
Hydrostatic force, on curved
        submerged surfaces, 76
    on plane submerged surfaces, 68
Hypersonic flow, 499

Ideal fluid, 243, 265
Ideal gas, 63, 486
Incompressible flow, 38, 45
Inertial control volume, 133
Inertial coordinate system, 133, 153
Intensive property, 116
Internal energy, 174, 486
Internal flow, 330
Inviscid flow, 37
Irreversible process, 488
Irrotational flow, 239
Irrotationality condition, 240
Irrotational vortex, 299
Isentropic flow, 515
    basic equations for, 515
        ideal gas, 523
    in converging nozzle, 529
    in converging-diverging nozzle,
        536
    tables for computation of,
        528, 621
Isentropic process, 489
Isentropic stagnation properties,
        501
    for ideal gas, 505

Jet pump, 209
Journal bearing, 341

Kinematic similarity, 318
Kinematic viscosity, 35, 615

Kinematics of fluid motion, 235
Kinetic energy flux coefficient, 359

Lagrangian method of description,
        8
Laminar flow, 44, 332
    between parallel plates, 334
    in pipe, 346
Laminar flow element (LFE), 456
Laplace's equation, 241
Laplacian, 658
Lift, 436
Lift coefficient, 437
    airfoil, 438
    spinning sphere, 445
Linear momentum, 114
Local acceleration, 249
Loss, major, 354
    minor, 354
Loss coefficient, 367
Lubricating oil, 617

Mach angle, 500
Mach cone, 498
Mach number, 45, 316
Magnus effect, 445
Major loss, see Head loss
Manometer, 59
Measurement of, flow, 447
    pressure, 283
    velocity, 284
    viscosity, see Viscometer
Mechanical energy, 360
Mechanical flowmeter, 460
Meniscus, 313
Meter, flow, 449
    elbow, 447
    float-type, 460
    mechanical, 460
    nozzle, 453
    orifice plate, 452

turbine, 460
venturi, 455
Mile, nautical, 662
Minor loss, *see* Head loss
Minor loss coefficient, 367
Model studies, 317
Molecular mass, 486, 618
Moment of momentum, 114, 172, 183
Momentum, angular, *see* Moment of momentum
   linear, *see* Newton's second law of motion
Momentum equation, differential form, 255
   inviscid flow, 266
Momentum flux coefficient, 466
Momentum integral equation, 408
   for zero pressure gradient flow, 410
Moody diagram, 364

Nautical mile, 662
Navier-Stokes equations, 257
Network, pipe, 391
Newton, 14
Newtonian fluid, 33
Newton's law of viscosity, 35
Newton's second law of motion, 8, 114
Noncircular duct, 397
Noninertial reference frame, 163
Non-Newtonian fluid, 33
Normal shock, 577
   basic equations for, 577
   effects on properties, 581, 582
   tables for computation of, 586, 637
   Ts diagram, 582
Normal stress, 32
Non-slip condition, 2, 25, 39

Nozzle, 521
   choked flow in, 531, 537
   converging, 529
   converging-diverging, 536, 592
   design conditions, 538
   normal shock in, 592
   overexpanded, 539
   underexpanded, 538

One-dimensional flow, 24
Orifice plate, 452
Overexpanded nozzle, 539

Pascal, 19
Pathline, 26
Pelton wheel, 217
Physical properties, 609
Pipe, aging, 366
   compressible flow in, 544
   head loss, 359
   networks, 391
   relative roughness, 363, 365
   standard sizes, 378
Pi theorem, 308
Pitot tube, 284
Pitot-static tube, 285
Planform area, 437
Poise, 35
Polar plot, 440, 443
Potential flow theory, 243
Potential function, 240
Potential velocity, 241
Power-law velocity profile, 355
Pressure, 54, 178
   absolute, 66
   dynamic, 284
   gage, 66
   isentropic stagnation, 501
   stagnation, 283
   static, 283
   thermodynamic, 190

Pressure coefficient, 317
Pressure distribution, airfoil, 436, 439
  converging nozzle, 530
  converging-diverging nozzle, 537, 593
  diffuser, 421
  entrance length of pipe, 369
  sphere, 430
Pressure drag, 43, 426
Pressure force, 53
Pressure gradient, 43, 55
  effect on boundary layer, 418
Pressure recovery coefficient, 372, 422
  ideal, 423
Pressure tap, 283
Primary dimension, 13
Profile, velocity, 24
Properties, fluid, 609
Propulsive efficiency, 220
Pump, 173

Rayleigh line flow, 562, 565
  basic equations for, 562
  choking, 568
  effects on properties, 566, 567
  maximum heat addition, 568
  tables for computation of, 572, 632
  Ts diagram, 566
Relative roughness, 363, 365
Reversible process, 488
Reynolds' experiment, 332
Reynolds number, 44, 316
  critical, 366, 430
Rigid-body, motion of fluid, 89, 269
Rotation, 236, 237
Roughness, relative, 363, 365

SI units, 14, 661

conversion factors, 662
  prefixes, 661
STP (Standard Temperature and Pressure), 21
Scalar product, 651
Secondary dimension, 13
Secondary flow, 373
Second law of thermodynamics, 115, 196
Separation, 43, 418
Shaft work, 188
Shear stress, 1, 32
Shear work, 189
Shock, normal, *see* Normal shock
Similarity, dynamic, 318
  geometric, 318
  kinematic, 318
Similitude, 321
Siphon, 278
Skin friction coefficient, 413
Slug, 14
Sluice gate, 140, 279
Solid, 1
Sound, speed of, 496
Span, 441
Specific gravity, 16, 609, 610
Specific heat, constant pressure, 487, 618
  constant volume, 486, 618
Specific heat ratio, 487, 618
Specific volume, 16, 486
Specific weight, 16
Speed of sound, 496
  ideal gas, 496
Stability, 87
Stagnation enthalpy, 517
Stagnation point, 42
Stagnation pressure, 284
  isentropic, 501
Stagnation pressure probe, 284
Stagnation properties, *see*

Isentropic stagnation
    properties
Stagnation state, 500
Stagnation temperature, 501
Stall, wing, 437
Standard atmosphere, 65
    properties of, 612
Standard pipe sizes, 378
State, equation of, 486
    thermodynamic, 500
Static pressure, 283
Static pressure probe, 283
Static pressure tap, 283
Steady flow, 23
Stoke, 35
Stokes' drag law, 428
Streakline, 26
Stream function, 232, 234
Streamline, 26
    equation of, 232
Streamline coordinates, 270
Streamline curvature, 272
Streamlining, 43, 434
Stream tube, 146, 287
Stress, 30, 32
    Newtonian fluid, 256
    normal, 32, 265
    notation, 32
    shear, 1, 32
    sign convention, 33
Stress field, 28, 32, 33
    inviscid flow, 265
Subsonic flow, 498
Substantial derivative, 249
Sudden expansion, 371
Supersonic flow, 498
Surface force, 28
Surface tension, 611
System, 5
Systems of dimensions, 13
Systems of units, 14, 661

Tables and charts, 609
Tds equations, 489
Thermodynamic pressure, see
    Pressure
Three-dimensional flow, 24
Throat, nozzle, 522
Total head tube, 285
Trailing vortex, 441
Transition, 45, 399
Transonic flow, 498
Turbine, impulse, 172
    reaction, 172
Turbine flowmeter, 460
Turbomachine, 172
Turbulent flow, 44, 332
Two-dimensional flow, 24

Underexpanded nozzle, 538
Uniform flow, at section, 25
Uniform flow field, 26
Units, 14, 661
Unit vector, 649
Universal gas constant, 486
Unsteady Bernoulli equation, 293
Unsteady flow, 24

Vector, differentiation of, 653
    identity, 658
    review of, 649
Vector product, 652
Velocity of approach factor, 451
Velocity coefficient, 451
Velocity field, 23
Velocity gradient, 40
Velocity measurement, 284
Velocity polygon, 174
Velocity potential, 241
Velocity profile, 24, 39
    in pipe flow, 355
Vena contracta, 279, 368, 449
Venturi flowmeter, 455

velocities (incomp. & steady flow)
p 226

Viscometer, capillary, 350
  concentric cylinder, 50
Viscosity, 35, 613
  absolute, 35, 614
  effective, 51
  kinematic, 35, 615
  physical nature of, 613
Viscous flow, 38
Viscous sublayer, 366
Volumetric flowrate, 126
Vortex, forced, 299
  free, 261, 299
  trailing, 441
Vortex generator, 444

Vortex shedding, 326
Vorticity, 239

Wake, 43
Wall shear stress, 357
Weber number, 322
Weight, 15
Wetted perimeter, 397
Work, 188
Work equ., 191
Yield stress, 33

Zone, of action, 500
  of silence, 500